# Narrow Gap Semiconductors 1995

## Other titles in the Series

The Institute of Physics Conference Series regularly features papers presented at important conferences and symposia highlighting new developments in physics and related fields. Recent titles include:

**129: Gallium Arsenide and Related Compounds 1992, papers presented at the 19th International Symposium, Karuizawa, Japan**
T Ikegami, F Hasegawa and Y Takeda

**136: Gallium Arsenide and Related Compounds 1993, papers presented at the 20th International Symposium, Freiburg im Braunschweig, Germany**
H S Rupprecht and G Weimann

**141: Compound Semiconductors 1994, papers presented at the 21st International Symposium, San Diego, CA, USA**
H Goronkin and U Mishra

## Related titles

**Narrow Gap Semiconductors, Proceedings of the NATO Workshop held in Oslo, 1991**
S Lovold and J B Mullin

**Narrow Gap Semiconductors 1992, Proceedings of the 6th International Conference, Southampton, UK**
R A Stradling and J B Mullin

# Narrow Gap Semiconductors 1995

Proceedings of the Seventh International Conference on Narrow Gap Semiconductors, Santa Fe, New Mexico, 8–12 January 1995

Edited by J L Reno

Institute of Physics Conference Series Number 144
Institute of Physics Publishing, Bristol and Philadelphia

CODEN IPHSAC 144 1–384 (1995)

*British Library Cataloguing in Publication Data*

A catalogue record for this book is available from the British Library.

ISBN 0 7503 0341 7

*Library of Congress Cataloging-in-Publication Data are available*

Published by Institute of Physics Publishing, wholly owned by The Institute of Physics, London
Institute of Physics Publishing, Techno House, Redcliffe Way, Bristol BS1 6NX, UK
US Editorial Office: Institute of Physics Publishing, The Public Ledger Building, Suite 1035, 150 South Independence Mall West, Philadelphia, PA 19106, USA

Printed in the UK by J W Arrowsmith Ltd, Bristol

# Contents

**Chapter 5: II–VI Growth**



**Chapter 6: III–V Characterization**

**Chapter 7: Infrared Photodetectors**

# Preface

This volume presents the papers from the Seventh International Conference on Narrow Gap Semiconductors, held in Santa Fe, New Mexico from 8 to 12 January 1995. It includes current results on the materials, characterization, and device aspects of a broad range of narrow gap semiconductors.

The conference consisted of 8 invited talks, 60 oral presentations, 20 poster papers, and 4 late news papers. This volume is organized by the sessions in which the papers were presented. Poster papers are included in the appropriate session with the oral presentations.

The editor wishes to especially thank the program committee, listed on the following page, for its hard work in selecting invited speakers and in deciding which abstracts should be accepted.

**John L Reno**
1995

## Organizing Committee

Ralph Dawson (*Chairman*), John Reno (*Program Chairman*), Steven Kurtz, David Seiler, Darryl Smith, Robert Wagner

## Program Committee

John Reno (*Chair*), Jose Arias, Gunther Bauer, Bob Biefeld, Tom Elliot, Steve Jost, Mike Kinch, Patrick McCann, Dale Partin, Shue-Chu Shen, Jerry Stringfellow, Robert Wagner, Arthur Willoughby

## International Advisory Committee

A Appelbaum, J W Baars, G Bauer, R Biefeld, C T Elliot, H Heinrich, V Ivanov-Omskii, R G Galazka, M Kinch, F Koch, D Kushev, G Landwehr, B Levine, C Littler, H Maier, Y Marfaing, T McGill, J Meyer, N Miura, T Nguyen Duy, G N Pain, D Partin, W Paul, C R Pidgeon, J L Robert, S C Shen, A Stradling, G Stringfellow, I Sunagawa, K Takita, W Tennant, H Watanabe, A M White, A Willoughby, W Zawadski

The conference organizers would like to thank the following exhibitors for their support:

Air Products (K Wong), Aixtron Inc. (T Lovis), EPI (D Reamer), Instruments SA (C Badgett, P Dargan), Johnson Matthey (G Gruska, C Parfeniuk), Lehighton Electronics (T Taylor), MCP Wafer Technology Ltd (K Lamb, M Lamb, L Ramirez)

*Inst. Phys. Conf. Ser. No 144: Section 1*
*Paper presented at 7th Int. Conf. on Narrow Gap Semicond. Santa Fe 9–12 January*



# GaSb-based semiconductor lasers in the 4-$\mu$m band*

H. K. Choi, G. W. Turner, and H. Q. Le

Lincoln Laboratory, Massachusetts Institute of Technology, Lexington, MA 02173-9108, U.S.A.

**Abstract.** The current status of GaSb-based semiconductor lasers emitting at ~4 $\mu$m is described. Double-heterostructure (DH) InAsSb/AlAsSb diode lasers have exhibited pulsed operation up to 170 K and the characteristic temperature is 20 K. Broad-stripe devices have operated cw up to 105 K, with single-ended cw power of 30 mW at 70 K. Quantum-well diode lasers consisting of a strained InAsSb/InAlAsSb active region have operated pulsed up to 165 K, with a characteristic temperature of 30 K up to 120 K. Broad-stripe devices have operated cw up to 123 K, with cw power at 80 K of 30 mW/facet. Ridge-waveguide lasers have operated cw up to 128 K and the cw threshold current at 80 K is 35 mA. Optical pumping of InAsSb/AlAsSb DH lasers using 0.94-$\mu$m InGaAs/AlGaAs diode arrays have exhibited peak power of 2 W and average power of 0.24 W at 92 K.

## 1. Introduction

High-performance semiconductor lasers emitting in the mid-infrared (IR) band (2 – 5 $\mu$m) are very desirable for many applications. A variety of molecules including all hydrocarbons have strong absorption lines in this band. Single-frequency mid-IR lasers are ideally suited for highly sensitive detection of trace gases for such applications as pollution or toxic gas monitoring and industrial process monitoring. In addition, there are three atmospheric transmission windows in this band. High-power lasers emitting in these windows can be very useful for laser radars, target designation, and illumination.

A number of material systems such as III-V [1]-[11], lead-salt [12],[13], and II-VI [14] have been investigated for fabricating mid-IR semiconductor lasers. Lead-salt lasers emitting at wavelengths longer than 3 $\mu$m have been commercially available for a number of years. These lasers with emission wavelengths of ~4 $\mu$m have exhibited pulsed operation at temperatures up to 290 K [12], and cw operation up to 200 K [13]. Yet they are not expected to produce high output power because of the low thermal conductivity of lead-salt alloys and their susceptibility to damage. Recently, HgCdTe double-heterostructure (DH) diode lasers emitting between 2.9 and 5.3 $\mu$m were reported, with pulsed operation below 90 K. However, it is questionable whether these lasers can be practical because of low thermal conductivity of HgCdTe and the high mobility of Hg atoms.

It is thus very desirable to develop mid-IR semiconductor lasers from III-V alloys, because this material offers good metallurgical and thermal properties and because better-quality substrates are readily available. Most of the III-V diode laser work has utilized antimonide-containing compounds grown on GaSb or InAs. At ~2 $\mu$m, GaInAsSb/AlGaAsSb quantum-well (QW) lasers grown on GaSb have exhibited excellent room-temperature performance, with pulsed threshold current density of 143 A/cm$^2$ and cw power of 1.3 W from a 300-$\mu$m aperture [3]. However, the performance of lasers emitting at ~4 $\mu$m is still very inferior, mainly because of considerably higher Auger recombination rates at longer

*This work was sponsored by Phillips Laboratory, the Department of the Air Force.

wavelengths. In this paper, we report recent progress in GaSb-based semiconductor lasers emitting at ~4 $\mu$m. Performance of both InAsSb/AlAsSb DH and InAsSb/InAlAsSb QW diode lasers will be described. In addition, high-power operation of InAsSb/AlAsSb DH lasers under optical pumping will be reported.

## 2. Properties of Antimonide-Based Material Systems

Figure 1 plots the room-temperature values of bandgap energy $E_g$ and corresponding wavelength $\lambda_g$ of lattice-matched $Ga_xIn_{1-x}As_ySb_{1-y}$ on GaSb as a function of x. The bandgap decreases steeply from 0.73 to 0.30 eV (with a corresponding increase in wavelength from 1.7 to 4.2 $\mu$m) as x decreases from 1.0 to 0.4. However, $E_g$ stays nearly constant for x < 0.4, and has a minimum value of 0.28 eV (4.4 $\mu$m) at x = 0.22. For 4-$\mu$m emission, we have chosen InAsSb for the active layer because it is stable at the growth temperature and is easier to grow than metastable $Ga_xIn_{1-x}As_ySb_{1-y}$ with x > 0.25 [15].

For efficient laser structures, cladding/barrier layers must have not only a smaller refractive index and a larger bandgap energy than the active layer, but also proper conduction- and valence-band positions to confine carriers in the active layer. The valence-band position of $InAs_{0.91}Sb_{0.09}$ is slightly lower than $AlAs_{0.08}Sb_{0.92}$, the extreme composition for the lattice-matched AlGaAsSb alloy. For DH lasers, this negative valence-band offset is not detrimental, as will be discussed later. However, it is a very serious problem for QW structures. In these structures, we have used InAlAsSb for the barrier layer because it has a lower valence-band position than InAsSb.

The epilayers were grown on (100)-oriented *n*-type GaSb substrates in a solid-source molecular beam epitaxy (MBE) system. A valved As source was used to permit almost instantaneous changes in either the cracked $As_2$ or uncracked $As_4$ flux. Two conventional effusion cells were used to provide various combinations of uncracked $Sb_4$ fluxes. We used Be and GaTe as the *p*- and *n*-type dopants, respectively. The growth rate and temperature for AlAsSb are ~0.8 $\mu$m/h and 520°C, respectively. The growth temperature for InAsSb is somewhat lower. Because relatively thick cladding layers are required, precise lattice matching of AlGaAsSb is extremely important. It is also very important to have the proper sequencing of shutters to maintain abrupt interfaces. Details of the growth procedure can be found in Ref. 16.

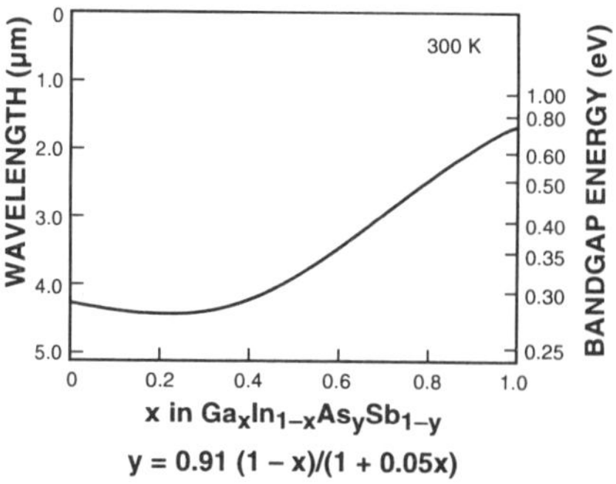


Fig. 1. Room-temperature bandgap and corresponding wavelength of lattice-matched GaInAsSb as a function of Ga content.

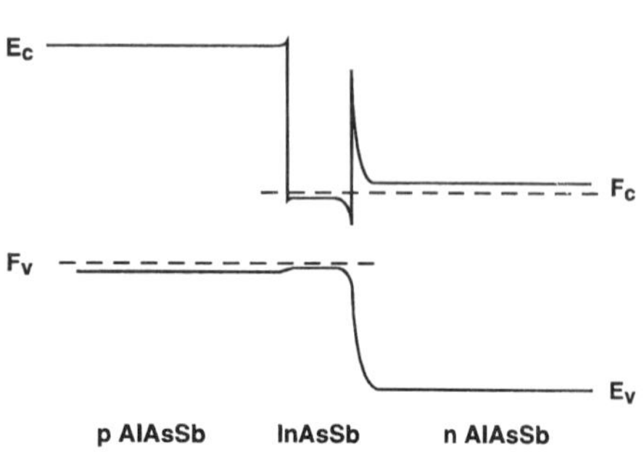


Fig. 2. Schematic energy band diagram for InAsSb/AlAsSb double-heterostructure (DH) laser under forward bias.

## 3. Diode Lasers

### *3.1. InAsSb/AlAsSb DH lasers*

The laser structure consists of the following layers: 0.5-$\mu$m-thick $n$-GaSb buffer, 3-$\mu$m-thick $n$-$AlAs_{0.08}Sb_{0.92}$ cladding, 0.8-$\mu$m-thick nominally undoped InAsSb active, 3-$\mu$m-thick $p$-$AlAs_{0.08}Sb_{0.92}$ cladding, and 0.075-$\mu$m-thick $p^+$-GaSb cap. Double-crystal x-ray diffraction measurement of the laser structure showed that the lattice match was better than $2.5 \times 10^{-3}$. No cross-hatch patterns were visible during inspection of the surface using Nomarski optical microscopy. The photoluminescence measured at 4.5 K had a strong peak at 3.78 $\mu$m (0.328 eV), with a full width at half-maximum (FWHM) of only 8 meV.

Broad-stripe lasers 60 or 100 $\mu$m wide were fabricated by using $SiO_2$ patterning. Ohmic contacts to the $p^+$- and $n$-GaSb were obtained by depositing Ti/Pt/Au and Au/Sn/Ti/Pt/Au, respectively, and alloying at 300°C. Individual devices 500 $\mu$m long were mounted $p$-side up on Cu heatsinks for pulsed measurements and $p$-side down on diamond submounts on Cu for cw measurements. They were then installed in a dewar for low-temperature operation.

Figure 2 shows the schematic energy band diagram for the laser structure under forward bias. Because of the huge conduction-band offset between InAsSb and AlAsSb, there is a large potential barrier for electrons between the active and the $n$-cladding layer, which increases the turn-on voltage and series resistance. On the other hand, this large built-in potential confines the holes in the active layer very effectively, even though the valence-band offset is negative.

Figure 3 shows the forward voltage vs current curves of a 100-$\mu$m-wide device at several temperatures. The turn-on voltage at room temperature is ~0.6 V, and it increases gradually to ~1.2 V at 70 K. Since the bandgap energy of InAsSb is only ~0.3 eV, the high turn-on voltage is indicative of the large potential barrier between the active and the $n$-cladding layer. The fact that the series resistance does not increase appreciably at low temperatures shows that the carrier freezeout at 70 K is not significant for AlAsSb.

Figure 4 shows the pulsed threshold current density $J_{th}$ of a 100-$\mu$m-wide device for temperatures between 60 and 170 K. At 60 K, the value of $J_{th}$ is 36 A/cm$^2$. The characteristic temperature $T_0$ for the entire temperature range is 20 K. The fact that a single value of $T_0$

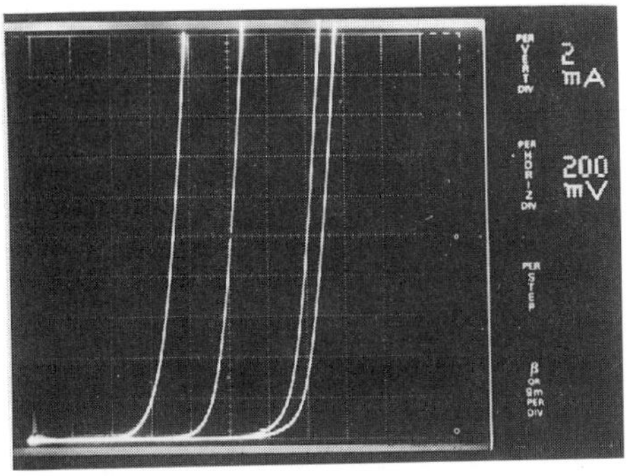


Fig. 3. Voltage vs current curves for 100 × 500-$\mu$m device at 300, 200, 100, and 70 K (from left to right).

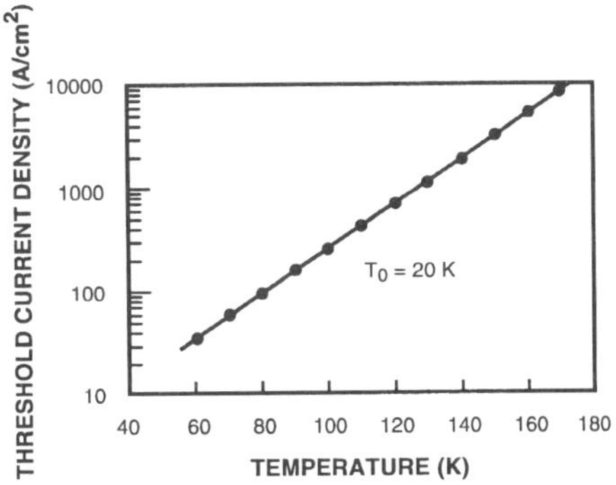


Fig. 4. Pulsed threshold current density vs temperature of 100 × 500-$\mu$m device.

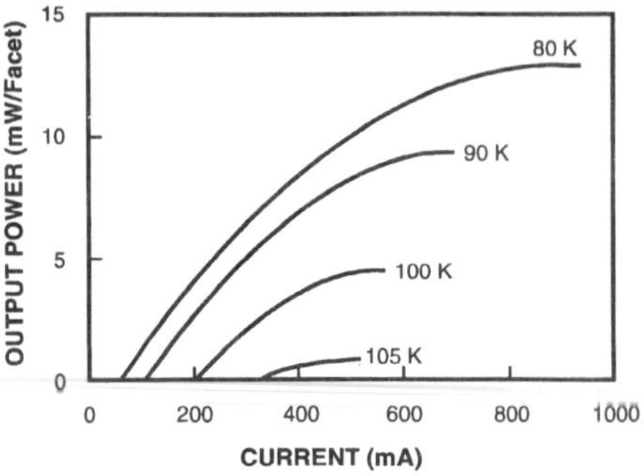


Fig. 5. CW power vs current for 60 × 500-μm device at several heatsink temperatures.

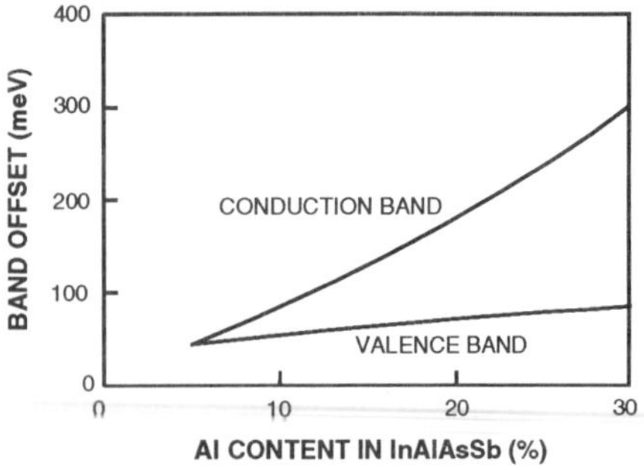


Fig. 6. Calculated conduction- and valence-band offset between InAsSb and InAlAsSb with a constant strain of $5 \times 10^{-3}$ as a function of Al content.

describes the temperature dependence of $J_{th}$ indicates that one dominant mechanism, most likely Auger recombination, is responsible. At the maximum operating temperature of 170 K, the value of $J_{th}$ is 8.5 kA/cm$^2$.

Figure 5 shows the cw output power vs current curves for a 60-μm-wide device at heatsink temperatures from 80 to 105 K. These curves were obtained by measuring the output with an InSb detector in the linear regime and calibrating it with a thermopile detector. The maximum power at 80 K is 12.5 mW/facet, with an initial differential quantum efficiency of 19% from both facets. Another device 100 μm wide was coated to have high (~90%) and low (~10%) reflectivity on the back and front facet, respectively. The maximum power from this device at 70 and 80 K is 30 and 24 mW, respectively. These lasers have been thermally cycled only a few times and operated for a few hours, but during this time the output power has been very stable.

### *3.2. InAsSb/InAlAsSb QW lasers*

As discussed in section 2, InAlAsSb is a suitable barrier material for the InAsSb active wells because it has a lower valence-band position than InAsSb. Figure 6 shows the calculated conduction-and valence-band offsets as a function of Al content in InAlAsSb with a fixed tensile strain of $5 \times 10^{-3}$ with respect to $InAs_{0.85}Sb_{0.15}$. The valence-band offset increases very gradually with the Al content, while the conduction-band offset increases rapidly.

The following laser structure was grown by MBE: 1-μm-thick *n*-GaSb buffer, 3-μm-thick *n*-$AlAs_{0.08}Sb_{0.92}$ cladding, active region consisting of ten 15-nm-thick $InAs_{0.85}Sb_{0.15}$ active wells and eleven 30-nm-thick $In_{0.85}Al_{0.15}As_{0.86}Sb_{0.14}$ barriers, 3-μm-thick *p*-$AlAs_{0.08}Sb_{0.92}$ cladding, and 50-nm-thick $p^+$-GaSb cap. Double-crystal x-ray diffraction measurement of the laser structure showed some broadening of the satellite peaks, indicating imperfect quality of the interfaces.

Broad-stripe lasers 100 μm wide were fabricated by using $SiO_2$ patterning. Lasers were mounted junction-side up on Cu heat sinks using In, and loaded into a dewar for low-temperature measurements. Pulsed $J_{th}$ of a 1000-μm-long device for temperatures between 80 and 165 K is plotted in Fig. 7. At 80 K, the value of $J_{th}$ for the QW device is 78 A/cm$^2$, compared to 100 A/cm$^2$ for the DH devices. The characteristic temperature $T_0$ up to 120 K is

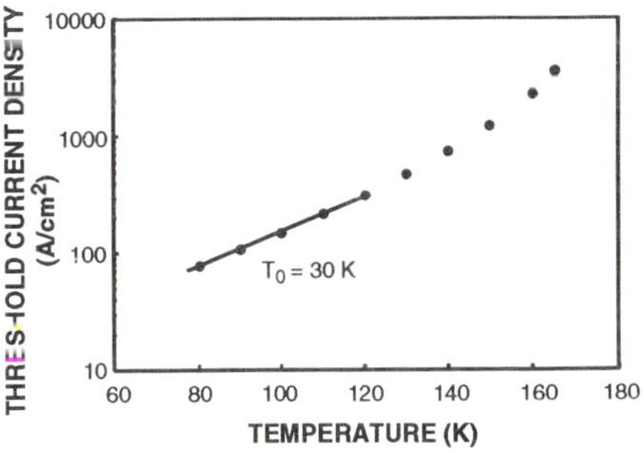


Fig. 7. Pulsed threshold current density of InAsSb/InAlAsSb quantum-well (QW) laser as a function of temperature.

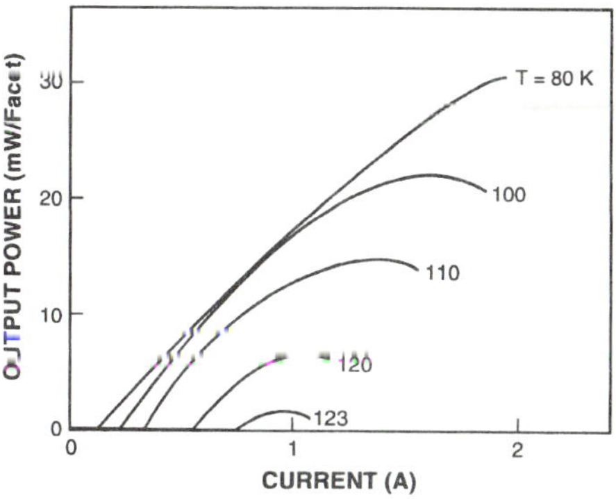


Fig. 8. CW power vs current of InAsSb/InAlAsSb QW laser at several temperatures.

30 K, which is larger than the 20 K observed for the DH lasers emitting at the same wavelength. Both the lower $J_{th}$ and higher $T_0$ for the QW devices indicate that the Auger recombination, believed to be the limiting mechanism for 4-$\mu$m lasers, has been decreased by the employment of the strained QW structure. As the temperature is increased above 120 K, $T_0$ becomes smaller. At the maximum operating temperature of 165 K, the value of $J_{th}$ is 3.5 kA/cm$^2$.

Figure 8 shows the cw power vs current for the same device. The maximum power at 80 K is 30 mW/facet, which is much higher than 24 mW obtained for a DH laser with high/low facet coatings and mounted junction-side down. The highest cw operating temperature is 123 K, which is higher than 105 K obtained for the 60-$\mu$m-wide DH device.

Ridge-waveguide lasers 15 $\mu$m wide were fabricated. The cw threshold current of a 1000-$\mu$m-long device is 35 mA at 80 K, and the maximum cw operating temperature is 128 K. The far-field pattern in the transverse direction has an FWHM of ~60°. In the lateral direction, the far-field pattern shows a single lobe with an FWHM of ~15°.

## 4. Optically Pumped Lasers

The high operating voltage of the InAsSb/AlAsSb diode lasers is one of the major reasons for low efficiency and output power. At 80 K, the operating voltage of the laser is 2–3 V depending on the amount of current flow. Because the photon energy at 4 $\mu$m is only ~0.3 eV, the high operating voltage is a significant factor for the low power efficiency.

One approach to increase the efficiency and output power is to employ optical pumping because pumping with 1-$\mu$m lasers is equivalent to 1.24 V. In the past, optical pumping was used mostly for diagnostic purposes because high-power lasers were very bulky. However, the availability of compact, efficient high-power diode lasers at ~1 $\mu$m has made optical pumping practical.

The pump source is a 1-cm$^2$ five-bar diode array stack emitting at 0.94 $\mu$m and capable of 250-W peak power at 15% duty cycle. A lens system focuses the stack output to a stripe 0.2–0.3 mm wide and 2–3 mm long on a sample inside the dewar. Details of the pumping geometry can be found in Ref. 17.

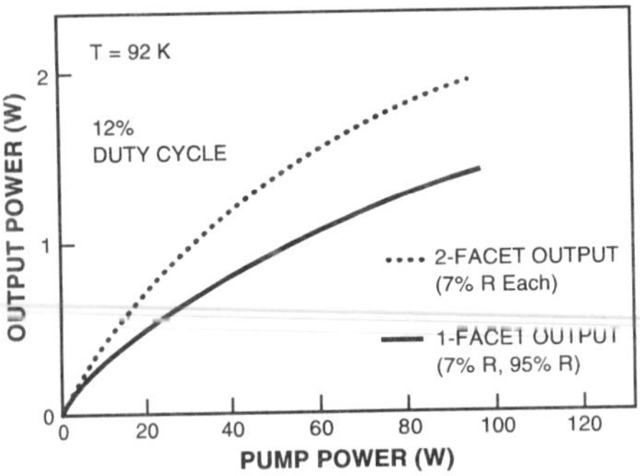


Fig. 9. Output power vs input power for InAsSb/AlAsSb DH laser pumped by 11-$\mu$s pulses from 0.94-$\mu$m diode laser array.

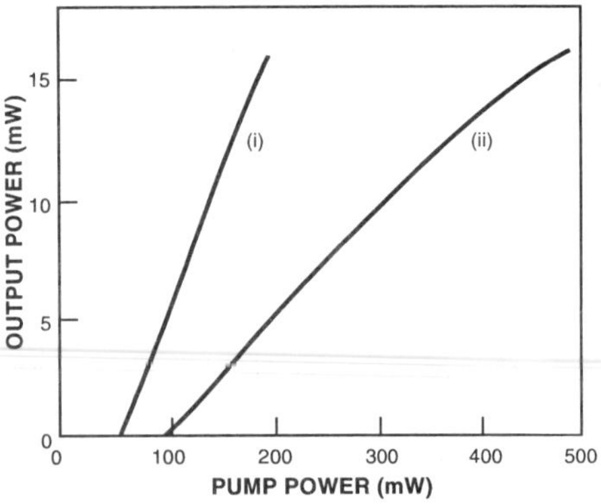


Fig. 10. CW output power at 86 K of InAsSb/AlAsSb DH laser pumped by (i) 2.1 $\mu$m and (ii) 1.06 $\mu$m lasers.

Figure 9 shows the output power vs input power for an InAsSb/AlAsSb DH laser pumped by 11-$\mu$s pulses at 92 K. With both facets coated for 7% reflectivity, a peak power of 2 W is obtained from both facets at a duty cycle of 12%, corresponding to an average power of 0.24 W. The initial power conversion efficiency is ~4%. With the back facet reflectivity increased to 95%, a peak power of 1.4 W is obtained from the front facet. These are the highest powers obtained for any semiconductor laser emitting at ~4 $\mu$m.

A much higher efficiency can be obtained by pumping with a 2-$\mu$m laser, which has an equivalent voltage of only 0.6 V [18]. Figure 10 compares cw output power at 86 K of InAsSb/AlAsSb DH lasers pumped by a 2.1-$\mu$m Ho:YAG laser and a 1.06-$\mu$m Nd:YAG laser. The input beam was focused to a 400 × 25-$\mu$m stripe. For 2.1-$\mu$m pumping, the slope efficiency is 11%, which corresponds to a differential quantum efficiency of 22%. The maximum cw power is 16 mW limited by the pump power. For comparison, the slope efficiency of the 1.06-$\mu$m pumped laser is ~5%. The same laser was pumped by the Q-switched (140 ns) Ho:YAG laser. The laser operated up to 211 K, the highest temperature for any 4-$\mu$m III-V laser.

## 5. Conclusions

Even though InAsSb/AlAsSb DH and InAsSb/InAlAsSb QW lasers still operate at low temperatures, the performance of 4-$\mu$m lasers has been substantially improved. The improvement in $J_{th}$ and $T_0$ observed in strained QW lasers is very encouraging. By optimizing the laser structure and growth conditions, it should be possible to obtain much higher performance. The operating voltage of the diode lasers needs to be reduced for higher power and efficiency, which may be accomplished by grading the heterointerfaces. Optical pumping will continue to play an important role in obtaining high output power at 4 $\mu$m. In particular, development of high-power 2-$\mu$m diode laser arrays will significantly increase the power efficiency.

**Acknowledgements**

We would like to thank D. R. Calawa, J. W. Chludzinski, M. K. Connors, L. Krohn, W. L. McGilvary, and J. V. Pantano for technical assistance, K. J. Challberg for manuscript editing, and I. Melngailis, D. L. Spears, and B-Y. Tsaur for encouragement and support.

**References**

[1] Bochkarev A E, Dolginov L M, Drakin A E, Eliseev P G and Sverdlov B N 1988 *Sov. J. Quantum Electron.* **18** 1362-1363

[2] Aidaraliev M, Zotova N V, Karandashev S A, Matveev B A, Stus' N V and Talalakin G N 1990 *Sov. Tech. Phys. Lett.* **15** 600-601

[3] Choi H K, Turner G W and Eglash S J 1994 *IEEE Photon. Technol. Lett.* **6** 7-9

[4] Choi H K, Eglash S J and Turner G W 1994 *Appl. Phys. Lett.* **64** 2474-2476

[5] Choi H K, Turner G W and Liau Z L 1994 *Appl. Phys. Lett.* **65** 2251-2253

[6] Baranov A N, Imenkov A N, Sherstnev V V and Yakovlev Yu P 1994 *Appl. Phys. Lett.* **64** 2480-2482

[7] Lee H, York P K, Menna R J, Martinelli R U, Garbuzov D Z, Narayan S Y and Connolly J C September 1994, presented at *14th Int. Semicond. Laser Conf.*, Maui, Hawaii, paper PD6

[8] Zhang Y H, Miles R H, Chow D H September 1994, presented at *14th Int. Semicond. Laser Conf.*, Maui, Hawaii, paper PD7

[9] Faist J, Capasso F, Sivco D L, Sirtori C, Hutchinson A L and Cho A Y 1994 *Electron. Lett.* **30** 865-866

[10] Le H Q, Turner G W, Eglash S J, Choi H K and Copetta D A 1994 *Appl. Phys Lett.* **64** 152-154

[11] Kurtz S R, Biefeld R M, Dawson L R, Baucom K C and Howard A J 1994 *Appl. Phys. Lett.* **64** 812-814

[12] Spanger B, Schiessel U, Lambrecht A, Bottner H and Tache M 1988 *Appl. Phys. Lett.* **53** 2582-2583

[13] Feit Z, Kostyk D, Woods R J and Mak P 1991 *Appl. Phys. Lett.* **58** 343-345

[14] Arias J M, Zandian M, Zucca R and Singh J 1993 *Semicond. Sci. Technol.* **8**, S255

[15] Tournie E, Lazzari J L, Mani H, Pitard F, Alibert C and Houllie A 1990 *Proc. SPIE* **1361** 641-656

[16] Turner G W, Choi H K, Calawa D R, Pantano J V and Chludzinski J W 1994 *J. Vac. Sci. Technol. B* **12** 1266-1268

[17] Le H Q, Goodhue W D, Maki P A and Di Cecca S 1993 *Appl. Phys. Lett.* **63** 1465-1467

[18] Le H Q, Turner G W, Ochoa J R and Sanchez A 1994 *Electron Lett.* **30** 1944-1945

*Inst. Phys. Conf. Ser. No 144: Section 1*
*Paper presented at 7th Int. Conf. on Narrow Gap Semicond. Santa Fe 9–12 January*



# 300K light emitting devices for the 3 - 10μm band from arsenic rich InAs/InAs$_{1-x}$ Sb$_x$ strained layer superlattices

M.J. Pullin, P.J.P. Tang, S.J. Chung, C.C. Phillips, R.A. Stradling. *Physics Department, Imperial College, London SW7 2BZ*

A.G. Norman, Y.B. Li, and L. Hart . *Interdisciplinary Research Centre for Semiconductor Materials, Imperial College, London SW7 2BZ*

**PACS numbers 78.60.Fi, 73.20.Dx, 78.65.Fa.**

**Abstract**

Arsenic rich InAs/InAs$_{1-x}$Sb$_x$ strained layer superlattices (SLS's) grown on GaAs substrates by Molecular Beam Epitaxy (MBE) are studied for their potential application as infrared emitters. The long wavelength emission (4-11μm) is highly sensitive to superlattice design parameters and is accounted for by a large type II band offset, greater than in previously studied antimony rich InSb/InAs$_{1-x}$Sb$_x$ SLS's. High internal PL efficiencies (>10%) and intense luminescence emission were observed at these long wavelengths despite large dislocation densities. Initial unoptimised InAs/InAs$_{1-x}$Sb$_x$ SLS LED's gave ≈200nW of emission at 300K, with an internal quantum efficiency >0.05% in the 4.2μm $CO_2$ absorption band. LED's with similar efficiencies are demonstrated at wavelengths extending out to 10μm, the longest yet reported for a 300K device.

In(AsSb) strained layer superlattices (SLS's) offer superior metallurgical stability and compatibility with existing III-V technology when compared with other infrared materials like (HgCd)Te and lead salts. Although lead-salt lasers already exist, they operate only below 200K [1]. To date, because of the narrow band gaps of the alloy constituents, Sb-rich InSb/InAs$_{1-x}$ Sb$_x$ SLS's have mainly been studied for long wavelength (≥10μm) photodiode [2] and emitter applications. Their attraction arises from a type-II band offset which gives a superlattice structure with an effective band gap lower than either of the alloy constituents.

The necessity of growth on non lattice-matched substrates however leads to threading dislocation densities of ~3 x $10^8$ $cm^{-2}$ which produce generation-recombination centres which degrade Sb-rich alloy opto-electronic devices. In contrast, Walukiewicz [3] has shown that InAs is unique among the III-V's in that point-like lattice defects produce electronic states above the conduction band edge, in agreement with the existence of an electron accumulation layer [4] which is formed by InAs surface states producing energy states above the conduction band minimum. Observations of high efficiency excitonic PL from InAs [5] and As rich In(AsSb) alloys [6] grown on GaAs substrates also suggests that the dislocations do not produce non-radiative recombination centres in the bandgap in these materials.

All the SLS wafers here were grown by MBE on (100) GaAs substrates. A 0.25-0.3μ m GaAs buffer layer was followed by ~1 μm of strain-symmetrising In(AsSb) alloy buffer [8] and the well and barrier thicknesses were equal. All alloy concentrations, and layer thicknesses were determined by X-ray diffraction modelling and, partly because of the small Sb concentrations used, phase separation effects were not seen. The 3 trial LED structures ( table I ) had undoped SLS regions grown between 1μm thick n- and p- type alloy layers doped to $10^{19}cm^{-3}$ The alloy composition for these doped layers was the average of the superlattice for strain symmetrisation. The diodes were fabricated into unpassivated mesa devices using conventional wet-chemistry photolithography.

**Table I.**

| Sample number | SLS (~2μm thick ) | SLS period | buffer (≈1μm thick) | Doping type | Approx. $T_0$ | Valence band offset (meV) for $e^-$ wells in In(As,Sb) | InAs |
|---|---|---|---|---|---|---|---|
| IC 389 | $InAs/InAs_{0.68}\ Sb_{0.32}$ | 30nm | $InAs_{0.84}\ Sb_{0.16}$ | Undoped | 97K | 830 x | 780 x |
| IC 481 | $InAs/InAs_{0.73}\ Sb_{0.27}$ | 23.5nm | $InAs_{0.86}\ Sb_{0.14}$ | Undoped | 72K | 790 x | 620 x |
| IC 494 | $InAs/InAs_{0.86}\ Sb_{0.14}$ | 16nm | $InAs_{0.93}\ Sb_{0.07}$ | Undoped | 55K | 880 x | 940 x |
| IC 495 | $InAs/InAs_{0.86}\ Sb_{0.14}$ | 24.8nm | $InAs_{0.93}\ Sb_{0.07}$ | Undoped | 46K | 880 x | 940 x |
| IC 499 | $InAs/InAs_{0.61}\ Sb_{0.39}$ | 21.5nm | $InAs_{0.8}\ Sb_{0.2}$ | Undoped | 61K | 710 x | 631 x |
| IC 493 | $InAs/InAs_{0.728}\ Sb_{0.274}$ | 20.4nm | $InAs_{0.863}\ Sb_{0.137}$ | p-n diode | - | - | - |
| IC 500 | $InAs/InAs_{0.905}\ Sb_{0.095}$ | 24.2nm | $InAs_{0.953}\ Sb_{0.047}$ | p-n diode | - | - | - |
| IC 501 | $InAs/InAs_{0.897}\ Sb_{0.103}$ | 16.4nm | $InAs_{0.949}\ Sb_{0.051}$ | p-n diode | - | - | - |
| IC 516 | $InAs/InAs_{0.875}\ Sb_{0.125}$ | 24 nm | $InAs_{0.937}\ Sb_{0.063}$ | p-n diode | - | - | - |
| IC 106 | InAs control sample | 3.3μm | - | Undoped | 16K | - | - |

**Table I.** Structures of the SLS wafers (all grown on GaAs substrates at 450°C). Except for IC 516 all the periods and alloy compositions were determined by X-ray diffraction. Valence band offset values are unstrained figures (in terms of x, the alloy composition difference ) obtained from numerical modelling of the PL peak energies by assuming that the electron wells lie in the In(As,Sb) or InAs material respectively.

Photoluminescence (PL) spectra were obtained using a double modulation technique [7] with a ~0.98μm laser diode as the exciting source. The sample was mounted in a closed cycle cryostat and the PL/EL was detected with a Bomem DA-3.02 Fourier transform spectrometer with a MCT detector sensitive up to λ≈20μm.

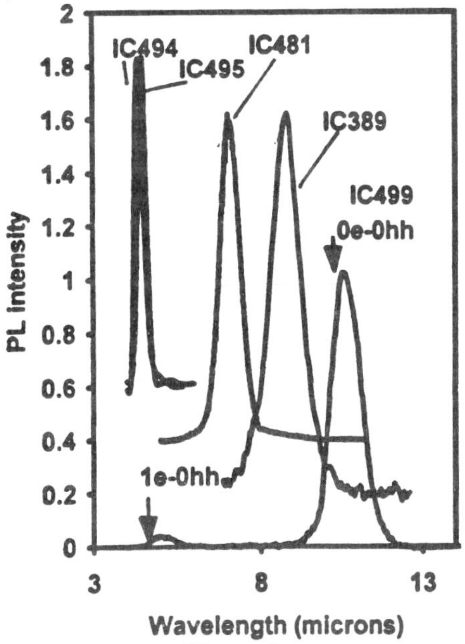


Figure 1.

Fig.1 summarises the PL spectra for the undoped SLS's listed in table I. For all samples PL features are observed at energies lower than the band gaps of either of the superlattice alloy constituents. A numerical model [9] incorporating the strain shifts of the energy bands [10,11] and non-parabolicity effects [12] was used to calculate the lowest electron-heavy hole transition energies. The unstrained alloy band gaps, spin-split off energies and momentum matrix elements were taken from previous magneto-absorption studies of the bulk alloys [13,14] leaving the valence band offset as the only free fitting parameter.

The low PL energies observed can only be accounted for by a large type-II band offset but in view of the anomalous band bowing in this alloy system it is *a proiri* unclear whether the electron wells lie in the InAs or the alloy. Table 1 lists the unstrained offset values resulting from the computer fit

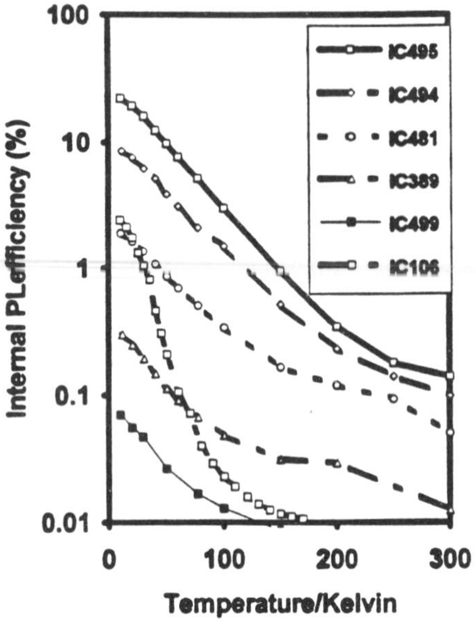


Figure 2.

corresponding to these two possibilities. The best fit to the spectrum of IC499 however (where 2 SLS transitions were tentatively identified, fig.1) corresponded to the alloy acting as the electron well. Studies of Sb-rich InSb/In(As,Sb) SLS's indicate that the layer with the lower Sb concentration is the electron well but the opposite assumption gives a tighter spread of modelled unstrained valence band offset values from these As-rich samples. The latter also agrees with magneto-absorption data [15] from IC389 and IC481 which suggest that the holes are confined to the InAs layers and electrons to alloy wells.

IC494 and IC495 have the same compositions but different periods and show a clear quantum size effect (a 9meV shift in PL energy) arguing for a band-to-band and against a defect-related origin of the PL. Similarly samples IC481, IC 389 and IC499 form a series with increasing Sb content and decreasing PL energy, showing the expected variation of effective bandgap with mean alloy composition. Irrespective of assumptions about the sign of the band offset, the fitted offset values (620-940 x) are all significantly greater than any values previously found for Sb-rich InSb/In(As,Sb) SLS [2].

Except for IC499 and IC 389 (the most Sb-rich of the samples) all superlattices had 10K internal PL efficiencies of >1%, several decades higher than for any Sb-rich SLS (on either InSb or GaAs substrates) that we have previously studied. The integrated PL efficiencies are also much more temperature stable than for the bulk InAs control sample IC106 (fig.2). We attribute this to the reduction in the ratio $m^*_{hh}/m^*_e$ caused by the effects of a) strain lightening the valence band and b) non-parabolicity and the type-II nature of the potential increasing $m^*_e$ above the value one would expect for the alloy with the same effective gap as the SLS. We believe that this reduction in the $m^*_{hh}/m^*_e$ ratio strongly quenches the CCCH Auger recombination channel which is the dominant non-radiative recombination mechanism in bulk alloys at close to 300K.

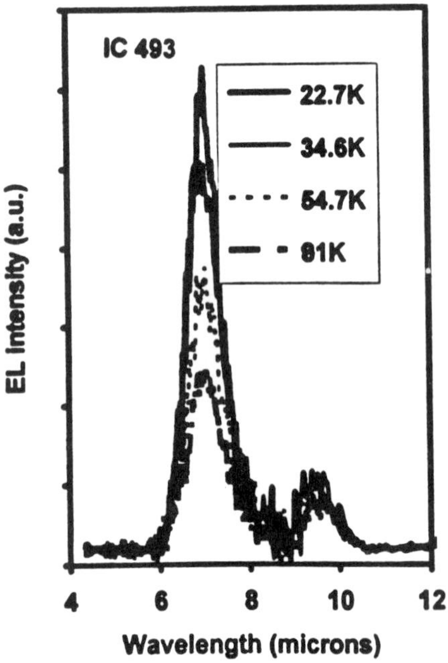


Figure 3.

If it is assumed that the PL temperature dependence arises only from a local competition between radiative and non-radiative recombination (i.e. ignoring

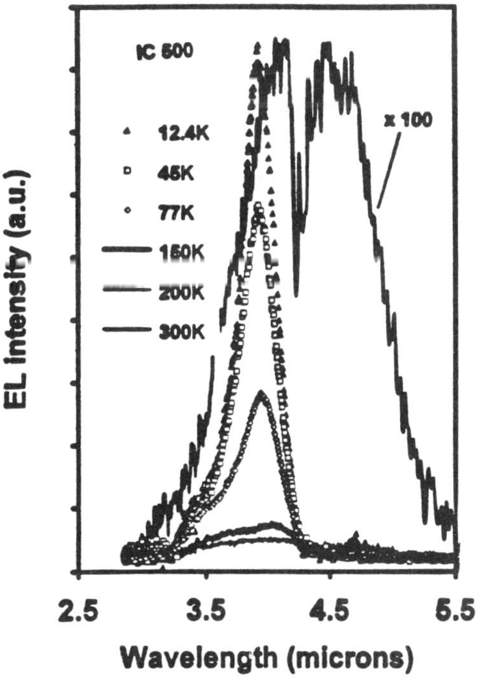


Figure 4.

for the moment the possibile role of the SLS potential in blocking carrier migration to the heavily dislocated substrate/epilayer interface) then the data of fig.2 yield estimates (table I) for the $T_o$ threshold current temperature parameters which would apply to lasers made from these materials.

These high PL efficiencies (fig.2) in the presence of high threading dislocation densities again suggest that point defects for As-rich alloys are resonant with the conduction band [3,5] as opposed to forming non-radiative defect levels within the band gap. The SLS series shows the same general trend as we observe in bulk $InAs_{1-x}Sb_x$ alloy samples of decreasing PL efficiency with increasing Sb concentration (e.g. we find a PL efficiency in $InAs_{0.5}Sb_{0.5}$ ~20 times less than $InAs_{0.84}Sb_{0.16}$ [5] ).

The EL spectra for the IC493 LED (fig.3) show principle peaks with widths and energies which match IC481, its undoped counterpart. The EL efficiencies for both LED's IC493 and IC500 (fig. 4) are remarkably stable against rising temperature; only a x 2.2 reduction is seen for a temperature increase from 22.7K to 81K . Both LED's IC493 and IC500 show weak long wavelength EL features (at 9.5μm and 4.6μm respectively) of uncertain origin but probably related to the doped strain-symmetrising alloy buffer layers. The 300K EL spectrum for IC500 (fig.4) extends beyond 5μm, spanning the $CO_2$ absorption line (in this case due to ≈50 cm of laboratory air) around 4.2μm hence demonstrating its potential for 300K uncooled infrared gas/pollutant sensor applications.

IC 500 (with no passivation or index-matching lens) has an external quantum efficiency of 0.0007% at 300K, only ~30 times less than at 20K (fig.5) and gives 200nW of optical power at 100mA, sufficient for gas sensing operations at room temperature over several metres.

Assuming only reflection and refraction losses this implies an internal 300K conversion efficiency of ~0.05% . In practice, due to a "soft" diode characteristic and finite in-plane conductivity some of the EL is generated beneath the opaque top contact and the real optical losses are probably higher,

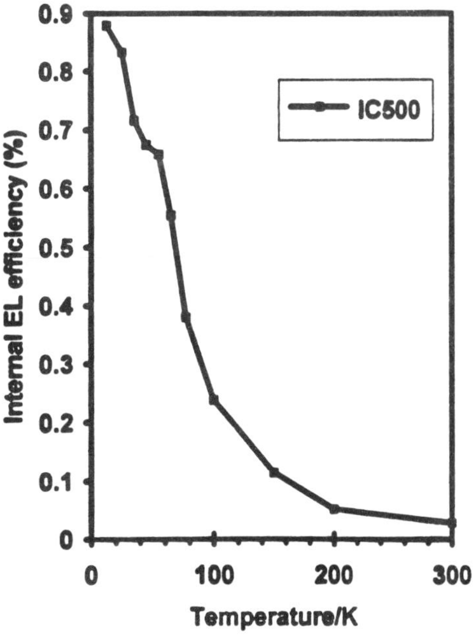


Figure 5.

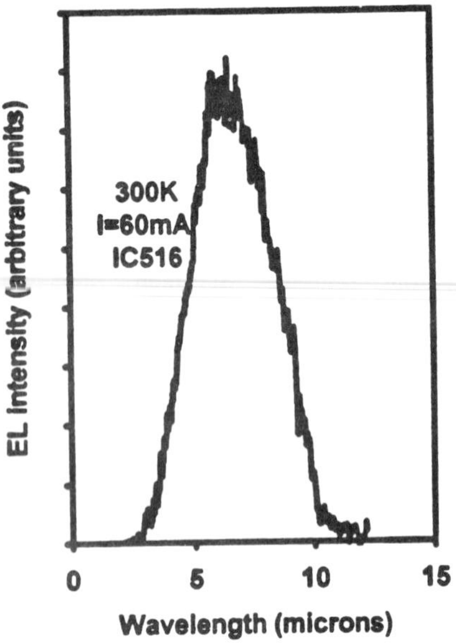


**Figure 6.**

implying that the actual internal conversion efficiency is greater than 0.05%.

The doping levels for these initial diodes are high, possibly resulting in increased Auger recombination rates; the integrated 12.4K PL intensity of IC500 was 600 times less than its undoped SLS counterpart IC495 for example. Overall we are confident that the external EL efficiencies can be greatly enhanced with improved diode contact design and with band potential profiles optimised for carrier confinement.

Also shown is a very recent LED EL spectrum (fig.6) from a sample with nominally the same well widths as IC 500 but with higher Sb concentration. The 300K quantum efficiency and output power is similar to the IC 500 LED but the EL spectrum now extends out to 10μm, to our knowledge a world record for a 300K device.

To conclude, intense photoluminescence has been observed from As rich In(As,Sb) strained layer superlattices covering a wavelength range out to 11 μm at 300K for the first time. This long wavelength emission has been achieved with relatively wide band gap (InAs)Sb alloys with large valence band offsets giving a type-II SLS structure. We have also demonstrated the first uncooled SLS LED with emission around 5 and out to ~10μm. All of our SLS structures have been grown on mismatched GaAs, offering integration with existing III-V technologies.

References

[1] see e.g. H Prier. Semicond Sci. and Technol. **5**, S12, (1990).
[2] S.R. Kurtz and R.M. Biefeld, Phys. Rev. **B 44**, 1143 (1991).
[3] W.L. Walukiewicz, Mat. Res. Soc. Symp. Proc. **148**, 137 (1989).
[4] P.D. Wang, S.N. Holmes, Tan Le, R.A. Stradling, I.T. Ferguson and A.G. de Oliveira, Semicond. Sci. and Technol. 7, 767 (1992) .
[5] P.J.P. Tang, C.C. Phillips and R.A. Stradling, Semicond Sci. and Technol. **8**, 2135, (1993).
[6] P.J.P. Tang, PhD thesis, London University (1994).
[7] N.L. Rowell, Infrared Phyics **28** (1), 37 (1988).
[8] I.T. Ferguson, A.G. de Oliveira and B.A. Joyce, J. Crys. Growth **121**, 267, (1992).
[9] R.L. Krönig, W.G. Penny, Proc. R. Soc. London, **A130**, 499, (1930).
[10] F.H. Pollak, "Semiconductors and Semimetals" edited by R.K. Willardson and A.C. Beer, pub. Academic, New York, **32**, 17, (1990).
[11] Landolt-Börnstien, "Numerical Data and functional Relationships in Science and Technology, New series III" **17a**, Berlin: Springer-Verlag, (1982).
[12] G. Bastard and J.A. Brum, IEEE J. Quant. Elec. **QE-22** (9), 1625, (1986).
[13] S.N. Smith, C.C. Phillips, R.H. Thomas, R.A. Stradling, I.T. Ferguson, A.G. Norman, B.N. Murdin and C.R. Pidgeon, Semicond. Sci. Technol. 7, 900 (1992).
[14] R.H. Thomas, PhD Thesis: London University (1992).
[15] Y.B.Li, R.A. Stradling, A.G. Norman, P.J.P. Tang and C.C. Phillips, Proceedings of the 22nd International Conference on The Physics of Semiconductors, Vancouver Canada,1994.(to be published by World Scientific, Singapore).

*Inst. Phys. Conf. Ser. No 144: Section 1*
*Paper presented at 7th Int. Conf. on Narrow Gap Semicond. Santa Fe 9–12 January*



# Interfaces in InAsSb/InGaAs strained-layer superlattices grown by MOCVD for use in infrared emitters*

R. M. Biefeld, D. M. Follstaedt, S. R. Kurtz, and K. C. Baucom

Sandia National Laboratory, Albuquerque, NM 87185-0601 USA

**Abstract.** We have prepared InAsSb/InGaAs strained-layer superlattices (SLS's) using metal-organic chemical vapor deposition (MOCVD). X-ray diffraction was used to determine lattice matching as well as composition and structure of the SLS's. The presence of an InGaAsSb interface layer was indicated by broadened x-ray diffraction peaks for samples grown under non-optimized conditions. Two types of interfacial layers apparently due to a difference in composition at the interfaces.were observed with transmission electron microscopy (TEM). The width of the x-ray peaks can be explained by a variation of the interfacial layer thicknesses. Optimized growth resulted in SLS's with narrow x-ray peaks.

## 1. Introduction

In our recent work we prepared heterostructures with biaxially compressed, As-rich $InAs_{1-x}Sb_x$ as active regions for midwave infrared emitters[1]. $InAs_{1-x}Sb_x$ / $In_{1-y}Ga_yAs$ ($x \approx y \approx 0.1$) strained-layer superlattices (SLSs), nominally lattice matched to InAs, were grown using metal-organic chemical vapor deposition. An SLS light-emitting diode was constructed which emitted at 3.6 µm with 0.06% efficiency at 77 K. An optically pumped, single heterostructure SLS / InPSb laser which emitted at 3.9 µm with a maximum operating temperature of approximately 100 K was also fabricated[1]. These results indicate the potential of this system for making infrared devices and the need for improved devices able to operate in this wavelength range at higher temperature. We are continuing to explore the growth of $InAs_{1-x}Sb_x/In_{1-y}Ga_yAs$ SLS's for their possible use in mid-wave, 2-6 µm infrared optoelectronic and heterojunction devices. This system was chosen because the compositions span the 2-6 µm wavelength range and they can be grown lattice matched to InAs with type I band offsets. Our previous studies of the InAsSb ternary system have demonstrated accurate composition control through the use of a thermodynamic model [2,3]. We have made high quality infrared detectors in our laboratory from doped strained-layer superlattices (SLS's) grown by MOCVD in the Sb rich end of this ternary[2,3]. We have focused our recent efforts on the growth of $InAs_{1-x}Sb_x/In_{1-y}Ga_yAs$ heterostructures on InAs with emphasis on the growth and characterization of the InAsSb layers due to the importance of this material in the active devices [1,4]. We have reported detailed growth conditions for $InAs_{1-x}Sb_x/In_{1-y}Ga_yAs$ SLS's and $InAs_{1-x}Sb_x$ alloys [4]. The efficiency of emitters in the 2-6 µm range is usually limited by Auger processes [1,4]. The increased electron-hole effective mass ratio of biaxially compressed InAsSb reduces the Auger transition rates and increases the infrared emission. In the present work, we have examined the structure of the $InAs_{1-x}Sb_x/In_{1-y}Ga_yAs$ SLSs in detail using x-ray diffraction and transmission electron microscopy. Herein we discuss observation of interface layers in these SLSs and possible models which can explain the observed x-ray diffraction patterns.

## 2. Experimental

This investigation was carried out in a previously described horizontal MOCVD system [5]. The sources of In, Sb, Ga, and As were trimethylindium (TMIn), trimethylantimony (TMSb), trimethylgallium (TMGa), and 100 percent arsine ($AsH_3$), respectively. Hydrogen was used as the carrier gas at a total flow of 8 SLM. The V/III ratio was varied from 2.7 to 10.3 over a temperature range of 475-525 $^{o}C$, at total growth pressures of 200 to 660 torr and growth rates of 0.75 to 3.0 μm/h. The group V molar fraction of TMSb in the vapor phase $[n_{TMSb}/(n_{TMSb} + n_{AsH3})]$ was varied from 0.06 to 0.09.

The growth was performed on (001) InAs substrates. which were cleaned by degreasing in solvents and deionized water. It was then etched for 20 to 30 seconds in a 50:1 mixture of sulfuric acid and hydrogen peroxide, rinsed with deionized water and blown dry with nitrogen.

X-ray diffraction patterns for the $InAs_{1-x}Sb_x$ SLS's were obtained by x-ray diffraction using a Cu x-ray source and a four crystal Ge monochromator. The (004) reflection was used to measure the lattice constant normal to the growth plane and the (115) or (335) reflections were used to determine the in-plane lattice constant for some cases [5]. In this way the composition determination is corrected for partial strain relaxation by misfit dislocations. The diffraction simulations were done using the Bede RADS program.

The optical properties of these SLS's were determined by infrared photoluminescence and absorption measurements. Infrared photoluminescence was measured at 14 K using a double-modulation technique which provides high sensitivity, reduces sample heating, and eliminates the blackbody background from the infrared emission spectra [6].

A key SLS specimen in this study (CVD 1172) was chosen for TEM examination. Cross-sectional specimens were prepared as described elsewhere [7,8]. Both transmission electron diffraction (TED) patterns and TEM images were obtained using a Philips CM20T (200 keV) microscope.

## 3. Results And Discussion

The TEM images clearly showed the layering of the SLS structure in all images. Interfacial layers could be identified in (004) two-beam images and high-resolution (lattice) images, but the latter gave the clearest determinations of their nature; see Figure 1. The lattice periodicity also gave an internal standard for magnification and allowed the overall periodicity to be accurately determined: 22.7±0.5 nm. Some individual periods varied from the average by about this same uncertainty. Lower magnification images indicated a constant periodicity throughout the 1.135 μm layer (50 periods). In addition, superlattice satellites extending from the zincblende reflections in the [00±1] directions give a periodicity of 23.0+0.5 nm, in good agreement with direct imaging. The diffraction patterns also showed the presence of CuPt order in the InAsSb layers, which is believed to be responsible for the reduction in bandgap of these layers from the expected 330 meV to 270 meV [4,8,9].

The first layer grown in the sequence was InAsSb; these layers could be identified in the SLS by their being thinner (~10 nm vs. ~12 nm for InGaAs) and by their slightly larger lattice constant in the growth direction. The contrast of the interfacial layers has a regular pattern for the specimen area imaged in high resolution, which appears to have fairly uniform thickness. A light interface is seen on going from InAsSb to InGaAs along the growth direction, and a dark interface on going from InGaAs to InAsSb. These interfacial contrasts occur between two layers with very similar contrast and lattice patterns, and are repeated within each period; thus we believe they indicate the presence of interfacial layers which differ from the two layers on either side. We believe that the interfaces have different chemical compositions from the layers. However, the nature of the compositional change at the interfaces cannot be specified without detailed modeling of the lattice image. The interfacial layer contrasts vary in width along individual interfaces and between different ones: the light interface width varies from 0.6-1.3 nm, while the dark one varies from 1.0-1.6 nm. These thickness variations can be used as described below to model the observed x-ray diffraction pattern.

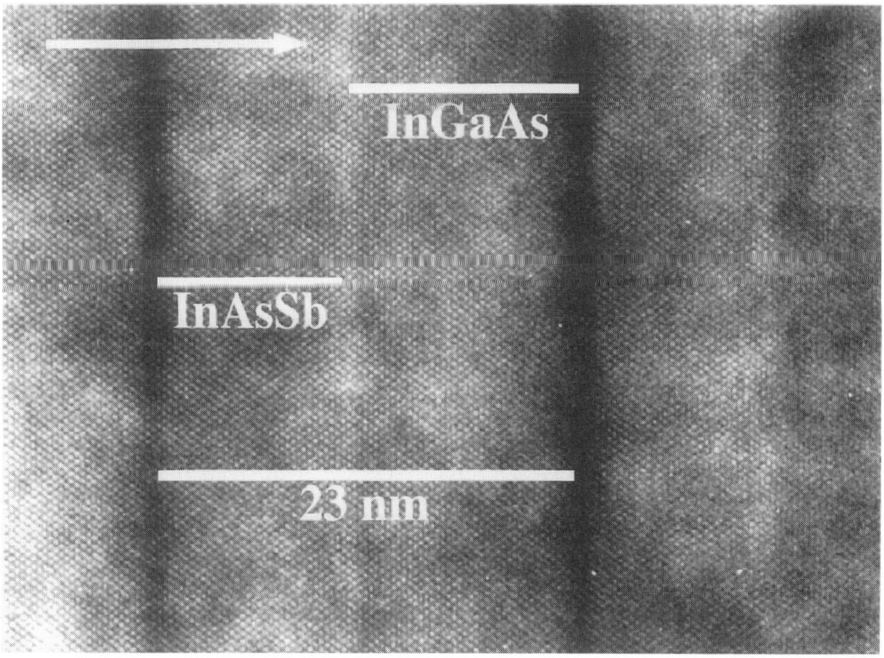


Figure 1. A high resolution lattice image showing the alternating light and dark interface regions of an $InAs_{1-x}Sb_x/In_{1-y}Ga_yAs$ SLS ( CVD1172). The growth direction is from left to right as indicated by the arrow in the upper left hand corner.

Both measured and calculated x-ray diffraction patterns of the (004) reflection of a selected $InAs_{1-x}Sb_x/In_{1-y}Ga_yAs$ SLS (CVD1172) grown on InAs are shown in Figure 2. The patterns show the intense InAs peak located at 0 arc seconds and the satellites due to the SLS. This sample was grown at atmospheric pressure with 30 second purges between layers with uninterrupted $AsH_3$ flow. The calculated pattern shown in curve C is for an $InAs_{0.9}Sb_{0.1}/In_{0.1}Ga_{0.9}As_{0.9}Sb_{0.1}/In_{0.88}Ga_{0.12}As\ /In_{0.95}Ga_{0.05}As_{0.8}Sb_{.2}$ SLS with 100/10/110/6 Å layer thicknesses. This structure was chosen assuming that the lighter interface has a lower average atomic number while the darker, second interface has more In and Sb. The compositions of the interfaces are not a unique choice and other compositions could be used and still yield a similar pattern with small adjustments to the compositions of the thick layers.

The calculated pattern shown in curve C indicates the ideal peak widths for a structure with interface layers that are repeated with no variation in thickness or composition. As one can see, the observed pattern is much broader than the ideal pattern illustrated by curve C. A better match to the observed pattern can be obtained by introducing thickness variations in the interfacial layers. Curve B was generated by using the compositions $InAs_{0.9}Sb_{0.1}/In_{0.2}Ga_{0.8}As_{0.9}Sb_{0.1}/In_{0.88}Ga_{0.12}As\ /In_{0.95}Ga_{0.05}As_{0.8}Sb_{.2}$, and varying the individual layer thicknesses while keeping the total period thickness constant at 222 Å. The layer thicknesses were varied from 99/8/110/5 Å to 90/17/101/14 Å in three steps of 2 Å each and a fourth step of 3 Å. This modeling procedure would give the same x-ray diffraction pattern as one in which the individual layer thicknesses were varied horizontally within the same period but keeping the same total vertical period. Again it should be emphasized that this is not a unique solution to the observed pattern but it is one that is plausible based on the observed structure from the high resolution micrograph. Other possible broadening mechanisms include a total period thickness or composition variations within the layers which are not supported by the TEM micrographs and the presence of

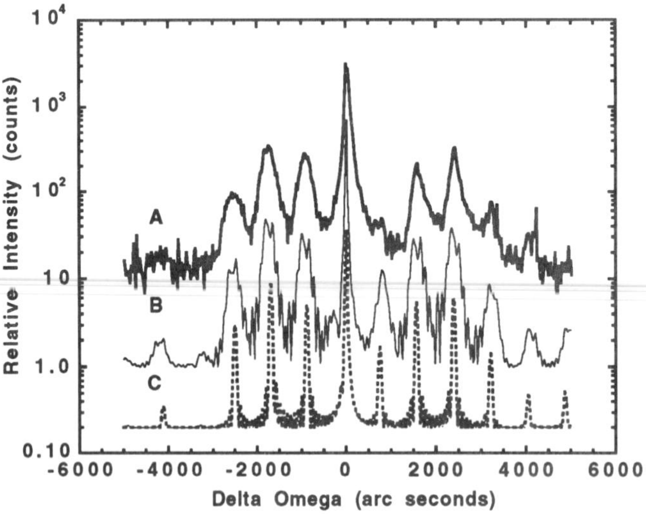


Figure 2. Curve A is the observed x-ray diffraction pattern of the (004) reflection of CVD1172. It is compared to calculated patterns in the lower curves to illustrate the improved match when variations in the thickness of the interface layers are included in the x-ray simulation B when compared to a simulation with no variations of thickness in the interface layers shown in C.

dislocations which are seen in TEM. The order of the layers also makes no difference to the calculated pattern. This simulated pattern, curve B, is a good match to the observed pattern, curve A, and is support for the existence of an interface layer of varying thickness as the explanation for the width observed in the x-ray diffraction pattern.

A variety of growth conditions were explored in an effort to improve the sharpness of the x-ray diffraction pattern. The optimum x-ray diffraction pattern was obtained at 200 torr with a 10 second growth interrupt at 475 $^{o}C$, a V/III ratio of 7, a growth rate of 1.8 μm/hour, and no reactants flowing during the interrupt. An x-ray diffraction pattern of an SLS grown using the optimum growth technique was simulated without using interface layers and no thickness variations and a good match between the observed and calculated pattern was obtained. These patterns are illustrated in Fig. 3. This improvement is due to the minimization of exposure of the InAsSb surface to $AsH_3$ and/or As. Extended $AsH_3$/As exposures are known to cause interface roughening in the InAs/GaSb system as well as asymmetrical interfacial layers [9,10]. The presence of Ga in these layers could be explained by the faster gas phase diffusion of TMGa compared to TMIn. Alternate explanations would be the roughness of the surfaces which enables mixing of Ga into the InAsSb layers at the interface or the diffusion of Ga into the InAsSb during growth.

In conclusion, we have established growth conditions for $InAs_{1-x}Sb_x/In_{1-y}Ga_yAs$ SLS's that minimize the width of the x-ray diffraction peaks. Asymmetric interfacial layers which are observed by TEM have been modeled using excess As and Ga at the interfaces. We anticipate that with these improvements in material quality, we should soon realize a reduction of Auger rates and higher temperature operation of midwave emitters with biaxially compressed InAsSb.

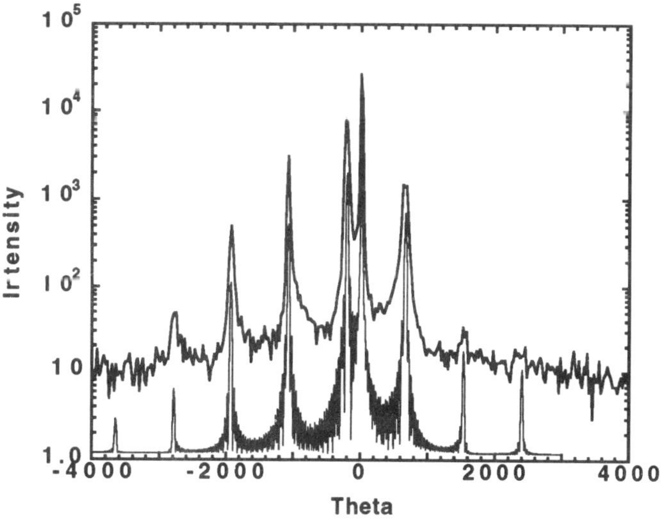


Figure 3. The upper curve is the observed x-ray diffraction pattern of the (004) reflection of CVD1479 compared to a calculated pattern in the lower curve to illustrate the improved diffraction pattern obtained using the optimum growth conditions.

**Acknowledgments**

We wish to thank J. A. Bur, M. P. Moran, and M. W. Pelczynski for technical assistance. This work was supported by the US DOE under Contract No. DE-AC04-94AL85000.

**References**

[1]Kurtz S R, Biefeld R M, Dawson L R, Baucom K C and Howard A J 1994 Appl. Phys. Lett. **64** 812-4.
[2] Biefeld R M 1986 J. Crystal Growth **75** 255-63.
[3] Biefeld R M, Kurtz S R and Casalnuovo S A 1992 J. Crystal Growth **124** 401-8.
[4] Biefeld R M, Baucom K C, Kurtz S R and Follstaedt D M 1994 Mat. Res. Soc. Symp. Proc., **325** 493-8 .
[5] Biefeld R M, Hills C R and Lee S R 1988 J. Crystal Growth **91** 515-26.
[6] Kurtz S R and Biefeld R M 1991 Phys. Rev. B **44** 1143-9 .
[7] Follstaedt D M, Biefeld R M, Kurtz S R and Baucom K C 1994 submitted to J. Electronic Materials.
[8] Follstaedt D M, Biefeld R M, Kurtz S R, Dawson L R and Baucom K C to be published elsewhere in this proceedings.
[9] Chow D H, Miles R H, Soderstrom J R and McGill T C 1990 J. Vac. Sci. Technol. B **8** 710-4.
[10] Feenstra R M, Collins D A, Ting D Z-Y, Wang M W and McGill T C 1994 J. Vac. Sci. Technol. B **12** 2592-7.
[11] Wei S-H and Zunger A. 1991 Appl. Phys. Lett. **58**, 2684-6

*Inst. Phys. Conf. Ser. No 144: Section 1*
*Paper presented at 7th Int. Conf. on Narrow Gap Semicond. Santa Fe 9–12 January*




# Heterostructures and Infrared Emitters with Compressed InAsSb Layers (Invited)

**Steven R. Kurtz and Robert M. Biefeld**
**Sandia National Laboratories, Albuquerque, NM 87185**

**Abstract:** An overview is presented of strained InAsSb heterostructures and infrared emitters. InAsSb/InGaAs strained-layer superlattices (SLS) and InAsSb quantum wells were characterized using magneto-photoluminescence and compared with unstrained InAsSb and InAs alloys. In heterostructures with biaxially compressed InAsSb, large quantum confinement energies were observed, and the holes exhibited a decrease in effective mass, approaching that of the electrons. This study demonstrates that the electrons and holes in the InAsSb heterostructures are confined in the InAsSb layers, and the band offsets are type I. A large increase in the Auger-1 threshold energy should accompany the strain-induced change in valence-band symmetry of the InAsSb layers. Correspondingly, the InAsSb heterostructures display high radiative efficiencies and increased activation energies for nonradiative recombination compared with the unstrained alloys. LEDs and lasers with InAsSb heterostructure active regions are described. InAsSb/InGaAs SLS LEDs operating at 300K at wavelengths ≤ 5 μm have been demonstrated. Optically pumped InAsSb/InGaAs SLS lasers, with InPSb cladding, had a maximum operating temperature of 100K.

## 1. Introduction

Frequently, the Auger-1 process (i.e. An electron and hole recombine by scattering a second electron up into the conduction band.) dominates radiative recombination in narrow bandgap III-V semiconductors, and as a result, the wavelength of diode lasers operating at room temperature has been limited to ≤ 2.1-2.3 μm.[1] In a biaxially compressed III-V layer, the |3/2,±3/2> hole ground state can increase the in-plane, electron-hole effective mass ratio ($m^*_e/m^*_h$) over that found in bulk material. In the 2-dimensional limit, this effective mass ratio will result in an increased threshold energy for Auger-1.[2-4] Therefore, midwave infrared (2-6 μm) emitters with biaxially compressed, InAsSb active regions may exhibit improved performance and higher temperature operation.[5] In this paper, we will describe the optical characterization of InAsSb/InGaAs strained-layer superlattices (SLSs) and InAsSb quantum wells with InAs barriers. We determine the electronic properties of these heterostructures and present evidence indicating that a large increase in Auger-1 threshold energy results from the strain-induced valence-band symmetry of the InAsSb layer. LEDs and lasers with InAsSb heterostructure active regions are described.

## 2. Optical Studies

A. Experimental Details

InAsSb and InGaAs alloys and heterostructures were grown by metal-organic chemical vapor deposition (MOCVD) on InAs substrates. SLS and ternary compositions, layer thicknesses, and lattice constants were determined from both (004) and (115) or (335) x-ray rocking curves. In this study, photoluminescence spectra for an InAsSb/InGaAs SLS and InAsSb quantum wells with InAs barriers are compared with those for unstrained $InAs_{0.93}Sb_{0.07}$ and InAs alloys. The SLS characterized in the optical study was an $InAs_{0.91}Sb_{0.09}$ / $In_{0.87}Ga_{0.13}As$ SLS (90 Å/ 130 Å layer thicknesses), nominally lattice matched to the InAs substrate. The multiple quantum well sample consisted of 318, 159, 106, and 53 Å thick quantum wells of $InAs_{0.91}Sb_{0.09}$, separated by 500 Å thick, InAs barriers. The sum of the four quantum well thicknesses is less than the critical layer thickness (≈ 1000 Å), and the quantum wells are pseudomorphic.

Throughout our studies of As-rich, InAsSb (5-50% Sb), the bandgaps of our InAsSb alloys were smaller than accepted values [6], and the unexpectedly low photoluminescence energies of As-rich, InAsSb heterostructures is traceable to the bandgap anomaly observed for the InAsSb alloys. Electron diffraction results indicate compositional ordering and phase separation may occur in the As-rich, InAsSb grown at low temperatures by vapor phase epitaxy.[7] As demonstrated in the following discussion, the InAsSb material displays "single phase, random-alloy-like" optical properties except for the bandgap anomaly, and we assume that the strain imposed by the InAs heterostructure dominates any internal strain occurring within domains of the InAsSb.

Midwave infrared photoluminescence was measured by operating an FTIR spectrometer in a double-modulation mode. In the magneto-photoluminescence experiments, a fluoride optical fiber was used to transmit infrared light in the magnet cryostat. The photoluminescent light was collected by the fiber and analyzed with the FTIR equipped with an InSb photodiode. All measurements were made in the Faraday configuration with the magnetic field parallel to the growth, (001) direction of the sample.

B. Magneto-Photoluminescence

Low temperature, zero-field photoluminescence spectra for the multiple quantum well, SLS, unstrained InAsSb, and InAs samples are shown in Figure 1. For each sample, the photoluminescence spectra consists of a single peak with a linewidth of approximately 10 meV. Emission from single quantum wells is clearly resolved in Fig 1(c), with the quantum well thickness indicated in the figure. (InAs barriers, 500 Å thick, separate the quantum wells in Fig. 1(c).) The photoluminescence energies of the thick (318 Å) quantum well and the InAsSb alloy provide estimates of the bandgap energies of the MOCVD alloys, and both alloys exhibit the bandgap anomaly. A large quantum size shift is observed for the quantum wells. Analysis of quantum size effects for InAsSb/InGaAs and InAsSb/InAs heterostructures indicates that both have type I band offsets. Contrary to our experimental results, a type II band offset would result in negligible quantum size shifts for the InAsSb/InAs heterostructure with equal InAs barrier thicknesses. The bandgap and conduction band offset are sensitive to ordering or phase separation occurring in the InAsSb; the valence band offset is insensitive to these effects.

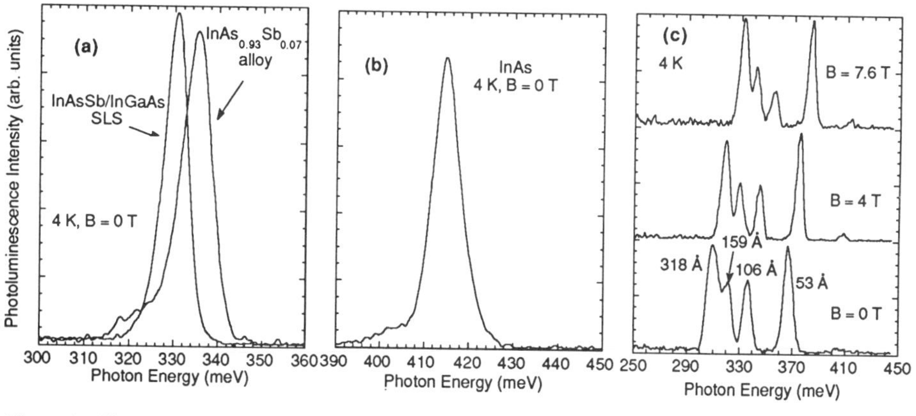


Figure 1 - Photoluminescence spectra of (a) $InAs_{0.91}Sb_{0.09}$ / $In_{0.87}Ga_{0.13}As$ SLS and $InAs_{0.93}Sb_{0.07}$ alloy, (b) InAs, and (c) $InAs_{0.91}Sb_{0.09}$/InAs quantum wells.

The magnetic field dependence of the photoluminescence energies for each of these samples is shown in Figure 2. For each photoluminescence line, the reduced mass, $\mu$, corresponding to the slope obtained from the free electron-hole approximation ($\frac{dE}{dB} = \mu_B\left(\frac{1}{m_e^*} + \frac{1}{m_h^*}\right) = \frac{\mu_B}{\mu}$; $\mu_B$ is the Bohr magneton), is indicated in the figure. The reduced masses for the unstrained alloys ($\mu_{InAsSb}$= 0.037 and $\mu_{InAs}$=0.031, in units of free electron mass) are significantly larger than those observed in the strained heterostructures ($\mu$=0.015-0.024) due to the low in-plane hole mass associated with the heavy-hole ( |3/2,±3/2> ) ground state of the biaxially compressed, InAsSb layers.

Excitonic behavior is revealed in the magneto-photoluminescence results. For all samples, the photoluminescence peak energy is insensitive to magnetic field for B< 2T, characteristic of a diamagnetic exciton, and in the linear region observed at higher fields, the reduced mass values obtained from the free carrier approximation are consistently too large, due in part to the binding energy of the exciton. Also, note the increase in reduced mass (or decrease in slope) as the quantum wells become thinner (Fig. 2(b)).

Using measured parameters for InAs and semi-empirical expressions for nonparabolicity and magneto-exciton energies, we can estimate exciton binding energies and correct the reduced mass values.[8] Assuming a 3-dimensional exciton, we obtain corrected reduced mass values, $\mu$=0.010, 0.026, and 0.023, for the 318 Å thick quantum well, unstrained InAsSb alloy, and InAs, respectively. The exciton binding energies for the 318 Å well and the alloys were 1.0 meV and 1.8 meV, respectively. Examining the thinnest (53 Å) quantum well, we estimate an exciton binding energy of 2.5 meV for a reduced mass, $\mu$=0.016, and assuming a 3-dimensional exciton. The change in slope in Fig 2(b), associated with quantum confinement, indicates an increase in exciton binding energy of the correct magnitude.[8]

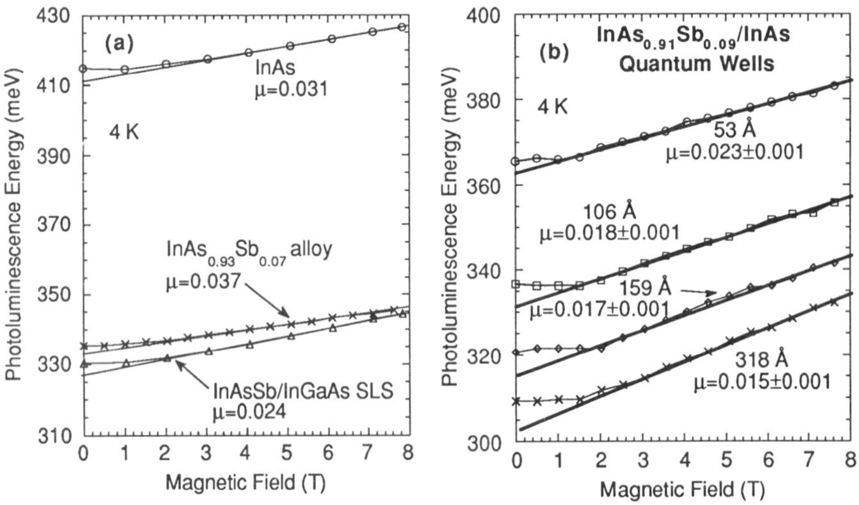


Figure 2 - Magnetic field-induced shift of the photoluminescence energy for (a) the InAsSb/InGaAs SLS and the InAsSb and InAs alloys, and (b) the InAsSb/InAs quantum wells.

C. Modification of Auger-1 Threshold Energies

In the 2-dimensional limit, a large increase in the Auger-1 threshold energy accompanies this strain-induced change in valence-band symmetry. We examined the temperature dependence of the photoluminescence intensity for SLS and bulk, $InAs_{0.93}Sb_{0.07}$ and InAs samples. (see Figure 3) At > 100 K, the radiative efficiency decreases exponentially for all samples, and the activation energy for nonradiative recombination ($\Delta E$) in the SLS displayed a marked increase compared with that of the unstrained alloys ($\Delta E$=0.26 $E_{gap}$ vs $\Delta E$=0.06 $E_{gap}$). Other InAsSb/InGaAs SLSs that we examined also displayed large activation energies. Due to variations in optical alignment and sample doping and defects, the differences in the relative radiative efficiency between samples may not be accurate, but generally, the SLSs have higher radiative efficiencies than the unstrained alloys.

Approaching room temperature, Auger-1 will be the dominant nonradiative process. The radiative efficiency for extrinsic, n-type material is

$$\eta = \frac{\tau_A}{\tau_R} \propto (1/n_0)\cdot \exp\left(\frac{\Delta E}{kT}\right) \quad (1) \qquad \text{and} \qquad \Delta E = \left(\frac{m_e^*/m_h^*}{1+(m_e^*/m_h^*)}\right)\cdot E_{gap} \quad (2),$$

where $n_0$ is the electron density, $\tau_R \propto (T/n_0)$ is the radiative lifetime, and

$\tau_A \propto (T/n_0^2)\exp(\Delta E/kT)$ is the Auger-1 lifetime (in 2 dimensions).[9] Eq. 1 is based on simple energy-momentum conservation for isotropic, parabolic bands, and the expression for activation energy, ΔE (ΔE = Auger threshold - Bandgap), is valid in 2 or 3-dimensions.[2] In the 2-dimensional limit, the SLS in-plane effective masses determine ΔE,[3,4] and therefore, the SLS activation energy is larger than that of the unstrained alloys due to the decreased hole mass in biaxially compressed layers of the SLS.

Electron-hole effective mass ratios for each sample are compared with theoretical values obtained from k·p calculations and with experimental values obtained from magneto-photoluminescence and temperature dependence studies. In each case there is good agreement between theoretical and experimentally determined mass ratios. The effective mass ratios ($m^*_e / m^*_h$) determined by the 3 methods for the SLS were in the range, 0.3-0.5 , and the effective mass ratios for the unstrained alloys were 0.06±0.02.[9] Comparing the SLS with the alloys, the SLS consistently displayed the predicted effects of the valence band under biaxial compression.

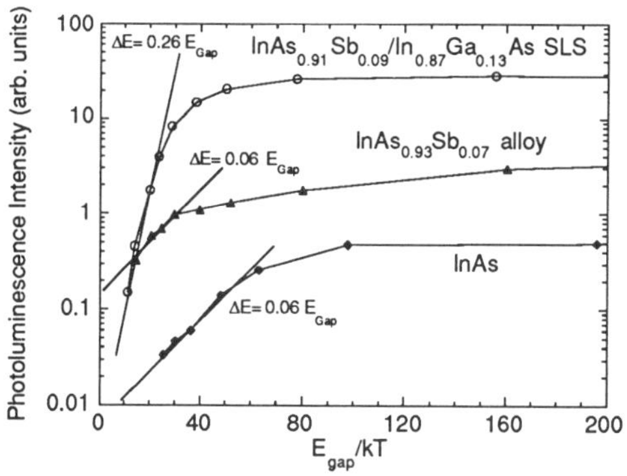


Figure 3 - Temperature dependence of the photoluminescence intensity (radiative efficiency) for the InAsSb/InGaAs SLS and InAsSb and InAs alloys. Activation energies (ΔE) are indicated in the figure.

## 3. Midwave (4 μm) Emitters

A. Broadband LEDs

For many chemical sensing applications, a broadband LED infrared source is preferred over a diode laser. We have constructed several MOCVD grown, LEDs as infrared sources and laser precursors. These devices demonstrate n and p-type doping of both InAsSb heterostructure and InPSb cladding materials. The LEDs were 380 μm diameter mesa devices with $InAs_{1-x}Sb_x$ / $In_{1-x}Ga_xAs$ SLS (x≈0.1) active regions. Typically, LED and laser emission occurs at the active region photoluminescence energy, shifted to slightly higher energy (≈ 10 meV) by band filling. In devices where the active region and cladding lattice constants are well matched to that of the substrate, defect-related nonradiative recombination is minimized, and the temperature dependence of the LED emission (Fig. 4(a)) resembles the temperature dependence of the SLS radiative efficiency (see Fig. 3). The LED in Figure 4(a) had a low temperature external efficiency of 0.06% at a wavelength, 3.6 μm. Correcting for reflection losses, the internal efficiency may be as high as 2%. As shown in Figure 4(b), these devices operate at room temperature and span the 4-5 μm range necessary for CO and $CO_2$ detection. The total external power output of this device at 300 K was 180 nW with 100 mA drive current.[5]

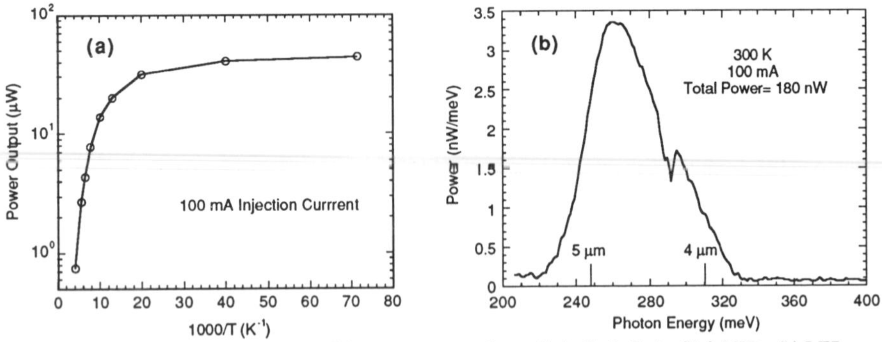


Figure 4 - (a) Temperature dependence of the power output for an InAsSb/InGaAs SLS LED. (b) LED emission spectrum at 300K.

B. Lasers

A laser structure was realized by growing a 2 µm thick layer of $InP_{0.65}Sb_{0.35}$, followed by a 1.5 µm thick SLS active region, on an InAs substrate. Finally, a 100 Å thick InPSb cap layer was grown on top of the SLS to reduce surface recombination. The structure of the active region determined by x-ray analysis was $InAs_{0.87}Sb_{0.13}$ / $In_{0.93}Ga_{0.07}As$ ( 95 Å/ 133 Å layer thicknesses). The buried InPSb layer and the space

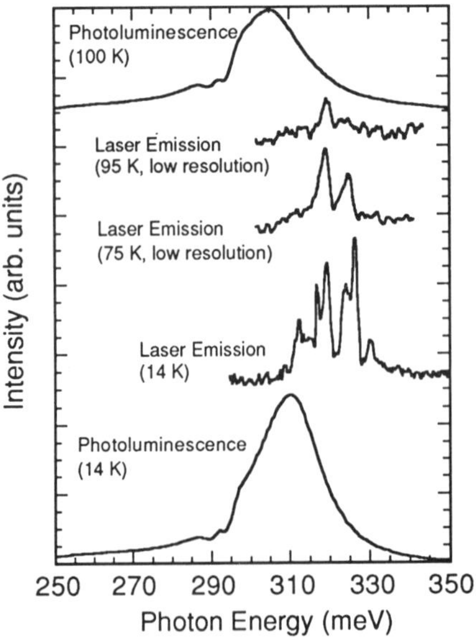


Figure 5 - Photoluminescence and lasing spectra for a SLS/InPSb "single heterostructure" up to 100 K.

above the SLS produced optical confinement within the active region. Laser emission at 3.9 µm was observed by optical pumping cleaved bars. Laser and photoluminescence spectra for the SLS-InPSb structure are shown in Figure 5. The maximum lasing temperature was 100

K.[5] To date, we have not demonstrated electrically injected MOCVD grown, SLS lasers with InPSb claddings operating at ≥ 4 μm. Shorter wavelength, 3.06 μm, electrically injected InGaAsSb/InPSb double heterostructure lasers were marginal and had a maximum operating temperature of 35 K.[10] MBE-grown, electrically injected lasers with InAsSb/InAlAs SLS active regions and AlAsSb claddings have been demonstrated with pulsed operating temperatures up to 85 K.[11] "More work is needed to prove that the strained quantum well structure improves 4 μm laser performance."

## 4. Conclusions

Using infrared photoluminescence and magneto photoluminescence, we have examined the electronic properties of a series of heterostructures with biaxially compressed InAsSb. Quantum size studies show that electrons are confined to the InAsSb layer, and magnetic field studies demonstrate that holes are confined to the heavy hole (small in-plane effective mass) ground state of the biaxially compressed InAsSb layers. The band offsets for these heterostructure are type I. Analysis of the magneto-excitonic behavior indicates that the hole mass may be quite small ( $\mu$=0.010), and exciton binding energy increases with quantum confinement. The photoluminescence efficiency versus temperature revealed that an increased activation energy for nonradiative recombination accompanies the decreased hole mass in the SLS. Photoluminescence efficiency activation energies for the SLS and unstrained alloys agree with estimated Auger-1 threshold energies, with the activation energy for the SLS approaching the Auger-1 value in the 2-dimensional limit. Although the material properties of the MOCVD-grown InAsSb are non-ideal, these heterostructures display Auger-1 threshold energies, effective masses, band offsets, and "random alloy-like" properties that are desirable for active regions in midwave infrared diode lasers.

Infrared emitters were constructed with MOCVD-grown, compressed InAsSb active regions. InAsSb/InGaAs SLS LEDs may have internal efficiencies as high as 2% at low temperature. Without device modifications to minimize ohmic heating or reflection loss, these LEDs emit ≈ 200 nW over the 4-5 μm range at room temperature. Optically pumped, InAsSb/InGaAs SLS lasers have been demonstrated with a maximum operating temperature of 100 K. Although strained heterostructures hold the promise of reduced Auger-1 rates, 4 μm lasers with these active regions have not yet demonstrated improved performance over lasers with unstrained bulk active regions.

The work presented in this overview resulted from collaborations and conversations with Sandia colleagues; L. R. Dawson, A. J. Howard, M. H. Crawford, D. M. Follstaedt, H. J. Hjalmarson, J. S. Nelson, K. C. Baucom, S. K. Lyo, and E. D. Jones. Further details may be found in their contributions to these proceedings. This work was performed at Sandia National Laboratories, supported by the U. S. Department of Energy under contract No. DE-AC04-94AL85000.

**References**

[1] Eglash S J and Choi H K 1990 Appl. Phys. Lett. **57** 1292; 1992 Appl. Phys. Lett. **61** 1154.
[2] Dutta N K 1983 J. Appl. Phys. **54** 1236
[3] Adams A R 1986 Elect. Lett. **22** 249
[4] Yablonovich E and Kane E O 1988 IEEE J. Lightwave Tech. **6** 1292
[5] Kurtz S R, Biefeld R M, Dawson L R, Baucom K C, and Howard A J 1994 Appl. Phys. Lett. **64** 812
[6] Fang Z M, Ma K Y, Jaw D H, Cohen R M, and Stringfellow G B 1990 J. Appl. Phys. **67** 7034, and references therein.
[7] Follstaedt D M, Biefeld R M, Baucom K C, and Kurtz S R (accepted for publication, J. Elect. Mat.)
[8] Kurtz S R and Biefeld R M (accepted for publication, Appl. Phys. Lett.)
[9] Kurtz S R, Biefeld R M, and Dawson L R (accepted for publication, Phys. Rev. B)
[10] Menna R J, Capewell D R, Martinelli R U, York P K, and Enstrom R E 1991 Appl. Phys. Lett. **59** 2127.
[11] Choi H K, Turner G W, and Le H Q (to be published).

*Inst. Phys. Conf. Ser. No 144: Section 1*
*Paper presented at 7th Int. Conf. on Narrow Gap Semicond. Santa Fe 9–12 January*



# High-power diode-laser-pumped midwave infrared HgCdTe/CdZnTe quantum well lasers

H. Q. Le and A. Sanchez

Lincoln Laboratory, Massachusetts Institute of Technology
Lexington, MA 02173-9108, USA

J. M. Arias, M. Zandian, and R. R. Zucca

Rockwell International Science Center
Thousand Oaks, CA 91358, USA

Y.-Z. Liu

Fermionics Corporation
Simi Valley, CA 93063-3480, USA

**Abstract.** High-power midwave infrared HgCdTe/CdZnTe broad-stripe quantum well lasers were studied using 0.94-μm diode arrays for pumping. Laser power, threshold, and efficiency were studied for devices of various lengths and at different temperatures. At 88 K, a device yielded 1.3-W peak power and 105-mW average power per facet at 3.2 μm. The highest operation temperature was 154 K. Possibility of compact devices with closed-cycle cryo-cooler is discussed.

## 1. Introduction

Mid-wave and long-wave infrared (MWIR and LWIR) lasers are useful for spectroscopic detection of trace gases. Low power (~ 1 mW) commercial Pb-salt diode lasers have been used in instruments for point sampling or long-path sampling detection with retroreflectors [1]. In these applications, laser detection techniques are effective but not unique. Non-laser approaches, *e. g.* Fourier transform infrared spectroscopy or ion mass spectroscopy are comparably effective. With higher power (> 1 W), lasers can be uniquely useful for remote sensing, which has the ability of rapid surveying of a large area. Presently, only solid state or gas lasers have sufficient power for remote sensing that relies on aerosol or topographic backscattering for the return signal. However, semiconductor lasers with ~ 1-W power level can also be sufficient for short to medium-range (~ 100 m) remote sensing with topographic backscattering. Such a power level is possible at ~ 80 K [2,3], and cryo-cooler such as Stirling cooler presently has sufficient capacity for such an approach.

This paper describes our efforts to develop high-power HgCdTe/CdZnTe MWIR lasers. The majority of recent activities in MWIR semiconductor lasers focuses on III-V materials: InAs, GaSb, or others [4-11]. While II-VI HgCdTe material has been developed extensively for IR detectors, it has not been fully exploited for laser development. Only a few demonstrations of HgCdTe and HgCdMnTe diode lasers have been reported [12-13], and other laser works employed optical pumping, in part because of difficult p-type doping [14-20]. With the availability of compact high-power diode lasers, we show that diode-pumped HgCdTe lasers can yield high power, and that diode-pumping can be a cost-effective approach to use this material system. This paper reports results on MBE-grown 3.2-μm HgCdTe/CdZnTe quantum well lasers and discusses issues relevant to practical device development.

## 2. Technical approaches

### *2.1. Materials*

Laser structures designed for optical pumping are generally different from those for injection [22]. Structure for optical pumping can be designed with a large vertical optical mode, with a matching gain profile, and without doping which can help reduce free carrier absorption. In this work, a simple QW structure is studied and sketched in Fig. 1. The structure consists of two 1-μm-thick $Hg_{0.5}Cd_{0.5}Te$ epilayers sandwiching an active region of three 20-nm wide $Hg_{1-x}Cd_xTe$ QWs with $x$ nominally = 0.38. The $Hg_{0.5}Cd_{0.5}Te$ epilayers serve both as pump absorbers and carrier and optical confinement layers. The entire structure is cladded on one side by the CdZnTe substrate and on the other side by a ~ 4-μm-thick polycrystalline CdTe capping layer.

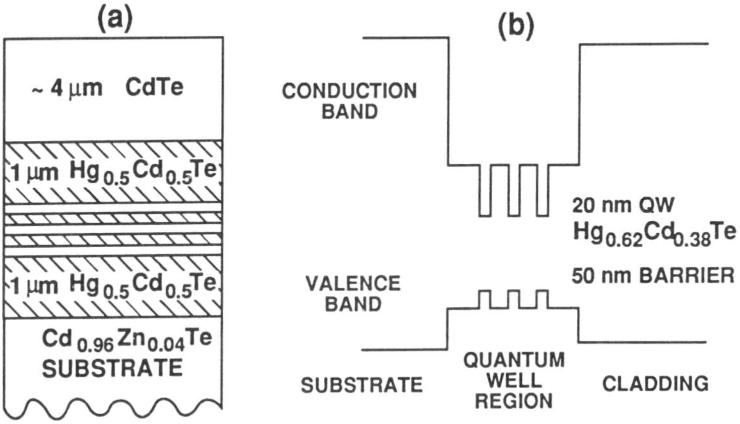


Fig. 1 (a) Laser wafer structure and (b) the bandgap diagram

Two wafers were grown at 190 C in a RIBER 2300 MBE system on (211)B-oriented $Cd_{0.96}Zn_{0.04}Te$ substrates. The (211)B orientation was chosen because it results in adequate HgCdTe surface morphology and has been used to produce injection diode lasers [13] and high-performance photodiodes [21]. Growth was carried out with effusive sources that contain Hg, $Te_2$ and CdTe. Nominal beam fluxes for these sources were in the $10^{-5}$-, $10^{-6}$- and $10^{-7}$-torr beam equivalent pressure ranges, as measured with a nude ion gauge in the growth position and corrected for atomic and molecular weight as well as ionization efficiency. Under these conditions, growth rates were 3 μm/h. The optimum Hg/$Te_2$ flux ratio necessary to maintain monocrystalline HgCdTe growth was determined using in situ reflection high-energy electron diffraction analysis.

### *2.2. Device processing and experiments*

The wafers were thinned and polished to ~ 100 μm, antireflection coated for 0.94 μm (pump wavelength) on the substrate side, metallized on the epi side, sectioned into 10 x 1- to 2-mm bars, and cleaved into devices of length from 0.5 to 3.5 mm. Each device was In-soldered on a Cu heat sink by exposure to a temperature of 160°C for less than 3 min. The heat sink was mounted on a cold finger in a Dewar with a temperature variable from 80 to 300 K.

The optical pump source was a 1-cm$^2$ 5-bar diode array stack emitting at 0.94 μm and capable of up to 250-W peak power at 15% duty cycle. A lens system focused the stack output into a stripe ~ 0.3 mm wide and ~ 4 mm long on a sample inside the Dewar. Narrower stripe width can be obtained with a pair of variable light concentrators mounted on the sample submount. Details of the pumping geometry have been described elsewhere [22]. The average power output from the sample was measured outside the Dewar with a thermoelectric power meter. The collection optics included an f/0.75 ZnSe lens system and a Ge filter, all of which were 3-5 μm antireflection coated. The measured power was corrected only for the transmittance of the $BaF_2$ Dewar window and not for the collection finite aperture. Since the output beam divergence was observed to exceed the Dewar window aperture, the actual power output was probably slightly higher than the figure reported here. The pulse peak power was derived by dividing the average power by the duty cycle. The pulse height and width were determined with a fast InSb detector.

## 3. Results and discussion

### *3.1 General consideration*

From the system point of view, as most cooling systems have rapidly decreasing heat pumping efficiency for decreasing temperature in the cryogenic range, achieving optimal laser power efficiency is crucial. The exponential rise of threshold for increasing temperature is certainly a major issue in MWIR and LWIR lasers. However, even in the temperature range where threshold is not excessive, it will be shown that the drop in efficiency is the main impediment to obtaining useful laser output power at high temperature. In the present study, using 0.94-μm pump diode to obtain 3.2-μm photons is not efficient because the excess energy of 0.93 eV per photon results mostly in heat. However, the study can be useful to assess the present efficiency problem and to project further improvement. In particular, pumping with 2-μm diode lasers can be very attractive [23].

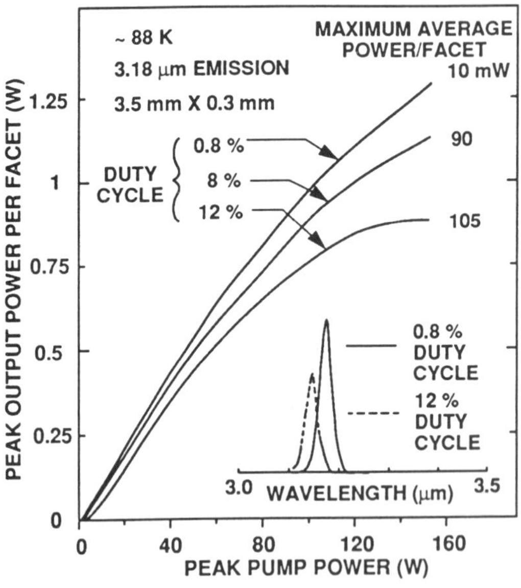


Fig. 2 Power output of 3.2-μm HgCdTe/CdZnTe QW lasers pumped with 0.94-μm diode array with various duty cycles at 88 K

### *3.2. Power output*

Result with the highest power output at 88 K is shown in Fig. 2. The device was 3.5-mm long and emitted at 3.18 μm. The pump pulse length was 8 μs, and the repetition rate was varied up to 15 kHz. At low duty cycle, peak power up to 1.3 W/facet was observed. At 12% duty cycle, a maximum average power of 105 mW per facet was obtained, but with lower peak power because of heating. Based on the observed blue shift in the emission spectra [inset of Fig. 2], the temperature rise was inferred to be ~ 26 K.

*3.3. Threshold and efficiency:*

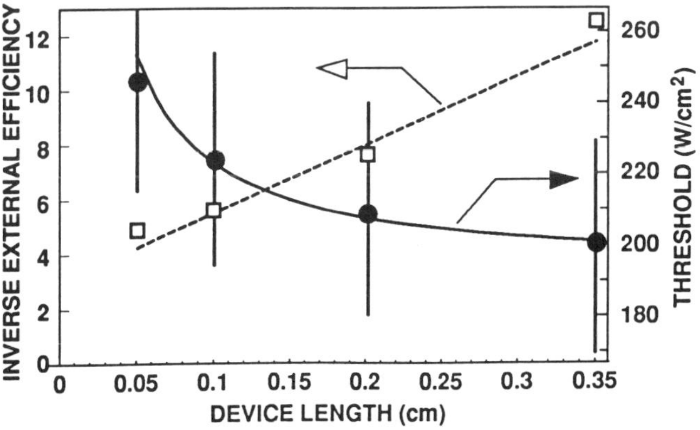


Fig. 3 Output efficiency and threshold at 83 K for devices of various lengths.

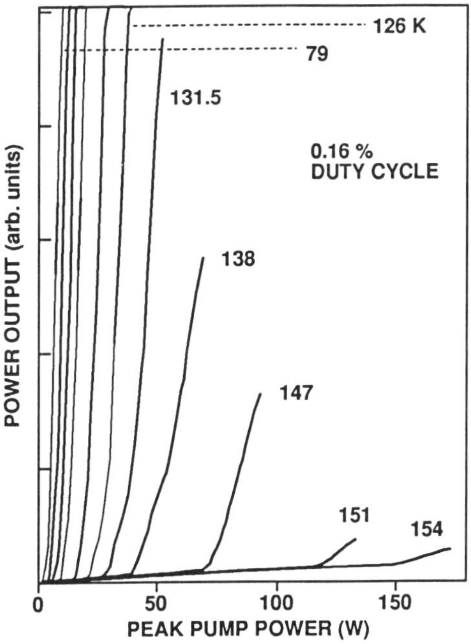


Fig. 4 Power output at different temperature.

At low lasing power and temperature, threshold and efficiency appear to have simple behaviors. The results at ~ 83 K with 0.8 % pump duty cycle for devices with different lengths excited with the same ~ 100-μm stripe width are shown in Fig. 3. The efficiency was measured for pump power just above the threshold. From the behavior of the efficiency, an internal loss of ~ 12 $cm^{-1}$ can be inferred. For the threshold, the lowest value observed, 200 $W/cm^2$ is equivalent to a current of 150 $A/cm^2$. Although the measurement uncertainty on the threshold in terms of absolute physical unit is large, its relative magnitude was determined with higher precision, and the variation vs. device length in Fig. 3 was clearly observed. The solid and dashed curves in Fig. 3 are simple model fits. In this linear model, the gain is expressed as $g = G_o(n - n_o)$, and the free carrier absorption loss is $\alpha = an$, where $G_0$ and $a$ are parametrized coefficients, $n$ is the population, and $n_0$ is the population at transparency. The dependence of the threshold population $n_{th}$ on the device length $L$ is given by $g_{th} - \alpha_{th} = \ln(1/R_1R_2)/2L$, where $R_1$, $R_2$ are the two facet reflectivity. The external efficiency is expressed $\eta_e = \eta_i[\ln(1/R_1R_2)]/[\ln(1/R_1R_2) + 2\alpha L]$ where $\eta_i$ is the internal efficiency, independent of device length. The relation between the absorbed pump photon flux $N_p$ and the population $n$ at or below the threshold can be expressed as $N_p - n(A + Bn + Cn^2) = 0$ where $A$, $B$, $C$ are the nonradiative, radiative, and Auger recombination rate coefficients, respectively. At threshold $B$ is negligible and at low temperature $C$ can also be neglected, then the relation between $N_p$ and $n$ is linear. The solid and dashed curves in Fig. 3 were obtained with $G_o n_o = 90$ cm$^{-1}$, $a n_o = 12$ cm$^{-1}$, and $R_1 = R_2 = 0.237$.

At higher temperature or higher lasing intensity, the behaviors of threshold and efficiency are more complex, and the linear model appears inadequate. Figure 4 shows a typical power output for temperature increasing from 79 to 154 K. The operation temperature of 154 K is the highest reported for any HgCdTe laser emitting at 3.1 μm or longer wavelengths. The duty cycle was fixed at 0.16% to avoid heating. The threshold and efficiency vs. temperature for samples of different lengths are shown in Fig. 5. Evidently, even with attainable threshold at higher temperature, the rapid drop in efficiency is even more detrimental to obtaining useful power output.

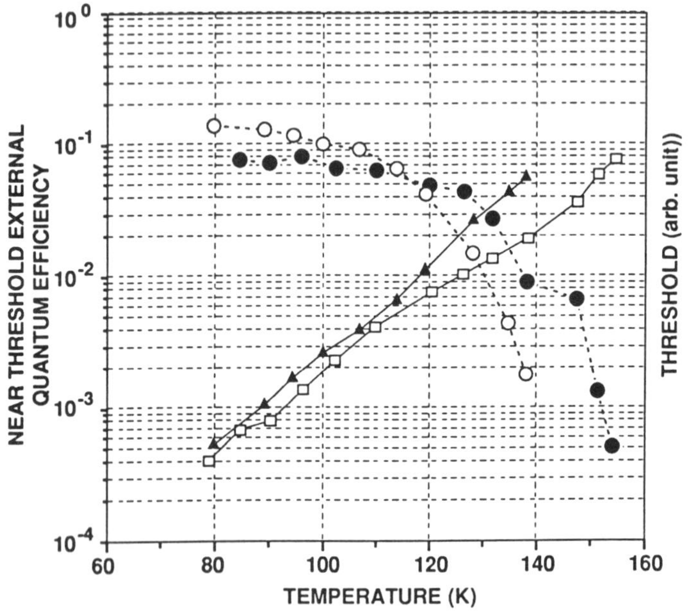


Fig. 5 Efficiency and threshold vs. temperature of 1- and 3.5-mm long devices. Triangles: threshold, open circles: efficiency of 1-mm device; squares: threshold, filled circles: efficiency of 3.5-mm device.

The difference between the two samples is most noticeable at high temperature. The threshold behavior is consistent with a lower gain parameter $G_o$, caused by thermally broadened carrier energy distribution, as well as by a higher carrier recombination rate $A + Bn_{th} + Cn_{th}^2$, especially the Auger rate $Cn_{th}^2$. The efficiency of the shorter sample starts out higher at low temperature owing to a lower internal loss, but decreases to a lower level at higher temperature, indicating that the internal efficiency $\eta_i$ and not internal loss is the dominant factor at high temperature. In Fig. 6, the threshold and efficiency for 1-, 2-, and 3.5-mm long samples are parametrically plotted against each other. The device length dependence becomes less significant, indicating a parametric relation between the two variables.

*3.4 Discussion:*

The above results suggest the following consideration in device development. In order to extract higher power efficiency at low temperature, *e. g.* 80 K, it is useful to increase the output coupling efficiency by reducing the facet reflectivity with low reflection coating, as shown in the calculation in Fig. 7. However, if the device is to be operated at higher temperatures, *e. g.* 110-130 K, it is advantageous to reduce the threshold for higher efficiency, as evidenced in Figs. 5 and 6 with longer device or higher reflectivity facet coating.

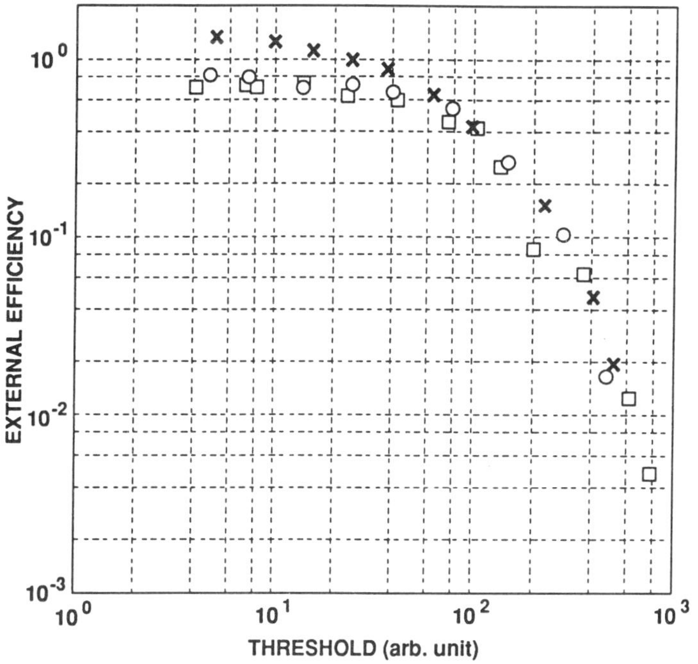


Fig. 6 Parametric plot of efficiency vs. threshold for various devices of different lengths: crosses: 1 mm, circles: 2 mm, and squares: 3.5 mm.

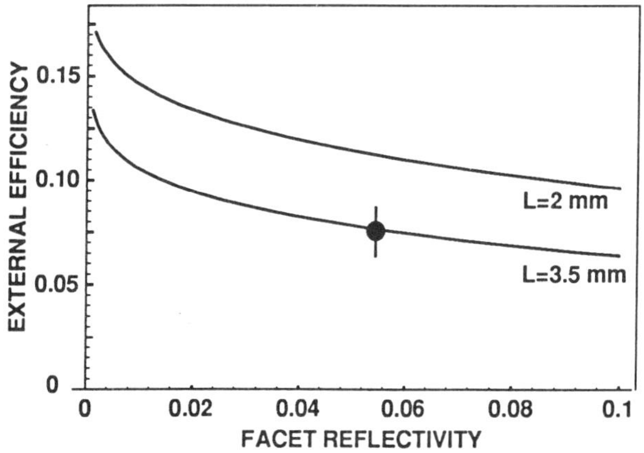


Fig. 7 Calculated laser output efficiency vs. facet reflectivity for two different device lengths. The circle indicates the experimental result with uncoated facets.

The present results have been obtained from only one wafer, whose extrinsic properties such as defect density and the type and density of unintentionally doped majority carriers are not known. It is premature and insufficient to analyze the wafer performance with intrinsic model. However, it can be expected that with future works and more experimental data, a clearer understanding of HgCdTe QW lasers can be obtained to help further improvement.

## 4. Summary

In summary, diode-laser-pumped HgCdTe quantum well lasers operated at 88 K yielded approximately 1-W-level peak power and 0.1-W-level average power/facet with a net optical-to-optical efficiency ~ 2%. With commercially available compact Stirling coolers having a few W heat load capacity at 77 K, compact laser devices with few W peak and few tens of mW average power are feasible. There is no evidence indicating that the efficiency at ~ 80 K observed here is fundamentally limited, and improvement can be expected.

We thank C. Cook and J. L. Daneu for expert assistance. This work was sponsored by Phillips Laboratory, Department of the Air Force.

## References

[1] See, for example, *Monitoring of Gaseous Pollutants by Tunable Diode Lasers*, R. Grisar, G. Schmidtke, M. Tacke, and G. Restelli (Eds.), Kluwer Academic Publishers, Massachusetts, USA, (1989).

[2] H. Q. Le, G. W. Turner, S. J. Eglash, and H. K. Choi, *Appl. Phys. Lett.* **64**, 152 (1994)

[3] H. Q. Le, J. M. Arias, M. Zandian, R. R. Zucca, and Y.-Z. Liu, *Appl. Phys. Lett.* **65**, 810 (1994)

[4] H. K. Choi *et al.*, presented at *Int. Semiconductor Laser Conf.* Paper M3.6, Maui, Hawaii, September 19-23, 1994.

[5] M. Aydaraliev, N. V. Zotova, S. A. Karandashov, B. A. Matveev, N. M. Stus', and G. N. Talalakin, *Semicond. Sci. Technol.* **8**, 1575 (1993)

[6] A. N. Baranov *et al., Appl. Phys. Lett.* **64**, 2480 (1994).

[7] H. Mani and A. Joullie, *Proc. SPIE* **1362**, 38 (1990).

[8] R. Martinelli *et al.*

[9] Y. H. Zhang, presented at *Int. Semiconductor Laser Conf.* Paper PD7, Maui, Hawaii, September 19-23, 1994.

[10] S. R. Kurtz *et al., Appl. Phys. Lett.* **64**, 812 (1994).

[11] J. Faist *et al., Electron. Lett.* **28**, 14 (1994).

[12] P. Becla, *J. Vac. Sci. Technol.* A **6**, 2728 (1988).

[13] M. Zandian, J. M. Arias, R. Zucca, R. V. Gil, and S. H. Shin, *Appl. Phys. Lett.* **59**, 1022 (1991).

[14] I. Melngailis and A. J. Strauss, *Appl. Phys. Lett.* **8**, 179 (1966).

[15] T. C. Harman, *J. Electron. Mater.* **8**, 191 (1979).

[16] K. K. Mahavadi, J. Bleuse, X. Chu, and J. P. Faurie, *Appl. Phys. Lett.* **55**, 1285 (1989).

[17] N. C. Giles, J. W. Han, J. W. Cook, Jr., and J. F. Schetzina, *Appl. Phys. Lett.* **55**, 2026 (1989).

[18] A. Ravid, A. Sher, G. Cinader, and A. Zussman, *J. Appl. Phys.* **73**, 7102 (1993).

[19] A. Million, T. Colin, P. Ferret, J. P. Zanatta, P. Bouchut, G. L. Destefanis, and J. Bablet, *J. Cryst. Growth* **127**, 291 (1993).

[20] J. Bleuse, N. Magnea, J.-L. Pautrat, and H. Mariette, *Semicond. Sci. Technol.* **8**, S261 (1993).

[21] J. M. Arias, J. G. Pasko, M. Zandian, S. H. Shin, G. M. Williams, L. O. Bubulac, R.-E. DeWames, and W. E. Tennant, *Appl. Phys. Lett.* **62**, 976 (1993).

[22] H. Q. Le, W. D. Goodhue, P. A. Maki, and S. Di Cecca, *Appl. Phys. Lett.* **63**, 1465 (1993).

[23] H. Q. Le, G. W. Turner, J. R. Ochoa, and A. Sanchez, *Electron. Lett.* **30**, 1944 (1994).

*Inst. Phys. Conf. Ser. No 144: Section 1*
*Paper presented at 7th Int. Conf. on Narrow Gap Semicond. Santa Fe 9–12 January*



# 3-4μm laser diodes based on GaInSb/InAs superlattices

**R H Miles, D H Chow, T C Hasenberg, A R Kost and Y-H Zhang**

Hughes Research Laboratories, Malibu, California 90265, USA

**Abstract.** Mid-wave infrared diode lasers based on GaInSb/InAs superlattice active layers have been demonstrated for the first time. The results follow significant improvements in the optical quality of these structures resulting from choices of group-V species and overpressures during growth by molecular beam epitaxy and use of an *in situ* anneal. Devices fabricated to date lase in the 3-4μm spectral range up to temperatures of approximately 150K, with typical 77K threshold current densities of 130-400 A/cm$^2$. The structures employ InAs/AlSb cladding layers and multi-quantum well active layers consisting of GaInSb/InAs superlattice wells and GaInAsSb barriers.

## 1. Introduction

Recent calculations[1] have suggested considerable promise for mid-wave infrared (MWIR) laser diodes based on GaInSb/InAs superlattice active layers. Predicted suppression of Auger recombination processes, which has been confirmed at long wavelengths,[2] would effectively eliminate these nonradiative decay channels for a variety of MWIR GaInSb/InAs superlattices.[3] Large in-plane oscillator strengths coupled with a low valence band density of states would yield unusually low threshold current densities. Further, the valence band edge energy of this class of superlattices is typically appreciably higher than that of candidate cladding layer materials such as AlGaAsSb, allowing confinement of both holes and electrons in GaInSb/InAs active layers for arbitrarily long emission wavelengths.

Application of these structures in infrared detectors or light emitters has until recently been hindered by the poor optical quality of $Ga_{1-x}In_xSb$/InAs superlattices grown by molecular beam epitaxy (MBE) and the inability to grow them by other techniques for $x>0$. However, appropriate choice of Sb species during MBE growth and subsequent use of an *in situ* anneal have now been shown to greatly improve the radiative efficiencies of these superlattices, leading to the demonstration of room temperature MWIR emission and low temperature optically pumped stimulated emission.[4]

In this paper we outline MBE growth techniques yielding superlattices of improved optical quality and report results from the first MWIR laser diodes to employ GaInSb/InAs active layers.

## 2. MBE growth and optical properties of GaInSb/InAs superlattices

Samples were grown in a VG V80 MBE machine equipped with a valved, cracked As source and an EPI 175 cracked Sb source. Only (100) GaSb substrates were used; high dislocation densities have been found to greatly degrade the optical quality of structures grown on GaAs. All $Ga_{0.75}In_{0.25}Sb/InAs$ active layers were grown at substrate temperatures of approximately 380°C, which has been found to yield superlattices of high structural quality.[5] 5s Sb soaks were employed at each InAs-GaInSb interface, yielding predominantly InSb-like interfacial bonds.[6] Growth of each superlattice was followed by a 30min, 500°C anneal, performed *in situ* in an Sb flux. This has been shown to greatly improve the optical efficiencies of these structures.[4]

Laser diodes were grown employing an Sb cracker zone temperature of 900°C. As shown in figure 1, the intensity of luminescence from these superlattices has been found to be very sensitive to cracking zone temperature.[4] The figure shows relative intensities of photoluminescence (PL) from 240Å thick 35Å/17Å $Ga_{0.75}In_{0.25}Sb/InAs$ superlattices sandwiched between carrier-confining AlGaSb layers. Samples were nominally identical except for the Sb cracking zone temperature employed during growth of the superlattice. Spectra were accumulated under pulsed AlGaAs laser diode photoexcitation, at an average power density of less than $10W/cm^2$. Data plotted in the figure are for samples cooled to 100K; results are similar for a temperature of 200K. While luminescence linewidths and peak positions are very similar in each case, PL is brightest at a cracking zone temperature of 900°C and drops markedly even for temperature changes as small as 50°C.

Analysis of the relative fluxes of $Sb_4$, $Sb_2$, and Sb produced by another cracker cell of the same type shows that approximately 80% of the antimony coming from the cracker at 900°C is in the form of monomers.[7] Based on these data it appears likely that monomeric Sb is the species critical to reducing densities of Shockley-Read recombination centers in these superlattices, and that the drop in optical efficiency at higher cracking zone temperatures is related to contamination. Secondary ion mass spectrometry has revealed high concentrations of Cr and Fe in samples grown at a cracking zone temperature of 1200°C, while these elements have not been detected in samples grown at 900°C.[4] As outgassing of Fe and Cr is characteristic of hot stainless steel, improved thermal isolation of the cracking zone should

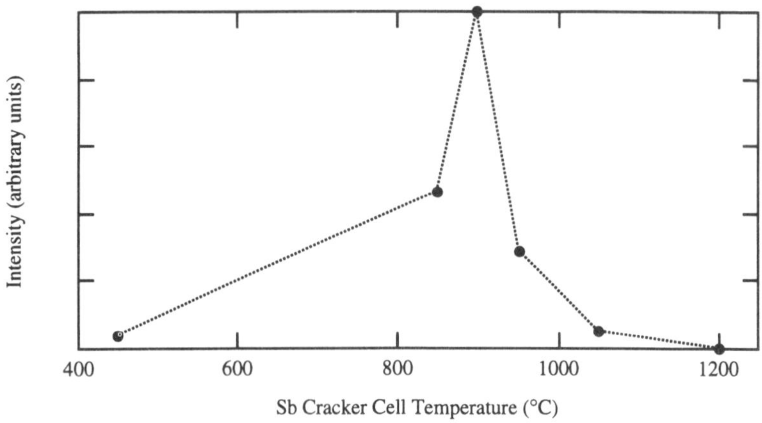


**Figure 1.** Relative intensities of 100K photoluminescence from 240Å thick 35Å/17Å $Ga_{0.75}In_{0.25}Sb/InAs$ superlattices grown with different antimony cracker cell temperatures. Inset is a typical 100K PL spectrum.

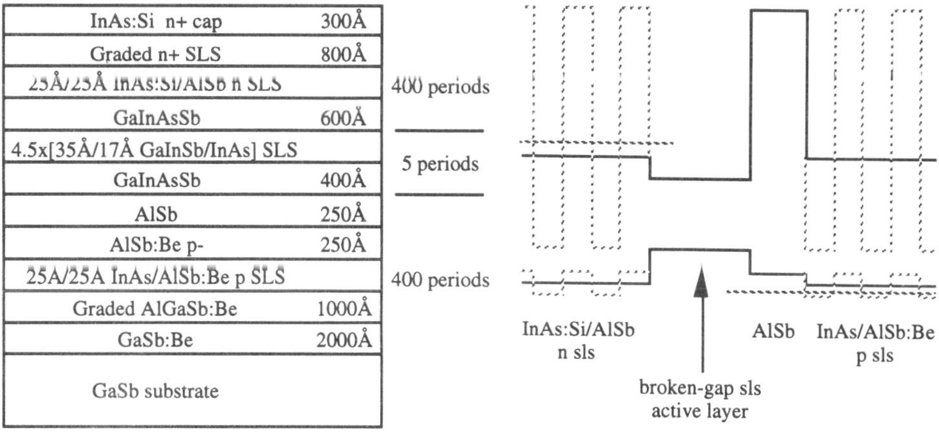


**Figure 2.** Schematic of a typical laser structure and simplified band edge diagram. In the band edge diagram, solid lines reflect conduction and valence band edges. Heavy dotted lines denote Fermi energies and light dotted lines Γ-point band edges of the AlSb and InAs layers comprising the cladding superlattices.

reduce these contaminants, likely yielding a monotonic increase in PL efficiency with cracking zone temperature over the range considered here.

Subsequent to growth of the device structures reported here, it was found that the radiative efficiency of MWIR $Ga_{0.75}In_{0.25}Sb/InAs$ superlattices is further improved by reducing the As overpressure to a point just sufficient to maintain a group-V stabilized surface during growth of the InAs layers. This reduction results in a reflection high energy electron diffraction pattern which slowly changes (over a time of approximately 30s) from 1x3 to 1x2 during growth of the InAs layers. We note in passing that x-ray diffraction data reveal no measurable changes in Sb cross-incorporation upon raising the As flux from that barely sufficient to yield a group-V stabilized InAs surface to one four times as great.

## 3. Laser design and characteristics

A schematic and simplified band edge diagram of a typical laser structure are shown in figure 2. The active layer is a multi-quantum well structure which in this case consists of five superlattice wells, each comprised of five 35Å thick $Ga_{0.75}In_{0.25}Sb$ layers and four 17Å thick InAs layers. This particular set of layer thicknesses and compositions was chosen for high optical transition strength and low Auger recombination rates, for a particular MWIR emission wavelength (in this case, about 3.5μm). Tailoring these three parameters expends the three available degrees of freedom in the superlattice (two sets of layer thicknesses and one composition). In general, Auger recombination in this system can be reduced by increasing the InSb fraction. Constraining the $Ga_{1-x}In_xSb$ composition to a value unlikely to result in misfit dislocations (x=0.25 for this particular structure), the Auger rate is typically found to be reduced for several discrete ranges of superlattice periods (for which superlattice minigaps fall an energy gap above or below the conduction or valence band edges, respectively). The choice between these ranges and the particular GaInSb:InAs thickness ratio are dictated by the desired emission energy and need for high optical efficiency. In general, the emission

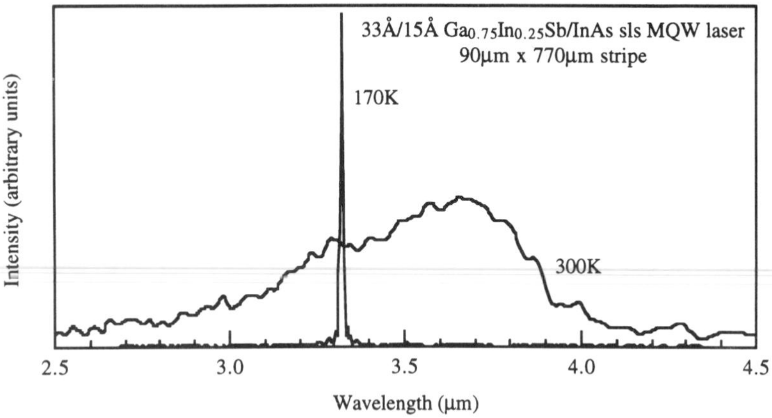


**Figure 3.** Room temperature electroluminescence and 170K emission above threshold for a 3.3μm laser.

wavelength of the broken gap superlattice is particularly sensitive to the choice of InAs layer thickness whereas the optical recombination rate is more dependent upon the fraction of InAs in the structure, since these layers are essentially optically dead. The particular choice of a 35Å/17Å $Ga_{0.75}In_{0.25}Sb$/InAs superlattice is advantageous because

(i) minigaps in both the heavy hole and conduction band growth-direction dispersion relations fall an energy gap away from the band extrema in this superlattice, and

(ii) the strong localization of holes in GaInSb layers usually raises optical transition rates for high GaInSb:InAs layer thickness ratios, such as are found in this structure.

An electrical injection laser can be formed by sandwiching the GaInSb/InAs -containing active region between InAs/AlSb superlattice cladding layers. These structures can be doped n- or p-type by codepositing Si during growth of the InAs layers or Be during the AlSb layers, respectively.[8] These clads avoid use of Te to achieve n-type layers and are more readily grown than conventional quaternaries, although their comparatively high index of refraction results in less effective waveguiding. As the cladding superlattices are several microns thick, constituent InAs and AlSb layer thicknesses must be almost equal, to balance alternating compressive and tensile stresses and avoid formation of misfit dislocations. However, the energies at which electrons and holes are injected into the active region (or the magnitude of the barrier blocking their escape) may be chosen by thinning or widening both sets of layers simultaneously. As shown in the figure, we employ an additional 500Å thick AlSb layer to block electron escape from the active region. Use of such a layer is facilitated by the choice of the GaInSb/InAs active layer; while the AlSb valence band energy is higher than that of InAs ($\Delta E_V \approx$ -110meV), the valence band edge of the superlattice is considerably higher than either.

Figure 3 illustrates room temperature electroluminescence and 150K pulsed emission above threshold for a laser structure similar to that illustrated in figure 2. The device is a broad area laser diode fabricated by a standard photolithographic process. Facets are cleaved and uncoated. The laser was driven at a 1% duty cycle, at a peak current of 1.5A. The shift from broad emission peaked at 3.7μm at room temperature to a sharp peak at 3.3μm at 170K is consistent with the expected temperature dependence of the superlattice energy gap. The device has a threshold current density of 134A/cm$^2$ at 77K. Measurements of $T_0$ have not yet been made on this or other lasers.

In addition to the 3.3μm laser of figure 3, we have demonstrated devices with 35Å/17Å and 35Å/21Å $Ga_{0.75}In_{0.25}Sb$/InAs superlattice active layers.[11] The lasing wavelengths of 3.47μm and 3.7μm, respectively, are in good agreement with theory. Threshold currents of 390A/cm$^2$ and 308A/cm$^2$ are measured for the devices at 77K. These values are appreciably greater than those calculated to be sufficient for an estimated gain of 30cm$^{-1}$,[3] and somewhat higher than the best reported in the literature; 77K threshold current densities of 158A/cm$^2$ are reported for a 4μm laser[9] and 85A/cm$^2$ for 3.9μm. We anticipate that use of high reflection coatings and refined device design will reduce thresholds, as will growth at lower As overpressures, which we have found to reduce extrinsic nonradiative recombination rates.

## 4. Conclusions

Significant improvements in the optical quality of GaInSb/InAs superlattices have been realized by employing a cracked Sb source and a low As overpressure during growth, and by use of a post-growth anneal. The first lasers fabricated from these structures display threshold current densities somewhat higher than the current state-of-the-art. However, calculations of Auger rates and threshold carrier concentrations suggest these to be the most promising of the active layers currently being considered for higher temperature MWIR operation, based on high values of $T_0$.[3] Realizing these benefits through demonstrated improvements in growth and refined device design is the aim of ongoing work.

## Acknowledgments

We gratefully acknowledge many helpful discussions with C. H. Grein of the University of Illinois at Chicago and H. Ehrenreich of Harvard University. We thank H. Dunlap, C. Haeussler, and L. Warren for expert technical assistance. This work was supported in part by the Air Force/Phillips Laboratory under Contract No. F29601-93-C-0037.

## References

[1] Grein C H, Young P M and Ehrenreich H 1994 *J. Appl. Phys.* **76** 1940
[2] Youngdale E R, Meyer J R, Hoffman C A, Bartoli F J, Grein C H, Young P M, Ehrenreich H, Miles R H and Chow D H 1994 *Appl. Phys. Lett.* **64** 3160
[3] Grein C H, Flatte M, Ehrenreich H and Miles R H in preparation
[4] Miles R H, Chow D H, Zhang Y-H, Brewer P D, and Wilson R G submitted to *Appl. Phys. Lett.*
[5] Chow D H, Miles R H, Söderström J R and McGill T C 1990 *J. Vac. Sci. Tech.* B **8** 710
Campbell I H, Sela I, Laurich B K, Smith D L, Bolognesi C R, Samoska L A, Gossard A C and Kroemer H 1991 *Appl. Phys. Lett.* **59** 846
[6] Chow D H, Miles R H and McGill T C 1991 *J. Cryst. Growth* **111** 683
[7] Brewer P D in preparation
[8] Chow D H, Zhang Y-H, Miles R H and Dunlap H L to appear in *J. Cryst. Growth*
[9] Eglash S J and Choi H K 1994 *Appl. Phys. Lett.* **64** 833
[10] Choi H K, Turner G W and Liau Z L 1994 *Appl. Phys. Lett.* **65** 2251
[11] Hasenberg T C, Chow D H, Miles R H, Kost A R and West L submitted to *Electr. Lett.*
Chow D H, Miles R H, Hasenberg T C, Kost A R and Zhang Y-H in preparation

*Inst. Phys. Conf. Ser. No 144: Section 1*
*Paper presented at 7th Int. Conf. on Narrow Gap Semicond. Santa Fe 9–12 January*



# Mid infrared lasers grown on InAs by modulated-molecular-beam epitaxy

**Y H ZHANG[1], H Q LE[2], D H CHOW[1], R H MILES[1]**

[1] Hughes Research Laboratories, Malibu, CA 90265
[2] Lincoln Laboratory, Massachusetts Institute of Technology, Lexington, MA 02173

**Abstract.** Stimulated emission has been observed from $InAs/AlAs_{0.16}Sb_{0.84}$ double-heterojunction and $InAs/InAs_xSb_{1-x}$ type-II superlattice (SL) laser structures grown on InAs under cw and pulsed optical pumping. For the $InAs/AlAs_{0.16}Sb_{0.84}$ double heterojunction lasers, the operating wavelength is 3.0 μm at low temperatures. The maximum cw operating temperature is 95 K. For the $InAs/InAs_xSb_{1-x}$ type-II SL lasers, the maximum temperatures are 95 K for cw operation and 195 K for pulsed operation. At 90 K these lasers yield a peak power of 1.3 W/facet and an average power of 150 mW/facet at 3.4 μm. The longest operation wavelength achieved is 3.65 μm at 81 K.

## 1. Introduction

High power midwave infrared (MWIR) (2 - 5 μm) semiconductor lasers have great potential in applications such as gas trace monitoring and military countermeasures. Three semiconductor material systems, *i.e.*, Pb salts, II/VI, and III/V compound semiconductors, have been used to make these lasers. The most developed III/V semiconductor MWIR lasers are based on GaSb substrates [1]. Room temperature or nearly room temperature operation of quaternary InGaAsSb/AlGaAsSb laser diodes at wavelengths shorter than 3.3 μm [2] and InAsSb/AlAsSb lasers with operating wavelength up to 4 μm have been reported [3]. In addition to these structures based on GaSb, InAs(Sb, P)/InAsSbP double heterojunction lasers grown on InAs by LPE [4] have also been explored. All these laser structures possess type-I (spatially direct) bandedge structure. Therefore, the longest operation wavelength of these lasers is limited by the narrowest bandgap of the constituent materials. Recently, MWIR lasers possessing GaInSb/InAs [5] and InAs/InAsSb [6] type-II superlattice (SL) active regions have been reported with cw operation at low temperatures. One of the most important advantages of these type-II SL lasers is that their operation wavelengths are not limited by the narrowest bandgaps of the constituent materials and could potentially cover a broad IR spectrum including longwave infrared (8 - 12 μm).

In the present paper we report our latest experimental results on MWIR lasers based on InAs substrates. Two types of laser structures are studied. The first type of laser structures consists of an InAs/AlAsSb double heterostructure and the second type has an InAs/InAsSb type-II SL active region. Under cw optical excitation lasing has been observed from these structures. Under pulsed optical pumping high power output is obtained from discrete devices. Our preliminary experimental results demonstrate practical MWIR lasers based on InAs substrates.

## 2. Sample Structures and Growth

In the present study a conventional molecular beam epitaxy (MBE) system was used to grow the samples. InAs wafers were mainly used as substrates although thick (>1 μm) strain-relieved InAs buffers grown on GaAs substrates were also occasionally used for calibration of growth conditions at early stages of the study. Two cracking effusion cells were used to provide $As_2$ and $Sb_2$ beams. The growth temperatures for InAs and

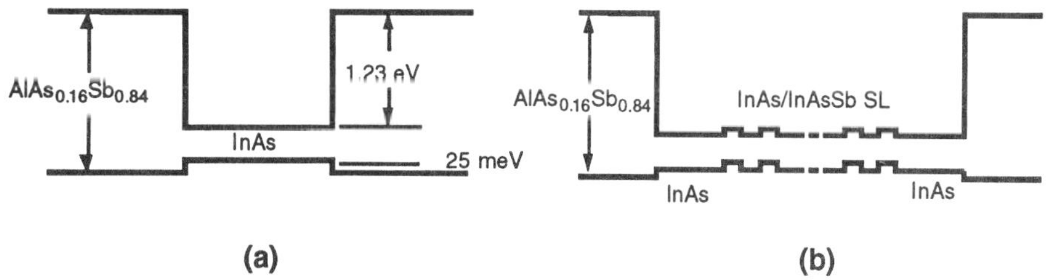


Fig 1 Bandedge diagrams (a) for studied InAs/AlAsSb double heterojunction laser structures and (b) for InAs/InAsSb type-II SL laser structures.

InAs/InAsSb SL layers were 480 to 500 °C, while those for $AlAs_xSb_{1-x}$ were 500 to 520 °C. The substrate temperature was measured by both a calibrated pyrometer and a thermocouple. The growth rates for InAs and $AlAs_xSb_{1-x}$ were kept around 0.50 µm/hr and 0.48 µm/hr, respectively. *In situ* reflection high energy electron diffraction (RHEED) was used to monitor surface reconstruction. After the growth, samples were characterized by using x-ray diffraction and photoluminescence (PL) techniques when applicable.

The laser structures used for the present study were designed for optical pumping. They consist of a 0.2-µm thick InAs buffer, followed by a 1.0 - 1.5-µm thick $AlAs_{0.16}Sb_{0.84}$ cladding layer, an active region, a second $AlAs_{0.16}Sb_{0.84}$ cladding layer, and finally a 300 Å or 0.1 µm thick InAs cap layer. For the double heterojunction lasers, the active region is a 1 µm thick InAs layer, while that for type-II SL laser is a 30-period 23Å-InAs/77Å-$InAs_{0.926}Sb_{0.074}$ SL sandwiched by two 0.4-µm thick InAs layers which function as optical absorbers and waveguide. The $AlAs_{0.16}Sb_{0.84}$ cladding layers provide both optical and carrier confinement. The schematic bandedge diagrams of the investigated samples are shown in Fig. 1(a) and (b) for the InAs/AlAsSb double heterojunction lasers and the InAs/$InAs_xSb_{1-x}$ type-II SL lasers, respectively. Our experimental results have shown that the crystalline and optical qualities of the $AlAs_{0.16}Sb_{0.84}$ cladding layers is very critical to the successful achievement of cw operation of the optical pumped lasers. Closely lattice matched $AlAs_xSb_{1-x}$ is thus desired to avoid misfit dislocations. A previously published study of the growth of $AlAs_xSb_{1-x}$ alloy has shown that these group-V alloys grown by Modulated MBE (MMBE) have superior structural and optical properties compared to those grown by conventional MBE [7]. The MMBE growth procedures for these samples are as follows. During growth of group-V alloys the shutter of Al effusion cell is kept open, while the As and Sb shutters are alternately opened and closed. The average AlAs mole fraction is proportional to the As-shutter duty-cycle [$As_{shutter-time}/(As_{shutter-time} + Sb_{shutter-time})$]. Similar procedures were also used for the growth of $InAs_xSb_{1-x}$ layers.

For optical pumping experiments, a cw Ti:sapphire laser pumped by an $Ar^+$ laser is used. The Ti:sapphire laser had 500 mW maximum power output at a wavelength around 800 nm. Two conventional cylindrical lenses were used to focus the excitation light on to the laser samples. The edge emitted light from the laser structures was analyzed by a Bomem Fourier transform infrared (FTIR) spectrometer. The pulse pumping light source was a laser-diode-array stack [1] with a 250 W peak power at 940 nm. A special lens system was used to focus the light output of the laser-array stack into a stripe of ~ 0.3 mm wide and 3.2 mm long on a sample inside a cryogenic Dewar. The average power output of the sample was measured outside the Dewar with a thermoelectric power meter. The pulse power was inferred by dividing the average power by the duty cycle.

## 3. Results and Discussion

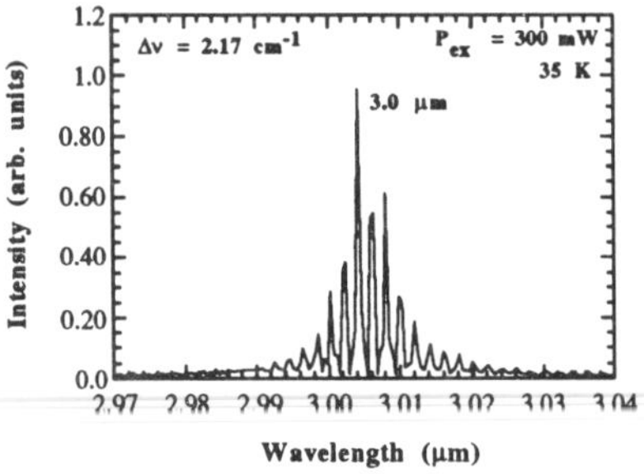


Fig. 2 CW lasing spectrum of an InAs/AlAsSb double heterojunction laser at 35 K under optical pumping. The equivalent threshold current density is ~ 280 $A/cm^2$.

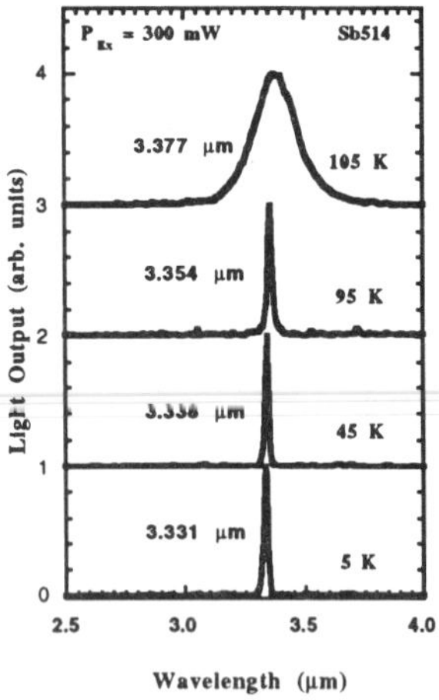


Fig. 3 Temperature dependent optical spectra of an InAs/InAsSb type-II SL laser under cw optical pumping of 300 mW.

## 3.1 CW Optical Pumping

Fig. 2 shows the lasing spectrum of an InAs/AlAsSb double heterojunction laser at 35 K under cw optical pumping. Assuming a 50 μm × 600 μm excitation stripe, the equivalent injected current density is estimated to be 375 $A/cm^2$ for 300 mW optical excitation power at 800 nm. At this wavelength InAs has an absorption coefficient of $1 \times 10^5$ $cm^{-1}$ [8]. If the internal quantum- and light-collection-efficiencies are taken to be ~1 and the 300 Å InAs cap layer for this sample absorbs 25% of the excitation light, the actual equivalent injected current density for the active region is around 280 $A/cm^2$. It is interesting to note that InAs/AlAsSb double heterostructure lasers were proposed more than a decade ago [9]. However, due to the difficulty of the growth of AlAsSb cladding layers, to the best of our knowledge, no successful laser demonstration has been reported before. We believe that our success in realizing cw operation of this type of laser is due to the improvement of the crystalline and optical quality of the AlAsSb cladding layers.

As shown in Fig. 2, only one set of longitudinal modes is observed in the lasing spectrum. No splitting of the longitudinal modes is visible, which would be indicative of coexisting of higher lateral mode along the growth direction. This characteristic is also seen in spectra taken under various excitation densities. The wavenumber separation (Δν) of the longitudinal modes is measured to be 2.17 $cm^{-1}$ with a cavity length (L) of ~ 0.6 mm. The equivalent index ($n_e$) is estimated to be 3.8 by using the simple formula $n_e = 1/(2\Delta\nu L)$. The spectrum center wavelength is 3.0 μm at 35 K. The maximum cw operating temperature for this device is 95 K under an excitation of 300 mW. At room temperature, strong single peak luminescence centered at 3.5 μm is observed from this sample.

Fig. 3 shows temperature dependent optical spectra of the edge emitted light from an $InAs/InAs_{0.926}Sb_{0.074}$ type-II SL laser structure under cw optical pumping of 300 mW. The optical transitions in the type-II SL are spatially indirect, from extended electron states to localized hole states in the $InAs_{0.926}Sb_{0.074}$ layers [6]. The overlap between the electron and hole wavefunctions is relatively large. By using the Kronig-Penney model and taking strain effects [10] into account, the calculated optical transition wavelength for the specific sample structure is 3.29 μm at 95 K [11]. The measured lasing wavelength at the same temperature is 3.35 μm. The discrepancy between the two values can be attributed to the uncertainty of the valence band offset between InAs and InSb. As the sample temperature is increased, the lasing wavelength increases monotonically. At 95 K it reaches 3.354 μm. At this temperature, since lasing is still achievable, the equivalent injected current density is estimated to be 56 $A/cm^2$ with a cavity length of 1.5 mm. At 105 K, this injection density is below threshold and only luminescence with a characteristic broad linewidth is observed. A higher cw operating temperature is expected under higher excitation. The characteristic temperature $T_0$ for this device is estimated to be around 32 K.

## 3.2 Pulsed Optical Pumping

In order to obtain high power output, a broad area device with ~ 0.3-mm × 3.1-mm was used for the pulsed optical pumping experiments. It is necessary to point out that the specific sample used has a 0.1 μm InAs cap layer, which absorbs the excitation light strongly. Fig 4 gives the light output power per facet of an $InAs/InAs_{0.926}Sb_{0.074}$ type-II SL laser vs. absorbed pumping power at 90 K. The pumping pulse length is 10 μs and the duty-cycle is 1%. A least square fitting of the experimental data in Fig. 4 reveals that the external differential quantum efficiency is 2.5%/facet for optical pumping power less than 20 W. A kink is observed around a pump power of 25 W. Above the kink, the linear behavior of the output power vs. pumping power remains up to the maximum power measured, but the quantum efficiency drops to 2%/facet. The peak power measured is 1.3W/facet under the maximum absorbed pumping power of 60 W. At maximum pumping power, neither deviation from the linear behavior of the power output vs. pumping power nor catastrophic degradation of the device is observed. Therefore, it is reasonable to believe that a higher maximum peak power could be achieved by using even higher

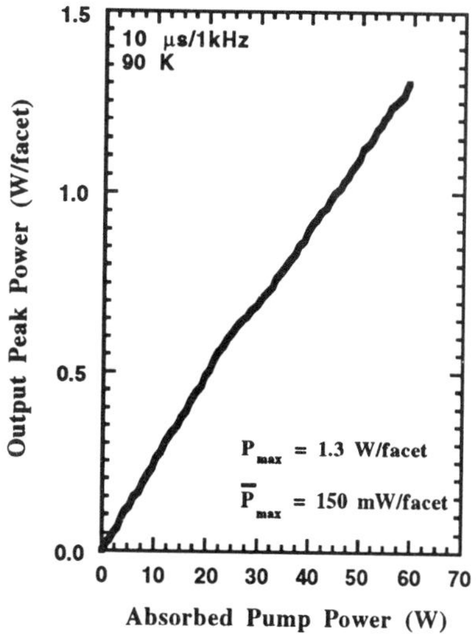


Fig. 4 Peak power vs. absorbed pumping power for an InAs/InAsSb type-II SL lasers under pulse optical pumping at 90 K.

pumping power. Furthermore, the structure used for the present study has not been optimized for high power lasers. A proper design of the laser structure such as increase the cladding layer thickness to increase optical confinement will certainly improve the device performance both in maximum power output as well as in quantum efficiency.

By increasing the optical pumping duty cycle, a 150 mW/facet average power is achieved at 95 K. For this case, the external differential quantum efficiency is also increased to 8%/facet. With the 60 W absorbed pump power, the maximum pulse operation temperature is 195K. With another $InAs/InAs_xSb_{1-x}$ type-II SL laser sample possessing a similar structure but different composition in the $InAs_xSb_{1-x}$ layers achieved an operating wavelength of 3.65 μm at 81 K under pulsed optical pumping. This is the longest lasing wavelength which has ever been achieved for the InAs/InAsSb type-II SL laser structures.

## 4. Conclusion

We have demonstrated cw operation of two types of MWIR laser structures, *i.e.* InAs/AlAsSb double heterostructure lasers and InAs/InAsSb type-II SL lasers, based on InAs substrates. The peak power of the second type of lasers has reached 1.3W/facet at 90 K with a cw operation temperature of 95 K and pulsed operation temperature of 195K. The longest lasing wavelength achieved is 3.65 μm at 81 K. These results suggest that InAs based MWIR lasers are very promising candidates for high performance MWIR lasers.

## Acknowledgment

The authors thank W. Clark for his careful reading of the manuscript and useful suggestions. We also appreciate the excellent technical support from L. Warren and C. Haeussler.

## References

[1] Choi H K, Turner G W, and Liau Z L, *Appl. Phys. Lett.* **65**, 2251-2253 (1994);
Le H Q, Turner G W, Eglash S J, Choi H K, and Coppeta D A, *Appl. Phys. Lett. 64*, 152-154 (1994) and references therein.

[2] Choi H K, Eglash S J, Turner G W, *Appl. Phys. Lett.* **64**, 2474-2476 (1994).

[3] Eglash S J, Choi H K, *Appl. Phys. Lett.* **64**, 833-835 (1994).

[4] Baranov A N, Danilova T N, Ershov O G, Imenkov A N, Sherstnev V V, and Yakovlov Yu P, *Sov. Tech. Phys. Lett.* **18**, 725-726 (1992);
Baranov A N, Imenkov A N, Sherstnev V V, and Yakovlov Yu P, *Appl. Phys. Lett.* **64**, 2480-2482 (1994).

[5] Miles R H, Chow D H, Zhang Y H and Brewer P D submitted to *Appl. Phys. Lett.;*
Miles R H, Chow D H, Hasenberg T C, Kost A R and Zhang Y H, the 7th Int. Conf. on Narrow Gap Semiconductors, Santa Fe, 1995.

[6] Zhang Y H, *Appl. Phys. Lett.* **66**, 118 (1995).

[7] Zhang Y H and Chow D H, *Appl. Phys. Lett.* **65**, 3239 (1994)

[8] Landolt-Börnstein, *Numerical Data and Functional Relationships in Science and Technology,* New Series III/17a, edited by O. Madelung, M. Schulz, and H. Weiss (Springer, Heidelberg, 1982).

[9] Casey H C Jr. and Panish M B. *Heterostructure Lasers* (Academic, NY, 1978)

[10] Kurtz S R, Biefeld R M, Dawson L R, Baucom K C, and Howard A J, *Appl. Phys. Lett.* **64**, 812-814 (1994).

[11] Zhang Y H, Miles R H, and Chow D H, to appear in *IEEE Quantum Electronics.*

*Inst. Phys. Conf. Ser. No 144: Section 2*
*Paper presented at 7th Int. Conf. on Narrow Gap Semicond. Santa Fe 9–12 January*



# Luminescence and Optical Gain from Free, Localized and Confined Excitons in Narrow-Gap $Hg_{0.7}Cd_{0.3}Te$

**J W Tomm, T Kelz and W Hoerstel**
Humboldt-Universität zu Berlin, Institut für Physik, FET, Unter den Linden 6. D-10099 Berlin, Germany
**T K Tran, B K Wagner, R G Benz II and C J Summers**
Quantum Microstructure Laboratory, Georgia Tech Research Institute, Atlanta GA 30332-0800, USA

**Abstract:** The results of a photoluminescence (PL) study of $Hg_{0.7}Cd_{0.3}Te$ at high excitation levels are reported. Bulk samples as well as superlattices (SL) with $Hg_{0.7}Cd_{0.3}Te$-wells were investigated at moderate and high excitation levels. For both kinds of samples excitonic and electron-hole plasma (e,h) transitions were identified. Optical gain spectra have been obtained from emission spectra under the variation of the excitation stripe length.

## 1. Introduction

The investigation of the elementary excitations in narrow-gap materials and structures at high excitation levels is of basic interest, but also the design of diode laser structures [1,2] demands knowledge about this subject.

Narrow-gap Hg based II-VI SL's have been widely investigated and a recent review is given in [3]. However, nearly all results discussed in these papers are obtained from HgTe/CdTe- or $HgTe/Hg_{1-x}Cd_xTe$-SL, i.e. structures with semimetallic HgTe wells. This structures are mainly devoted to detector applications.

An earlier study on highly Iodine-doped $Hg_{0.7}Cd_{0.3}Te/Hg_{0.15}Cd_{0.85}Te$ SL's [4] showed substantially enhanced luminescence intensities for these structures which have semiconducting wells compared to HgTe. Therefore we consider $Hg_{0.7}Cd_{0.3}Te/Hg_{0.15}Cd_{0.85}Te$ SL's or multiple quantum wells (MQW) as being extremely promising materials for the forming of optically active layers for II-VI infrared (IR) diode lasers designed for applications in the 3-5 μm wavelength range.

An additional reason, which makes both bulk and heteroepitaxial $Hg_{0.7}Cd_{0.3}Te$ to interesting objects is that if one applies a magnetic field one can - additionally to the well known Landau quantization and spin splitting - basically influence the band edge emission behaviour by changing the exciton binding energy ($R_x$) from $R_x$=1.2 meV at B=0 T to $R_x$= 4.1 eV at B=7 T. The latter number approaches already the corresponding value in GaAs.

In this paper we present the first joined study of the emission and gain mechanisms in highly excited $Hg_{0.7}Cd_{0.3}Te$ and $Hg_{0.7}Cd_{0.3}Te/Hg_{0.15}Cd_{0.85}Te$ SL's inclusively by magnetoluminescence techniques.

## 2. Experimental

IR PL spectra were recorded with a classical dispersive setup using a Nd:YAG laser (hν=1.17 eV) as an excitation source. The laser was used either with Q-switching (pulsed, FWHM of

Table 1: Structural parameters of the SL structure

| SL structure | SL25 | SL27 |
|---|---|---|
| well mole fraction | 0.3 | 0.3 |
| well thickness (nm) | 20 | 10 |
| barrier thickness (nm) | 10 | 5 |
| number of periods | 150 | 200 |

pulses about 80 ns, 1 kHz repetition rate) or without (cw) to achieve various excitation density ranges. An $Hg_{1-x}Cd_xTe$ photodiode was used for signal detection. For magnetoluminescence experiments an Spectromag 4000 system (B up to 7 T) was added. For *PL measurements* the optics were arranged to collect the emission perpendicular to the sample surface. The *gain measurements* were performed in a 90 degree configuration as used for optical pumping of semiconductor lasers. The variation of the illuminated sample length was realized by a micrometer drive for moving and positioning of the cryostat (cf. reference [5]).

As bulk reference samples we used $Hg_{1-x}Cd_xTe$ ($0.3<x<0.4$) samples grown by the travelling heater method (THM). Equilibrium carrier concentrations at T = 77 K were between $p_0=10^{15}cm^{-3}$ and $p_0=10^{16}cm^{-3}$ with mobilities in the 300-400 $cm^2/(Vs)$ range.

The SL were grown by photoassisted MOMBE (50 $mWcm^{-2}$, 488 nm) on (211)B CdTe substrates at a temperature of 448 K and a growth rate of 1 µm/h. Di-isopropyl tellurium, dimetylcadmium and elemental Hg were used as the deposition sources with the Te and Cd organometalles precracked in high temperature injectors. A more detailed description of the SL's is given in reference [4]. For this study we have chosen two structures with a very different emission behaviour at high excitation level. The stuctural data verified by X-ray analysis are collected in Table 1.

Transport measurements indicated good sample quality: The dependence of the SdH oscillation position ratio (for fixed Landau quantum numbers N) for both SL's exhibits a cosine - like dependence on the tilt angle indicating the 2D character of the carrier gas. From the SdH oscillation we determined the sheet concentrations ($3.22\times10^{11}cm^{-2}$ for SL25 and $3.64\times10^{11}cm^{-2}$ for SL 27) whereas the effective ground state electron masses ($0.035m_0$ for SL25 and $0.041m_0$ for SL27) were obtained from the temperature dependence of the amplitudes. The mean volume concentration calculated from the sheet concentration agrees very well with the Hall concentrations ($1.04\times10^{17}cm^{-3}$ for SL25 and $3.3\times10^{17}cm^{-3}$ for SL27) confirming the consistency of our analysis. The mobilities were of the order of $1.8\times10^4$ $cm^2/(Vs)$. The magnetic field dependence of the conductivity $\sigma_{xy}$ for SL25 shows well pronounced quantum Hall plateaus identified with the quantum numbers i = 1,2,3,4. The absolute value of $\sigma_{xy}$ shows steps varying by $(145+/-5)\times e\times 2/h$. So nearly all of the SL wells contribute to the quantum Hall effect in this structure. Thus, we can state that the transport data indicate that the SL's investigated in this study are very homogeneous.

## 3. Results and discussion

### *3.1. Spontaneous luminescence and gain mechanisms in bulk $Hg_{0.7}Cd_{0.3}Te$*

We have discussed the nature of the optical transition in bulk $Hg_{1-x}Cd_xTe$ ($0.3<x<0.4$) already in previous papers (cf. [6] and references there). Therefore, here we only briefly review the arguments for the assignments of the relevant near-to-edge emission lines. Impurity lines and donor-acceptor-pair emission are not considered. The edge emission of all high quality bulk samples is related to excitonic transitions in (nearly) every case. The excitonic contributions are relatively easy to identify if

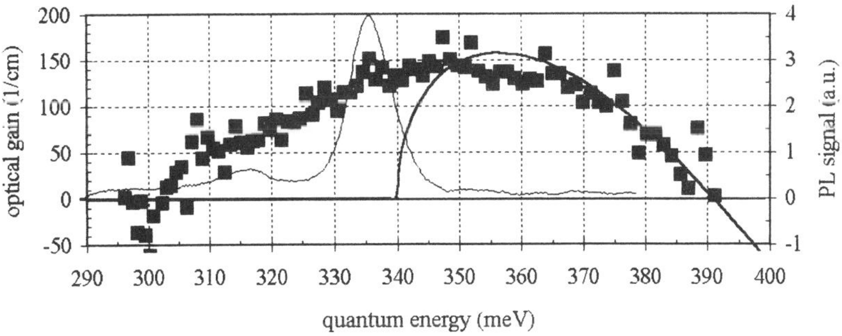


Fig. 1 Photoexcited optical gain spectrum (full squares) and fit taking into account (e,h) plasma transitions for a $Hg_{0.63}Cd_{0.37}Te$ sample. The temperature is T=6 K, the excitation density 100 $kWcm^{-2}$. The effective carrier temperature was estimated from PL (cf. reference [10]) to (280+/-20) K. For comparision a PL spectrum measured at T=6 K (P=8 $Wcm^{-2}$) was added, which was assigned to bound exciton transitions.

(i) they exhibit the well-known bound exciton behaviour (nearly temperature-independent line position and thermal quenching below T=20 K cf. [7] ) or if

(ii) they show the behaviour of intrinsic excitons localized in potential fluctuations with the thermal delocalization, cf.[8].

We analyzed both cases with magnetoluminecence and found, that for (i) the magnetic field shift of the PL peak is extremely weak, whereas for (ii) the observed shift is stronger but less than the expected value for free excitons. An even stronger magnetic field depencence of the PL line positions - e.g. the shift expected for (e,h) transitions as analysed in magnetotransmission experiments - was not observed for high-quality $Hg_{0.7}Cd_{0.3}Te$.

For high excitation levels in the order of 100 $kWcm^{-2}$ two different gain mechanisms were identified. For samples mentioned under (i) we assume exciton-exciton scattering as a gain mechanism, since the observed IR shifts of up to 18 meV can not be explained exclusively by band gap renormalization. The interpretation, that optical gain can arise from bound exciton transitions was made already by Ivanov-Omskii et al. [9]. For heavily doped samples we assigned the (e,h)-plasma transitions as the gain mechanism [10]. Fig. 1 displays a gain spectrum (full squares) from a $Hg_{0.63}Cd_{0.37}Te$ sample ($E_g$ = (340+/-3) meV) showing bound exciton PL under moderate excitation. It was determined by recording emission spectra under the variation of the excitation stripe length. The fit (full line) was made assuming exclusively (e,h) plasma transitions without taking into account contributions below $E_g$, which empirically can be described by a 'tailing' of the absorption edge. Obviously an exclusive assignment to (e,h) transitions is impossible. Due to the pumping with 10 $kWcm^{-2}$ the carrier gas is heated to about 280 K (cf. [10]) and distinct excitonic contributions are not expected. However, the gain below $E_g$ indicates remarkable excitonic contributions.

### *3.2 Spontaneous luminescence and gain mechanisms in $Hg_{0.7}Cd_{0.3}Te/Hg_{0.15}Cd_{0.85}Te$ superlattices*

Until now SL luminescence from Hg-containing II-VI SL's was exclusively interpreted in the framework of (e,h) plasma (band-to-band) recombination. The only exception is the paper by Harris et al. [11] who claimed the observation of acceptor bound exciton recombination ($A^0$,X) at energies below $E_g$. In [4] we have demonstrated, that the SL magnetoluminescence data is fitted better with excitonic corrections than without and consider this as one additional

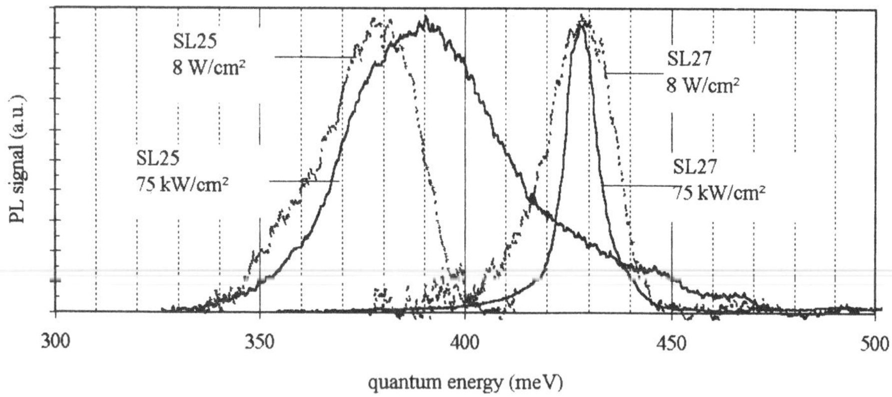


Fig. 2 PL spectra at moderate (cw, dashed) and high excitation levels (pulsed, full lines) from SL25 (left spectra) and SL27 (right spectra) at T=6 K. All spectra are scaled to unit.

argument for the presence of excitonic effects in narrow-gap II-VI SL's. The present new data at high excitation levels for both SL structures indicate a very different behaviour and can only be explained by assuming that in narrow-gap SL's optical transitions of different physical origin can occur and the exclusive consideration of (e,h) transitions fails.

Varying the excitation density from 600 $mWcm^{-2}$ (cw) to 100 $kWcm^{-2}$ (pulsed) we found that the FWHM for SL25 *increases* from 25 meV to 42 meV, whereas for SL27 a *decrease* from 20 meV to 9 meV occurs. The peak position shift also shows opposite tendencies: With increasing excitation density (cf. above) for SL25 the peak shows a blue shift by about 11 meV, whereas for SL 27 the corresponding a weak IR shift was observed. The excitation density ($I_e$) dependence of the integrated PL signal magnitude ($I_{PL}$, 'area below the spectrum') for both structures followed relations like

$$I_{PL} = I_e^{\alpha} \tag{1}$$

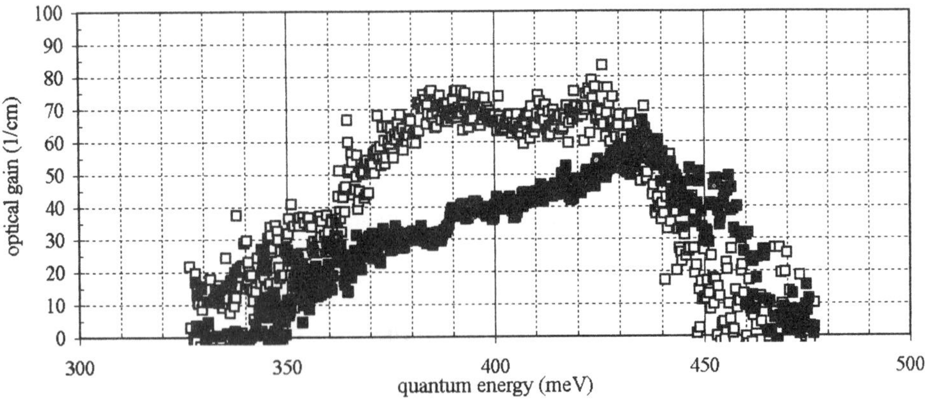


Fig. 3 Photoexcited optical gain spectra from SL25 at T=6 K for two excitaion densities 50 $kWcm^{-2}$ (full squares) and 100 $kWcm^{-2}$ (empty squares).

were the exponent α is equal to 0.6 for SL25 and 0.8 for SL27. The value of α for SL27 does not change if one analyzes the PL signal amplitude at peak intensity instead of the integral. Doing this for SL25 the value for α becomes smaller and approaches α=0.5. This indicates the presence of contributions from more than one optical transition with different power dependencies. In order to illustrate the statements given above, Fig. 2. displays spectra for different excitation levels for both structures. The signal magnitude was scaled to unit.

Fig. 3 displays two gain spectra from SL25. As already assumed from the conventional PL two transitions occur and both of them also contribute to the optical gain and obviously they have different power dependencies. However, a bandstructure calculation for this structure (cf. [4]) predicts only a single ground state. The microscopic nature of this mechanisms will be investigated further.

## 4. Conclusions

We present the results of a study of the luminescence and optical gain mechanisms in bulk $Hg_{0.7}Cd_{0.3}Te$ and $Hg_{0.7}Cd_{0.3}Te/Hg_{0.15}Cd_{0.85}Te$ SL's at optical pumping intensities.

For the bulk material we observed optical gain also below the energy and assigned this gain contribution to an exciton exciton scattering mechanism, whereas the above band gap contribution comes from the (e,h) plasma.

It was found for one SL structure that different optical transitions form the PL line and that also optical gain can arise from more than one mechanism.

**Acknowlegement:**

The bulk reference samples were grown by P. Gille. The authors wish to thank W. T. Masselink for helpful discussions and S. Halpern for technical assistance.

## References

[1] Becla P 1988 *J. Vac. Sci. Technol. A* **6** 2725-7

[2] Zandian M, Arias J M, Zucca R, Gil R V and Shin S H 1991 *Appl. Phys. Lett.* **59** 1022-4

[3] Myers T H, Meyer J R and Hoffman C A 1993 *Semiconductor Quantum Wells and Superlattices for Long-wavelength Infrared Detectors/Infrared Detectors* (Norwood: Artech House, Inc.)

[4] Tomm J W, Kelz T, Hoerstel W, Herrmann K H, Tran T K, Bicknell-Tassius R N, Wagner B K, Benz II R G and C. J. Summers 1995 accepted for publication in *Semicond. Sci. Technol.*

[5] Shaklee K L, Nahory R E and Leheny R F 1973 *J. Luminescence* **7** 284-309

[6] Tomm J W, Herrmann K H, Hoerstel W, Lindstädt M, Kissel H and Fuchs F 1994 *J. Crystal Growth* **138** 175-81

[7] Hunter A T and McGill T C 1982 *J. Vac. Sci. Technol.* **21** 205-7

[8] Lusson A, Fuchs F and Marfaing Y 1990 *J. Crystal Growth* **101** 673-7

[9] Ivanov-Omskii V I, Kurbanov K R, Rustamov R B, Smirnov V A and Yuldashev Sh U 1984 *Fiz. Tekh. Poluprov.* **18** 1509-11

[10] Herrmann K H, Tomm J W and Lindstädt M 1994 CIRP6 accepted for publication in *Infrared Physics*

[11] Harris K A, Yanka R W, Mohnkern L M, Myers T H, Yang Z, Yu Z, Hwang S and Schetzina J F 1992 *J. Vac. Sci. Technol. B* **10** 1574-81

*Inst. Phys. Conf. Ser. No 144: Section 2*
*Paper presented at 7th Int. Conf. on Narrow Gap Semicond. Santa Fe 9–12 January*



# Investigation of Impurity/Defect Level in $Hg_{1-x}Cd_xTe$ Alloys Using Infrared Magneto-Photoconductivity

**P.L.Liu, W.Lu, X.F.Lu, H.M.Gong*, X.N.Hu*, Y.M.Mu, J.H.Chu, and S.C.Shen**
National Laboratory for Infrared Physics, Academia Sinica, 420 Zhong Shan Bei Yi Road, Shanghai, 200083, People's Republic of China

* Shanghai Institute of Technical Sciences, Academia Sinica, 420 Zhong Shan Bei Yi Road, Shanghai, 200083, People's Republic of China

**Abstract.** The infrared photoconductivity spectra of un-intentionally doped $n-Hg_{1-x}Cd_xTe$ with x=0.2-0.4 were measured at 4.2K in the magnetic field up to 10 Tesla in Faraday configuration. Both the intrinsic and the extrinsic photoconductivity were observed. The results from these samples provide direct evidence that two trap levels in the energy gap edge and two resonant defect levels in the conduction band and an acceptor shallow level with activation energy about 7meV for these x-value samples.

## 1. Introduction

The ternary semiconductor Mercury Cadmium Telluride (MCT) is one of the most important material for the construction of the intrinsic infrared detector. The presence of impurity or defect levels in this material often has a strong effect on the quality of device performance.[1] For example, defects or impurities create energy levels lying deep in the forbidden energy gap region or in the conduction band, acting as effective recombination centers affecting the carriers lifetimes.[2,3] These levels also provide centers that enhance interband tunneling, giving rise to dark currents in the intrinsic infrared detector based on p-n diode or metal-insulator-semiconductor device. It is therefore highly desirable to know the energy level positions of these centers in $Hg_{1-x}Cd_xTe$.

## 2. Experiment and Results

The magneto-photoconductivity were measured on a Bruker 113v Fourier Transform Infrared Spectrometer and an Oxford Superconducting Magnet-Cryostat combined by an Optical Transportation System, which transmits the unpolarized infrared radiation, i.e. the interference optical beam from the spectrometer to the sample mounted in the center of magnet [4]. The experiments were carried out on MCT single crystal bulk grown from Te-rich solution by a modified Travelling Heater Method, and examined by x-ray Tomography, infrared transmission, Hall coefficient and resistivity measurement after annealing. The samples were cut, polished, etched, and had gold contacts. The alloy composition parameters x of samples MCT-23, MCT-30 reported here are 0.23, 0.31, the energy gap Eg=119meV, 270meV, the free electron concentration $n = 8 \times 10^{14}\ cm^{-3}$, $7 \times 10^{14}\ cm^{-3}$ at 77K, and Fermi level are about 3 meV, 1.4 meV above the bottom of conduction band, respectively.

Fig.1 shows the photo-conductivity of sample MCT30 in the far infrared range. It is an extrinsic photo-conductivity since the energy is small in comparison with the energy gap for x=0.30. Two pronounced peaks can be resolved from fig.1 for different magnetic field. A long high-energy photo-conductivity tail is also observed, which is typical to impurity or defect ionization spectra.

Fig.2 shows the photo-conductivity of sample MCT23 in the mid-infrared range. It is a typical intrinsic photo-conductivity with the energy gap at 121meV indicating the composition x=0.23 in good agreement with the nominal composition value x=0.24. However, with the increasing of the magnetic field, the evaluation of the low energy peak and high energy peak is different. The low energy peak becomes weaker and sharper with the increasing of magnetic field in contract with the high energy peak which develops quickly to a strong and broad band. Moreover, one weak peak is observed just below the intrinsic photo-conductivity.

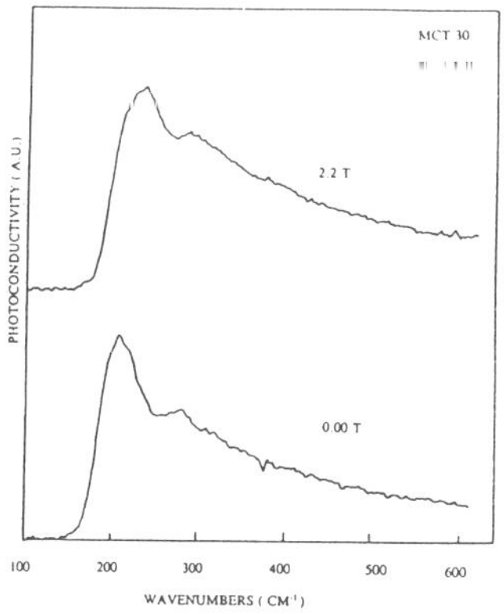


Fig.1 Magneto-photoconductivity for sample MCT30 at 4.2K in far infrared region.

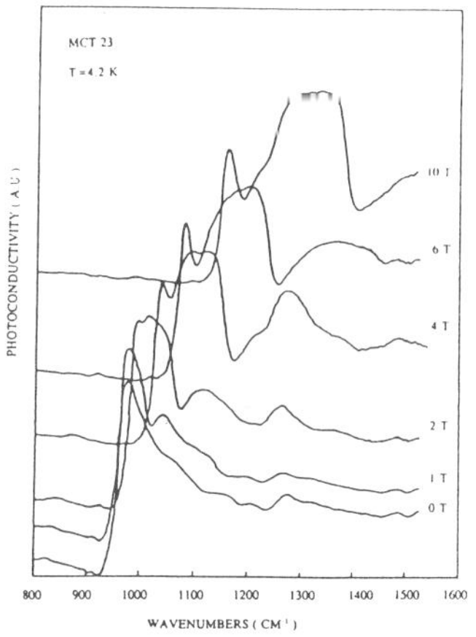


Fig.2 Magneto-photoconductivity for sample MCT23 at 4.2K in mid-infrared region.

For the sample MCT30 we not only observed the intrinsic photo-conductivity rising from $2150cm^{-1}$ and two peaks ( at $2206cm^{-1}$ and $2256cm^{-1}$ ) overlapping on the spectrum, but also observed a rather weak peak at 2115 $cm^{-1}$ just below the threshold of photo-conductivity.

## 3. Discussion

### 3-1. The Impurity or Defect Levels in the Energy Gap

Both of our studied samples are n-type material, so that we believe that the photo-conductivity results from the electrons in the conduction band. Following this way, the final state of all observed impurity or defect levels related transition will be in the conduction band. Fig.3 shows the dependence of the transition energy on the magnetic field for the trap levels to the conduction band. One of these two levels is at the position of 35meV below the conduction band edge in agreement with that in another work[5]. Another level is at the position of about 26meV above the valence band edge. This one is the first observation. Notice the almost the same slope of the magnetic field dependence of their energy of photo-conductivity peak for $E_c$-26meV level in comparison with the interband transition. So that this level is almost independent on the magnetic field showing a typical deep level behavior. The magnetic field dependence of the transition energy for the level ($E_c$-35meV) to the conduction band has much lower slops in comparison with the interband transition. This behavior is different with that in reference 5. But we shall note that our sample is n-type instead of the p-type in ref.5. So we think although almost the same level position was obtained in our work and the work in ref.5, the $E_c$-35meV level observed in this work may be different from that in ref.5.

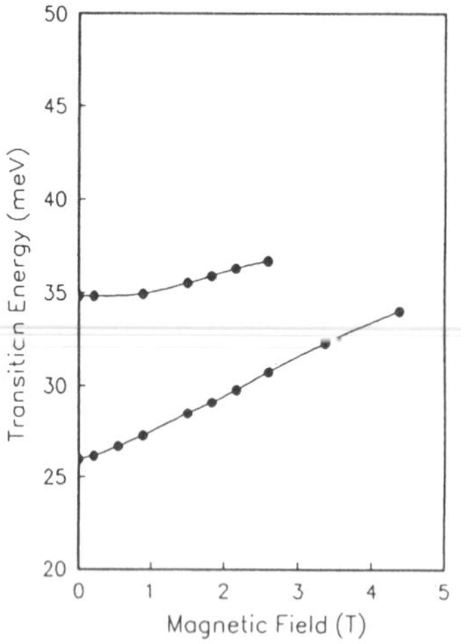


Fig.3 Magnetic field dependence of the transition energy from the deep level to conduction band edge for sample MCT30 at 4.2K. Dots represent the experimental data and lines are just for connection.

The open circles in fig.4 and fig.5 show the extrinsic photo-conductivity in both sample MCT23 and MCT30. In agreement with the experimental results in other works[5,6], we identify this structure as resulting from electron transitions from a shallow acceptor to the lowest conduction band Landau level. The activation energy is about 7meV.

## 3-2. The Resonant Levels in the Conduction Band

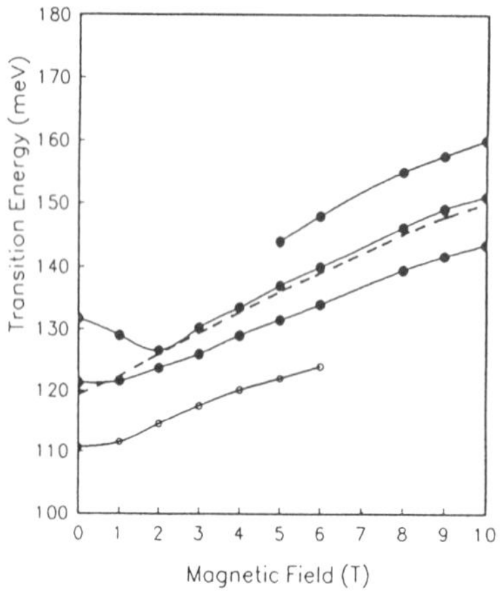


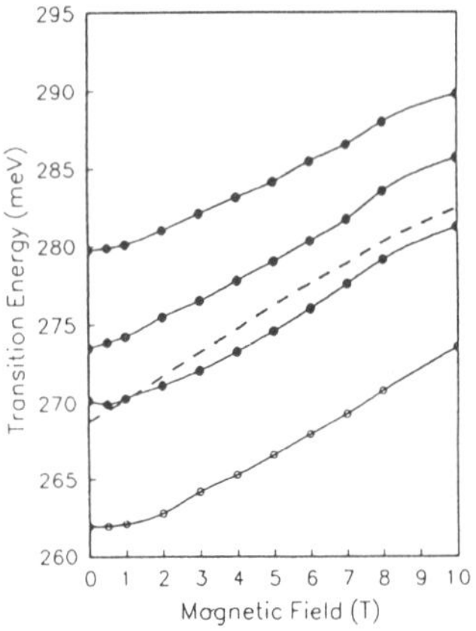


Fig.4 Magnetic field dependence of the transition energy from the valence band edge to the resonant level and the conduction band lowest Landau level (dots) and from shallow accept level to conduction band edge (circle) for sample MCT23 at 4.2K. Lines are just for connection.

Fig.5 Same as Figure 4 but for sample MCT30.

The dots in fig.4 and fig.5 present the experimental data taken from the photo-conductivity.

The lines are just used for connection. We have calculated the theoretical magnetic field dependence of the interband transition by using the energy band parameters for these two samples.[7] In the calculation the non-parabolic property of the energy band has been included. The theoretical curves are shown in fig.4 and fig.5 as dashed lines. No best fit is made, but the agreement between the experimental data and the theoretical curves is reasonable. In fig.1 the interaction between the resonant state and the Landau level happened around 2 Tesla. Around this magnetic field range, a clear repulse behavior is demonstrated between the low lying resonant level and the lowest Landau level. Such an interaction also makes the intensity of the interband transition and the valence-band to the resonant level transition mixed each other as shown in fig.2. So in this sample the low lying resonant level evaluates to a non-resonant level located below the conduction band edge as the magnetic field exceeds 3 Tesla. So that a rather broad resonant state related photo-conductivity band at a low magnetic field (for example, 1 Tesla) evaluated into a non-resonant level related to sharp peak at high magnetic field ( for example, 6 Tesla). The high lying resonant state has almost the same slope of the magnetic field dependence of the energy in comparison with the interband transitions. This may indicate that the wavefunction of the resonant states have included a large factor of the conduction band state wavefunctions.

It is the same for the sample MCT30 as shown in fig.5. However, comparing the two such kind of slopes for a low lying and high lying resonant levels, the high lying resonant state has low slope. It can be understood as that the high lying resonant level includes more behavior of high energy states in conduction band in comparison with the low resonant level. So that it has a low slope due to larger effective mass value for the higher energy states in the conduction band.

## 4. Conclusion

We have investigated the impurity or defect levels in $Hg_{1-x}Cd_xTe$ by using the Magneto-Photoconductivity Spectroscopy Technique, that include: (1) the observation of the transitions from the shallow acceptor levels and the deep levels to the conduction band; (2) the observation of the transition between the valence band and the resonant level in the conduction band.

## Acknowledgments

The authors wish to thank Prof. Shen Jie, Prof. Xu Guosen for the crystal and sample preparation, and Prof. M. von Ortenberg for considerable discussions.

## References

[1] Jones C E, Nair V, Lindquist J, and Polla D L, 1982 *J. Vac. Sci. Technol.* **21** 187
[2] Littler C L, Maldonado E, Song X N, Yu Z, Elkind J L, Seiler D G, and Lowney J R, 1992 *J. Vac. Sci. Technol.* **B10** 1466
[3] Chu J H, Mi Z Y, Sizmann R, Koch F, Wollrab R, Ziegler J, and Maler H 1992 *J. Vac. Sci. Technol.* **B10** 1569
[4] Liu Pulin, Shi Guoliang, Wang Peigang, Chen Minhui, Lu Wei, Zhu Jinbin, Liu Weijun, and Shen Xuechu 1993 *Chin. J. of Infrared and Millimeter Wave* **12** 308
[5] Kecura Z, Hlidek P, Hoschl P, Koubele V, Prosser V, and Zvara M 1990 *Phys. Stat. Sol.* **(b) 158** K173
[6] Seiler D G, Loloee M R, Milazzo S A, Durkin A J, and Littler C L 1989 *Solid State Commun.* **69** 757
[7] Liu Pulin, Lu Xiaofeng, Lu Wei, Chu Junhao, Shen Xuechu, Michael von Ortenberg 1993 *Chin. J. of Infrared and Millimeter Wave* **12**, 122

*Inst. Phys. Conf. Ser. No 144: Section 2*
*Paper presented at 7th Int. Conf. on Narrow Gap Semicond. Santa Fe 9–12 January*



# Magnetophonon Oscillations in the Longitudinal Magnetothermal EMF of Narrow Gap $Hg_{1-x}Cd_xTe$

J.Baars, C. L. Littler*, D. Brink, and M. Bruder**

Fraunhofer-Institut für Angewandte Festkörperphysik
Tullastrasse 72, D-79108 Freiburg i. Br.
Germany

*Permanent Address: Department of Physics, University of North Texas, Denton, Tx 76203
**AEG-Aktiengesellschaft, Infrarotmodule, D-74072 Heibronn, Germany

The magnetophonon(MP) effect, as manifested in the longitudinal magnetothermal emf (LME), has been observed for the first time in narrow gap samples of $Hg_{1-x}Cd_xTe$. The MP oscillations, seen in the magneto-Seebeck coefficient, have been described in terms of unresolved MP transitions involving both longitudinal optical (LO) HgTe and CdTe phonons for transitions with $\Delta l > 1$. The MP resonances attributable to the different LO phonons become resolved for $\Delta l = 1$. To interpret the data, a phase shift of $\pi/2$ to lower fields is applied to the calculated MP resonance magnetic field positions. The interpretation of the MP data involving both the HgTe and CdTe LO phonons is found to satisfactorily explain not only the present data but also past results in $Hg_{1-x}Cd_xTe$ obtained from both longitudinal and transverse magnetoresistance measurements. Previously, past results were described only in terms of the longitudinal optical HgTe phonon scattering and contributions from other Landau levels of differing spin states. Our results thus support the premise that the HgTe and CdTe LO phonons are also involved and possibly dominate in the scattering process which gives rise to MP resonances observed in narrow gap $Hg_{1-x}Cd_xTe$.

## 1. Introduction

The magnetophonon (MP) effect has been studied in a number of semiconductors in order to determine various features of the band structure, transport properties, and the nature of the electron-phonon interaction in these materials[1,2]. However, in the solid solution semiconductor $Hg_{1-x}Cd_xTe$, only a limited number of observations have been made[3-7]. In these past observations, the MP effect was observed in the longitudinal and transverse magnetoresistance. In InSb[8], the MP effect is observed as resonant maxima in the transverse magnetoresistance(TMR) and resonant minima in the longitudinal magnetoresistance(LMR); however differing theoretical explanations of this shift are still open to question. [2] This shift from maxima to minima appears to also hold true for $Hg_{1-x}Cd_xTe$, as seen in the comparison of LMR[3] and TMR [6,7]data. However, there exists additional confusion in the literature as to whether the minima or maxima in the longitudinal magnetothermoelectric(LME) voltage represent the MP resonances. This is, in part, due to the fact that experimentally neither the maxima nor the minima of the LME voltage in InSb occur at the magnetic field positions of the MP resonances[9,10]; each is well-displaced from the expected MP resonance field positions. It has been shown that[11], in the limit of no Landau-level broadening and non-polar materials, one should observe resonant maxima in the LME voltage at magnetic field positions slightly higher than those expected for the MP effect. However, in the limit of large Landau level damping[12], it has also been shown that the MP resonances should occur as weak minima in the LME voltage, displaced to the low field side of the expected MP field positions by a phase shift of $\pi/2$. For the measurements reported here, where strong damping is expected, the $\pi/2$ phase shift is applied and shown to describe the MP resonances in terms of both unresolved and resolved transitions involving both longitudinal optical (LO) HgTe and CdTe phonons.

## 2. Theory

In order to obtain a quantitative description of the MP resonances, it is necessary to take into account the nonparabolicity of the conduction band of $Hg_{1-x}Cd_xTe$ and the polaron correction to the MP effective mass. Corrections to the MP resonance field positions that result from the thermal population of higher Landau levels are neglected in our initial analysis; aspects concerning population effects on the observed MP positions will be discussed later. The resonance condition for magnetophonon transitions is given by

$$E_{l'}^{\pm} - E_{l}^{\pm} = \hbar\omega_{LO}, \tag{1}$$

where the $\pm$ refers to the different spin states, $l$ ($l'$) refer to the Landau quantum numbers, and $\omega_{LO}$ is the longitudinal optical phonon frequency. An approximate analytical expression for the Landau level energies given by Weiler[13], along with the appropriate band parameters[14] and phonon energies, was used to calculate the MP transition energies given by Eq. 1. The effective mass, m*, is influenced by the polaron effect, introducing a multiplicitive factor of $(1 + \eta\alpha/2)$ to m*. For $Hg_{1-x}Cd_xTe$, $\eta = 0.83$ and $\alpha = 0.107$ [6]; thus the polaron effect produces a small but noticeable shift of the MP resonances to higher magnetic fields.

Barker[12], using a resolvent perturbation analysis of the Kubo-Luttinger formula, has shown that the magnetophonon structure in the longitudinal magneto-Seebeck coefficient is strongly modified by Landau level damping. Under the conditions $\hbar\omega_c/(k_BT) \geq 1$ and $\hbar\omega_{LO}/(k_BT) > 1$ (conditions for our experiment), where $\omega_c$ is the cyclotron frequency and $\omega_{LO}$ is the phonon frequency, the oscillatory part of the thermoelectric power is given by,

$$Q_1 = -Q_o \cos(2\pi\, \omega_{LO}/\omega_c - \pi/2), \tag{2}$$

generating weak minima displaced to the low field side of the MP resonance field positions $\omega_{LO}/\omega_c$. Thus, one would expect that the MP resonances would manifest themselves as weak minima in the measured LME of $Hg_{1-x}Cd_xTe$ which occur on the low field side of the expected MP resonances.

## 3. Experimental

For this study, n-type bulk crystals and epitaxial layers of $Hg_{1-x}Cd_xTe$ with $x \approx 0.21$ were used. The bulk crystals were grown by solid state recrystallization from quenched stoichiometric melts and 0.5 mm-thick slices were cut from the bulk crystals parallel to the (111) surfaces. The epitaxial layers were deposited from Te-rich solutions on (111)B surfaces of lattice-matched CdZnTe crystals using the dipping technique. To convert the as-grown p-type material to n-type, the bulk slices and epitaxial layers were annealed in sealed quartz ampules under saturated Hg vapor pressures at temperatures of 225 C for 3 to 4 weeks. Subsequently, the samples were chemo-mechanically polished using a 0.5% bromine-methanol solution. Gold leads were bonded to indium contact pads evaporated at the edges of the rectangular-shaped specimens measuring $\approx 5 \times 12\ mm^2$. The bulk and epitaxial material were of uniform composition and had donor densities in the range of low $10^{14}\ cm^{-3}$, as verified from temperature-dependent Hall-effect and infrared transmission. The magneto-thermoelectric effect was measured in the longitudinal configuration ($\mathbf{B} \parallel \nabla T$) by employing the lateral gradient method[15]. A temperature difference of 2 to 4 K across the sample was used for determining the LME between 50 and 150K.

## 4. Results and Discussion

The magnetophonon spectra from a bulk sample of $Hg_{1-x}Cd_xTe$(x = 0.215) is shown in Figure 1. Plotted in this figure is the calculated 2nd derivative of the LME voltage versus magnetic field. Here the resonant minima appear as peaks in the 2nd derivative spectra. Note in this figure that a doublet appears in the MP resonance at ≈ 12 kG. This doublet can be described in terms of magnetophonon (MP) transitions involving both longitudinal optical (LO) HgTe and CdTe phonons, as seen in Figure 2, which shows the comparison of theory to data as a function of temperature. The solid lines are the calculated MP transition energies as a function of quantum number difference and temperature for both the HgTe and CdTe phonons, and the solid dots are the field positions of the MP resonances. To correctly describe the observed MP LME data , a phase shift (to lower fields) of $\pi/2$ was applied to the field positions of the LME resonances calculated using Eq.'s 1-4 and the values of 17.1 meV and 19.4 meV for the HgTe and CdTe phonon energies, respectively. A good fit of theory to data, as shown in Fig. 2, is clearly obtained. It can further be seen that, for most temperatures and quantum numbers, the MP transitions resulting from interactions of electrons with both the HgTe and the CdTe phonons are unresolved. However, at higher fields and temperatures, the $0^+ \rightarrow 1^+$ and $0^+ \rightarrow 2^+$ MP transitions become resolved. Note also that an additional MP resonance, resulting from the CdTe $0^+ \rightarrow 4^+$ and the HgTe $0^+ \rightarrow 3^+$ MP transitions, is seen. This suggests that the CdTe phonon plays a stronger role in the scattering than has been previously believed. This extra resonance cannot be explained by invoking additional MP transitions from higher spin states.

Past magnetoresistance results[3-7] for low x-value (x < 0.3 ) samples have been described only in terms of LO phonon scattering involving HgTe phonons. In one of these studies[3], resolution of the $0 \rightarrow 1$ MP resonance was observed . Kahlert and Bauer [3] interpreted their LMR data as arising from HgTe LO-phonon scattering alone involving both spins, i.e. scattering between the $0^+$ and $1^+$ and $0^-$ and $1^-$ Landau levels.

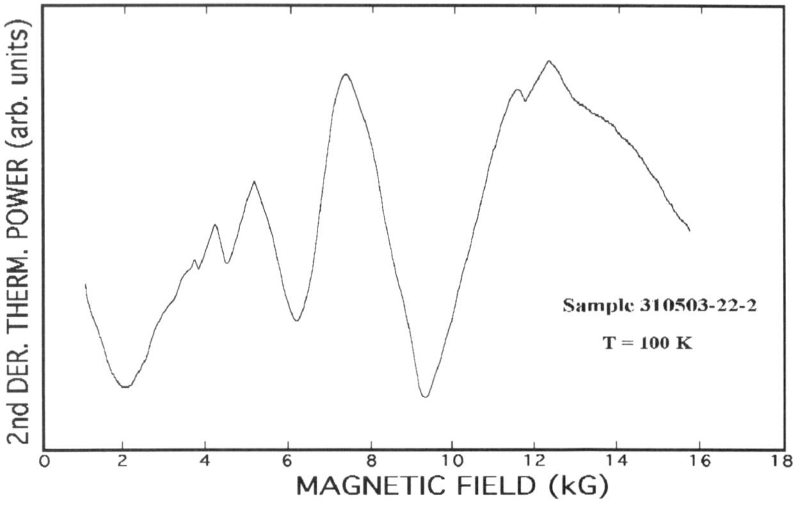


Figure 1: Second derivative($d^2U/dB^2$) of the longitudinal magnetothermoelectric voltage U(B) vs. B at T = 100 K. The resonant peaks represent minima in U(B).

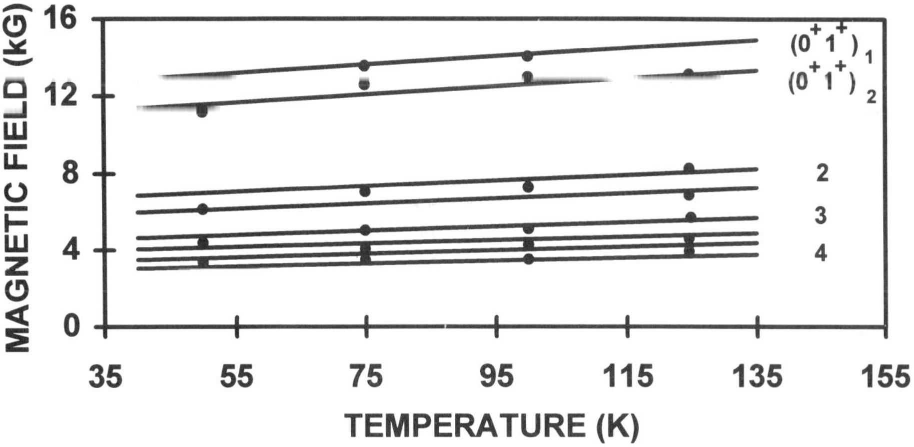


Figure 2: Comparison of the measured MP magnetic field positions (•) with the predictions of theory(—) vs temperature for spin conserving transitions $(0^+l^+)_i$ involving CdTe-like LO phonons (i=1) and HgTe-like LO phonons (i=2). The numbers 2, 3, and 4 refer to the final Landau quantum state of the MP transitions.

Figure 3 shows the result of both our analysis and that of Ref. [3], applied to the data of Ref. [3]. Here, the solid lines refer to the calculated $0^+ \rightarrow l^+$ MP transition energies and the dashed lines refer to the $0^- \rightarrow l^-$ MP transition energies. It is seen that the data are described reasonably well by each approach, with agreement being somewhat better if both phonons are included in the analysis. A similar fit to data from an x = 0.195 LPE sample and the data of Ref. [6] both yield good agreement when both phonons are included. Some observations should be made at this point. At low temperatures and moderate fields the $0^-$ level is not expected to be populated. However, as the temperature increases, the $0^-$ will become populated, even for the highest fields reached in this experiment. Thus one would expect the contribution of the $0^- \rightarrow l^-$ HgTe and CdTe transitions to the observed MP resonances to increase with increasing temperature. This, in turn, would lead to a shift to higher fields of the observed MP resonances with increasing temperature, which is in agreement with temperature dependence of the data seen in Figure 2 and 3. Thus, it appears that at the higher temperatures all possible MP transitions are present, whereas the MP transitions from both phonons involving the $0^+$ level contribute only at the lower temperatures. Recall also that the additional resonance noted in Fig. 2, resulting from a $0^+ \rightarrow 4^+$ CdTe- phonon MP transition and a $0^+ \rightarrow 3^+$ HgTe-phonon MP transition , cannot be explained by invoking the $0^- \rightarrow l^-$ HgTe MP transitions. Finally, resolution of the $0^+ \rightarrow 1^+$ MP transition involving the CdTe phonon has been seen in an x = 0.332 sample.[5] In this case, the resonance resulting from scattering with the CdTe phonon was approximately equal in strength to that resulting from the HgTe phonon. Thus, it seems clear that scattering involving both phonons, as well as thermal population effects, must be considered in order to explain the observed MP resonances. It is possible, considering the apparent strength of the CdTe MP transition seen in our data, that the involvement of the CdTe phonon may be larger than would normally be expected from the expected oscillator strengths. Future work will focus on the aquisition of LME data on other x-value samples to help confirm the strength and nature of the contribution of the CdTe phonon to the MP spectra.

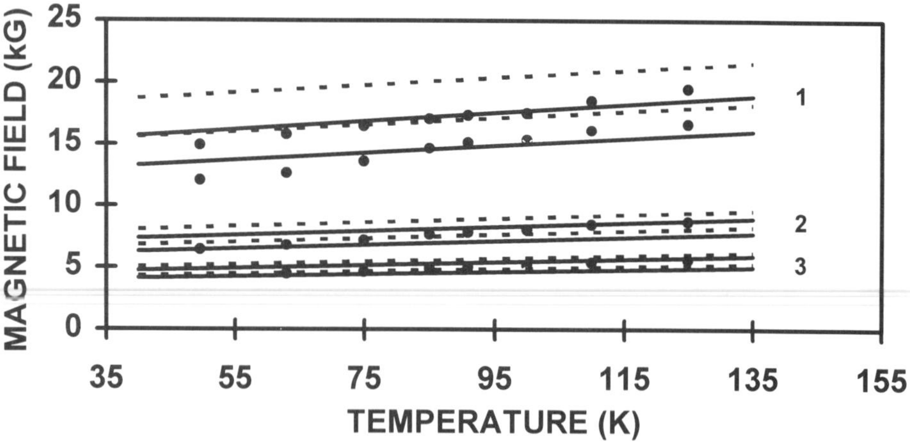


Figure 3: Comparison of the data (x = 0.212) of Ref . [3] (•) with theoretical preditions for the spin-conserving MP transitions $0^+ \rightarrow l^+$(—) and $0^- \rightarrow l^-$(---) vs temperature. The numbers refer to the final Landau quantum state for each spin and phonon.

## 5. Conclusions

The magnetophonon(MP) effect has been observed for the first time in the longitudinal magnetothermal emf of narrow gap samples of $Hg_{1-x}Cd_xTe$. The resonances have been described in terms of unresolved MP transitions involving both longitudinal optical (LO) HgTe and CdTe phonons for transitions with $\Delta l > 1$ which become resolved for $\Delta l = 1$. To interpret the data, a phase shift of $\pi/2$ to lower fields was applied to the calculated MP resonance magnetic field positions. The interpretation of the MP data involving both the HgTe and CdTe LO phonons was found to satisfactorily explain not only the present data but also past results in $Hg_{1-x}Cd_xTe$. The results support the concept that the HgTe and CdTe LO phonons are involved and possibly dominate in the scattering process which gives rise to MP resonances observed in narrow gap $Hg_{1-x}Cd_xTe$.

## 6. References

[1] Peterson, R. L. 1975 *Semiconductors and Semimetals* (New York: Academic) vol. 10, p. 221
[2] Firsov, Yu. A. , Gurevich, V. L. and Parfeniev, R. V. 1991 *Landau Level Spectroscopy* (Amsterdam: North Holland) vol. 2, p. 1183
[3] Kahlert, H. and Bauer, G. 1973 *Phys. Rev. Lett.* **30** 1211
[4] Amirkhanoy, Kh. I. , Bashirov, R. I. , Gadzhieva, R. M., and Elizarov, V. A. 1976 *Sov. Phys. Solid State* **18** 1360
[5] Mozhaev, E. A., Ivanov-Omskii, V. I., Mal'tseva, V. A., Mashovets, D. V., and Parfen'ev, R. V. 1977 *Sov. Phys. Semicond.* **11** 1260
[6] McClure, S. W., Seiler, D. G., Littler, C. L., and Goodwin, M. W. 1985 *J. Vac. Sci. Tech.* A**3** 271
[7] Takita, K., Suzuki, A., and Masuda, K. 1986 *Solid State Commun.* **58** 209
[8] Seiler, D. G., Joseph, T. J., and Bright, R. D. 1974 *Phys. Rev. B***9** 716
[9] Puri, S. M. and Geballe, T. H. 1963 *Bull. Am. Phys. Soc.* **8** 309
[10] Muzhdaba, V. M., Parfen'ev, R. V., and Shalyt, S. S. 1966 *Sov. Phys. Solid State* **7** 1922
[11] Arora, V. K. and Peterson, R. L. 1974 *Phys. Rev. B* **9** 4323
[12] Barker, J. R. 1973 *J. Phys. C: Solid State Phys.* **6** L52
[13] Weiler, M. H. 1979, *J. Magn. Magn. Mater.* **11** 131
[14] Seiler, D. G. and Littler, C. L. 1992 *Semiconductors and Semimetals* (New York:Academic) vol. 36, p. 294
[15] Baars, J. Brink, D., and Ziegler, J. 1991, *J. Vac. Sci. Tech.* **B9** 1709

*Inst. Phys. Conf. Ser. No 144: Section 2*
*Paper presented at 7th Int. Conf. on Narrow Gap Semicond. Santa Fe 9–12 January*



# Optical properties and interdiffusion of (001) $HgTe/Hg_{1-x}Cd_xTe$ superlattices

**F Goschenhofer, V Latussek, S Einfeldt, M O Möller, C R Becker and G Landwehr**

Physikalisches Institut der Universität Würzburg, Am Hubland, 97074 Würzburg, Germany

**Abstract.** Interdiffusion in $HgTe/Hg_{1-x}Cd_xTe$ superlattices has been investigated by means of changes in their optical properties. The energy of the lowest interband transition, H1→E1, which is calculated using a finite $HgTe/Hg_{1-x}Cd_xTe$ interface width, is compared with the experimental value in order to ascertain the width and shape of the interface. The results, which can be explained better if interdiffusion is linear rather than nonlinear, are strongly dependent on the sample. Moreover, the observed interdiffusion in the superlattice with the smallest diffusion coefficient is larger than the interdiffusion which occured during growth: the absorption coefficient of as-grown superlattices is independent of time spent at the growth temperature contrary to predictions using experimentally determined interdiffusion results. Apparently, the results of the interdiffusion experiments depend on the thermal history of the sample.

## 1. Introduction

The optical [1] and electrical [2] properties of Hg based heterostructures depend strongly on the abruptness of their interfaces and consequently on interdiffusion in these interfaces during growth and during subsequent technological processes after growth. Previous diffusion experiments on $HgTe/Hg_{1-x}Cd_xTe$ superlattices have been primarily based on x-ray diffraction methods [3–5] as well as on chemical mapping [6]. Even though optical absorption depends to a large extent on the interdiffusion of these interfaces as demonstrated by Leopold *et al.* [1], optical absorption has not previously been employed to quantitatively determine diffusion in $HgTe/Hg_{1-x}Cd_xTe$ superlattices.

The absorption coefficient of $HgTe/Hg_{1-x}Cd_xTe$ superlattices displays step-like features corresponding to subband transitions which depend on various parameters, among which are period, well and barrier thicknesses, composition of the wells and barriers, valence band offset, strain and the concentration profile of the interfaces. If these parameters with the exception of the interface profile are known with sufficient accuracy, then a theoretical determination of the absorption coefficient should result in the concentration profile of the interfaces. This method has been employed here to determine the temporal dependence of this concentration profile and hence interdiffusion in these interfaces.

## 2. Growth and parameters

The HgTe/$Hg_{1-x}Cd_xTe$ superlattices were grown on (001) $Cd_{0.96}Zn_{0.04}Te$ substrates by molecular beam epitaxy (MBE), RIBER 2300, as has been described in detail elsewhere [7, 8]. The substrates were degreased, etched in HCl, loaded into the MBE system and then heat treated at temperatures up to 320°C. The substrate temperature was measured by means of a thermocouple which was in contact with a molybdenum substrate holder and was calibrated at the melting points of indium and tin. The day to day uncertainty in the temperature is $< 2$°C. Before the superlattice was grown, a 500 Å thick CdTe buffer was grown at 270°C. The superlattices were then grown at 180°C. A polycrystalline CdTe cap layer with a thickness of about 600 Å was grown at 35°C either directly after growth or in one case after several weeks.

The superlattice period as well as the well thickness and hence that of the barrier were determined by means of high resolution x-ray diffraction [7,9]. This is done by taking advantage of the large HgTe structure factor compared to that of CdTe for the (002) Bragg reflection. In general, the investigated superlattices are of good quality as indicated by the full width at half maximum of their (004) Bragg reflection, i.e. as low as 22 arcsec, and the large number of satellites. [7, 8] The composition of the initial barrier material was determined to be $0.68 \pm 0.02$ by means of transmission and reflection measurements on thick test layers of (001) $Hg_{0.32}Cd_{0.68}Te$ grown under identical conditions. Knowing the well and barrier thicknesses we have been able to confirm this value for the composition of the barriers by annealing several superlattices and then measuring the composition of the resulting alloy.

## 3. Optical absorption coefficient

Optical transmission and reflection measurements were carried out in the middle and near infrared with a Fourier transform spectrometer, IFS88, from Bruker Instruments Inc. A pyroelectric detector was employed rather than a HgCdTe detector because of its better linearity. The aperature was kept as small as possible for the same reason, i.e. a diameter of 2 to 3 mm. The absorption coefficient was determined by fitting the experimental transmission and reflection spectra to a theoretical description of the multi-layer system using the standard matrix procedure, e.g. Ref. [10]. Minimizing the difference between experimental and theoretical values of merely the transmission at a constant index of refraction resulted in the same absorption coefficient within experimental uncertainity. Therefore, in most cases this procedure was carried out for only the transmission using an appropriate constant refractive index.

The electronic band structure was calculated with an envelope function method [11] which differs formally from two previous methods [12, 13] but combines the essential aspects of both approaches. The bands of both bulk HgTe and CdTe are described by Kane's four-band model ($8 \times 8$ $\mathbf{k} \cdot \mathbf{p}$) including second order far-band contributions, but neglecting the small anisotropy of the valence bands. The effects of strain due to lattice mismatch were also taken into consideration [2]. The numerical values for the band parameters were deduced from measurements on bulk HgTe and $Hg_{1-x}Cd_xTe$ according to Weiler [14]. The valence band offset between HgTe and CdTe was taken to be 550±50 meV which we have found to be necessary in order to correctly describe subband transition energies and the absorption coefficient [15]. The valence band offset for $Hg_{1-x}Cd_xTe$ was then assumed to vary linearly with $x$ [16]. The initial $x$ value for the barrier in conjunction with a concentration profile due to subsequent interdiffusion of the interface was integrated into the theory.

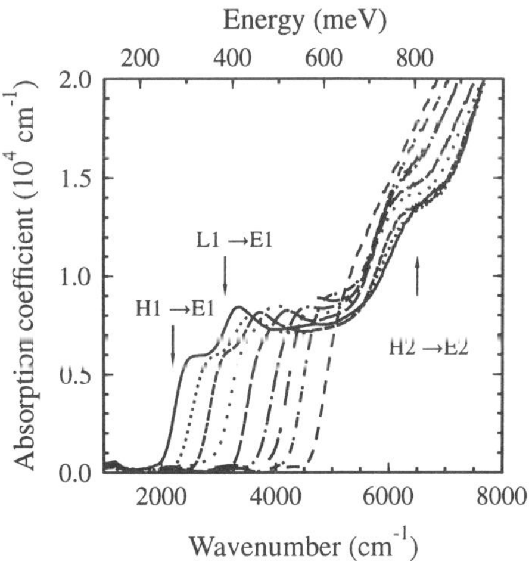


**Figure 1.** The experimental optical absorption coefficient as a function of wavenumber ($cm^{-1}$) and energy (eV) at various diffusion times for the superlattice Q424 at 300 K. The diffusion temperature was 180°C.

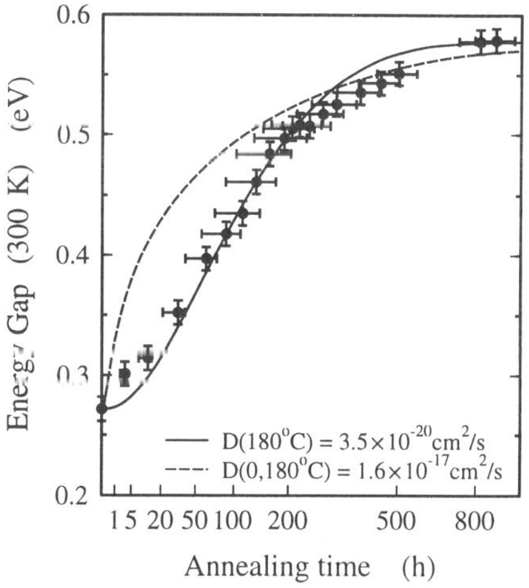


**Figure 2.** The shift in the $E_0$ band gap, i.e. H1→E1, of Q424 as a function of diffusion time for a diffusion temperature of 180°C. The solid curve corresponds to a linear fit and the dashed curve to a nonlinear fit.

Diffusion experiments were carried out on the superlattice structures discussed above in vacuum, an Ar atmosphere or oil at 160 and 180°C, i.e. at temperatures similar to those used during growth. After each diffusion step the absorption coefficient was determined and occasionally x-ray diffraction of the sample was measured.

## 4. Results and discussion

The experimentally determined optical absorption coefficients before and after some of the annealing steps at 180°C for the superlattice Q424 are shown in Fig.1. After the final annealing step, i.e. after 880 hours, the superlattice has become an alloy. Theoretically determined energies of various interband transitions before annealing, which are indicated by arrows, correspond to the observed structure in the absorption coefficient. As shown in Fig.2 the observed shift of the band gap can be much better described by linear than by nonlinear diffusion. Using a Cd dependence similar to that of Tardot et al. [5], nonlinear diffusion can be expressed by

$$D(x,T) = D(0,T)\exp\left(\frac{-0.50\,x\ \ eV}{kT}\right)$$

where $x$ is the Cd concentration of the barrier. All of the diffusion results in this investigation are listed in Table I and the nonlinear results are shown in Fig.3 together with published results. As can be seen, our results are extremely dependent on sample. They vary from sample to sample by a factor of 50. Some of our nonlinear results are much higher and some much lower than an extrapolation of the published results to 160 and 180°C. The large scatter in our data can at present not be correlated with any growth condition or sample parameter. However, it is worthwhile mentioning that the diffusion in the first four samples in Table I can be

TABLE I. The number of periods, the period and layer thicknesses(Å), the linear diffusion coefficient, $D(T)$, and the concentration independent component of the nonlinear coefficient, $D(0,T)$.

| | Periods | $\bar{d}_P$ | $\bar{d}_w$ | $\bar{d}_b$ | Diffusion | Diffusion coefficient | |
|---|---|---|---|---|---|---|---|
| | | $\pm 0.5$ | $\pm 1$ | $\mp 1$ | Temperature | $D(T)$ | $D(0,T)$ |
| | | Å | Å | Å | °C | $\times 10^{-19}$ cm$^2$/s | $\times 10^{-16}$ cm$^2$/s |
| Q766 | 100 | 62.0 | 23.7 | 38.3 | 180 | 3.0 | 1.0 |
| | | | | | 160 | 1.5 | 0.5 |
| Q769 | 100 | 74.4 | 31.7 | 42.7 | 180 | 5.0 | 2.0 |
| | | | | | 160 | 2.5 | 1.5 |
| Q784 | 100 | 80.3 | 30.9 | 49.4 | 180 | 33 | 6.5 |
| | | | | | 160 | 8 | 1.6 |
| Q785 | 100 | 93.0 | 32.8 | 60.2 | 180 | 16 | 8.0 |
| | | | | | 160 | 8 | 4.0 |
| Q424 | 120 | 110.7 | 34.0 | 76.7 | 180 | 0.35 | 0.16 |

described by a slow initial diffusion and then by a subsequent faster diffusion as has been reported by Arch et al. [4]. The presence of a cap layer was shown to have a relatively small effect on interdiffusion in Q766.

The expected Cd concentration profile for the superlattice Q424 corresponding to an $E_0$ band gap of 400 meV for both linear and nonlinear diffusion is shown in Fig.4. In addition, the as-grown profile should be dependent on position in the superlattice. In other words, this profile should depend on the length of time spent at the growth temperature, i.e. 180°C. A shift

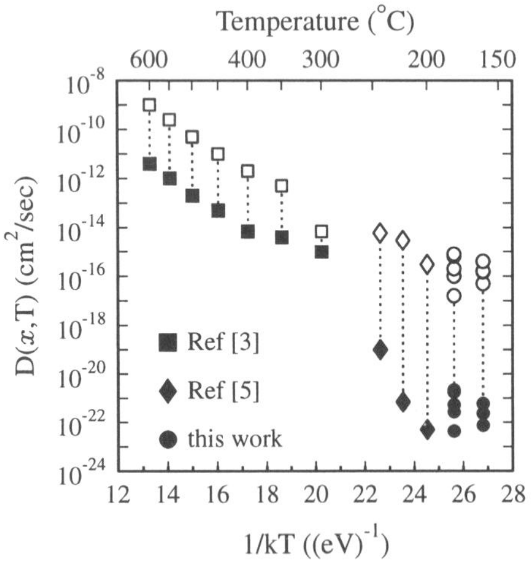


**Figure 3.** An Arrhenius plot of nonlinear diffusion coefficients in CdTe/HgTe. The upper symbols (empty) represent diffusion in HgTe, D(0,T), and the lower symbols (filled) diffusion in CdTe, D(1,T).

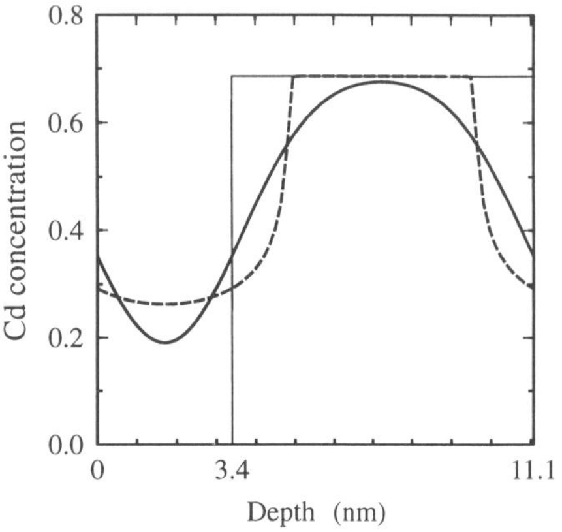


**Figure 4.** The expected Cd concentration profile for the superlattice Q424 for both linear (solid line) and nonlinear diffusion (dashed line) after an annealing time which corresponds to a $E_0$ band gap of 400 meV.

in the band gap across a superlattice of approximately 6 meV is expected at this temperature for a typical growth time of two hours.

In order to corroborate the expected results of diffusion during growth, Q424 was etched to reduce its thickness and then the absorption coefficient of the resulting superlattice was measured. This was repeated four times until only 80 of the original 120 periods were present. Within experimental error of less than 1 meV, no shift was observed. This means that an upper limit for linear diffusion in Q424 is given by $D(180°C) < 1\times10^{-20}$ $cm^2/s$ which is at least a factor of four smaller than the experimental value for this coefficient. The difference between diffusion during growth and after growth is not understood at the present. However, we speculate that diffusion depends on the number of dislocations which are generated due to differences in thermal expansion coefficients of substrate, wells, barriers and cap layer. The low value for diffusion during growth would correspond to an average interface width of approximately two monolayers which is a factor of three narrower than that determined by Kim et al. [6].

In conclusion, we have shown that interdiffusion in HgTe/ $Hg_{1-x}Cd_xTe$ superlattices can be best explained by linear diffusion. The resulting diffusion constant varies by a factor of 50 depending on the sample. Some of the diffusion coefficients are appreciably smaller and some appreciably larger than the literature values. However, the diffusion coefficient for Q424, which is the smallest measured value, is nevertheless larger than that for interdiffusion which occurs during growth by at least a factor of four. We speculate that this is due to the thermal history and differences in thermal coefficients of expansion.

The support of this project by the Bundesministerium für Forschung und Technologie (contract TK 0369) and the Deutsche Forschungsgemeinschaft is gratefully acknowledged.

## References


[1] Leopold D J, Broerman J G, Peterman D J and Wroge M L 1990 *Appl. Phys. Lett.* **52** 969-71

[2] Simon A, Bertho D, Boiron D and Jonanin C 1990 *Phys. Rev. B* **42** 5221-8

[3] Tang M-F S and Stevenson D A 1987 *Appl. Phys. Lett.* **50** 1272-4

[4] Arch D K, Staudermann J L and Faurie J P 1986 *Appl. Phys. Lett.* **48** 1588-90

[5] Tardot A, Hamoudi A, Magnea N, Gentile P and Pautrat J L 1993 *Appl. Phys. Lett.* **62** 2548-50

[6] Kim Y, Ourmazd A, Bode M and Feldman R D 1989 *Phys. Rev. Lett.* **63** 636-9

[7] Becker C R, He L, Regnet M M, Kraus M M, Wu Y S, Landwehr G, Zhang X F and Zhang H 1993 *J. Appl. Phys.* **74** 2486-93

[8] Becker C R, Latussek V, Heinke H, Regnet M M, Goschenhofer F, Einfeldt S, He L, Bangert E, Kraus M M and Landwehr G 1993 *Growth and Characterization of Materials for Infrared Detectors* R E Longshore and J W Baars, Editors Proc. SPIE 2012 p 138-48

[9] Möller M O, Bicknell-Tassius R N and Landwehr G 1992 *J. Appl. Phys.* **72**, 5108-16

[10] Fasold D, Hehl K and Jetschke S 1984 *phys. stat. sol. (a)* **86**, 125

[11] Bangert E, Boege P, Latussek V and Landwehr G 1993 *Semicond. Sci. Technol.* **8** S99-101

[12] Ram-Moham L R, Yoo K H and Aggarwal R L 1988 *Phys. Rev. B* **38** 6151-9

[13] Johnson N F, Ehrenreich H, Hui P M and Young P M 1990 *Phys. Rev. B* **41** 3655-69

[14] Weiler M H 1981 *Semiconductors and Semimetals* **16** editors Willardson R and Beer A C (New York: Academic) 119-89

[15] von Truchseß M, Latussek V, Sizmann R, Helgesen P, Goschenhofer F, Becker C R, Landwehr G and Batke E submitted to *Phys. Rev. B*

[16] Shih C K and Spicer W E 1987 *Phys. Rev. Lett.* **58** 2594-7

# EFFECT OF Hg PRESSURE ON THE DIFFUSION OF Hg IN CdTe

**M U Ahmed[1], E D Jones[1], J B Mullin[2] and N M Stewart[3]**

[1]Department of Applied Physics, Coventry University, Priory Street, Coventry, CV1 5FB, UK
[2]EMC, "The Hoo", Brockhill Road, West Malvern, WR14 4DL, UK
[3]BT Research Laboratories, Martlesham Heath, Ipswich, IP5 7RE, UK

**Abstract.** In this paper the results of the diffusion of Hg into CdTe as a function of Hg overpressure in closed tube diffusions at 328°C and 400°C are reported. Concentration profiles were measured using both radiotracer sectioning and secondary ion mass spectrometry and a mathematical function composed of the sum of two complementary error functions gave the best fit to the profiles. Results obtained are consistent with diffusion occurring via a vacancy mechanism.

## 1. Introduction

CdTe is used extensively in the fabrication of infra red detectors using the narrow band-gap II-VI ternary compound (HgCd)Te (MCT). Previous work on the diffusion of Hg in CdTe has been carried out by Jones et al [1], Takita et al [2] and Hogg et al [3]. All reported work has been on the temperature dependence of the diffusion. In addition to knowing the rate of diffusion of Hg in CdTe it is important to have an idea of the mechanism by which the diffusion occurs. One possible method which will give information on this is to carry out Hg pressure dependency measurements at constant temperatures.

There are two ways of carrying out such measurements in sealed capsules: (a) isothermal diffusions where the pressure is controlled at values less than the saturated vapour pressure (SVP) of elemental Hg by controlling the mass of Hg, (b) diffusion in a two - zone furnace where one zone is used to maintain the pressure at SVP values using temperatures which are lower than that used for the diffusion anneal and the CdTe slice is held at the required diffusion temperature in the other zone.

In this paper the results of the diffusion of Hg into CdTe as function of Hg overpressure at two temperatures, 328°C and 400°C are reported. The diffusion times were 5 days and 1 day respectively.

## 2. Experimental Techniques

Slices of bulk grown single crystal CdTe of dimensions, 8mm x 5mm, were mechanically polished using successively finer grades of diamond pastes, 6, 3, 1, 0.1 μm and then chemically polished using 1% $Br_2$/ methanol solution. In this process a total thickness of 300 μm was removed from the surface of the slice. Each polished CdTe slice, together with radioactive Hg (or non - radioactive Hg) was sealed in a silica ampoule under high vacuum conditions ( typically 1 x $10^{-6}$ torr).

Diffusions were carried out using method (b) described above. The range of Hg pressure, $P_{Hg}$, used was 0.0037 - 2 atmospheres. In addition two CdTe slices were preannealed with Cd at 800°C for 1 day under SVP conditions prior to the pressure dependency measurements.

After completion of the Hg diffusion, there was no noticable change in the physical appearence of the slices and no measurable change in the weight of the slice was observed. The concentration profiles for the diffusions carried out at 400°C were measured using Secondary Ion Mass Spectrometry (SIMS) [4], while those annealed at 328°C were measured using a radiotracer sectioning technique (RTS) [ 5].

### 3. Results

The majority of the diffusion profiles decayed by over four decades before flattening out to give a signal due to the natural background of the instrument and as can be seen in figure 1 the profile consists of two distinct parts. A mathematical function of the form:

$$C = C_{01}\ erfc(x/b) + C_{02}\ erfc(x/d) + C_{03}. \qquad (1).$$

which was fitted using a Mathematica software package gave the best fit to the profiles, where parameters *b* and *d* gave two values øf the diffusivity, $D_1$ and $D_2$ respectively, $C_{01}$ and $C_{02}$ are the corresponding surface concentrations and $C_{03}$ is the constant background level.

The absolute concentration of Hg atoms for the SIMS measurements were calibrated from measurements of the Hg concentration on the surface of the diffused slice obtained using Energy Dispersive X- ray Analysis (EDAX) on a Scanning Electron Microscope (SEM). The concentration obtained using this method was termed $C_{0s}$.

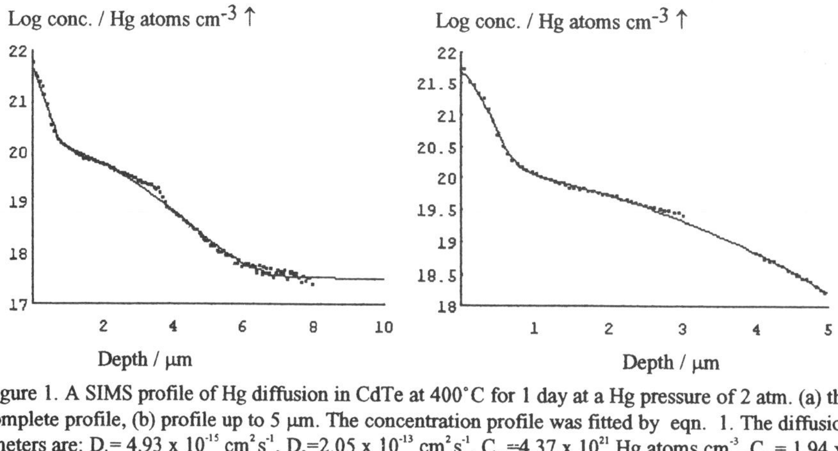


Figure 1. A SIMS profile of Hg diffusion in CdTe at 400°C for 1 day at a Hg pressure of 2 atm. (a) the complete profile, (b) profile up to 5 µm. The concentration profile was fitted by eqn. 1. The diffusion parameters are: $D_1$= 4.93 x $10^{-15}$ $cm^2s^{-1}$, $D_2$=2.05 x $10^{-13}$ $cm^2s^{-1}$, $C_{01}$=4.37 x $10^{21}$ Hg atoms $cm^{-3}$, $C_{02}$= 1.94 x $10^{20}$ Hg atoms $cm^{-3}$, and $C_{03}$=3.44 x $10^{17}$ Hg atoms $cm^{-3}$. The kink in the profile at a depth of approximately 3 µm is due to a change in the sensitivity range of the instrument and has been ignored in the curve fitting.

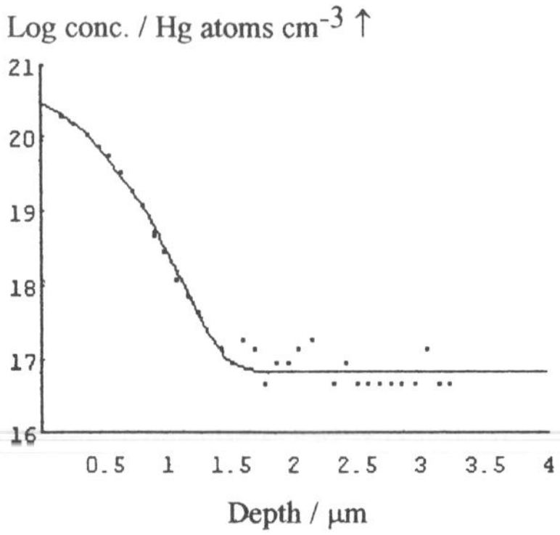


Figure 2. A SIMS profile of Hg diffusion in CdTe at 400°C for 1 day at a Hg pressure of 2 atm, in which the CdTe had been preannealed with Cd. The concentration profile was fitted by eqn. 2. The diffusion parameters are: $D_1 = 1.89 \times 10^{-15}$ cm$^2$s$^{-1}$, $C_{01} = 3.70 \times 10^{20}$ Hg atoms cm$^{-3}$ and $C_{03} = 9 \times 10^{16}$ Hg atoms cm$^{-3}$.

The concentration profiles of slices which had been preannealed with Cd ( figure 2 ), showed only one diffusion component and were fitted best by:

$$C = C_{01}\ erfc(x/b) + C_{03} \qquad (2).$$

A graph of diffusivities versus the Hg pressure is shown in figure 3 where it can be seen that the results for both sets of measurements (328°C and 400°C) show the same behaviour. The use of different techniques for measuring the diffusion profiles enhances confidence in the results. The slow diffusing component, $D_1$, increases with $P_{Hg}$ with an approximate $D_1 \propto P_{Hg}^{0.5}$ dependence up to a pressure of 0.4 atmospheres, and then $D_1$ decreases with further increase in pressure with an approximate $D_1 \propto P_{Hg}^{-1.2}$ dependence. The fast diffusing component, $D_2$, increases with $P_{Hg}$ over the whole pressure range with an approximate $D_2 \propto P_{Hg}^{0.15}$ dependence.

The corresponding values of $C_0$ for the diffusions carried out at 400°C are shown in figure 4 where it can be seen that the values increases systematically with increasing $P_{Hg}$. The surface concentrations, $C_{0s}$, obtained using EDAX and used to calibrate the SIMS profile are also plotted in figure 4. As can be seen there is a systematic discrepency between the two sets of results, ie between $C_{01}$ and $C_{0s}$. The reason for this is that the results obtained using EDAX gives an average value over a depth of 1 μm into the slice, the depth to which the electron beam of the SEM penetrates the CdTe.

Another important feature displayed by the results in figure 4 are the low values obtained for $C_{0s}$ for the CdTe slices preannealed with Cd and then diffused with Hg, a reduction of a factor of 100 at 0.065 atmospheres and a factor of 10 at 2 atmospheres.

## 4. Discussion and Conclusion

The occurrence of two component profiles when group II elements diffuse into II-VI compounds is common and well documented and reasons have been proposed for the second faster component.

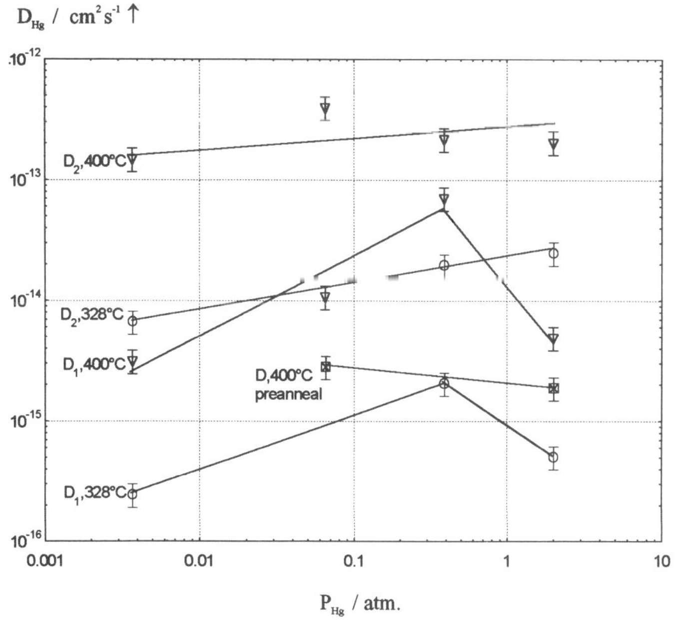


Figure 3. A graph of the Hg diffusivities, $D_1$ and $D_2$, versus the Hg pressure, $P_{Hg}$, for diffusion at 400 °C and 328°C.

Because of the improved depth resolution used in the measurements reported here the functions used in equation 1 gave much better fits than other possible functions, eg ones that describe dislocation and subgrain boundary diffusions.

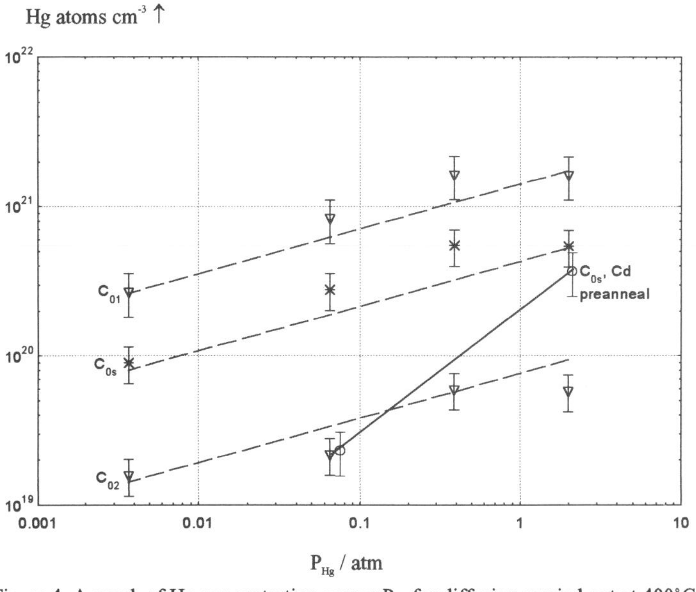


Figure 4. A graph of Hg concentration versus $P_{Hg}$ for diffusion carried out at 400°C.

For the diffusion of Hg in CdTe, under certain physical conditions there is evidence that rate limited diffusion is occurring [4] but that is not the case for the measurements reported here. Another feature in these measurements is that when the CdTe slices were preannealed with Cd equation 2 gave the best fit to the profiles.

One possible explanation for the behaviour shown in figure 3 and 4 is to assume that diffusion occurs by some kind of vacancy mechanism. As $P_{Hg}$ increases from low values up to 0.4 atmospheres the concentration of Hg atoms at the surface increases and occupy more of the vacant sites on the Cd sublattice, this is consistent with the increase in the surface concentration and the diffusivity. At pressures above 0.4 atmospheres where the concentration continues to increase and the diffusivity decreases rapidly one of two possibilities occur; (a) the vacant sites on the Cd sublattice are filled with Hg atoms completely, (b) a Hg - Cd alloy forms on the surface of the CdTe slice which hinders the diffusion of Hg into the slice. Either effect could explain the results obtained.

This possible explanation is supported by the results obtained from the slices which have been preannealed with Cd. In this case the excess vacancies on the Cd sublattice have been filled with Cd atoms which in turn reduces both the Hg concentration in the slice and the corresponding value of the diffusivity. Furthermore, the occurrence of one component profiles imply that the defect type responsible for the second component was removed during the preanneal with Cd.

CdTe can contain Te precipitates which can affect the rates of diffusion of Hg. It was not possible to observe any such precipitates in the CdTe slice on the micrographs obtained using infra red microscopy prior to the diffusion. Neither was it possible to observe any precipitates in CdTe slices which had been diffused at 400°C. Te profiles obtained using SIMS were, to within limits of experimental error, perfectly uniform and micrographs obtained from SIMS ion imaging of Te also gave uniform spatial distributions. Such micrographs were taken over a 150 μm diameter at the surface of the CdTe slice and at depths of 1, 3 and 8μm into the material. There was no physical evidence of Te evaporating from the surface of the CdTe slice during the diffusion and condensing on the inner wall of the capsules. From these observations the presence of Te precipitates in the material, which would have affected the Hg diffusion, can be ruled out.

The spatial distribution of Hg at the surface of the slice was uniform for all the samples, but deeper into the slice ( > 5 μm ) some slight non-uniformities became apparent for all samples. The ion image for Hg taken at a depth of 8μm did not reveal any Hg at all.

## Acknowledgements

The authors would like to thank The Defence Research Agency (DRA), Malvern, UK and the Engineering and Physical Science Research Council (EPSRC), UK for financial support, GEC - Marconi Infrared Ltd, Southampton, UK, for providing the CdTe slices used in this work and Mr. N. Shaw of the DRA for valuable discussions.

## References

[1] E D Jones, V Thambipillai and J B Mullin 1992 *J. Crystal Growth,* **118** 1

[2] K Takita, K Murakami, H Otake, K Masuda, S Seki and H Kudo 1984 *Appl. Phys. Lett.* **44(10)** 996

[3] J C H Hogg, A Bairstow, G W Matthews, D Shaw and J D Stedman 1993 *Mater. Sci. Eng.* **B16** 195

[4] M U Ahmed, E D Jones, J B Mullin and N M Stewart 1995 *Mater. Sci. Forum* at press.

[5] N M Stewart, E D Jones and J B Mullin 1992 *J. Mater. Sci.* **3** 211

*Paper presented at 7th Int. Conf. on Narrow Gap Semicond. Santa Fe 9–12 January*

# Studies on the Pressure-dependency of the Diffusion of Iodine into CdTe

E. D. Jones[1], J. Malzbender[1], N. Shaw[2] and J. B. Mullin[3]

1 School of Natural and Environmental Sciences, Coventry University, Coventry, CV1 5FB, UK

2 DRA Electronics Division, RSRE, St Andrews Road, Great Malvern, Worcester, WR14 3PS, UK

3 Electronic Materials Consultant, The Hoo, Brockhill Road, West Malvern, Worcester, WR14 4DL, UK

**Abstract.** An investigation on the pressure dependency of the diffusion of iodine into CdTe at temperatures of 220°C, 360°C and 500°C is reported. The diffusion profiles were found to be composed of four components which were characterised by effective diffusivities $D_1$, $D_2$, $D_3$ and $D_4$. For each component the diffusivity remained constant indicating that the diffusivity was independent of the $CdI_2$ pressures except at low pressures at 500°C. In the case of the corresponding surface concentrations, $C_0$, initially each component showed a decrease, went through a minimum and increased sharply before either gadually falling again or becoming constant as the $CdI_2$ pressure was reduced systematically. These features are discussed in this paper.

## 1. Introduction

Iodine has been used on a regular basis as an n-type dopant in epitaxial material in infrared devices [1], but recent measurements have shown that iodine is a fairly fast diffusant if diffused from an external source [2], producing diffusivities of $10^{-14}$ cm$^2$ s$^{-1}$ at room temperature [3]. Such measurements have produced complex diffusion profiles which could be fitted empirically by a function composed of the sum of four complementary error functions, giving four effective values of the diffusivity for each profile. This proved to be a satisfactory method to indicate how the diffusion of iodine was affected by parameters such as temperature, time and pressure. The reasons for the occurrence of the four component profiles are still not fully explained.

In general, it is possible to estimate the most probable defect mechanism by which the diffusion occurs by measuring the diffusivity as a function of the iodine partial pressure in the capsule. The partial pressure of the diffusant in an enclosed system can be varied, whilst keeping the sample temperature consant, by two methods:

(a) Reducing the mass of the diffusion source to a level below the mass needed for saturation vapour pressure.

(b) By reducing the temperature of the diffusion source in order to reduce the related saturation vapour pressure [4].

Method (a) is usually carried out under isothermal conditions in a single zone furnace, whereas in method (b) a furnace with two temperature zones is used, with the CdTe slice placed in the higher temperature zone and the diffusion source in the lower one. Because of the temperature gradient along the length of the capsule in method (b), the CdTe slice may lose material from the surface during the diffusion anneal and the possibility of this occurring should always be checked. In fact, loss of material from the surface is possible in both methods but in this study it is more likely to occur in method (b) due to the transport properties of iodine [5].

## 2. Experimental

Wafers of bulk grown CdTe, supplied by GEC Marconi Infrared Limited, Southampton, which had been cut with a diamond saw from large boules containing a few grains were used in this investigation. Single crystal slices (~ 8 mm × 8 mm) were cleaved out and chemically polished with 1 % bromine in methanol to remove cutting damage. In this process a total thickness of between 100 and 200 $\mu$m was removed from the surface of each slice.

In the series of measurements described here the dependence of diffusivity and $C_0$ values on the $CdI_2$ partial pressure was determined at 220°C, 360°C and 500°C. The concentration profiles were measured using radiotracer sectioning techniques [6]. Again four component profiles were obtained and the diffusivities, $D_1$ - $D_4$, ($D_1$ was the slowest and $D_4$ the fastest component) and the corresponding values of the surface concentrations, $C_{01}$ - $C_{04}$, are plotted in Fig. 1 - 3. For the results obtained at 220°C, which are plotted in Fig. 3 the low radioactivity of the tracer did not allow the analysis of the fastest profile part $D_4$ to be measured.

One measurement was carried out using method (a), which acted as a check on the bulk of the measurements carried out using method (b) (see Fig. 1, points labelled A, temperature of the slice 500°C). Close agreement was obtained between the results obtained using the two methods and so the temperature gradient had little effect on the results.

The range of $CdI_2$ pressure used for the three temperatures 500°C, 360°C and 220°C was $3 \cdot 10^{-5}$ to 10 mm Hg, $10^{-14}$ to 0.032 mm Hg and $3 \cdot 10^{-13}$ to $10^{-4}$ mm Hg, respectively. The maximum pressure in each case was given by the saturated vapour pressure under isothermal conditions. The corresponding minimum temperatures of the diffusion sources were 200°C, 20°C and 37°C, respectively.

During the diffusions the diffusion source was not allowed to come into direct contact with the CdTe slice and so only molecules from the gas phase could enter the slice. At the end of each diffusion no weight change in the CdTe slice was detected and it was not possible to observe any change in its physical appearance.

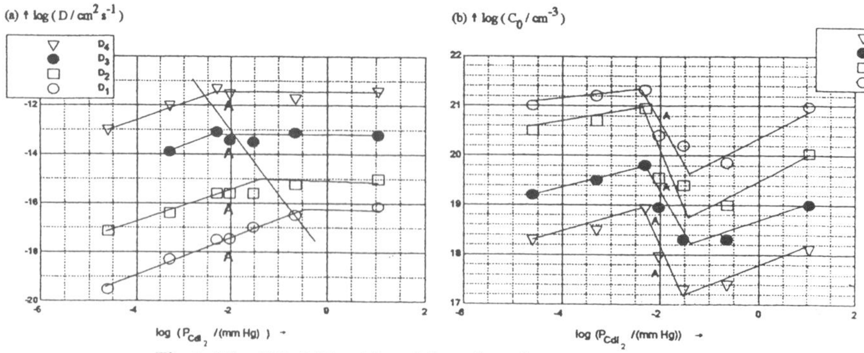


Fig. 1: The diffusivities (a) and $C_0$ values (b) as a function of $CdI_2$ partial pressure for a temperature of 500°C.

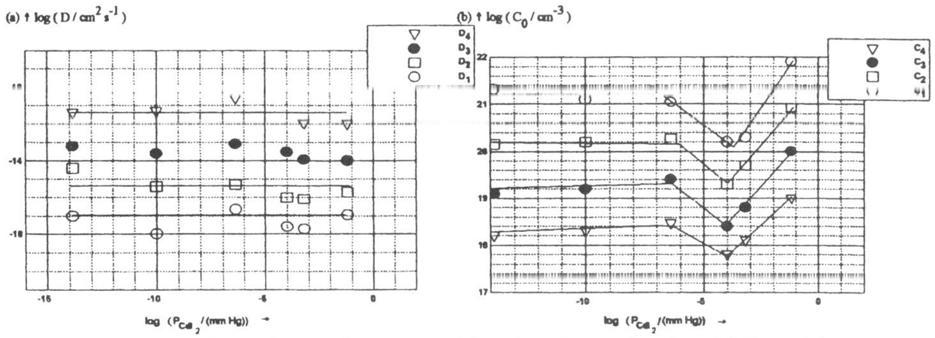


**Fig. 2:** The diffusivities (a) and $C_0$ values (b) as a function of $CdI_2$ partial pressure for a temperature of 360°C.

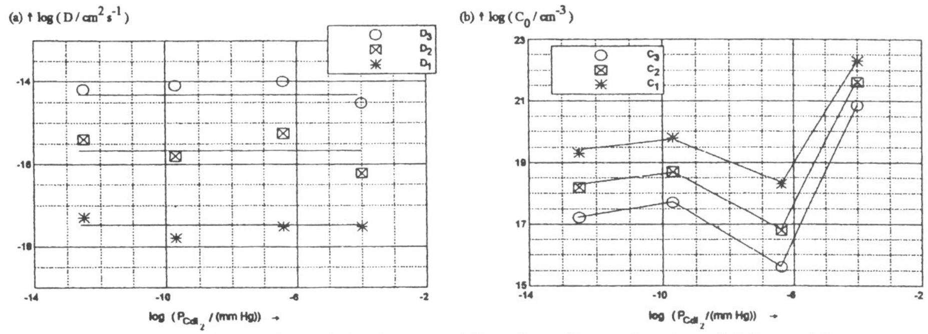


**Fig. 3:** The diffusivities (a) and $C_0$ values (b) as a function of $CdI_2$ partial pressure for a temperature of 220°C.

## 3. Results

The diffusivity for all four components remained constant, indicating that it was independent of the $CdI_2$ partial pressure, except possibly at low pressures at 500°C (Fig. 1). In this case all diffusing components showed a pressure dependency with: $D \propto p_{CdI_2}^{0.75}$, but the change from

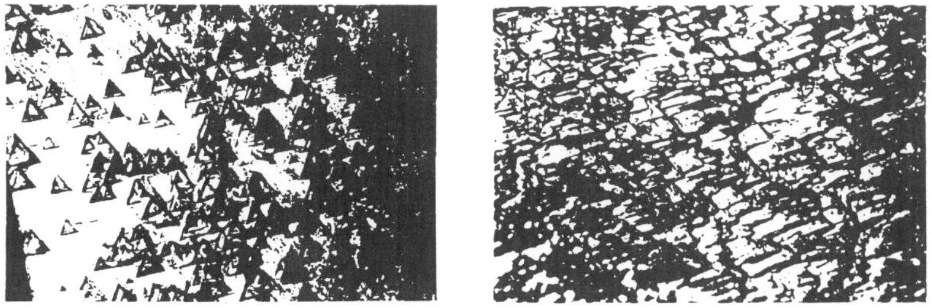

Fig.4: Micrographs showing the surface of a slice diffused at 500°C under a $CdI_2$ overpressure of $3 \cdot 10^{-5}$ mm Hg for 5 days. Different patterns can be seen which might be related to different orientations of the CdTe slice ( × 320).

pressure independent diffusion to pressure dependent conditions occurred at different values for the four components, increasing gradually from the fastest diffusing component (0.0025 mmHg) down to the slowest (0.25 mm Hg).

The results for $C_0$ for the four components are not as consistent as the diffusivities, but all three sets of graphs exhibit the same behavior. As the pressure of the $CdI_2$ is reduced from its highest level, the values of $C_0$, for all four components, initially show a decrease, go through a minimum and then increase sharply before either gradually falling again or becoming constant.

Micrographs showing the surface of a slice diffused at 500°C under a $CdI_2$ overpressure of $3 \cdot 10^{-5}$ mm Hg are shown in Fig. 4, where a pyramidal structure can be seen, which was probably due to preferential evaporation from the surface of the slice. The different morphologies that can be seen in the micrographs are probably due to different orientations of the subgrains in the CdTe slice. The density of the pyramids is similar to the dislocation density in good quality CdTe ($10^5$ cm$^{-2}$). At a slice temperature of 500°C these features were less significant at high $CdI_2$ pressure than at low and were not be observed at all for slices annealed at lower temperatures.

## 4. Discussion and Conclusions

Rather complicated behaviours were obtained here, which can be related possibly either to the response of the surface of the slice to the surrounding gas phase or to the defect chemistry after equilibrium has been reached in the capsule.

All crystalline solids left in vacuum will undergo vaporisation, but the actual rate of vaporisation will depend on the temperature, the surrounding ambient and the chemical composition of the vaporising species [7]. Preferential evaporation will occur at crystal edges, screw and edge dislocations and areas where surface nucleation has occurred. In the latter case, the vaporisation may be influenced by the presence of impurity atoms. A variety of vaporisation morphologies for different types of materials have been observed and lattice defects have been established as origins of several of them [7]. Because of the particular crystal structure, different vaporisation rates can be expected from opposite faces of a slice of a compound like CdTe, which will lead to features similar to those as observed with chemical etching [7].

Such evaporation features have been observed in this investigation for a CdTe slice possessing different twin structures, which was diffused at 500°C for 5 days under a $CdI_2$ overpressure of $3 \cdot 10^{-5}$ mm Hg (see Fig. 4). Features associated with different orientations of the slice, were observed, which are similar to observations made in MCT, where annealing in a Cd - Hg atmosphere caused the surface to reveal different defect structures [8]. The density of the pyramids on the surface of the CdTe slice (Fig. 4) used in this investigation was approximately the same as the dislocation density in good quality CdTe ($10^5$ cm$^{-2}$), which suggests a connection to the dislocation density.

Farrow et al [9] reported on the vapourisation of $Hg_{0.8}Cd_{0.2}Te$ and observed the evaporation of Cd from the surface of the slices above 450°C, but this cannot be related to the measuremants reported here at 500°C. The work of Farrow et al is related to evaporation into a dynamic vacuum and will proceed indefinitely whereas diffusions described here were carried out in a closed volume and evaporation will proceed until an equilibrium is reached. Any similarity between the results reported here to those reported by Farrow et al must have occurred early on in the diffusion before equilibrium conditions had been reached.

At the start of the diffusion the slice will lose both Cd and Te from the surface of the slice until it is in equilibrium with the Cd and Te partial pressures. In normal circumstances such an equilibrium is reached very quickly and consequently the diffusion must relate to these conditions in the capsule and cannot be associated with surface loss other than changes in stoichiometry. The pressure dependence of diffusivity at 500°C must be a genuine bulk phenomenon relating to a defect process and the pressure independent part to a different process. The diffusions carried out at the other temperatures are probably related to the second process over the whole pressure range.

As both the surface concentrations, Co, and the diffusivities exhibit a consistent but complex behavior with respect to the $CdI_2$ pressure, it is probable that these parameters are related in some way and can be used to explain the defect structure, but the experimental observations given below must be consistent with any model that is developed. These are:

1.The four diffusivities, $D_1$ to $D_4$, as well as the corresponding surface concentrations, $C_{01}$ to $C_{04}$, showed the same behaviour versus $CdI_2$ partial pressure, thus indicating that the mechanisms responsible for those profile parts are interrelated.

2.The value of the pressure at the change over point from pressure dependent to pressure independent diffusion at 500°C increasing gradually from the fastest diffusing component (0.25 mm Hg) down to the slowest (0.0025 mm Hg).

3.The occurrence of Te on the inside wall of the silica capsule in the pressure dependent diffusions at 500°C and not in any of the pressure independent diffusions.

4.The fact that the temperature at which the increase in the $C_0$ values occurs ($10^{-6}$ mm Hg at 220°C, $10^{-3}$ mm Hg at 360°C and $3 \cdot 10^{-2}$ mm Hg at 500°C, respectively) is close to the saturated vapour pressure of Te at these temperatures [10] ($10^{-6}$ mm Hg at 220°C, $10^{-2}$ mm Hg at 360°C and $7 \cdot 10^{-1}$ mm Hg at 500°C, respectively) suggests the involvement of the Te partial pressure.

## Acknowledgements

The authors wish to thank Dr P Capper, GEC Marconi Infrared, Southampton for supplying the CdTe slices.

## References

[1] Maxey C D, Whiffin P A C and Easton B C 1991 *Semicond. Sci. Technol.* **6** C26.

[2] Malzbender J, Jones E D, Mullin J B and Shaw N 1994 *J. Electronic Materials: Material in Electronics* accepted for publication.

[3] Jones E D, Clark J C, Malzbender J, Mullin JB, Shaw N and Brinkman A W 1994 *J. Electronic Materials.*

[4] Bailar J C, Emeleus H J, Nyholm Sir R and Trotman - Dickenson AF 1975 *Comprehensive Inorganic Chem.* vol 2 (Pergamon Press) p 261.

[5] Paorci C and Pelosi C 1977 *Appl. Phys. Rev.* **12** 155.

[6] Jones E D, Stewart N M and Mullin J B 1992 *J. Crystal Growth* **117** 244.

[7] Sangwal K and Rodriguez-Clemente R 1991 *Solid State Phenomena* **17 & 18** 213.

[8] Yang J, Yu Z, Lui J and Tang D 1990 *J. Crystal Growth*, **101** 281.

[9] Farrow R F C, Jones G R, Williams G M, Sullivan P W, Boyle W J O and Wotherspoon J T M 1979 *J.Phys.D* **12** L117.

[10] Honig R E 1962 *RCA Review* p 567.

*Inst. Phys. Conf. Ser. No 144: Section 2*
*Paper presented at 7th Int. Conf. on Narrow Gap Semicond. Santa Fe 9–12 January*



# The Study of Mercury Vacancies in $Hg_{1-x}Cd_xTe$ from Capacitance and Photoluminescence Spectroscopy

**Yong Chang, Junhao Chu, Wenguo Tang, Shiping Guo and Dingyuan Tang**
**National Laboratory for Infrared Physics, Academia Sinica, Shanghai 200083, China**

**Abstract.** The quantum capacitance and photoluminescence (PL) spectroscopy were measured for several $Hg_{1-x}Cd_xTe$ samples. A resonant defect level located at 45meV above the bottom of conduction band has been investigated from the capacitance spectroscopy measurement for p-type $Hg_{0.79}Cd_{0.21}Te$ bulk material under different annealing conditions. This defect state results from the Oxygen-occupied mercury vacancies. A deep donor state level due to the As-occupied mercury vacancies, which is located at about 8.5 meV below the conduction band, has been observed from the PL spectroscopy from 5K to 100K for a series of Arsenic implanted MBE-grown $Hg_{0.61}Cd_{0.39}Te$ samples. Furthermore two acceptor levels, which are located at about 14.5meV and 31.5meV above the valance band, respectively, have been also found from the PL measurements. They can be assigned as the mercury vacancies and As-occupied Tellurium-site, respectively.

## 1. Introduction

The mercury vacancies and their complexes with other impurity atoms play an important role in narrow gap semiconductor materials $Hg_{1-x}Cd_xTe$ by exerting an enormous influence on the characteristics of materials and devices. Some theoretical investigations[1,2] have already pointed out that the mercury vacancies could cause a series of deep or resonant levels in the gap or in the conduction band for $Hg_{1-x}Cd_xTe$ materials. The quantum capacitance spectroscopy and photoluminescence (PL) spectroscopy are powerful tools to investigate these states for narrow gap semiconductor $Hg_{1-x}Cd_xTe$.

In this paper, a resonant defect level resulting from the oxygen-occupied mercury vacancies in $Hg_{0.79}Cd_{0.21}Te$ and a deep donor level originating from As-occupied mercury vacancies as well as two acceptor levels in $Hg_{0.61}Cd_{0.39}Te$ have been investigated by using capacitance and PL spectroscopy.

## 2. Experiment

The samples investigated in quantum capacitance spectroscopy experiment are $Hg_{0.79}Cd_{0.21}Te$ bulk materials grown by the solid state recrystallization method. The surface of the samples were coated by ZnS and metallic gate to obtain the MIS structures. In measuring the voltage dependence of capacitance a bridge circuit with a high precision capacitance decade was applied for the differential capacitance measurement of C-V curve with a high sensitivity of 0.01pF/V. The measurement were performed at 4.2K and with a frequency of 233Hz.

PL measurements were performed by using an $Ar^+$ (514.5nm) laser for excitation and a Nicolet 800 Fourier transform infrared spectrometer in the temperature range of 4K to 295K. The samples for the PL experiment were cut from the same piece of As-implanted

$Hg_{1-x}Cd_xTe$ thin film grown by MBE. They were then annealed at different temperatures of 450°C, 400°C and 350°C for sample M1, M2 and M3, respectively. The above As-implanted samples were first investigated by PL measurements, then they were etched in HBr solution to remove the doping layer for the next measurement. The transmission measurements were also performed for all samples for comparison. The resolution of the measurements were set at 0.5meV.

## 3. Results and discussion

The p-type $Hg_{1-x}Cd_xTe$ (x=0.21) samples with acceptor concentration $N_A^* = 3\times10^{16} \sim 4\times10^{17}cm^{-3}$ were investigated by capacitance spectroscopy. It was found from the measured spectroscopy that an additional peak appears before the inversion onset for all the samples with $N_A^* \approx 10^{17}$[3]. In order to investigate the origin of this resonate defect state the capacitance spectroscopy measurements have been carried out for the MIS samples under different annealing and doping conditions. In Fig.1, a resonant peak can be observed obviously in the upper curve for the sample with $N_A^* = 1.65\times10^{17}cm^{-3}$. The fitting calculation shows that the concentration of the resonant states is $1.1\times10^{17}cm^{-3}$, and the energy level is $E_R$=41meV above the conduction band edge. The middle curve is the measured result for the sample after annealing at temperature 250°C for 10h. This shows that the doping concentration changes to $5.9\times10^{16}cm^{-3}$ and the resonant peak becomes much smaller than before. The fitting calculation tells us that the concentration of this resonant level located at 41meV decreases to $4\times10^{16}cm^{-3}$. The lower curve is the measured result for the Au-doped sample formerly with lower concentration, which was then annealed at 250°C for 10h in order to have Au impurities fully diffused. In this case the concentration of the acceptor becomes $6.95\times10^{17}cm^{-3}$, however, no resonant peak (other than the original peak) appears in the measured C-V curve. The fitting calculation shows that the number of resonant states is still equal to $4\times10^{16}cm^{-3}$; the same number as before Au-doping. This experiment reveals clearly that the resonant states depend on only mercury vacancies and not on other acceptor impurity doping. Furthermore, this also shows that the resonant states originate from oxygen atoms substituting for mercury vacancies, and the annealing experiment can be used as a tool to eliminate or reduce the resonant defect states. The theoretical calculation shows that S-like resonant levels produced by neutral and singly charged oxygen impurities substitutional for the cation Hg or Cd are about 50meV for $O_{Hg}^{o}$ and 20 meV for $O_{Hg}^{+}$ for $Hg_{0.79}Cd_{0.21}Te$[1]. This is coincident with the experiment result of this work.

The samples used for PL measurements have been described in section 1. Fig.2 shows the measured PL spectroscopy for sample M1 annealed at 450°C with surface composition $X_s$=0.394, that was determined by fitting the transmission spectroscopy[5]. The energy gap $E_g$ (eV) can be calculated from the formula of[6]: $E_g = -0.295+1.87x_s - 0.28x_s^2 + (6.0 - 14.0x_s + 3.0x_s^2)\cdot(10^{-4})\cdot T + 0.35x_s^4$

In Fig.2 four PL peaks labeled A, B, C and D can be clearly observed for sample M1. The exciting power $P_I$ dependencies on PL peak intensity $I_{P.PL}$ has been also investigated for the measured samples. The results are shown in Fig.3. It obeys the following rule: $I_{P.PL} = CP_I^{\nu}$ , where C is a constant and $\nu$ is a coefficient related to the recombination mechanisms. Schmidt et.al.[7,8] have established an effective model to calculate the range of $\nu$ for various kinds of recombinations of CdTe. It is assumed that the model is also appropriate to $Hg_{1-x}Cd_xTe$. That is : $\nu_{BX} > \nu_{FX} > \nu_{\text{Bound to Free}} > \nu_{D^oA^o}$ , and just like CdTe, the $\nu$ will be the same value when the exciting laser power is varied over less than two decades except for the $D^oA^o$ transition.

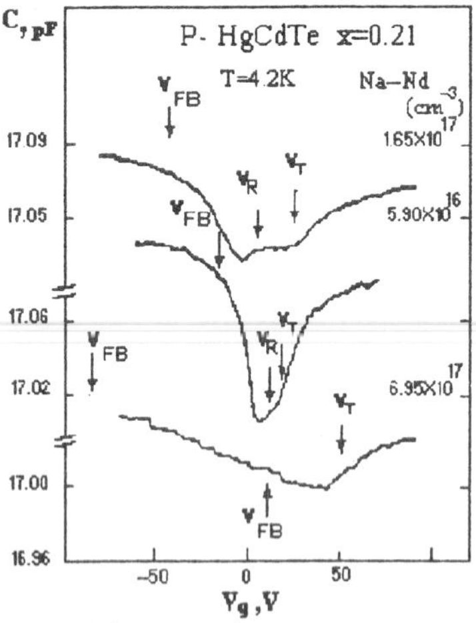


Fig.1. The capacitance spectroscopy for $Hg_{0.79}Cd_{0.21}Te$ ($N_A^* = 1.65 \times 10^{17} cm^{-3}$), and the C-V curves for the sample after annealing and Au-doping.

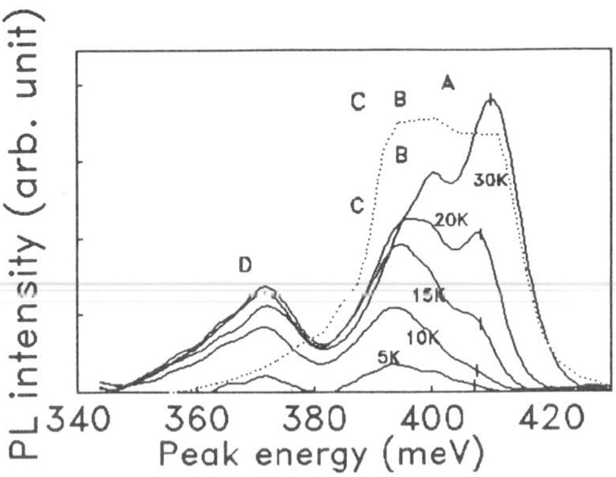


Fig.2. The PL spectra of sample M1 in several temperature before etched. (cw $Ar^+$ 514.5nm, laser output power is 300mW and the exciting power density is $3W/cm^2$.) The dotted line refers to the 30K PL spectra of sample M2 which was annealed about 50°C lower than sample M1. (The composition nonuniformity has been corrected.) As for the other samples which were annealed under lower temperatures, the PL can not be measured out.

Peak A is very near $E_g$. The index $\nu$ of this peak is the largest in all peaks and there exists no any other peaks in its upper energy side at all experiment temperatures. Therefore the peak A can be assigned as the band to band transitions at high temperature[9].

Fig.4 shows the PL spectra of sample M1 after etching, the doping layer was already removed by etching. It is easy to see that peaks B, D are obviously weakened while peak C is enhanced. The $\log(I_{P.PL})$ vs. $\log(P_I)$ before etched and after etched are shown in Fig.3. The peak energy position of peak C has the same temperature dependence with that of peak A and is not effected by exciting power. This means that peak C is due to the bound exciton(BX). The energy of peak C is at $16 \sim 17$ meV below peak A. The binding energy of exciton is about 1.5meV for $Hg_{1-x}Cd_xTe$ ($x \approx 0.39$), thus the impurity level involved is $14.5 \sim 15.5$ meV. In comparison with other literatures[10-12], we know that it originates from Hg vacancies. In fact, the influence of Hg vacancies is impossible to avert in $Hg_{1-x}Cd_xTe$.

The peak B and peak D have not been reported as far as we know. Analyzing from the exciting laser power and temperature dependence of peak energy and peak intensity, we attribute peak B to BX luminescence. Peak B is strong before etched and becomes weak after etched. This results demonstrated that peak B is As-implantation-related. This fact further manifests that peak B results from the donor level which comes from As-occupied mercury vacancies after annealing. As a result, it can also be deduced that this donor level is 8.5meV below conduction band.

The intensity of peak D is also strong before etched and becomes weak after etched just like peak B, it shows that peak D is related to As doping, too. The $\nu$ of this peak is not a constant and exhibits a saturation tendency as exciting laser power increases. The blue shift of this peak was observed when excitation is weak, moreover, the peak intensity decreases rapidly with the increase of temperature . These characteristics are typical of donor-acceptor($D^oA^o$) pair, therefore, it can be deduced that peak D is from $D^oA^o$ transition.

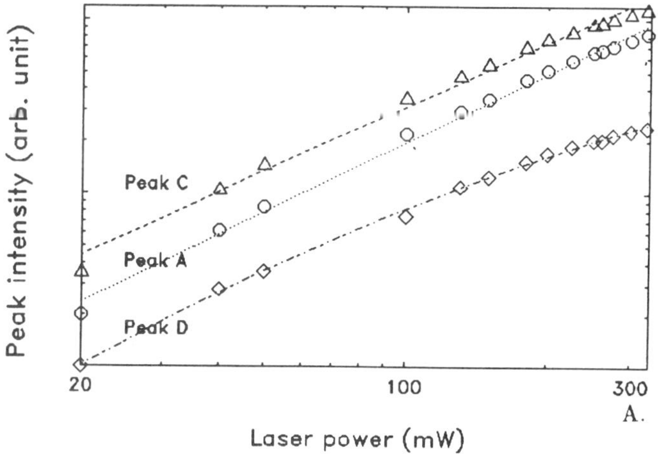


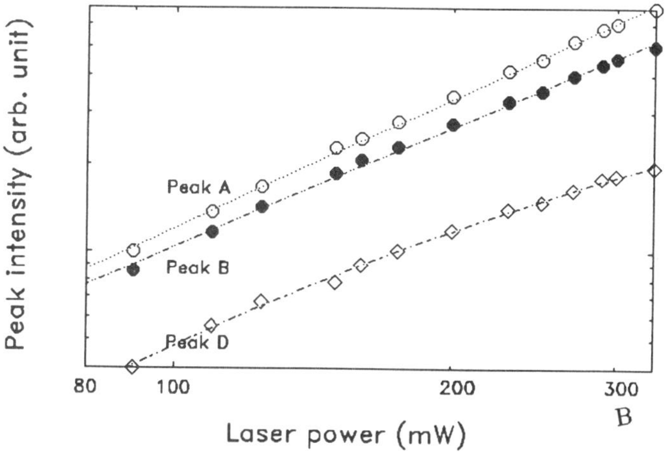


Fig.3. The main PL peak intensity of sample M1 vs. the power of the exciting laser. (a)35K after etched; (b)30K before etched. For the peak A, B and C, the slop of the fitted line represents $\nu$ in formula $I_{P.PL} = CP_I^\nu$, which are constants of 1.0 to 2.0. For peak D, the saturation tendency of $D^oA^o$ can be easily seen in (a) but not so obviously in (b), the explanation is given in text. The lines between data points of peak D are drawn only as a visual aid.

To investigate the origin of peak D, other measurements were performed for comparison. PL spectra of sample M2, which annealed under 400°C, were shown in Fig.2. There exists only three peaks (A, B, and C), the peak D can hardly be observed. This is due to the lower annealing temperature that is not high enough to let the As occupy the Tellurium site, in order to obtain the p type layer near the surface. Therefore, peak D is inferred as $D^oA^o$ pair and the acceptor is provided by As in anion site. As for the donor level involved, it is deduced that it is offered by the As in mercury site, since only donor-related peak B is dominated in Fig.2. Thus, from the energy position of peak D and the 8.5meV donor level coming from As in Hg site, we can infer that the acceptor level provided by As in anion site is about 31.5meV in $Hg_{0.61}Cd_{0.39}Te$. This result is in good agreement with the hydrogen atom analogy[13] of 28 meV and linearly extrapolation value of 30.25 meV using the data given in the literature[14,15] with $N_A \approx 1.0 \times 10^{17}$.

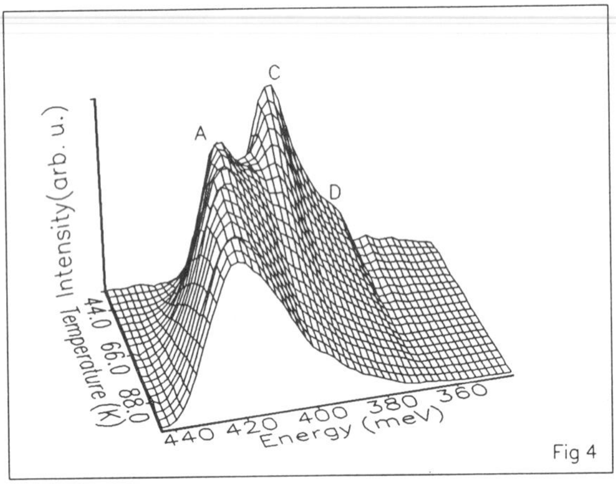


Fig.4. The PL spectra from sample M1 after etched in several experiment temperatures.

## 4. Conclusion

The experimental results are reported in this paper concerning the mercury vacancies and related complexes with such impurities as O, As, studied by means of capacitance spectroscopy and PL method for both bulk and MBE-grown materials of $Hg_{1-x}Cd_xTe$.

## References

[1] A.Kobayashi, O.F.Sankey, and J.D.Dow, *Phys. Rev. B* **25**, 6367(1982)
[2] W.C.Myles, *The 1987 U.S. Workshop on the Physics and Chemistry of of Mercury Cadmium Telluride*, Oct. 1987, New Orleans, Louisiana P.O/DI-31
[3] J.H.Chu, Z.Y.Mi, R.Sizman, F.Koch, R.Wollrab, J.Ziegler, and H.Maier, *J. Vac. Sci. Technol. B* **10**, 1569(1992)
[4] Junhao Chu and Z.Y.Mi *Chinese J. Infra. & Milli. Waves* **8**, 547(1990)
[5] Kun Liu, Junhao Chu, Biao Li, and Dingyuan Tang, *Appl. Phys. Lett.* **64**, 1(1994)
[6] Junhao Chu, Shiqiu Xu, and Dingyuan Tang, *Appl. Phys. Lett.* **43**, 1064(1983)
[7] T.Schmidt, G.Daniel, and K.Lischka, *J. Cryst. Growth* **117**, 748(1992)
[8] T.Schmidt, K.Lischka, and W.Zulehner *Phys. Rev. B* **45**, 8989(1992)
[9] F.Fuchs and P.Koidl *Semicond. Sci. Technol.* **6**, C71(1991)
[10] A.T.Hunter and T.C.McGill, *J. Appl. Phys.* **52**, 5779(1981)
[11] A.T.Hunter and T.C.McGill, *J. Vac. Sci. Technol.* **21**, 205(1982)
[12] J.W.Tomm, K.H.Herrmann, and A.E.Yunovich, *phys. stat. sol.(a)* **122**, 11(1990)
[13] C.L.Jones in:*Properties of Cadmium Mercury Telluride (EMIS Datareviews series, No.3)* Edit by J.C.Brice and P.Capper (London, INSPEC, IEE, 1987), p.151
[14] P.Capper, J.J.G.Gosney, C.L.Jones, and I.Kenworthy, *J. Cryst. Growth* **71**, 57(1985)
[15] I.Kenworthy, P.Capper, C.L.Jones, J.J.G.Gosney, and W.G.Coates, *Semicond. Sci. Technol.* **5** 854(1990)

*Inst. Phys. Conf. Ser. No 144: Section 2*
*Paper presented at 7th Int. Conf. on Narrow Gap Semicond. Santa Fe 9–12 January*



# Transformation of the photoluminescence properties of $Hg_{1-x}Mn_xTe$ for changing sign of the g-factor

Yu I Mazur[1], G G Tarasov[1] and J W Tomm[2]

[1] Institute of Semiconductors Physics, National Academy of Sciences, prospect Nauki 45, 252650 Kiev, Ukraine
[2] Institut fur Physik, Humboldt-Universitat zu Berlin, Sitz: Invalidenstr. 110, Unter den Linden 6, D-10099 Berlin, Germany

**Abstract.** Infrared photoluminescence (PL) has been investigated in $Hg_{1-x}Mn_xTe$ (MMT) single crystals for MnTe mole fractions (x) covering the range where the effective g-factor ($g_{eff}$) changes the sign (x=0.104-0.202). The behaviour of the PL feature assigned to electron-hole-plasma transitions was examined for temperatures in the 5-70 K range and for magnetic fields up to 7T for both Faraday and Voigt geometries. With increasing magnetic field the PL signal magnitude grows getting a maximum at $B=B_{max} \approx 2.3$ T and then decreases, approaching the noise level at about 5 T (for x=0.104). A strong narrowing of PL peak is observed for all samples investigated in the magnetic field. A modified Pidgeon-Brown model was successfully applied for describing the magnetic field dependencies of the emission lines for x=0.104 but fails for x=0.167 and 0.202. The exchange interaction parameters are obtained by fitting the data to be $N_0\beta = 0.9$ and $N_0\alpha = -0.45$.

## 1.Introduction

MMT belongs to the class of semimagnetic semiconductors (SMSC) [1]. This material gives the unique opportunity to study the effect of the exchange interaction between the spins of $3d^5$ electrons of the localized paramagnetic $Mn^{2+}$ ions with the free electrons. Due to this interaction a significant enhancement of gyromagnetic factor g takes place for both the $\Gamma_6$ and $\Gamma_8$ bands. The g-factor becomes strongly dependent on temperature and x and can change also the sign. The spin-splitting of the band electron states is directly proportional to the magnetization of the crystal. Therefore the change of the g-factor sign causes a change of the sequence of spin sublevels of the Landau doublet, which can be detected by means of magneto-optics [2] or PL [3]. For investigating the influence of the exchange interaction in a wider energy gap ($E_g$) range (from $E_g < 0$, if $x < 0.07$ to positive $E_g$ if $x > 0.07$) of course the manganese content must be changed. Since $E_g$ and the Mn concentration can not be varied independently in the MMT system a widening of the gap automatically leads to complication of magnetic properties of the SMSC due to possibility of a transition from a diamagnetic through a paramagnetic into a spin glass or an antiferromagnetic state [4]. Consequently a very different behaviour can be expected for MMT of various Mn concentrations in the range mentioned.

The effect of spin-spin exchange on Landau levels in MMT is usually calculated within the framework of a modified Pidgeon-Brown model where the contribution of the exchange interaction is taken into account by the addition of two new parameters $\alpha$ and $\beta$. The experimental data obtained by magneto-optical [2], magneto-transport [5], and PL measurements [3] strongly support the use of this approximation. It has been reported by Galazka et al. [3] that the PL spectrum of MMT ($0.12 \leq x \leq 0.15$) consists of two broad features, formed by plasma recombination as well as by transitions from conduction band to a acceptor level.

Here we present the results of a PL study for MMT samples with MnTe mole fractions of 0.104, 0.167 and 0.202. This samples allowed us to investigate the influence of the sign change of $g_{eff}$ on the behaviour of the interband PL as a function of both magnetic field and temperature.

## 2. Experiment

The sample set investigated consist of single crystals grown by a modified travelling heater technique. All samples were as-grown p-type. The typical acceptor concentration of $N_a \sim 10^{16}$ cm$^{-3}$ was produced by thermal annealing in Hg vapour. The accuracy for the mole fraction values is $\Delta x = \pm 0.003$ and was determined by X-ray

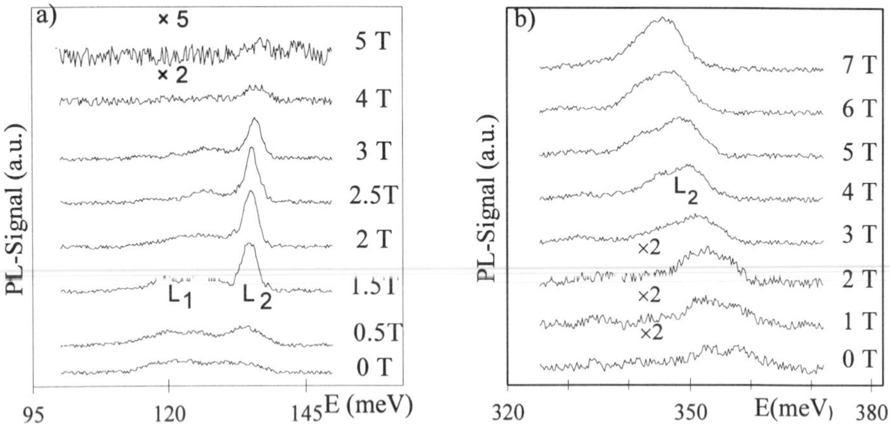


Fig.1. PL spectra of MMT samples vs. magnetic field (T=5 K, Voigt geometry):
a) x=0.104 and b) x=0.167. The electron-hole-plasma feature is labelled as $L_2$, low energy one is $L_1$.

microprobe. Before the optical measurements the samples were mechanically polished and etched in a 6% bromine-methanol solution. The PL was excited by a diode laser emitting 40 mW (cw) at 1.58 eV. All investigations were carried out with a IFS-88 Fourier spectrometer using a modified double modulation technique according to that reported in reference [6]. The spectral resolution was about 0.3 meV and the excitation densities were typically 1-2 W/cm$^2$. Magnetic field (0-7 T) and temperature (5-70 K) dependent measurements were performed with an Oxford Spectromag 4000 system. The studies in the magnetic field were carried out in both Voigt and Faraday geometries.

Fig.1 shows PL spectra for two samples with different mole fractions for Voigt geometry at T=5 K. The mole fractions chosen cause a small negative $g_{eff}$ for x=0.104 and a positive $g_{eff}$ for x=0.167. Two distinct features can be distinguished in the spectra of both samples. The low-energy line ($L_1$) is broad and evidently defect-related. A complicated structure is observed for this feature if a magnetic field was applied and we believe that this line is caused by conduction band-acceptor transitions. The second peak ($L_2$) is very strongly dependent on temperature and magnetic field and is the main goal of the investigations reported here. At zero magnetic field the peak $L_2$ corresponds to the direct $\Gamma_6$-$\Gamma_8$ transition, involving electronic states of conduction band $\Gamma_6$ and predominantly states of heavy-hole band $\Gamma_8$. The spectral peak position varies with temperature following $E_g$. These data well agree with those obtained in reference [7] from PL experiments in MMT alloys for temperatures between 10K and 300K. The variation of the magnetic field has an even stronger influence on the $L_2$ line: $L_2$ shifts slightly towards higher energy and its strength grows by a factor of 15 for increasing B up to the maximum at $B=B_{max}$=2.3 T for x=0.104. A further increase in B results in a $L_2$ PL magnitude decreasing. The detection limit is reached at about B=5 T.

## Theory

We use a modified Pidgeon-Brown model to describe the PL evolution in the magnetic field [2]. According to reference [2] the energy spectrum of the SMSC can be determined as an eigenvalue problem of the Pidgeon-Brown matrix, where the exchange interaction is incorporated by the introduction of additional terms. These terms include exchange integrals for interaction of the electrons and the holes with the $Mn^{2+}$ ions (parameters $\alpha$ and $\beta$, respectively). The Pidgeon-Brown matrix can be solved analytically only for the lowest spin sub levels of the conduction band ($\Gamma_6$) and for the upper levels of the hole band ($\Gamma_8$). In the case of the parabolic approximation holds

$$\Gamma_6\!:\begin{cases} E_a(0)=E_0+\hbar\omega_c/2+\frac{1}{2}g_{eff}\mu_B B \\ E_b(0)=E_0+\hbar\omega_c/2-\frac{1}{2}g_{eff}\mu_B B \end{cases} \qquad \Gamma_8\!:\begin{cases} E_a(-1)=\frac{\hbar\omega_0}{2}(\tilde{\gamma}-\gamma_1+k)-\tilde{B} \\ E_b(-1)=\frac{\hbar\omega_0}{2}(3k-\tilde{\gamma}-\gamma_1)-3\tilde{B} \end{cases} \tag{1}$$

where $\omega_c$ is the cyclotron frequency, $\omega_0$ is the free electron cyclotron frequency, $\mu_B$ is the Bohr magneton and $\tilde{\gamma}, \gamma_1$ and $k$ are modified Luttinger parameters.

$$g_{eff} = g^* + \frac{\alpha N_0 x \langle S_z \rangle}{\mu_B B} = g^* + 6\tilde{A}/\mu_B B, \quad \tilde{B} = \frac{1}{6}\beta N_0 x \langle S_z \rangle. \tag{2}$$

$g_{eff}$ includes g* as the g-factor determined by the band structure as well as the contribution of the s-d exchange interaction. The contributions of exchange interaction to the spin splitting for electrons (parameter $\tilde{A}$) and holes (parameter $\tilde{B}$) include the thermal average of the spin operator along the direction of the applied magnetic field $\langle S_z \rangle$ which is determined by B, the temperature (T) and the spin-spin interaction (usually antiferromagnetic) in the sub-system of localised spins, $N_0$ is the number of unit cells per unit volume of crystal and x is the MnTe mole fraction. The Luttinger parameters $\gamma_1 = 3, \tilde{\gamma} = 0.05$ and k=-1.65 are taken to be the same as for zero-gap MMT alloys and $g^* = -82$. Due to the wide variation of reported values for the exchange integrals ($-0.6 \leq N_0\alpha \leq -0.4$ eV, $0.6 \leq N_0\beta \leq 1.5$ eV) the following parameters were to be chosen to be close to those obtained from comparable experiments: $N_0\beta$ = 0.9 and $N_0\alpha$ = -0.45. To describe the temperature and field dependencies of $\langle S_z \rangle$ we use the following formalism [8]:

$$\langle S_z \rangle = -S_0 B_{5/2}\left(\frac{g_{Mn}\mu_B B}{k_B(T + T_0)}\right), \tag{3}$$

where $B_{5/2}$ is the Brillouin function for the spin of 5/2 and $g_{Mn}$ is the g factor for the $Mn^{2+}$ ions and is taken to be of 2. $S_0$ and $T_0$ are fitting parameters, which can be determined from temperature dependent measurements of the magnetic susceptibility. It can be shown that the $S_0$ values determined experimentally agree with those calculated if (assuming a random distribution of the $Mn^{2+}$ ions) the probability of the generation of various magnetic clusters is taken into account.

## 4. Experimental results and discussion :

According to the selection rules for interband transitions [11] one can find the following dominant sequences active in both absorption and PL: $a_{\Gamma_8}$(n-1)$\rightarrow b_{\Gamma_6}$(n) in Voigt geometry and $b_{\Gamma_8}$(n-1)$\rightarrow b_{\Gamma_6}$(n) in Faraday geometry with $n \geq 0$. Taking into account the thermal population of the states at low temperatures we believe that the short wavelength peak of the PL spectrum ($L_2$) is formed just by them. Starting with the parameters chosen above the energies of successive transitions were calculated. Fig.2 shows the experimental data and the fits using the modified Pidgeon-Brown model for MMT for two different x versus magnetic field. Reasonable agreement was obtained for both Voigt and Faraday geometries for the sample with x=0.104 using $S_0$=1 and $T_0$=8.4 K consistently. Mismatch between theory and experiment is found for x=0.167. The Faraday component is well described with $S_0$=0.6 and $T_0$=13 K whereas the Voigt geometry data obviously show a qualitatively different behaviour.

For the sample with x=0.104 the spin sub levels of the conduction band become higher and due to $g_{eff}$<0 the usual behaviour ($E_b(0) > E_a(0)$) is observed. Due to small value of $g_{eff}$ ($|g_{eff}| \approx 5$ at T=5 K and B=2 T) the spin splitting of the lowest Landau levels for the conduction band (n=0) is small enough. The dependencies ΔE(B) calculated for $b_{\Gamma_6}(0) \rightarrow a_{\Gamma_8}(-1)$ transitions corresponds to the magnetic field dependence of the $L_2$ peak position measured and displays a slight increase ($d\Delta E/dB$) = 0.4 meV/T. For a mole fraction of 0.167 $S_0$ decreases and $T_0$ increases. Interband magneto-absorption data have revealed a pronounced decrease of the Kane matrix element $E_p$ with increasing Mn concentration ($E_p$=15 eV) [2]. The calculation of the lowest spin-split Landau levels shows that the $E_b(0) \rightarrow E_a(-1)$ transition does not correspond to the $L_2$ peak in PL for the parameter set chosen above ($N_0\beta$=0.9 and $N_0\alpha$= -0.45). Thus we would have to set them outside of that what we believe to be reasonable. Therefore we conclude that there must exist an other mechanism which is not taken into consideration in the Pidgeon-Brown model (e.g., renormalisation of the energy spectrum caused by non-locality of potential, electron-phonon interaction).

The magnetic field dependencies of the PL lines (Fig.2) furthermore enable us to determine or to confirm MnTe mole fractions (as done here) giving evidence of the high sensitivity of the PL technique.

Fig.3 depicts the behaviour of the PL peak magnitude and its full width at half maximum (FWHM) when the temperature or the magnetic field (Voigt geometry) are varied (x=0.104). We additionally want to emphasise two peculiarities which are inherent to the SMSC alloys investigated:

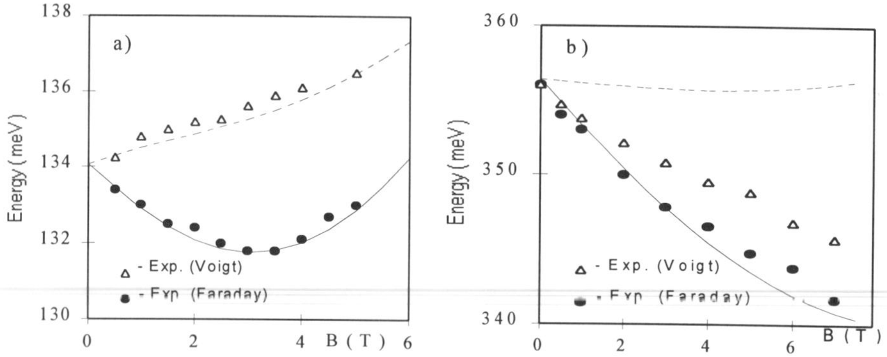


Fig.2. Spectral peak position of the $L_2$ line vs. magnetic field: a) x=0.104; b) x=0.167. Dashed and solid lines calculation for Voigt and Faraday geometries, respectively.

i) The non-monotonic behaviour of the peaks amplitude in the magnetic field and

ii) the drastic narrowing of the PL line to a value of about 2.5 meV at B= $B_{max}$ =2.3T. Note, that the magnetic field behaviour of the PL peak's FWHM correlates with the peaks amplitude. At $B_{max}$ the PL peak exhibits also the smallest FWHM.

For x=0.167 the evolution of the PL spectrum has been traced a) for increasing magnetic fields when the temperature is constant and b) for increasing temperature when the magnetic field remains constant. The general tendencies experimentally observed are that the magnetic field enhances the emission, at least, up to B=7T and that increasing temperature decreases the maximum intensity of the PL peak. We are not sure that the maximum of the PL signal (cf. Fig. 1) was already reached at B=7 T. Higher fields were not available with our apparatus. It can here only be assumed that its saturation takes place at higher magnetic fields. The influence of increasing temperature was investigated for B=7 T and T=5 K. The PL signal vanishes practically at T=30 K. Furthermore note the strong long-wavelength asymmetry of the PL peak. This asymmetry can be ascribed to a set of impurity states generated by structural disordering in this kind of crystals. These states are probably band-like and contribute also to the PL signal. The abrupt high-energy falloff of the PL emission is dominated by plasma-like transitions. The $L_2$ peak position at B=7 T is shifted with increasing temperature much more rapidly ($\partial\left(\hbar\omega_{max}\right)/\partial T = 0.6$ meV/K) than for increasing T without field. The PL peak becomes broader and weaker with increasing T. This observation can be assigned to a magneto polaron mechanism [9]. Through this mechanism it is possible for carriers to move together with the cluster of localised magnetic moments oriented by the carrier (magneto-polaron effect). The polaron mass is larger than the mass of free carriers and the polaron mobility is temperature dependent. For a hole in the degenerated valence band the spin can be taken as J=3/2. For increasing temperatures the decay of the polaron complex is activated resulting in a reduction of the recombination probability and thus in decreasing PL signals.

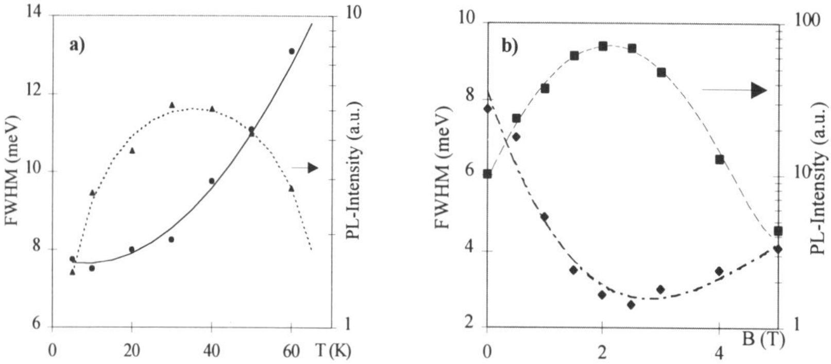


Fig.3. $L_2$ peak magnitude and FWHM for MMT vs. a) temperature (left) and b) magnetic field (right) in Voigt geometry.

For x=0.202 the tendency traced for x=0.167 is conserved . The magnitude of the PL line decreases and the role of structural disordering becomes more pronounced. The temperature coefficient $dE_g/dT$ approaches zero. This fact is in accordance with the known $E_g(x,T)$ behaviour for MMT alloys of x=0.13-0.22 reported by Piotrowski et al. [7] and contradicts Georgitse et al.[10]. Following them, the $Mn^+$ local level in MMT moves into the conduction band for x<0.20 and does not manifest itself in the radiative processes, whereas at x>0.20 this level emerges within the energy gap and participates in the radiative recombination. The same effect should be expected for increasing temperature. However, according to our data $dE_g/dT$ is smaller zero for MnTe mole fractions exceeding 0.2. Thus so far we did not observe the PL mechanism suggested in [10] which involves the antibonding states of $Mn^+$ ions.

## 5.Conclusions

The investigation of MMT samples with x from 0.1 to more than 0.2 allowed us to trace the evolution of the PL spectra assigned to transitions between spin split Landau levels of the conduction ($\Gamma_6$) and the heavy hole valence ($\Gamma_8$) bands for both a negative and a positive sign of the effective g-factor.

For x=0.104, we have obtained a very narrow PL peak with a non-monotonic magnetic field dependence of the amplitude and an extremely small FWHM of about of 2.5meV at B=2.3 T. To our knowledge this is the narrowest linewidth ever reported for spontaneous emission from a narrow gap semiconductor. The magnetic field dependencies of this peak were measured in Voigt or in Faraday geometries and its shift towards higher energy is explained by the influence of exchange interaction.

For a MnTe mole fraction x=0.167 the effective gyromagnetic factor becomes positive and an inverse ladder of spin sublevels for the lowest Landau levels of the conduction band is found. The description requires an additional modification of the Pidgeon-Brown model. It seems that a renormalization of energy should be taken into account. The PL peak was found to be strongly asymmetrical. We believe that high energy side of this peak is formed mainly by electron-hole-plasma transitions. For describing the temperature dependence of the PL the magneto-polaron mechanism was taken into account.

The tendency observed for x= 0.167 is also traced for the MnTe mole fraction of x=0.202.

This work was supported by the Volkswagen-Stiftung.

## References.

[1] Furdyna J K 1988 *J. Appl. Phys.* **64** 29-64

[2] Bastard G, Rigaux C, Guldner Y, Mycielski A, Furdyna J K and Mullin D P 1981 *Phys.Rev. B***24** 1961-75

[3] Galazka R R, Gelmont B L, Ivanov-Omskii V I, Postolaki I T and Smirnov V A 1986 *Solid State Commun.* **58** 311-14

[4] Brandt N B and Moshchalkov V V 1984 *Advances in Physics* 194-256

[5] Jonnson W B and Anderson I R 1984 *Phys. Rev. B* **29** 6679-87

[6] Fuchs F, Lusson A, Wagner J and Koidl P 1989 *Proc. SPIE* **1145** 323-6

[7] Piotrowski T, Tomm J W, Puhlmann N, Gensicke S, Schmidt H and Werner L 1990 *Acta Physica Polonica* A77 195-8

[8] Lyapilin I I and Tsidilkovskiy I M 1985 *Uspehi Fiz. Nauk* **146** 35-68

[9] Ryabchenko S M and Semenov Yu G 1983 *Zh. Eksp. Teor. F iz.* **84** 1419-31

[10] Georgitse E I, Ivanov-Omskii V I, Masterov V F, Pogorletskii V M, Piotrowski T and Smirnov V.A. 1990 *Sov. Phys. Semicond.* **24** 1339-42

*Inst. Phys. Conf. Ser. No 144: Section 2*
*Paper presented at 7th Int. Conf. on Narrow Gap Semicond. Santa Fe 9–12 January*



# CARS and Faraday Effect in Narrow Gap HgCdMnTe

W. Herbst[1], H. Pascher[1], F. Geist[1], M. Baran[2], T. Dietl[2], W. Dobrowolski[2], R. Szymczak[2],

[1]Experimentalphysik I, Universität Bayreuth, D-95440 Bayreuth,
[2]Institute of Physics, Polish Academy of Sciences, Warsaw, Poland

**Abstract.** In n- and p- doped narrow gap $Hg_{1-x-y}Cd_yMn_xTe$ ($0<x\leq0.16$; $E_g\approx200$ meV) Faraday rotation, spin resonant four wave mixing (CARS) and magnetisation measurements have been performed. Experiments were carried out in magnetic fields up to 7 T and at temperatures between 1.8 K and 30 K.

The bandstructure in a magnetic field was calculated using a Pidgeon & Brown **k·p** model with two 4x4 matrices describing the exchange interaction added to the original Hamiltonian. Compared to the commonly used formula which simply adds an exchange term to the effective g-factor this model also accounts for couplings of valence and conduction bands.

Using this model, from the experimentally obtained conduction band spin splittings and Faraday rotation exchange integrals and interband matrixelements were fitted.

## 1. Introduction

Effective g-factors of semimagnetic semiconductors are strongly dependent on temperature and on an external magnetic field, because of the exchange interaction between localised magnetic moments of the paramagnetic ions and the band electrons and holes. In II-VI-diluted magnetic semiconductors the exchange interaction can be described via two exchange integrals, $\alpha$ and $\beta$, for the conduction and valence band, respectively. In narrow-gap diluted magnetic semiconductors the magnetic contribution to the spin-splittings is comparable to the spin flip energy of a hypothetical non-magnetic semiconductor with the same energy-gap. This is in contrast to wide-gap materials, where the magnetic part dominates.

The bandstructure changes caused by the exchange interaction can be calculated either by adding terms proportional to the magnetisation of the sample to the effective g-factors or, more precisely, by modifying the Pidgeon & Brown Hamiltonian [1] with exchange matrices. This model is essentially the same as applied by Bastard et al. [2] and Kossut [3].

CARS spectroscopy is one of the most precise methods to measure spin-splittings in narrow-gap semiconductors in an external magnetic field. In II-VI materials one normally gets information on the conduction band only, because of the higher mobility of the electrons compared to the holes. In the present work, however, we observed an additional resonance, which arises from an intra-valence band transition.

Interband Faraday effect is very sensitive to the sum of spin-splittings in conduction and valence bands. The observed rotation can be described within the above mentioned extended Pidgeon & Brown model summing up allowed interband transitions.

## 2. Band structure model

In narrow gap semiconductors there is a strong mixing among heavy and light hole states and between those and the conduction band. Via this coupling a change in the valence band spin splittings due to exchange interaction also affects the conduction band and vice versa. In order to account for this mixing of states a modified Pidgeon & Brown model [1,2,3] was used to calculate the band structure. The interaction of the Mn-ions with the band carriers is taken into account by adding a Heisenberg like term to the original Pidgeon & Brown Hamiltonian. In II-VI materials two exchange parameters $N_0\alpha$ and $N_0\beta$ are describing the exchange interaction in conduction and valence bands, respectively. In Equ.1a,b the exchange matrices are given for a- and b-set, which are not coupled for $k_Z=0$ even in semimagnetic compounds.

**Set a:**
$$\mathbf{H} = \mathbf{H}_{\mathbf{P\&B}} + \frac{\mathbf{M}}{\mathbf{g_M \cdot \mu_B}} \begin{pmatrix} -\alpha/2 & 0 & 0 & 0 \\ 0 & -\beta/2 & 0 & 0 \\ 0 & 0 & \beta/6 & -\sqrt{2}\beta/3 \\ 0 & 0 & -\sqrt{2}\beta/3 & -\beta/6 \end{pmatrix} \tag{1a}$$

**Set b:**
$$\mathbf{H} = \mathbf{H}_{\mathbf{P\&B}} + \frac{\mathbf{M}}{\mathbf{g_M \cdot \mu_B}} \begin{pmatrix} \alpha/2 & 0 & 0 & 0 \\ 0 & \beta/6 & 0 & -\sqrt{2}\beta/3 \\ 0 & 0 & \beta/2 & 0 \\ 0 & -\sqrt{2}\beta/3 & 0 & \beta/6 \end{pmatrix} \tag{1b}$$

$H_{P\&B}$ denote the original Pidgeon & Brown Hamiltonians, $g_M \approx 2$, $\mu_B$ is Bohr´s magneton.

In the framework of a mean field approximation the temperature and the magnetic field dependence of the exchange interaction follows the magnetisation M of the sample, which, in turn is proportional to a modified Brillouin function [4] (see Equ.2.)

$$M = N_0 \cdot x \cdot g_M \cdot \mu_B \cdot S_0 \cdot B_{5/2}\left(\frac{5 \cdot \mu_B \cdot B}{k_B(T+T_0)}\right) \tag{2}$$

An effective temperature $T_0$ and an effective spin $S_0$ are introduced to consider a coupling of the Mn-ions among each other.

If the magnetisation and the energy gap $E_g$ are known, from measured conduction band g-factors and either Faraday rotation or intra- valence band transitions the interband matrixelement P and the exchange parameters can be calculated.

## 3. Experimental methods and results

The investigated samples are cut from bulk crystals $Hg_{1-x-y}Cd_yMn_xTe$. Mn-contents $0<x\leq 0.16$ were determined by x-ray micro analysis. The y-values of the different samples are adjusted such that the energy gaps of all samples are around 200 meV. The fundamental gaps are determined by infrared transmission measurements.

### 3.1 Magnetisation measurements

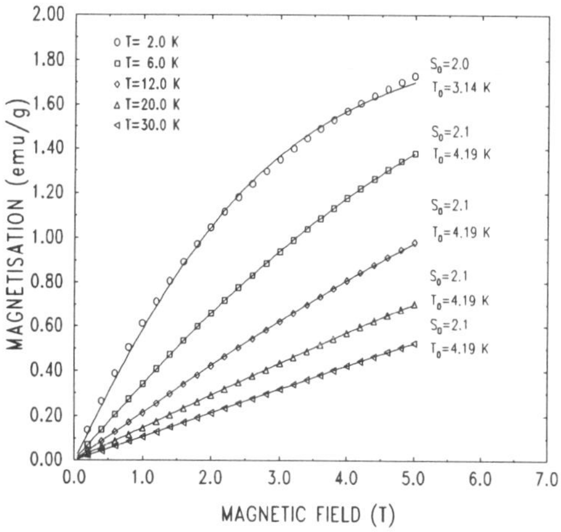


***Fig.1:*** *Magnetisation of $Hg_{1-x-y}Cd_yMn_xTe$, x=2.9%, y=16.3%*

The magnetisation of the samples was measured with a SQUID-magnetometer in external magnetic fields up to 5 T and at temperatures between 2 K and 30 K. Least square fits to Equ.2 were performed in order to determine $T_0$ and $S_0$. Figure 1 shows the results for a p-type sample with a Mn-content of 2.9%. The considerable high value of $T_0$ and the reduction of $S_0$ with respect to 5/2 clearly indicate some antiferromagnetic coupling of the Mn-ions even with relatively small Mn-content.

## 3.2 Coherent anti-Stokes Raman scattering

Coherent anti-Stokes Raman scattering (CARS) is a type of four wave mixing spectroscopy, where two laser beams with a frequency difference $\Delta\omega$ are superimposed in the sample. The intensity at $\omega_{AS} = \omega_{laser} + \Delta\omega$ is resonantly enhanced if $\Delta\omega$ is within the linewidth of a Raman allowed transition. Thus CARS is one of the most precise methods to measure spin splittings in narrow gap semiconductors over a wide range of the external magnetic field. In our experiments the intensity of radiation with frequency $\omega_{AS}$ was recorded for fixed frequency combinations of the two $CO_2$ lasers as a function of the magnetic field. Figures 2a,b show typical recordings for two temperatures.

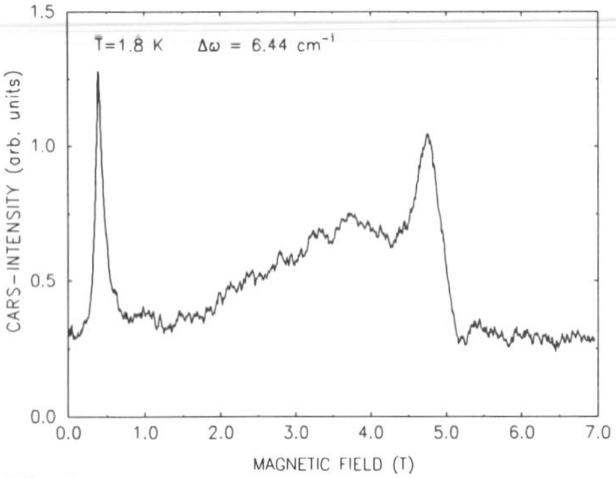


***Fig.2a:*** *Anti-Stokes intensity vs. magnetic field*
*T=1.8 K; n-*$Hg_{1-x-y}Cd_yMn_xTe$*; x=0.16, y=0*

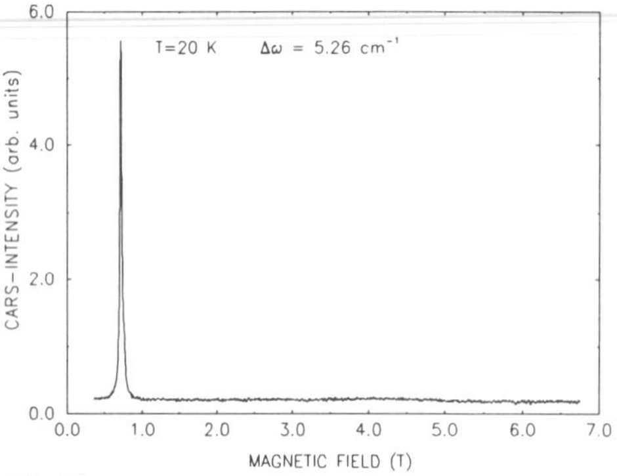


***Fig.2b:*** *Anti-Stokes intensity vs. magnetic field*
*T=20 K; n-*$Hg_{1-x-y}Cd_yMn_xTe$*; x=0.16, y=0*

The spin flip transition within the conduction band dominates the CARS-spectra of a II-VI narrow gap semiconductor in the mid infrared spectral region. This is also the case for p-type samples, because there is always a considerable number of electrons, photo excited by the radiation of the strong lasers via two-photon absorption. Due to the exchange interaction in HgCdMnTe the effective g-factor of the conduction band changes its sign at low temperatures and not too high magnetic fields. Therefore one, two or three electron spin resonances appear in the CARS-recording. The number of observed resonances depends on the magnitude of the chosen frequency difference. At higher temperatures ($T > 10$ K) the influence of the exchange interaction is reduced and the dependence of the conduction band spin splitting on the magnetic field becomes linear, like in non magnetic material. Therefore only one sharp resonance remains in the CARS spectrum.

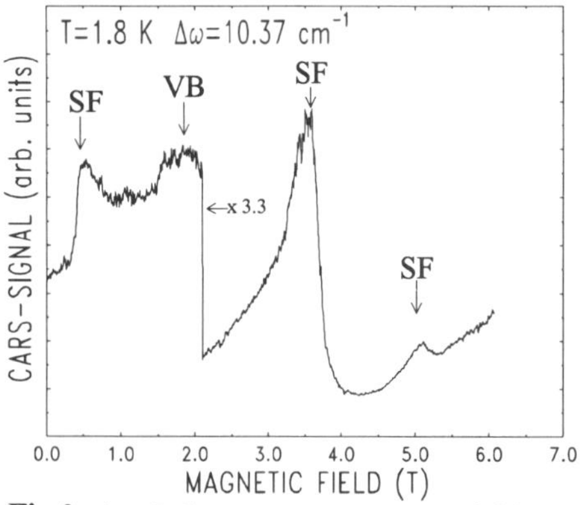


***Fig.3:*** *Anti-Stokes intensity vs. magnetic field*
*T=1.8 K; p-*$Hg_{1-x-y}Cd_yMn_xTe$*, x=2.9%, y=16.3%;*
*SF:electron spin resonances; VB: hole resonance*

In the p-type material for not too high magnetic fields and at low temperatures we found an additional resonance ("VB" in Fig. 3), which we identified as an intra valence band transition either from the level -3/2(1) to 1/2(0) or from -3/2(1) to 3/2(1) . For the first mentioned Raman process one of the virtual transitions is a non central transition which is only allowed for non zero momentum parallel to the magnetic field, the other one is allowed for all $k_z$. The observation of this transition is remarkable since apart from a few exceptions in zincblende semiconductors intra- valence band transitions are not observed in the Raman spectrum.

Recordings like shown in Figs 2,3 are taken for a variety of laser frequency differences. In this way the complete magnetic field dependence of the electron spin splitting and the VB-transition is obtained. Figure 4a shows the results for the p-type sample at low temperature, figure 4b for various temperatures and n-type material

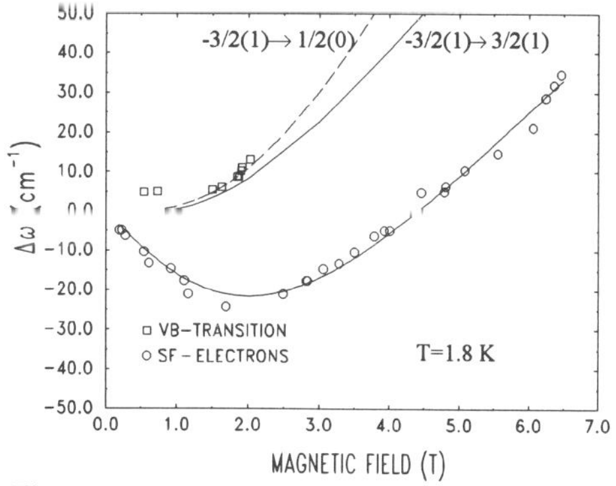


***Fig.4a:*** *Raman allowed intraband transition energies vs. magnetic field; T=1.8 K; p-$Hg_{1-x-y}Cd_yMn_xTe$,x=2.9%, y=16.3%.*

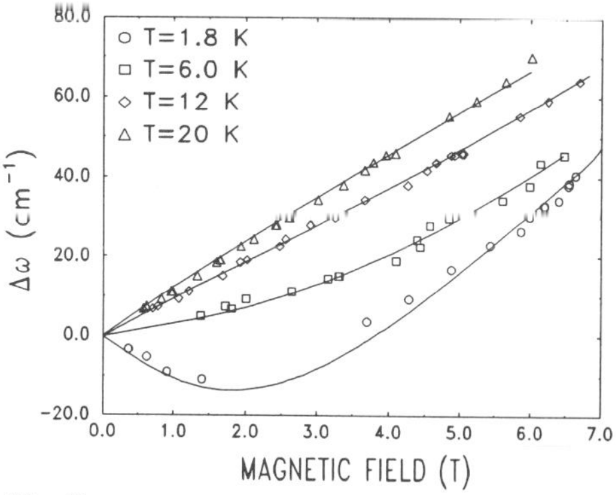


***Fig.4b:*** *Conduction band spin splitting vs. magnetic field for different temperatures; n-$Hg_{1-x-y}Cd_yMn_xTe$.x=3.0%, y=15.9%*

The lines in Figs 4a,b represent theoretical calculations according to the extended Pidgeon & Brown model, where the exchange parameters $N_0\alpha$, $N_0\beta$ and the interband matrixelement P were used as adjustable parameters. The energy gap was determined from infrared transmission data, and the other parameters were taken from literature. The complete set of band parameters is listed in table 1.

### 3.3. Interband Faraday effect

Interband Faraday rotation is very sensitive on the sum of the spin splittings of conduction and valence bands. Therefore the exchange interaction in a semimagnetic alloy causes a strong modification of the observed angle of rotation compared to a nonmagnetic semiconductor. The rotation $\Theta_F$ can be calculated, if all interband transitions, which are allowed in Faraday configuration are taken into account (see Equ.(3) [5]).

$$\theta_F = -\frac{\pi\cdot\hbar\cdot e^2\cdot L}{n_0\cdot c}\cdot\sum_{i,i'}\left[\frac{\left|(v_+)_{i,i'}\right|^2-\left|(v_-)_{i,i'}\right|^2}{\hbar^2\cdot(\omega_{i,i'}^2-\omega^2)}\cdot f_i\cdot(1-f_{i'})\right] \qquad (3)$$

A partial blocking of the interband transitions is considered by Fermi distribution functions $f_i$. For the calculation of the Landau levels and transition matrixelements $(v_{+/-})_{i,i'}$ the modified Pidgeon & Brown model was used. The interband matrixelement and the exchange integrals were fitted to the CARS and Faraday data, using the measured magnetisation of the samples.

We measured Faraday rotation with the radiation of a liquid nitrogen cooled CO-laser and a $CO_2$-laser, respectively. The angle of the plane of polarisation was detected with a fixed analyser after the sample.

Figure 5a,b shows the intensity behind the analyser for two different temperatures for a sample with 16% Mn. A comparison of the periods shows the decrease of specific rotation with increasing temperature. In Fig.5c the dependence of the angle of rotation on the magnetic field is plotted for a sample with x=2.9%. The Brillouin function like behaviour and the enhancement if the photon energy approaches the gap are clearly seen.

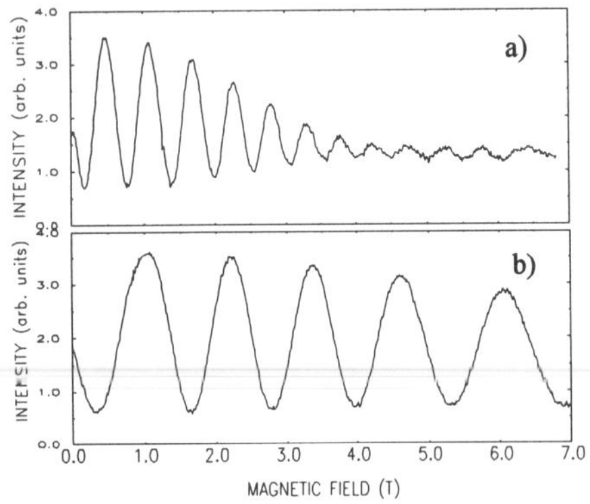


*Fig 5: Trace of Faraday rotation experiment, $Hg_{1-x-y}Cd_yMn_xTe$, x=16%, y=0.*
***a)** T=1.8K; **b)** T=20K*

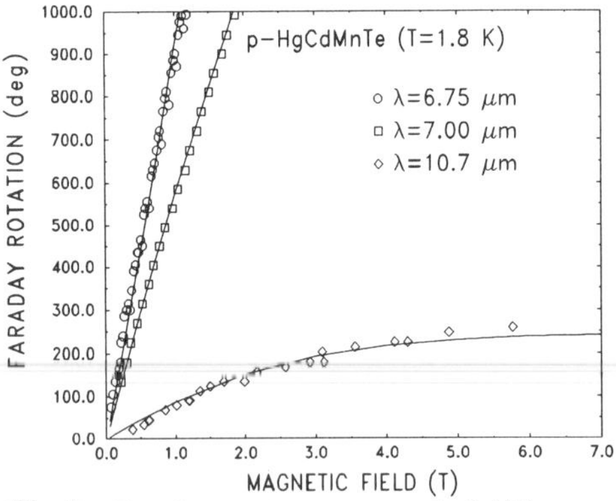


***Fig.5c:*** *Faraday rotation vs. magnetic field for three different wavelengths.*
*— calculated*

## 4. Results and discussion

The excellent matching of theory and experiment seen in Figs 4,5c serves as a proof for the validity of the obtained set of band parameters.

**Table 1:** Bandparameters and exchange integrals of $Hg_{1-x-y}Cd_yMn_xTe$ ($x \leq 0.16$, $y \leq 0.16$)

| P/eVm | $E_P$/eV | $N_0\alpha$/eV | $N_0\beta$/eV | $\gamma_1$[6] | $\gamma_2$[6] | $\gamma_3$[6] | $\kappa$[6] |
|---|---|---|---|---|---|---|---|
| $8.4 \cdot 10^{-10}$ | 18.6 | 0.27 | -0.60 | 1.5 | -1.3 | 0.5 | -0.2 |

The matrix element agrees with values published for HgCdTe with similar energy gaps. The valence band exchange integral is as given by Dobrowolska et al. [7]. However, the conduction band exchange is considerably smaller than results published up to now. Assuming $N_0\alpha = 0.4$ after Ref.[7], the conduction band CARS data can not at all be understood without a considerable change of the interband matrix element, which in turn, would be in contradiction to the observed intra valence band transition.

## References

[1] Pidgeon C R, Brown R N 1966 *Phys. Rev.* **146**, 575
[2] Bastard G, Rigaux C, Guldner Y, Mycielski J, Mycielski A 1978 *J. Phys. Paris* **39**, 87
[3] Kossut J 1988 in *Diluted Magnetic Semiconductors*, p 183ff, eds. Furdyna J K and Kussut J; *Semiconductors and Semimetals* vol 25, eds. Willardson R K and Beer A C, (Academic Press)
[4] Gaj J A, Planel R, Fishman G 1979 *Solid State Commun.* **29**, 435
[5] Jiménez-Gonzáles H J, Aggarwal R L, Favrot G 1994 *Phys. Rev. B*, **49**, 4571
[6] Pastor K, Grynberg M, Couder J in *Physics of Narrow-Gap Semiconductors:* ed. Kaczmarek E, Rauluszkiewicz J, Gorska M: Proc. III Int. Conf. Warszawa 1977 (Elsevier, Amsterdam 1978)
[7] Dobrowolska M, Dobrowolski W, 1981 *J.Phys/C* **14**, 5689

*Inst. Phys. Conf. Ser. No 144: Section 2*
*Paper presented at 7th Int. Conf. on Narrow Gap Semicond. Santa Fe 9–12 January*



# Galvanomagnetic properties of $(Cd_{1-x-y}Zn_xMn_y)_3As_2$

R. Laiho*, K.G. Lisunov+, V.N. Stamov+ and V.S. Zahvalinskii+

*Wihuri Physical Laboratory, University of Turku, 20500 Turku, Finland
+Institute of Applied Physics, 277028 Kishinev, Moldova

**Abstract.** Galvanomagnetic properties of the type $II_3V_2$ diluted magnetic semiconductor alloy $(Cd_{1-x-y}Zn_xMn_y)_3As_2$ are investigated in the region of $0 \le y \le 0.08$, $x + y = 0.3$. Well developed Shubnikov-de Haas single period oscillations are observed, showing a clear spin splitting in one of the samples (y = 0.02). From the data values of the cyclotron effective mass and the $g$-factor are determined. In crystals with a high Mn concentration the cyclotron mass depends strongly on the applied magnetic field.

## 1. Introduction

The $II_3V_2$ alloys $(Cd_{1-x-y}Zn_xMn_y)_3As_2$, briefly CZMA, are diluted magnetic semiconductors characterized by a small band gap, $E_g$, or a gapless state appearing at the composition $y = 0$, $x = 0.08$ [1]. It has been suggested that Zn and Mn atoms influence $E_g$ in the same manner, i.e. an increase of $y$ or $x$ in the region $x > x_0$ leads to an approximately linear increase of $E_g$ [2]. Another contribution of Mn to band parameters of CZMA is due to the s-d exchange interaction between electrons and localized magnetic moments. This can be described by an effective $g$-factor determining the spin-splitting of the Landau levels in a magnetic field.

Details of the band structure and the band parameters of heavily doped semiconductors can be investigated by observing Shubnikov-de Haas (SdH) oscillations. By analyzing the SdH oscillation patterns measured at different temperatures and magnetic fields it is possible to determine the effective $g$-factor and the cyclotron effective mass, $m_c$, of electrons. In diluted magnetic semiconductors the $g$-factor depends on the strength of the s-d exchange interaction while $m_c$ is not expected to depend on this effect [3].

SdH oscillations have been extensively investigated in CZMA for compositions around the gapless state, the value of $E_g$ being smaller than 0.1 eV [2,4,5]. In this work we investigate galvanomagnetic properties of CZMA and transformation of the band parameters in quantizing magnetic fields when the value of $E_g$ is increased by changing the composition in the region of $x + y = 0.3$, $y \le 0.08$. The band gap for these compositions is about 0.24 eV if the linear increase of $E_g$ with increasing of $x$ or $y$ is supposed [2].

## 2. Experimental details

Single crystals of CZMA were grown from stoichiometric amounts of $Cd_3As_2$, $Zn_3As_2$ and $Mn_3As_2$ by using a modified Bridgeman method (slow cooling of a melt in the presence of a temperature gradient). The Mn concentrations of the crystals were y = 0.02, 0.04, 0.06 and 0.08. The composition and homogeneity of the samples were determined by x-ray and microprobe analysis.

Magnetoresistance measurements were made in magnetic fields between 0 - 11 T by using the six-probe technique. The specimens were cut to the form of a rectangular prism having dimensions of 1 x 1 x 5 $mm^3$. The electrodes were fixed on them by soldering. For measurements the sample probe was inserted in a He exchange gas dewar where its temperature could be regulated with an accuracy of 0.5 %.

## 3. Results and discussion

In Table 1 are collected values of the Hall concentration $n_R$ and the Hall mobility $u_R$ for our CZMA samples. In the temperature region of the magnetoresistance measurements, $T$ = 4.2 - 31 K, the Hall coefficient, $R_H$, does not depend on the magnetic field, $B$, and both $R_H$ and zero field conductivity, $\sigma_0$, are temperature independent. Between 31 - 300 K $\sigma_0$ decreases monotonically with increasing temperature.

As shown in Fig. 1(a) well resolved single period SdH oscillations are observed in all our specimens. There is a strictly linear dependence between the inverse values of the magnetic field, $B_{max}$, and the quantum numbers, $N_{max}$, of the maxima of the SdH oscillations. The SdH periods, $P_{SdH}$, are independent from the magnetic field. The values of the SdH concentration, $n_{SdH}$ are determined from $P_{SdH}$ in the spherical and parabolic approximation. These data are presented in Table 1. The ratio $n_R/n_{SdH}$ is close to unit as expected (see Table 1).

Table 1. Parameters of the specimens investigated.

| $y$ | $n_R \cdot 10^{-18}$ | $u_R \cdot 10^{-4}$ | $n_R/n_{SdH}$ | $m_c(0)/m_0$ | $\alpha/m_0 \cdot 10^3$ | $T_D$ | $T_{Du}$ |
|---|---|---|---|---|---|---|---|
| | $cm^{-3}$ | $cm^2/V{\cdot}s$ | | | $T^{-1}$ | K | K |
| 0.00 | 1.8 | 4.5 | 1.4 | 0.044 | - | 44 | 2.2 |
| 0.02 | 1.0 | 3.5 | 1.0 | 0.032 | - | 24 | 3.6 |
| 0.04 | 3.0 | 1.4 | 0.9 | 0.040 | 12 | 33 | 7.9 |
| 0.06 | 4.0 | 1.2 | 1.1 | 0.028 | 8.1 | 39 | 13 |
| 0.08 | 5.5 | 0.5 | 1.3 | 0.042 | 9.4 | 32 | 21 |

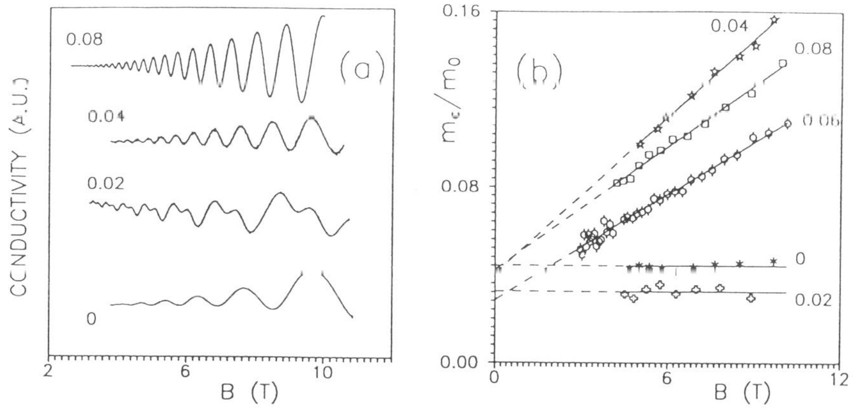


Fig. 1. Experimental traces of the SdH oscillations at 4.2 K (a) and magnetic field dependence of the cyclotron effective mass (b). The values of $y$ are shown near the curves.

The amplitudes of the SdH oscillations decrease monotonously with increasing temperature but without any shift of their positions. In the specimen with $y = 0.02$ a well resolved spin-splitting of the SdH maxima is observed at 4.2 K. This splitting is suppressed with increasing temperature and is not visible any more at $T \geq 20$ K.

The SdH amplitude, $A$ may be described by an equation [6],

$$A \sim B^{1/2} z/\sinh(z) \exp[-2\pi^2 m_c k_B T_D/(\hbar e B)] \cos(\pi \upsilon), \qquad (1)$$

where $z = 2\pi^2 m_c k_B T/(\hbar e B)$, $m_c$ is the cyclotron mass, $T_D$ is the Dingle temperature, $\upsilon = g m_c/(2m_0)$, and $m_0$, $\hbar$, e and $k_B$ have their usual meanings. Eq. (1) describes the amplitude of the first harmonics of the SdH oscillations which is the only component to be considered in our case since $A_2 \sim 0.01\ A_1$ ($A_1$ and $A_2$ are the amplitudes of the first and the second harmonics, respectively).

Analysis of the SdH oscillations in Fig. 1(a) gives no evidence for the field or temperature dependence of the $g$-factor. Usually $T_D$ is temperature independent at low values of $T$ [7]. Therefore, the cyclotron mass can be obtained from Eq. (1) by using values of the SdH amplitudes measured at two different temperatures in the same magnetic field. The SdH amplitudes have been determined after subtracting the monotonic part of the conductivity. As shown in Fig. 1(b) the cyclotron mass does not depend on $B$ at a low Mn concentration ($y \leq 0.02$) but at high concentrations ($y \geq 0.04$) it increases considerably with an increasing magnetic field, obeying a linear law, $m_c(B) = m_c(0) + \alpha B$. The values of $\alpha$ and $m_c(0)$ are collected in Table 1. Our result for $m_c(0)$ agrees well with that quoted in Ref. [8] for the case of $y = 0$, $x = 0.3$ and similar $n_R$ ($m_c/m_0 = 0.032$ - $0.038$ for $n_R = (1\text{-}2)\cdot 10^{18}$ cm$^{-3}$).

Using the observed spin-splitting of the SdH maxima at $y = 0.02$ the effective $g$-factor, $g = 24$ is obtained. It is independent from the magnetic field. At $y = 0$ and $x =$

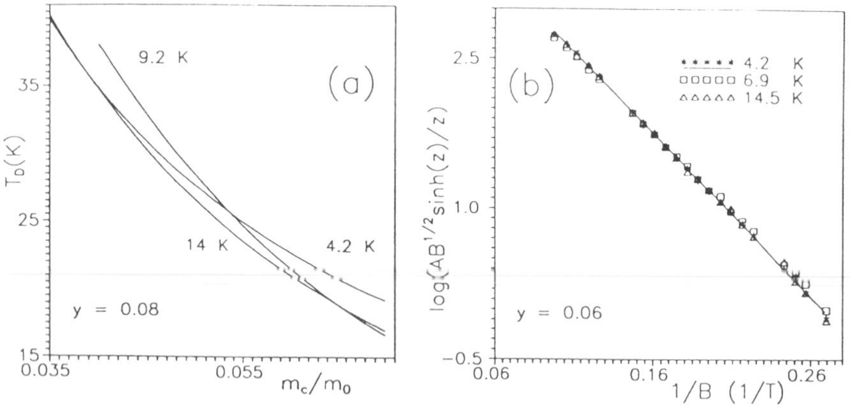


Fig. 2. The dependencies of $T_D$ on $m_c$ (a) and of the left hand side of Eq. (3) on $1/B$ (b) for different temperatures.

0.29 - 0.33 the values of $g$ found in Ref. [8] are in the interval 13 - 37, in agreement with our result.

A possible temperature dependence of $g$ is too weak to break the regularity of the SdH oscillations or to influence the temperature dependence of the SdH amplitude. At the same time omission of the factor $\cos(\pi\upsilon)$ in Eq. (1) would provide errors in the calculations of $m_c$ even if $g(T)$ is a weak function of $T$. In such a case we can use the method [9] described below if the independence of $m_c$ from $B$ is supposed. Eq. (1) may be rewritten as

$$\log[AB^{1/2}\sinh(z)/z] \sim \log[\cos(\pi\upsilon)] - 2\pi^2 m_c k_B T_D/(e\hbar B). \qquad (2)$$

Usually the dependence of $\log[\cos(\pi\upsilon)]$ on $B$ in diluted magnetic semiconductors can be neglected [9]. Then the right hand side of Eq. (2) is a linear function of $1/B$ and $T_D$ can be determined from the slope of the left hand side of Eq. (2) vs. $1/B$ for each value of $m_c$. In such a way we obtain three formal functions $T_D(m_c)$ for three different temperatures. These curves must intersect at one point giving the values of $T_D$ and $m_c$. Application of this procedure to the case of $y = 0$ gives the value of $m_c$ which is equal to that quoted in Table 1. For the sample with $y = 0.08$ this method leads to a 100% error of both $T_D$ and $m_c$, which are found to be within the intervals 19 - 37 K and $(0.038 - 0.065)m_0$, respectively (see Fig. 2(a)). In the cases of $y = 0.04$ and $y = 0.06$ this way of analysis cannot be used, because two out of the three curves $T_D(m_c)$ do not intersect at all.

Using the function of $m_c(B)$ as given above we can rewrite Eq. (2) in the form

$$\log[AB^{1/2}\sinh(z)/z] \sim \log[\cos(\pi\upsilon)] - 2\pi^2\alpha k_B T_D/(e\hbar) - 2\pi^2 m_c(0) k_B T_D/(e\hbar B). \qquad (3)$$

The right hand side of Eq. (3) must be a linear function of $1/B$ and independent from $T$.

Both these conditions are fulfilled with high accuracy (see Fig. 2(b)). This confirms the temperature independence of $T_D$ and the linear relationship between $m_c$ and $B$. Using the dependence of the left hand side of Eq. (3) on $1/B$ and the data of $m_c(0)$ in Table 1 the values of $T_D$ are obtained (see Table 1). In all specimens we find $T_D > T_{Du}$, where the parameter $T_{Du} = \hbar e/[\pi k_B m_c(0)u_R]$ determines the broadening of the Landau levels due to scattering of the electrons by lattice imperfections. The non-thermal broadening of the Landau levels must be attributed to some inhomogeneity of the distribution of the electron concentration in the crystal [7]. The situation is similar in the alloys with $y = 0$ [10].

## 4. Conclusions

Measurements of the SdH effect have been performed in CZMA alloys far from the gapless state. No phenomena connected with the influence of the s-d exchange interaction to the effective $g$-factor have been observed within the region of the compositions investigated. At the same time, a strong dependence of the cyclotron effective mass at the Fermi level on the magnetic field is found when the manganese concentration is increased. Therefore, this effect can be attributed to the presence of Mn in the alloys.

The observed dependence of $m_c$ on $B$ cannot be related to non-parabolicity of the band spectrum (see e.g. Ref. [11]). Instead, a possible reason may be an influence of the magnetic field to the band gap. If such influence is assumed a growth of $E_g$ by ~ 40 - 60 % would be expected when $B$ is increased up to 10 T.

## References

[1] Wagner R J, Palik E D and Swiggard E M 1969 Phys. Lett. **30 A** 175
[2] Lubczynski W, Cisowski J, Kossut J and Portal J C 1991 Semicond. Sci. Technol. **6** 619
[3] Kossut J 1988 Semiconductors and Semimetals vol 25 (London: Academic Press) p125
[4] Neve J J, Bouwens C J R and Blom F A P 1981 Solid State Commun. **38** 27
[5] Lubczynski W, Voiron J, Picoche J C, Cisowski J, Thullier J C and Zdanowicz W 1989 Semicond. Sci. Technol. **4** 223.
[6] Adams E and Holstein T 1959 Journ. Phys. Chem. Solids **10** 254.
[7] Tsidil'kovskii I M 1972 Electrons and Holes in Semiconductors (Moscow: Nauka)
[8] Arushanov E K, Gubanova A A , Knyazev A F, Lashkul A V, Lisunov K G and Sologub V V 1988 Fizika i Tehnika Poluprovodnikov (Sov. Physics and Technology of Semiconductors) **22** 338
[9] Takeyama S and Narita J 1986 J. Phys. Soc. Japan **55** 274
[10] Knyazev A F 1982 Ph.D Thesis (Kishinev)
[11] Dornhaus R and Nimtz G 1983 Springer Tracts in Modern Physics vol 98 (Berlin:Springer-Werlag) p 119

*Inst. Phys. Conf. Ser. No 144: Section 2*
*Paper presented at 7th Int. Conf. on Narrow Gap Semicond. Santa Fe 9–12 January*



# Optical and photoelectrical properties of $\beta$–$FeSi_2$ thin films*

**W.Z. Shen, S.C. Shen, W.G. Tang**
National Laboratory for Infrared Physics, Shanghai Institute of Technical Physics, Shanghai 200083, P.R.China

**L.W. Wang**
State Key Laboratory of Functional Materials for Informatics, Shanghai Institute of Metallurgy, Shanghai 200050, P.R.China

**Abstract.** Absorption, photocurrent, and far–infrared reflectivity spectra of the polycrystalline $\beta$–$FeSi_2$ thin films in relation with the Si substrate temperature are presented. The photocurrent spectra related with the fundamental interband, extrinsic defect transitions of $\beta$–$FeSi_2$ and the intrinsic transitions of Si substrates are observed. Both the absorption and photocurrent measurements show the direct gap characteristics. A simple recombination model is proposed to account for the photocurrent results. We show that $\beta$–$FeSi_2$ films with better quality can be achieved at higher substrate temperature during growth. The simplicity, sensitivity and reliability of the photocurrent measurements for studying $\beta$-$FeSi_2$ are well demonstrated.

## 1. Introduction

Recently, the particular interest in semiconducting $\beta$ phase of iron silicide ($\beta$–$FeSi_2$) on Si substrates is connected with its reported direct gap characteristics ($E_g$~0.85–0.89eV)[1–4], well suitable for the fabrication of infrared detectors or optoelectronic devices integrated in the silicon processing technology[5,6]. The possibility of growing high quality $\beta$–$FeSi_2$ thin films provides substantial motivation for the investigation of this material's optical and electrical properties. Optical techniques are very powerful tools for studying the fundamental properties of $\beta$–$FeSi_2$ structures. The reported optical measurements for the band gap and defects in $\beta$–$FeSi_2$ were performed mainly by using absorption[1–3] and photoluminescence (PL)[3,7]. However, the interband absorption measurements are bothered by the limitation of the sample thickness and the influence of the defects on the absorption edge, while the PL measurements have to be done at liquid helium temperatures[3,7]. Recently, Lefki *et al* [8,9] have first used the photocurrent (PC) method to study the electronic properties of this material and showed that the maxima of PC spectra correspond to the value of the band gap $E_g$ in the thicker silicide samples. Nevertheless, up to date, the photocurrent spectra of $\beta$–$FeSi_2$ in wider spectral range have been little reported.

On the other hand, an improvement of the film quality might be achieved by investigating in detail the film properties, including optical and electrical properties, in relation with the deposition conditions. Dimitriadis *et al* [3] have studied the electrical and optical properties of $\beta$–$FeSi_2$ films in relation with the annealing temperature and showed that the lowest annealing temperature (800°C) yields the best quality of the material. Peale *et al* [10] showed that a significant improvement in the epitaxial growth of $\beta$–$FeSi_2$ has been achieved by employing a template layer, codeposition of Fe and Si reactants, and strained Si substrates. We have grown polycrystalline $\beta$–$FeSi_2$ films with the annealing temperature of 800°C and investigated, in this paper, the dependence of the film quality on the Si substrate temperature during growth by absorption, photocurrent, and far–infrared reflection measurements.

## 2. Experiment

The polycrystalline $\beta$–$FeSi_2$ thin films were grown by either solid phase epitaxial (SPE) growth or reactive deposition–solid phase epitaxial (RD–SPE). 100nm high purity iron (99.99%) was deposited on the n–type (5 ~ 8 Ωcm) Si (100) substrates in ultrahigh vacuum chamber (~$10^{-8}$

mbar) using an electron beam. Sample No.1 was grown by SPE on a room temperature Si substrate, while samples No.2, No.3, and No.4 were grown by RD-SPE with the substrate temperature of 500, 550 and 600°C, respectively. Following deposition, all samples were annealed at 800°C in a flowing nitrogen atmosphere in order to form the silicide layers of about 310 nm in thickness[3]. Our Rutherford back-scattering spectroscopy (RBS) measurements show their thicknesses of about 300nm. The formation of the $\beta$ phase of iron silicide was verified by X-ray diffraction measurements. Absorption and photocurrent measurements were performed by a Nicolet 800 Fourier transform infrared spectrometer (FTIR) over the temperature range of 77K to room temperature (290K). Absorption spectra were detected by a InSb photodiode cooled with liquid nitrogen and with a tungsten lamp focused onto the samples. In-plane photocurrent spectra were recorded with zero bias voltage and also with a tungsten lamp focused onto the samples. The ohmic contacts of the samples, checked by the I-V characteristics, were simply formed by silver threads with conducting glue. The room temperature far-infrared reflection and transmission measurements were carried out with a PE-983 grating infrared spectrometer.

## 3. Results and discussion

The absorption spectra for a typical $\beta$-$FeSi_2$ sample (No.4) at room temperature and 77K are shown in Fig.1(a). The weak, low-energy features are believed due to extrinsic transitions related with defect states within the forbidden energy gap, while the strong absorption and edge-like behavior above 0.80eV indicate the onset of the fundamental interband transitions. Note that the gap energy increases with decreasing temperature, as expected. In order to confirm its direct transition characteristics, in this case, the quantity $(\alpha h\nu)^2$ is drawn as a function of ($h\nu$ - $E_g$), where $\alpha$ is the measured absorption coefficient, $h\nu$ is the photon energy and $E_g$ is the experimental energy gap, the result for the room temperature absorption spectrum is given in Fig.1(b). We first fit the lower energy defect feature with a linear shape function (dotted line) and substract it from the experimental data due to its continued contribution to the shape of the fundamental interband transitions. The steep and straight line (dashed line) indicates a direct transition of our $\beta$-$FeSi_2$ samples and give the energy gap of 0.848 eV at room temperature, 0.883 eV at 77K. The same result persists for the other samples.

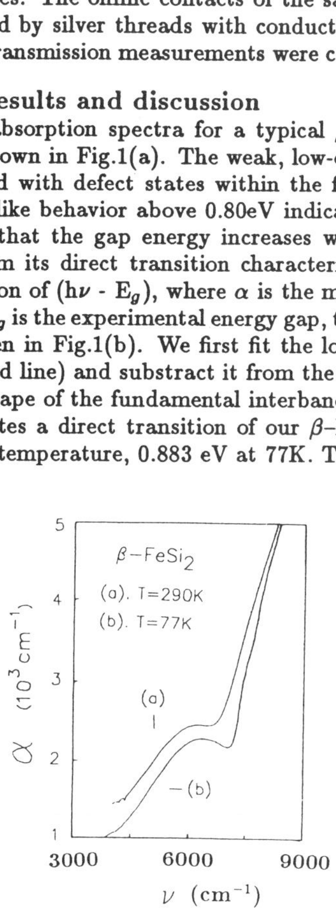


Fig.1(a). Optical absorption spectra of a typical $\beta$-$FeSi_2$ sample No.4 at 77K and 290K.

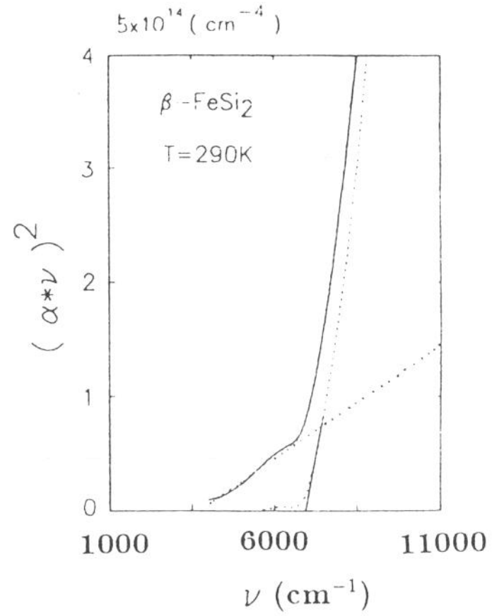


Fig.1(b). Fitting procedure for the typical sample at 290K.

Plotted in Fig.2 are the absorption spectrum and in-plane photocurrent (PC) spectrum at room temperature (290K) from another $\beta$-$FeSi_2$ sample (No.3). In comparison with the absorption spectrum, the three structures in the PC spectrum can be indentified without obscurity. The peaks B and C (at 0.841 and 0.464eV) correspond to the fundamental interband transitions and extrinsic defect transitions of $\beta$–$FeSi_2$, respectively. The peak positions obtained from the PC spectrum are in very good agreement with the absorption results. The strong PC peak labelled A obviously comes from Si substrate. To the best of our knowledge, this is the

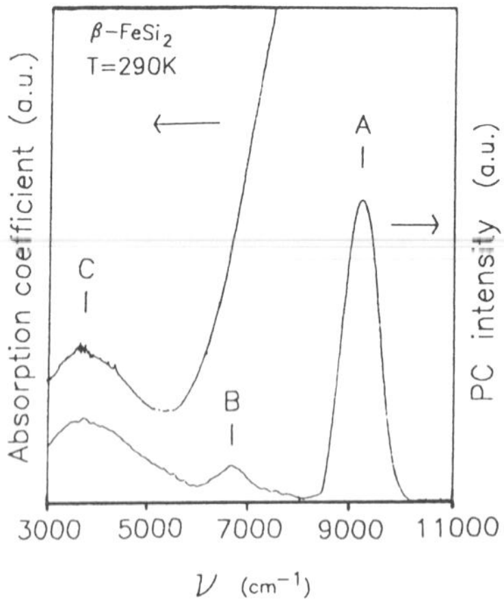


Fig.2 Optical absorption and in-plane photocurrent spectrum with zero bias voltage of another typical $\beta$–$FeSi_2$ structure (sample No.3) at 290K.

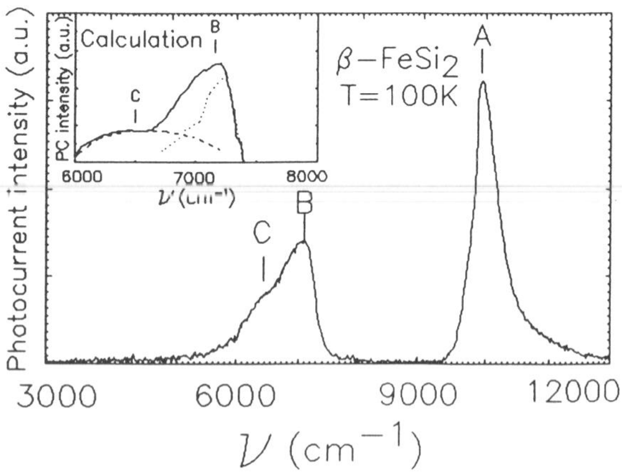


Fig.3 Photocurrent spectrum of the typical $\beta$–$FeSi_2$ sample No.4 at 100K with zero bias voltage. Shown in the inset are the theoretical PC line shapes of extrinsic defects (dashed curve), fundamental interband (dotted curve) and the total contribution (solid curve), based on a simple recombination model.

first observation of photocurrent signals of $\beta$–$FeSi_2$ related with the all transitions of the fundamental interband, extrinsic defect transitions of $\beta$–$FeSi_2$ and Si substrate intrinsic transitions. Based on the De Vore's model[8,11], the photocurrent signals can be analyzed as a diffusion–recombination process of the photogenerated minority carriers within the silicide. At steady state, we have:

$$\frac{d^2\Delta n(y)}{dy^2} - \frac{\Delta n(y)}{L_D^2} + \frac{\alpha\tau I_0 exp(-\alpha y)}{L_D^2} = 0$$

where $\Delta n(y)$ is the excess minority–carrier concentration in the $y$ direction ($0 \leq y \leq$ d, d being the silicide thickness), $L_D$ is the diffusion length, $\alpha$ is the absorption coefficient determined from the absorption measurements, $\tau$ is the bulk lifetime and $I_0$ is the flux of incident light. Using the boundary conditions and the assumption that the photogenerated current is proportional to the excess density $\Delta n(d)$, we now can fit the PC line shape of sample No.4 at 100K as given in Fig.3. The calculated line shape for fundamental interband and extrinsic defect transitions, together with their total contribution, are given in the inset of Fig.3 with the input parameters d = 276 nm and $L_D$ = $2\times10^{-5}$ cm. In the thicker silicide samples, the maxima of the photocurrent spectrum actually correspond to the band gap $E_g$ and defect states within the energy gap. In addition, note that this fitted silicide thickness d is in agreement with the RBS measurements of about 300 nm. As we know, the photogenerated current also depends on both the carrier mobility and recombination coefficient, and the carriers of $\beta$–$FeSi_2$ thin films have rather low mobility (e.g., $\sim$ 2 $cm^2/Vs$ at 290K)[3], far less than that of silicon (e.g., $\sim$ 1350 $cm^2/Vs$ at 290K), while the experimental PC intensity ratio of peak B and A is nearly 1:8 at 290K (see Fig.2) and 1:3 at 100K. This results confirm further the direct interband transition characteristics of $\beta$–$FeSi_2$.

The good agreement among the absorption, photocurrent experiments and theoretical fitting results provides strong evidence that the photocurrent measurement is a direct and reliable tool in revealing the energy band structure of $\beta$–$FeSi_2$. Fig.4 shows the in–plane photocurrent spectra at 77K for all four samples. Several messages can be drawn from the PC spectra. First, the photocurrent peaks shift to higher energy as the temperature decreases, as expected. The obtained variation of energy gap with measuremental temperature $E_g(T)$ from PC measurements, in good agreement with the absorption results, also agrees well with the temperature–dependent equation of $E_g(T)$ published by Giannini *et al*[12] except for $E_g(0)$=0.882 eV, which shows the strong effect of the electron–phonon coupling. The PC intensity of the fundamental interband transitions increases rapidly with decreasing experimental temperature (see together with Fig.2). Second, the position of the defect transition peak C, origining from the lattice mismatch between the Si substrates and silicide layers, shifts very rapidly to higher energy as the increase of substrate temperature, while the position of the interband transition peak B changes much slow (within 10 meV). This observation is further consistent with the results of absorption measurements. Finally, the interband transition peak becomes stronger and narrower for higher substrate temperature for growth, which is indicative of the best quality of the material under the highest substrate temperature of growth (600°C).

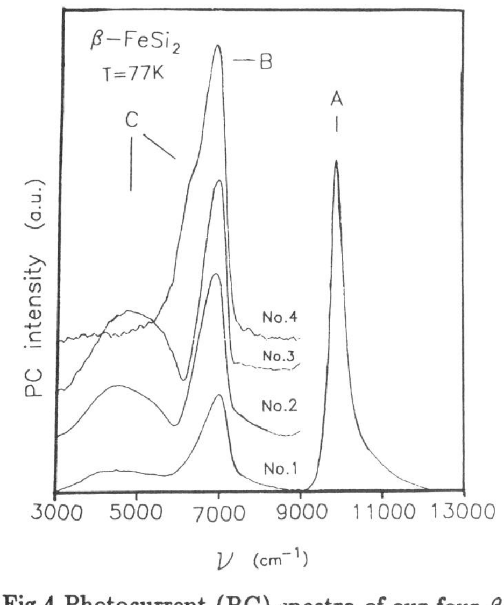


Fig.4 Photocurrent (PC) spectra of our four $\beta$–$FeSi_2$ samples at 77K with zero bias voltage. These samples were grown at different Si substrate temperatures: (1) $T_s$=27°C (sample No.1), (2) $T_s$=500°C (sample No.2), (3) $T_s$=550°C (sample No.3), (4) $T_s$=600°C (sample No.4). The peaks A of sample No.2, 3, and 4 were not shown here.

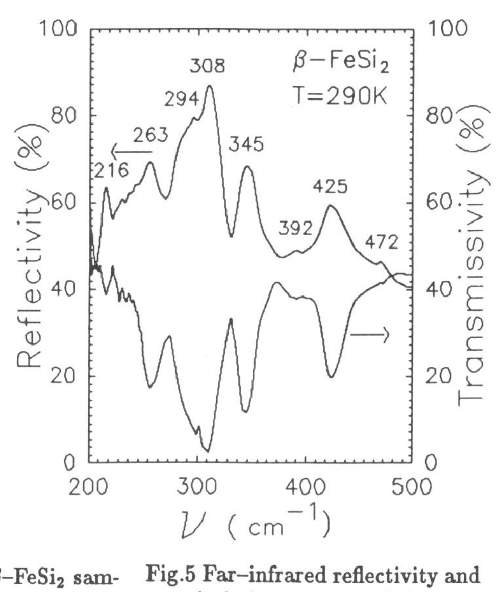


Fig.5 Far–infrared reflectivity and transimission spectra of the typical $\beta$–$FeSi_2$ sample (No.4) at 290K.

Finally, we show in Fig.5 the far–infrared reflectivity and transmission spectra of the sample No.4 of $\beta$–$FeSi_2$ in the wavenumber range from 200 to 500 $cm^{-1}$ measured at room temperature (290K). The six characteristic far–infrared reflection peaks of $\beta$–$FeSi_2$ reported in the literature[13–15] are clearly observed and their energy positions are in good agreement with that reported in the literature, as well as the transmission spectrum. Furthermore, the structures around 216 and 472 $cm^{-1}$ are first observed. We believe the presence of these two more characteristic far–infrared structures might be related with the improvement of the film quality, since the critical points for the phonon spectra should be more clear for the material with better structure. Confirmation of this association was obtained by the experimental facts that these two peaks in the reflectivity and transmission spectra gradually disappeared for the samples grown with lower substrate temperature, also in consistent with the PC results described above.

## 4. Conclusion

In summary, the polycrystalline $\beta$–$FeSi_2$ thin films were grown by either solid phase epitaxial (SPE) or reactive deposition–solid phase epitaxial (RD–SPE) and annealed at 800°C in a flowing nitrogen atmosphere. Absorption, photocurrent, and far–infrared reflectivity spectra of the $\beta$–$FeSi_2$ thin films in relation with the Si substrate temperature of growth are presented. Both the absorption and photocurrent measurements show the direct gap behavior of the material. The photocurrent spectra related with the fundamental interband, extrinsic defect transitions of $\beta$–$FeSi_2$ and the intrinsic transitions of Si substrates are observed. The transition energies obtained via photocurrent spectra are in good agreement with the results of the absorption measurements and the theoretical fitting calculation. We obtain the best quality $\beta$–$FeSi_2$ thin film under the highest Si substrate temperature of 600°C. The simplicity, sensitivity and reliability of the photocurrent measurements in revealing the energy band structure of $\beta$–$FeSi_2$ are demonstrated.

* This work was supported partly by Shanghai NSF No. 94JC14006.

## References

[1] Birkholz U and Schelm J 1968 *Phys. Stat. Sol.* **27** 413; 1969 **34** K177; 1970 **39** 197
[2] Bost M C and Mahan J E 1985 *J. Appl. Phys.* **58** 2696; 1986 *J. Vac. Sci. Technol. B* **4** 1336; 1988 *J. Appl. Phys.* **64** 2034
[3] Dimitriadis C A, Werner J H, Logothetidis S, Stutzmann M, Weber J, and Nesper R 1990 *J. Appl. Phys.* **68** 1726
[4] Lefki K, Muret P, Cherief N, and Cinti R C 1991 *J. Appl. Phys.* **69** 352
[5] Christensen N E 1990 *Phys. Rev. B* **42** 7148
[6] Mahan J E, Geib K M, Robinson G Y, Long R G, Xinghua Y, Bai G, Nicolet M, and Nathan M 1991 *Appl. Phys. Lett.* **56** 2126
[7] Hunt T D, Reeson K J, Gwillian R M, Homewood K P, Wilson R J, and Sealy B J 1993 *EMRS*, **Symposium E** LG17
[8] Lefki K and Muret P 1993 *J. Appl. Phys.* **74** 1138
[9] Muret P, Lefki K, Nguyen T T A, Cola A, and Ali I 1994 *Semicond. Sci. Technol.* **9** 1395
[10] Peale D R, Haight R, and Ott J 1993 *Appl. Phys. Lett.* **62** 1402
[11] De Vore H B 1956 *Phys. Rev.* **102** 86
[12] Giannini C, Lagomarsino S, Scarinci F, and Castrucci P 1992 *Phys. Rev. B* **45** 8822
[13] Birkholz J, Finkenrath H, Naegele J, and Uhle N 1968 *Phys. Stat. Sol.* **30** K81
[14] Borghesi A, Piaggi A, Stella A, Guizetti G, Nava F, and Santoro G 1990 in *Proceedings of the 20th International Conferences on Semiconductors*, edited by E.M. Anastassakis and J.D. Joaannopoulos (World Scientific, Singapore) p.2091
[15] Guizetti G, Marabelli F, Onda N, and Kanel H Von 1993 *Solid State Commun.* **86** 217

*Inst. Phys. Conf. Ser. No 144: Section 2*
*Paper presented at 7th Int. Conf. on Narrow Gap Semicond. Santa Fe 9–12 January*



# Influence of mercury pressure on composition profile of LPE $Hg_{1-x}Cd_xTe$ film

B. Li, J.H.Chu, X.Q.Chen, J.Y.Cao and D.Y.Tang

National Laboratory for Infrared Physics,
Shanghai Institute of Technical Physics
Shanghai 200083, China

## Abstract

This paper presents a quantitative calculation for the variations of the melt point and the mole fraction of the growth solution caused by the imbalance in the Hg pressure. The effect of Hg vapor pressure on the composition profile along the depth is also determined and compared with the experimental data. The results show that the non-equilibrium Hg pressure over the solution will result in changes of the average composition and the composition profile of the grown epitaxial film. Adequate Hg loss during the growth period can reduce the composition variation from the epilayer/substrate interface to film surface.

## 1. Introduction

The liquid phase epitaxial (LPE) growth of $Hg_{1-x}Cd_xTe$ from Te-rich solution is a main method for the production of quality materials. However, the difficulty of maintaining the equilibrium mercury vapor pressure during the LPE growth cycle is still a main troublesome factor though the Hg pressure over the Te-rich solution is manageably small. The imbalance in the Hg pressure will cause the average composition and composition profile of the growth layer to vary, that slide the cut-off wavelength and increase the fluctuation of characteristic among detector pixels. Many studies have discussed the effect of imbalance Hg pressure on the liquidus temperature and composition of HgCdTe liquid phase epitaxy[1-8], but no quantitative analysis is given as we known. In this paper, based on the associated solution model and phase diagram theory, a quantitative calculation for the average composition and composition profile change caused by the imbalance Hg pressure is presented and compared with the experimental data.

## 2. Calculation procedure

As we know, the concentration of Hg fraction in the whole solution, which is denoted by $C_i^0$, is constant at equilibrium with the vapor and can be derived from the associated solution model[8,9]. If the Hg reservoir drifts off its equilibrium temperature $T_{Hg} \rightarrow T_{Hg} + \Delta T_{Hg}$, the surface of the solution will change its composition to $C_i^f = C_i^0 + \Delta C_i$, where $\Delta C_i$ is a function of $\Delta T_{Hg}$[10]. Assuming that the transport of Hg atom in Te-rich (Hg,Cd)Te solution also obeys the diffusion rule as in liquid Te[11], then the variation of Hg concentration over the solution is:

$$\Delta C(x,t) = (C_i^f - C_i^0)\exp(-\mu x / \sqrt{t}) \qquad (1)$$

where parameter $\mu$ is the diffusion coefficient and independent of time. x, the distance from the surface in cm. t, the diffusion time in hour. From equation (1) one can get the Hg weight change during the growth cycle by integrating $\Delta C(x,t)$ along the height h and multiplying by area S for a growth solution. It generalizes to

$$\Delta W_i = 200.59 \cdot S \cdot \Delta C_i \cdot \sqrt{t} / \mu[1 - \exp(-\mu h / \sqrt{t})] \qquad (2)$$

Assuming that the thermodynamic components Hg, Cd, Te in the solution are numbered consecutively 1 through 3 and their number of gram-atoms denoted as $M_j$ (j=1,2,3), when the number of gram-atoms of Hg change from $M_1$ to $M_1+\Delta M$ due to the Hg pressure nonequilibrium, the liquid compositions of $(Hg_{1-z}Cd_z)_{1-y}Te_y$ will transform from (y, z) to (y', z'), correspondingly. The variation quantity of melting point caused by the growth solution change is

$$\Delta T = dT/dz \cdot (z' - z) + dT/dy \cdot (y' - y) \qquad (3)$$

where dT/dz and dT/dy are slope of liquidus surface.

As for Hg-Cd-Te system, Huang assumes that the mole fraction of each kind of atom in the solid $x_s$ is directly proportional to its net flux from liquid to solid, and all segregation atoms have the same proportional coefficient[12-14]. If the species Hg, Cd, Te are numbered consecutively 1 though 3, the interface segregation coefficient k* can be deduced as:

$$\begin{aligned} k^* &= \frac{1}{x_{L2}} \times \frac{J_{LS2} - J_{SL2} + fC_{L2}}{(J_{LS1} - J_{SL1} + fC_{L1}) + (J_{LS2} - J_{SL2} + fC_{L2})} \\ &= k_0 + \frac{1-(y+z)k_0}{(y+z)f + K_1/k_0} f \end{aligned} \qquad (4)$$

where $J_{LSi}$ and $J_{SLi}$ are unidirectional diffusition flux from liquid to solid and from solid to liquid, respectively. $C_{Li}$ is the liquid concentration of i-th specie, f is the growth velocity, $K_i$ is a function of molecular vibration parameters and can be called the absolute segregation coefficient, and $k_0$, the equilibrium segregation coefficient.

By the definition of the effective segregation coefficient $K_{eff}$[15], one can deduce following expression:

$$k_y = k_y^* \cdot [1 - \frac{(1+F)\Delta T}{(1-y)m}] \qquad (5a)$$

$$k_{eff} = k^* \cdot \frac{1 - \Delta T / [z(1-y)m]}{1 - (1+F)\Delta T / [(1-y)m]} \qquad (5b)$$

where parameter $F = (\frac{R}{x_s} - 1)\sqrt{\frac{D_2}{D_1}} = (\frac{R}{k_{eff} \cdot z} - 1)\sqrt{\frac{D_2}{D_1}}$ and $R = \frac{k^*(k_y^* - 1)}{k^* k_y^* - 1}$, $D_i$ is the diffusivity of i-th specie. $k_y$ is the effective segregation coefficient of HgCd, $\Delta T$, the temperature interval. The growth rate f and film thickness d can be derived from the mass transport equation[15]. In the supercooled LPE growth of $Hg_{1-x}Cd_xTe$, f and d are:

$$f = A\sqrt{D_2/\pi}\cdot(\Delta T_s\cdot t^{-1/2} + \alpha\cdot t^{1/2})$$
$$d = 2A\sqrt{D_2/\pi}\cdot(\Delta T_s\cdot t^{1/2} + \frac{2}{3}\alpha\cdot t^{3/2}) \quad (6)$$

respectively. Where $A = R/[(1-k_y^{*-1})k_y k_{eff} z(1-y)m]$ is a proportional constant., α is the cooling rate and ΔT=αt.

In equation (5b) is shown that $K_{eff}$ depends on supercooling, cooling rate, liquid composition and Hg pressure over the growth solution. If the complement Hg reservoir diverges its equilibrium temperature during the LPE process, the liquid composition and melting point will change that lead to the degrees of supercooling and composition profile of epitaxial film vary.

## 3. Result and discussion

To verify equations (1) to (3) of the composition variation versus Hg pressure change, we use the experimental result of Chiang et al about the film composition relative to the change of mercury reservoir temperature for slide LPE of $Hg_{1-x}Cd_xTe$[4] as for comparison. The calculation result agrees well with the measured data, that is, for one degree change of temperature at the mercury reservoir, the variation of x-value is approximately 0.007 for the $Hg_{0.8}Cd_{0.2}Te$ film[8].

Electron microprobe (EDAX) is used to characterize the composition depth profile of $Hg_{1-x}Cd_xTe$ epi-film grown by the LPE technique with Hg complement reservoir at 0 and 4°C lower than its equilibrium temperature as illustrated in figure 1. In the figure the theoretical curves of composition profile relative to various temperature deviations of Hg reservoir are also included. As for not considering the interdiffusion between film and substrate that exists in the LPE growth process, the slope of the measured composition profile is sharper than the calculated result, especially in the substrate/epilayer interface region. The gradient of composition profile becomes gentle as Hg loss from the solution, and sharp when Hg gained to the melt, while the composition of $Hg_{1-x}Cd_xTe$ epilayer deviates the designed value with Hg pressure change. It has been pointed out that Hg loss from the growth melt causes the ratio of Cd mole fraction and liquidus temperature increase with the growth time[8]. The former leads to the epilayer composition raise, while the latter offers a resistance of the effective segregation $k_{eff}$ decrease by slowing down the increase of supercooling with growth duration for supercooled LPE mode. Both help improve the composition uniformity of the epilayer.

On the contrary, Hg loss from the liquid increases the x-value and Hg vacancies of epi-film that are detrimental to the device characteristic. These weaknesses can be overcome by lowering the Cd mole fraction in the liquid to coincide with the cut-off wavelength of the detector and selecting the proper annealing conditions to decrease the hole concentration.

With a view of better understand the effect of Hg pressure on composition profile of $Hg_{1-x}Cd_xTe$ film, we further studied the dependence of longitudinal compositions on liquid compositions y and z of $(Hg_{1-z}Cd_z)_{1-y}Te_y$ as shown in figures 2(a) and (b). The results conform to the conclusion that adequate Hg loss from the solution improves the uniformity of composition profile. Another evidence is that the composition profile and film thickness are sensitive to the Te mole fraction y and insensitive to the Cd mole fraction z in the liquid.

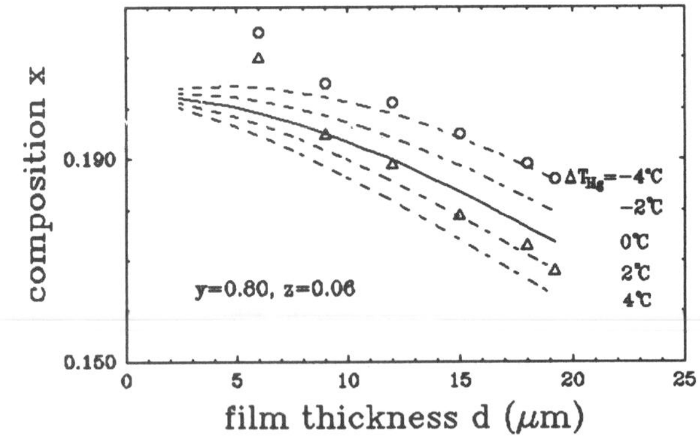


Figure 1: Composition depth profiles for $Hg_{1-x}Cd_xTe$ epilayer at various Hg pressure. Solid lines are calculated results with Hg reservoir temperatures -4, -2, 0, 2, 4°C above the balance value, respectively. Circles and triangles are EDAX analysis corresponding to epilayers grown with Hg temperature 0 and 4°C below its equilibrium value.

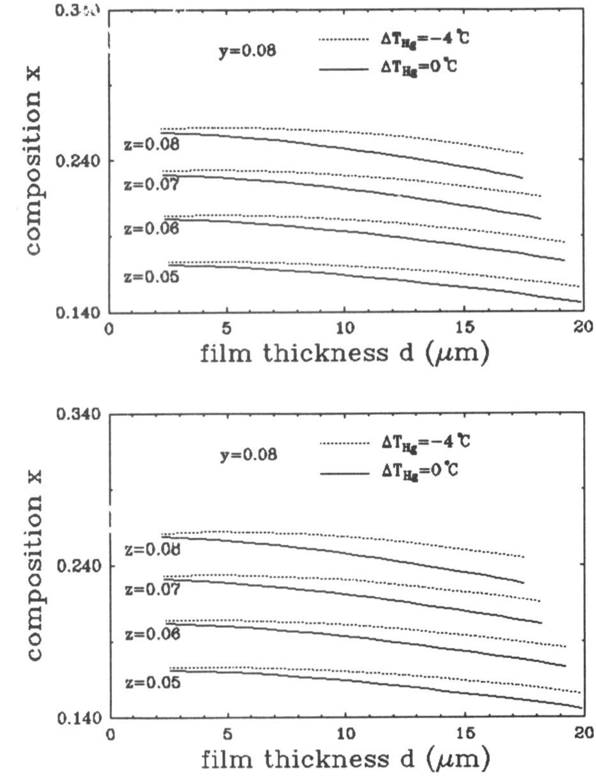


Figure 2: Dependence of composition profiles on liquid composiitons y [upper figure] and z [lower figure] of liquid $(Hg_{1-z}Cd_z)_{1-y}Te_y$. Solid lines refer to epilayers grown at balance Hg pressure, dashed lines to film grown with Hg temperature 4°C below equilibrium value.

Decrease y value provides a uniform composition curve with large thickness, while raise z value only deduce film thickness slightly.

The anticipated disadvantages of a small Te composition in the liquid include the increase in melting point and interdiffusion of the matrix elements at the surface, which deteriorate the compositional uniformity. Moreover, with y value decrease the liquid composition changes due to nonequilibrium Hg pressure raise[8]. Hence the proper y values of LPE growth melt may be in the range of 0.78 to 0.82.

## 4. Conclusion

Thermodynamic analysis has been applied to liquid phase epitaxial $Hg_{1-x}Cd_xTe$ from Te-rich solution. The sensitivity of the average composition and the composition profile of the deposited material is assessed as a function of Hg pressure. Adequate Hg loss and proper prescription of liquid composition can improve the composition uniformity in depth for $Hg_{1-x}Cd_xTe$ LPE film.

References:

[1] Harman T C 1980 *J.Electron. Mater.* **9** 945
[2] Brice J C 1986 *Prog. Crystal Growth and Character.* **13** 39
[3] Harman T C 1981 *J.Electron. Mater.* **10** 1069
[4] Chiang C D, Wu T B and Chung W C 1988 *J. Crystal growth* **87** 161
[5] Takami A, Kawazu Z, Takiguchi T, Mitsui K, Mizuguchi K and Murotani T 1992 *J. Crystal Growth* **117** 16
[6] Suh S H, Moon S W, Kim J S, Lim S W, Kwak N J, Kim H K and Kim J M 1992 *J. Crystal Growth* **121** 417
[7] Astles M G, Shaw N, Blackmore G and Hall R S 1992 *J. Crystal Growth* **117** 213
[8] Li B, Chu J H, Chen X Q et al, *J.Crystal Growth,* (accepted)
[9] Brebrick R F et al 1983 *Semiconductors and Semimetals,* Vol.19 (New York: Academic)
[10] Ruda H, Jedral L, Lagowski J and Gatos H C 1984 *J. Elecchem. Soc.,* **131** 1159
[11] Eger D, Yellin N and Dor L B 1986 *J. Electron. Mater.,* **15** 75
[12] Huang X L 1986 *Master Thesis* (Kunming Institute of Physics) 58
[13] Li Z, Wu C S and Huang X L 1987 *Infrared Technolgy,* **9** 41
[14] Li Z, Wu C S and Huang X L 1988 *Infrared Technolgy,* **10** 35
[15] Pamplin B R 1975 *Crystal Growth* (Pergamon) 777

*Inst. Phys. Conf. Ser. No 144: Section 2*
*Paper presented at 7th Int. Conf. on Narrow Gap Semicond. Santa Fe 9–12 January*




# Electronic and magnetic properties of diluted magnetic semiconductor $Hg_{1-x}Mn_xTe_{1-y}Se_y$

V. A. Kulbachinskii, P. D. Maryanchuk, and I. A. Churilov
Low Temperature Physics Department, Physics Faculty, Moscow State University, 119899, Moscow, Russia

M. Inoue, M. Sasaki, H. Negishi, and Y. Hara
Department of Materials Science, Faculty of Science, Hiroshima University, Higashi-Hiroshima 724, Japan

Electrical resistivity, Hall coefficient, and magnetoresistance for diluted magnetic semiconductors $Hg_{1-x}Mn_xTe_{1-y}Se_y$ crystals with $y$=0.01 and $x$=0.03, 0.06, 0.14, 0.30 have been measured in the temperature range 4.2-300 K and in magnetic fields up to 7 T. Negative magnetoresistances are observed at 4.2 K in all samples, which can be explained by the change of the acceptor activation energy in the magnetic field. The temperature and magnetic field dependences of the Hall coefficient and its sign reversal in magnetic field can be explained quantitatively by the coexistence of three group of carriers, electrons and two types of holes with different mobilities. Magnetic susceptibility measurements using a SQUID magnetometer in the temperature range 4.2-80 K reveal that these solid solutions show a paramagnetic behavior and sample with $x$=0.30 exhibits a spin-glass transition at a temperature lower than that of $Hg_{1-x}Mn_xTe$ with the same Mn content.

## 1. Introduction

Diluted magnetic semiconductors such as $Hg_{1-x}Mn_xTe$ or $Hg_{1-x}Mn_xSe$ behave in a similar fashion to their normal counterparts $Hg_{1-x}Cd_xTe$ and $Hg_{1-x}Cd_xSe$. By changing the composition $x$, we can modify the band structure, *i.e.*, the effective masses of electrons and holes, the energy gap, etc.[1,2] $Hg_{1-x}Mn_xTe$ and $Hg_{1-x}Mn_xSe$ are narrow gap semiconductors with magnetic properties, both showing spin-glass transitions at low temperatures.[3-6] From the Hall and Shubnikov-de Haas (SdH) experiments, it has been found that as-grown $Hg_{1-x}Mn_xTe$ is of p-type conductivity, while $Hg_{1-x}Mn_xSe$ is of n-type;[7] in the latter the electron concentration is high ($10^{17}$-$10^{18}$ $cm^{-3}$), which is almost independent of temperature. Thus it is interesting to synthesize mixed crystals $Hg_{1-x}Mn_xTe_{1-y}Se_y$ and to investigate their transport and magnetic properties. In this work the transport properties of single crystals $Hg_{1-x}Mn_xTe_{1-y}Se_y$ were measured in the temperature range 4.2-300 K and in magnetic fields up to 7 T. Magnetic susceptibilities were also measured in the magnetic field of 400 mOe using a SQUID magnetometer.

## 2. Experimental

Single crystals of $Hg_{1-x}Mn_xTe_{1-y}Se_y$ ($y$=0.01, $x$=0.03, 0.06, 0.14, 0.30) were grown by a Bridgeman technique. After synthesis the crystals were annealed in the vapor of Se at 250 °C for 200 hours. The value of Mn concentration $x$ was determined from the magnetic susceptibility at room temperature measured by a Faraday method at H=1000 Oe. Transport measurements were taken by a standard dc method with a sample current up to 50 mA. Ac magnetic susceptibility measurements were carried out using a SQUID magnetometer with the amplitude of ac magnetic fields 400 mOe and the frequency 1 Hz in the temperature range 4.2 -80 K.

## 3. Results and discussions

### A. Electrical properties

Of particular interest is the magnetic field dependence of Hall coefficients $R_H$, which vary with temperature, as illustrated in Fig. 1(a) for a typical sample (x=0.06). Points correspond to experimental results and solid lines are theoretical fitting (see later). At low temperatures $T$<20 K, $R_H$ has a positive sign, while at higher temperatures it becomes negative in low magnetic field. With increasing magnetic field some show a sign reversal from negative to positive. These behaviors are different from those for p-type $Hg_{1-x}Mn_xTe$ and n-type $Hg_{1-x}Mn_xSe$. Though not shown here, the transverse magnetoresistances show that with increasing magnetic field these are positive in low magnetic field, showing a positive peak, followed by negative magnetoresistance; the peak value decreases with increasing Mn content $x$. When the temperature is increased, the magnetoresistance becomes positive. These results are similar to those found for $Hg_{1-x}Mn_xTe$,[4] whereas $Hg_{1-x}Mn_xSe$ shows always a positive magnetoresistance over a wide temperature range.[7]

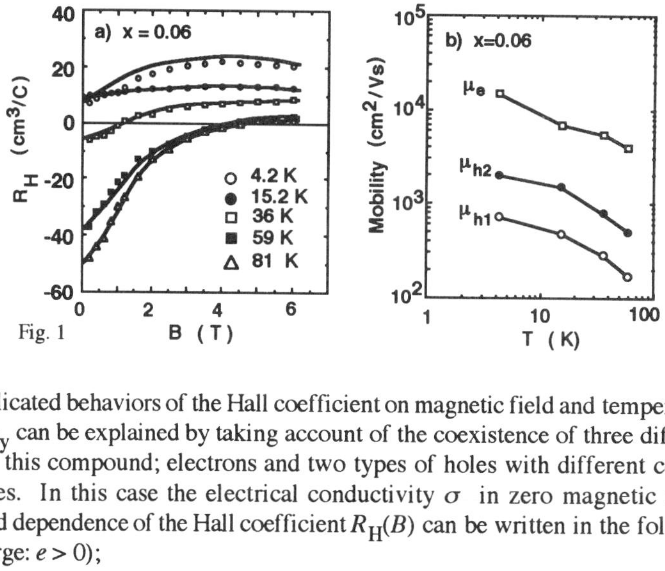

Fig. 1

The complicated behaviors of the Hall coefficient on magnetic field and temperature in $Hg_{1-x}Mn_xTe_{1-y}Se_y$ can be explained by taking account of the coexistence of three different groups of carriers in this compound; electrons and two types of holes with different concentrations and mobilities. In this case the electrical conductivity $\sigma$ in zero magnetic field and the magnetic field dependence of the Hall coefficient $R_H(B)$ can be written in the following forms (electron charge: $e > 0$);

$$\sigma = e(n\mu_e + \Sigma p_i \mu_{hi}), \quad (1)$$

$$R_H(B) = N/(A^2 + N^2B^2), \quad (2)$$

with

$$N = e[-n\mu_e^2/(1 + \mu_e^2B^2) + \Sigma p_i \mu_{hi}^2/(1 + \mu_{hi}^2B^2)],$$

$$A = e[n\mu_e/(1 + \mu_e^2B^2) + \Sigma p_i \mu_{hi}/(1 + \mu_{hi}^2B^2)],$$

where the Hall factor is set equal to 1 for all samples for simplicity. Here $n$ is an electron concentration, $p_i$ hole concentration of the $i$th hole ($i$=1 and 2), $\mu_e$ the corresponding electron mobility and $\mu_{hi}$ the hole mobility. We have calculated the magnetic field dependence of the Hall coefficient using Eqs. (1) and (2) with best-fit parameters of carrier concentration ($n$, $p_i$) and mobility ($\mu_e$, $\mu_{hi}$) to fit experimental data for each sample. The calculated curves of $R_H$ (solid curves) for sample with $x$=0.06 and the corresponding carrier mobilities at some fixed temperatures are shown in Figs. 1(a) and 1(b), respectively. The latter gives us the general trend of the temperature dependence of mobilities for electrons $\mu_e$ and two types of holes, $\mu_{h1}$ and $\mu_{h2}$.

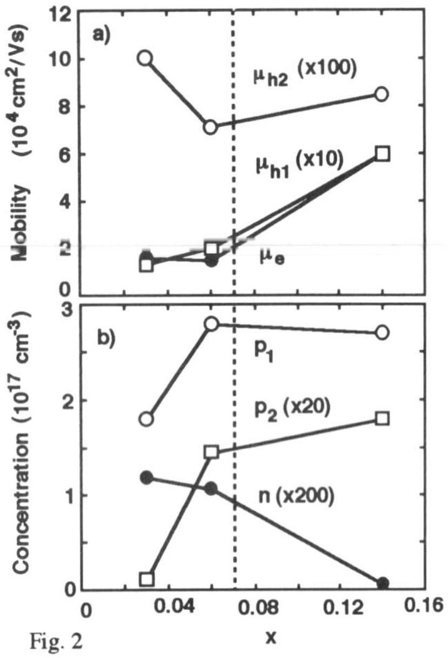

Fig. 2

Figure 2 shows the best-fit values of (a) the mobility and (b) carrier concentration at 4.2 K plotted against the Mn content $x$ . The dashed line marks the critical concentration $x_c$=0.07 for $Hg_{1-x}Mn_xTe$, below which it is gapless ($E_g$=0) and above which it has a small direct energy gap.[2] One notes that with increasing $x$ the electron concentration $n$ is decreased appreciably and the electron mobility $\mu_e$ increases at $x$=0.14. On the other hand, the hole concentrations $p_1$ and $p_2$ increase remarkably in the gapless region but change less above $x_c$=0.07, while the mobility $\mu_{h1}$ increases linearly but $\mu_{h2}$ shows a minimum at $x$=0.06. These results suggest that in our system the change from gapless to open gap may also occur between $x$=0.06 and 0.14.

We may compare our results with those for $Hg_{1-x}Mn_xTe$. In the latter, the main role in conductivity is due to the acceptor holes in the 'resonant acceptor impurity' band due to native vacancies, while the electron concentration is very low [$n \cong (2\text{-}4) \times 10^{14}$ cm$^{-3}$];[4] in contrast, $Hg_{1-x}Mn_xSe$ always has n-type conductivity and the Hall coefficient does not depend on magnetic field and temperature,[7] as mentioned earlier. Based on our experimental results and the existing band model for semimagnetic semiconductors,[2] we propose the possible band scheme for $Hg_{1-x}Mn_xTe_{1-y}Se_y$, as depicted schematically in Fig. 3. This band scheme is essentially the same as those for the component systems $Hg_{1-x}Mn_xTe$ and $Hg_{1-x}Mn_xSe$. But the difference is the position of the acceptor impurity band formed near the Fermi energy $E_F$. In $Hg_{1-x}Mn_xTe$, the activation energy of the impurity level $E_a$ is known to to be about 3-4 meV and coincides with the Fermi energy. Using well-known Kane's three band model and taking account of the matrix element of the momentum operator between the states $\Gamma_6$ and $\Gamma_8$ to be 7 × 10$^{-8}$ eVcm,[7] the Fermi energy for our system is calculated to be $E_F \cong 5$ meV, which is slightly higher than that for $Hg_{1-x}Mn_xTe$. In addition to electrons in the conduction

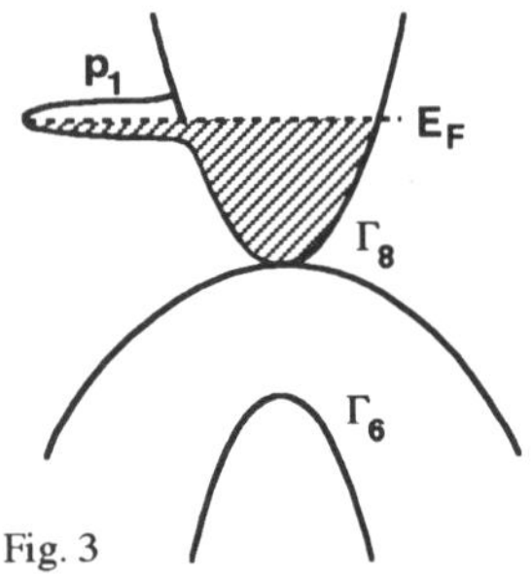

Fig. 3

band with the density $n$ and holes with the density $p_1$ in the acceptor impurity band, in our case the second type holes with the density $p_2$ exist, which may be ascribed to heavy holes in the valence band.

Moreover, the negative magnetoresistance observed at low temperatures for $Hg_{1-x}Mn_xTe_{1-y}Se_y$ can also be explained in the same manner as for $Hg_{1-x}Mn_xTe$, which is ascribed to an unusual dependence of the acceptor activation energy on magnetic field arising from the strong magnetic field dependence of the acceptor wave function.[1,2] That is, with increasing magnetic field the binding energy of a hole to acceptor site decreases, which causes the increase in the hole concentrations in the valence band, leading to a large decrease in resistance, *i.e.*, so-called magnetic boiloff effect.

## B. Magnetic properties

Typical results for ac magnetic susceptibilities $\chi$ of two samples with $x = 0.14$ and $0.30$ over the temperature range 4.2-80 K are illustrated in Fig. 4, where the inverse susceptibilities are plotted against temperature. At high temperatures $T$>20 K the temperature dependence of ac susceptibilities $\chi(T)$ follow the Curie-Weiss law, $\chi(T)= C/(T - \theta)$, where $C$ is a Curie constant, $\theta$ a Curie-Weiss temperature. As marked by arrow in the inset, the sample with $x$ =0.30 shows an indication of spin-glass transition at glass-temperature $T_g = 5$ K. The value of $T_g$ is plotted against Mn concentration $x$ for our sample in Fig. 5, together with those for $Hg_{1-x}Mn_xTe$[4,5] and $Hg_{1-x}Mn_xSe$[3,6] for comparison. As can be seen, $T_g$ of our sample is lower than those of other materials with the same Mn content ($x$ = 0.30).

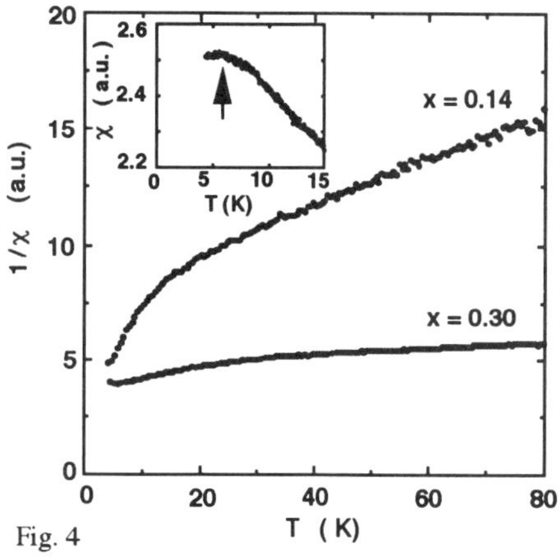

Fig. 4

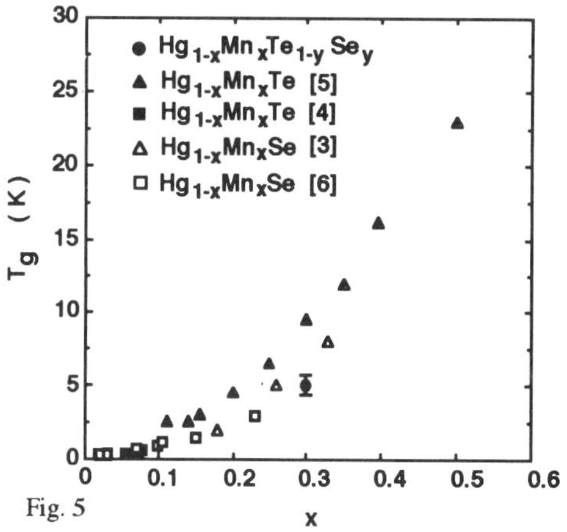

Fig. 5

In diluted magnetic semiconductor $Hg_{1-x}Mn_xTe$[4] or $Hg_{1-x}Mn_xSe$,[6] there is the formation of magnetic clusters with an antiferromagnetic interaction; the effective magnetic moment of Mn in these materials is $5.92\mu_B$ at higher temperatures and the Curie-Weiss temperature $\theta$ depends on Mn content. For our solid solutions $Hg_{1-x}Mn_xTe_{1-y}Se_y$, the absolute values of $\theta$ are higher compared to those of $Hg_{1-x}Mn_xTe$ with the same Mn contents. The nearest-neighbour exchange constant $J$, evaluated by the relation $\theta = gS(S+1)xzJ/3k_B$ with $g$=2, $S$=5/2 and $z$=12, decreases with $x$ from $J$=-10 K for $x$=0.03 to -7.6 K for $x$=0.30, and the value for $x$=1 can be extrapolated to be $J$=-7.5 K, which is reasonable magnitude; $J$=-7.15 K in $Hg_{1-x}Mn_xTe$ and $J$=-10.9 K in $Hg_{1-x}Mn_xSe$.[1] In $Hg_{1-x}Mn_xTe$ system there are always a number of native defects or vacancies of Hg and Te atoms, whereas in the solid solutions $Hg_{1-x}Mn_xTe_{1-y}Se_y$ the Se atoms fill substitutionally these vacancies, which reduces local cation-anion-cation distances around the Mn atoms, leading to the enhancement of the indirect superexchange interactions between Mn-(Te or Se)-Mn, and thus the increase of the Curie-Weiss temperature.

On the other hand, the spin-glass temperature $T_g$ (≈5 K) is lower in our system ($x$=0.30) compared to those for $Hg_{1-x}Mn_xTe$ and $Hg_{1-x}Mn_xSe$ ($x$=0.30), as mentioned earlier. In the host materials $Hg_{1-x}Mn_xTe$ and $Hg_{1-x}Mn_xSe$ (fig. 5b), $T_g$ increases with Mn concentration $x$, especially for the higher concentrations than the percolation critical concentration for nearest neighbour ($x_p$≈0.17-0.20), which means that the bigger the magnetic cluster size, the higher the glass temperature. Therefore the decrease of $T_g$ by substitution of Se atoms in the host lattices may suggest that the size of the magnetic clusters becomes smaller and/or the interaction between the clusters in our system becomes weaker compared to those for the host $Hg_{1-x}Mn_xTe$. This indicates that the fluctuation between the spins with antiferromagnetic nearest neighbour interaction can be increased remarkably by substitution of the small amount of anion species, leading to the modification of the sizes and distribution of the magnetic clusters, and $T_g$. These results are consistent with the fact that the anion radius becomes smaller in the order of B=Te, Se, and S in $A_{1-x}Mn_xB$ (A=Cd, Zn, and Hg), the exchange constant $J$ increases but $T_g$ becomes lower for the same Mn concentration.[1]

## 3. Conclusion

Solid solutions $Hg_{1-x}Mn_xTe_{1-y}Se_y$ formed from p-type $Hg_{1-x}Mn_xTe$ by mixing a small amount of n-type $Hg_{1-x}Mn_xSe$ have revealed interesting electronic and magnetic properties different from those of the constituent systems. Even though the band structure is essentially the same as those of the hosts, the Mn induced acceptor impurity band is enhanced in energy, the electron concentration is increased and the hole concentration in the impurity band remains almost unchanged, but the second type holes appears additionally in the solid solutions. Such three types of conduction carriers are found to be responsible for the temperature and magnetic field dependences of the observed Hall coefficients, in contrast to two carrier model valid for the host $Hg_{1-x}Mn_xTe$ or a single carrier model for $Hg_{1-x}Mn_xSe$. The effect of Se substitution into the $Hg_{1-x}Mn_xTe$ host lattices is also reflected in the magnetic properties, where the nearest neighbour exchange interactions between the spins become stronger, but the size of magnetic clusters formed by the Mn atoms is decreased compared to those in the host $Hg_{1-x}Mn_xTe$.

## References


1 J.K. Furdyna and J. Kossut, *Diluted Magnetic Semiconductors* (Academic, San Diego, 1988).
2 N.B. Brandt and V.V. Moshchalkov, Adv. Phys. **33**, 194 (1984).
3 G.D. Khattak , C.D. Amarasekara, S. Nagata, R.R. Galazka, and P.H. Keesom, Phys. Rev. B **23**, 3553 (1981).
4 N.B. Brandt, V.V. Moshchalkov, A.O. Orlov, L. Skrbek, I.M. Tsidilkovskii, and S.M. Chudinov, Zh. Eksp. Teor. Fiz. **84**, 1059 (1983) [Sov. Phys. JETP **57**, 614 (1983)].
5 A. Mycielski, C. Rigaux, M. Menant, T. Dietl, and M. Otto, Solid State Commun. **50**, 257 (1984).
6 N.B. Brandt, Zh.T. Ismailov, V.A. Kulbachinskii, S.M. Chudinov, N.P. Gavaleshko, and P.D. Maryanchuk, Fiz. Nizk. Temp. **12**, 215 (1986) [Sov. J. Low Temp. Phys. **12**, 125 (1986 )].
7 S. Takeyama and R.R. Galazka, Phys. Stat. Sol. (b) **96**, 413 (1979).

*Inst. Phys. Conf. Ser. No 144: Section 2*
*Paper presented at 7th Int. Conf. on Narrow Gap Semicond. Santa Fe 9–12 January*



# Optical transmission and infrared photoluminescence of $Hg_{1-x-y}Cd_xMn_ySe$ single crystals

Yu.I.Mazur[1], G.G.Tarasov[1], S.R.Lavoric[1], J.W.Tomm[2] and H.Kissel[2]

[1] Institute of Semiconductors Physics, National Academy of Sciences, prospect Nauki 45, 252650 Kiev, Ukraine
[2] Institut für Physik, Humboldt-Universitat zu Berlin, Sitz: Invalidenstr. 110, Unter den Linden 6, D-10099 Berlin, Germany

**Abstract.** Optical transmission in the range of fundamental absorption edge of $Hg_{1-x-y}Cd_xMn_ySe$ (x=0.09-0.11, y=0.02-0.03) single crystals is investigated at temperature varying from 5 K to 300 K. It is ascribed to band-to-band transitions with account of band-tail absorption. Infrared photoluminescence is detected for the first time in the mercury selenides compounds, flaring up at temperature increase. Energy gap and principal optical characteristics temperature dependences are determined for the compounds under investigation.

## 1. Introduction

Multicomponent solid solutions on the basis of binary or ternary compounds of narrow-gap semiconductors are of great interest for manufacturing of crystalline, layer or compositional structures used widely in FIR technique [1-2]. However their real potential for semiconducting devices is not obvious up to nowadays because of high structural defects, low degradation characteristics and technological problems. So $Hg_{1-x}Cd_xSe$ single crystals defeat in many aspects in comparison with $Hg_{1-x}Cd_xTe$ compounds due to their lower quality. To improve it the doping by ions of rear earth or transition elements was suggested [3].

Quaternary semimagnetic compounds $Hg_{1-x-y}Cd_xMn_ySe$ (MCMS) belong to narrow-gap semiconductors at x=0.1, 0<y<0.1. Electrical and optical investigations show that a continuous transformation of typically semi-metal properties for HgSe (x=0.0, y=0.0) into those one inherent to direct-band semiconductors of zinc blend structure takes place under increasing of x, y content [4-5]. For the set of samples investigated in this study a comparatively high mobility of free carriers in the range of $(1\text{-}3)\cdot 10^5$ $cm^2$/ Vs at low temperatures was found and infrared photoluminescence (IR PL) was observed at band-to-band excitation to the best of our knowledge for the first time in the mercury selenides. This makes the multicomponent MCMS an interesting object for examination of magnetic component influence on the electric and optical properties of ternary compound.

Previously the data on galvano-magnetic and optical measurements were published for these semiconductors [6-7].Here we present the results of extended analysis of optical properties of narrow-gap MCMS in the range of fundamental absorption edge.

## 2. Transmission of MCMS single crystals near the edge of fundamental absorption.

Mixed MCMS single crystals (x=0.09-0.11, y=0.02-0.03), synthesized by modified Bridgman technique [7], were as-grown of n-type with the electron concentration about of $n=(3\text{-}5)10^{17}$ $cm^{-3}$ . X-rays testing showed the homogeneous space distribution of manganese ions up to the y-value about of 0.005. Transmission (T) measurements were performed with Fourier-spectrometer in the spectral range from 400 $cm^{-1}$ to 2000 $cm^{-1}$. The accuracy of measurements was about of 2 $cm^{-1}$ over all the interval under investigation. The temperature was varied from 5 K up to 300K. A typical spectral dependence T(ν) is shown in Fig 1 for different temperatures. It can be seen an interference pattern in the spectral range below absorption edge. At the temperature increase the absorption edge shifts towards the higher energies and the transmission dependencies have more smooth behavior in the range of the shorter wavelengths.

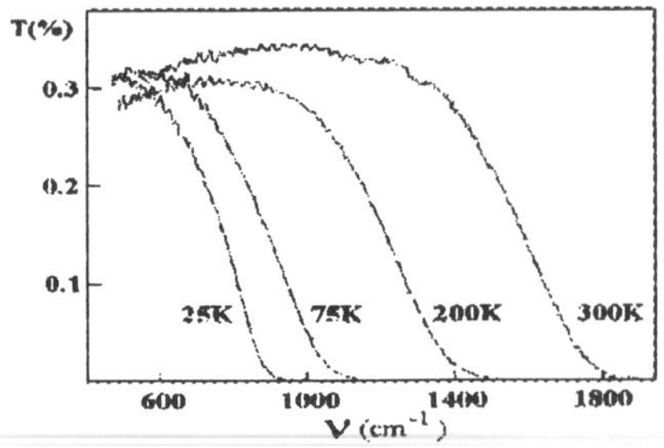


Fig.1 Transmission spectra of MCMS ($x=0.095$, $y=0.023$) at different temperatures. Crystal thickness $d=39\mu$, $n_e=8.8*10^{16}$ $cm^{-3}$

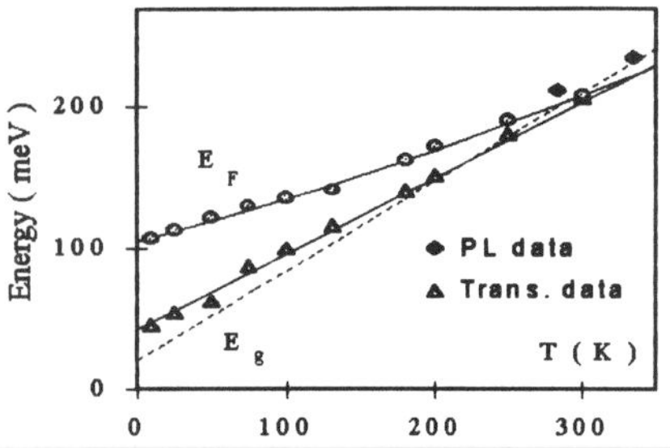


Fig.2 Energy gap $E_g$ and the Fermi energy $E_F$ vs temperature for the crystal, shown in Fig.1. Dashed line is $E_g$ calculated for virtual crystal.

To analyze the absorption coefficient $\alpha(\omega)$ for MCMS we have developed the calculation scheme more unified than that one suggested by Anderson [8]. It is applicable for wide class of narrow-gap semiconductors described in the Kane's approximation and takes into account the direct optical transitions between the bands of light (lh) and heavy (hh) holes and the conduction (c) one. Here the energy bands are determined by the equation [9]

$$E'\left[E'\left(E_g - E'\right)\left(E'+\Delta\right) + k^2P^2\left(E'+2\Delta/3\right)\right] = 0, E' = E(k) - \hbar^2k^2/2m_0 \qquad (1)$$

All energies are measured from the valence band edge, P denotes the interband matrix element, $\Delta$ is the valence band spin-orbit splitting, $k$ is the electron wave vector. Being the fourth order at every $k$-value , Eq.(1) gives four roots, satisfying the conditions:

$$E_1' = 0, \qquad E_2' + E_3' + E_4' = E_g - \Delta$$
$$E_2'E_3' + E_2'E_4' + E_3'E_4' = -\left(E_g\Delta + k^2P^2\right) \qquad E_2'E_3'E_4' = k^2P^2 2\Delta/3. \qquad (2)$$

For direct optical transitions the energy conservation law is hold,

$$E_3(k) - E_{1,2}(k) = \hbar\omega \qquad (3)$$

Subscripts 1, 2, 3 correspond to hh, lh and c notations.

Using Eq.(3) to fix the relation between the $E_2'(k)$ and $E_3'(k)$ we found all solutions ($E_2', E_3', E_4'$) and wave number $k_2$, actual for direct transition (the valence band of light holes - the conduction band), from the set of Eqs.(2)

$$E_2'(k_2) = \frac{1}{6}\left[p_2 + q_2\right], E_3'(k_2) = \frac{1}{6}\left[p_2 + q_2 + 6\hbar\omega\right], E_4' = \frac{1}{6}\left[p_2 - 2q_2 - 3\hbar\omega\right], \qquad (4)$$

$$k_2^2 = \frac{3E_2'E_3'E_4'}{2\Delta P^2},$$

where $$p_2 = 2\left(E_g - \Delta\right) - 3\hbar\omega, q_2 = \sqrt{p_2^2 + 12\left[E_g\Delta + k_2^2P^2 + \hbar\omega\left(p_2 + \hbar\omega\right)/2\right]}.$$

Eqs.(4) allow to calculate also the variation of the inter band matrix element P and , as a result, to determine the fraction of absorption caused by the light hole band to conductivity band transition with account of the Burstein-Moss absorption edge shift [8]. The same consideration for the heavy-hole band to conduction band transition gives the following set of solutions for Kane's equation (1)

$$E_2'(k_1) = \frac{1}{2}(p_1 + q_1), E_3'(k_1) = \hbar\omega - \frac{\hbar^2k_1^2}{2m_0}\left(1 + \frac{m_0}{m_1}\right), E_4' = \frac{1}{2}(p_1 - q_1), \qquad (5)$$

$$k_1^2 = \frac{3E_2'E_3'E_4'}{2\Delta P^2}, \text{ with } p_1 = E_g - \Delta - E_3', q_1 = \sqrt{p_1^2 + 4\left(E_3'p_1 + E_g\Delta + k^2P^2\right)}.$$

Taking into account that $E_j' = E_j - \hbar^2k^2/2m_0$, when $j\geq2$, and $E_1 = -\hbar^2k^2/2m_1$ we possess all the dependencies to calculate the absorption coefficient of MCMS, including the Burstein-Moss shift [8], for every frequency $\omega$ of absorbing photon even in the case of deep band-to-band excitation. The value of wave number

$k$, actual for the direct transition between the valence band and the band of conduction is calculated according to Eqs.(4) and (5) by solving the third order algebraic equation for $k^2$.

According to [8], $\alpha(\omega)$ is determined , in fact, by two parameters $E_g$ and $E_F$ and can be used for their calculation in the range of fundamental absorption edge. We have fitted the experimental dependencies $\alpha(\nu)$ by the theoretical ones varying only two parameters $E_g$ and $E_F$. The temperature dependences of $E_g$ and $E_F$ found from the fitting procedure are plotted in Fig.2. The linear approximation $E_g(x,y,T)=E_g(x,y,0)+\beta T$ is also shown. The value $E_g(0) = E_g(x,y,0) = 45$ meV corresponds to the x-content equals to 0.155 for the compounds $Hg_{1-x}Cd_xSe$. This means that the manganese presence opens the energy gap more effectively than the cadmium one:.$dE_g/dy>2.5dE_g/dx$ The coefficient $\beta$ of thermal extension for forbidden band equals to 0.54 meV/K being consistent with the $\beta$ value for $Hg_{1-x}Cd_xSe$ of cadmium content x the same as the total value $x + 2.52y$ for MCMS. The values of $E_g$ and $E_F$ demonstrate that MCMS with electron concentration of $8.8\ 10^{16}$ $cm^{-3}$ is degenerate up to $T \approx 300$ K. This explains a substantial Burstein-Moss shift of the optical absorption edge. At the temperature decrease the difference $E_F$-$E_g$ grows being an immediate manifestation of conduction band curvature dependence on the width of energy gap under the condition of slight dependence (or independence) of electron concentration on the temperature.

## 3. IR photoluminescence of $Hg_{1-x-y}Cd_xMn_ySe$.

The problem is its manifestation and identification for the mercury selenides of the composition under consideration. In narrow-gap semiconductors there exists a number of physical sources causing the decrease of the intrinsic quantum efficiencies. Analysis of experimental data shows [10] that decreasing of energy gap $E_g$, increasing of equilibrium carrier concentration as well as growth of pumping make the intrinsic non-radiative processes prevailing. This is the main restriction for manifestation of radiative transitions in narrow-gap semiconducting compounds. The experimental data on the IR PL for $Hg_{1-x}Cd_xSe$ are practically absent. The situation can be improved by doping it with manganese ions. Our data on the optical properties of $Hg_{1-x-y}Cd_xMn_yTe$ have shown [11] that the small amount of Mn can significantly affect them. This behavior was explained by the influence of low Mn-concentration on the degree of structural ordering in multicomponent compound. The same effect takes place evidently in MCMS. As a result, a detectable photoluminescence is developed here at least at room temperatures. It is interesting to note that the IR PL signal gradually decreases and disappears at the noise level when the temperature decreases from T=300K down to T=150K. This experimental fact has various grounds for explanation. It can be the decrease of lifetimes in the band-to-band transition at the temperature increase, caused by strong electron-phonon coupling. This effect has been detected experimentally for $Hg_{1-x}Cd_xSe$ [12].

Fig.3 presents the IR PL spectrum of MCMS ($x$=0.093, $y$=0.023, $n_e=1.05*10^{17}cm^{-3}$) single crystal, measured at two temperatures. The visible dip at the central part of luminescence spectra is ascribed to the $CO_2$ absorption. General behavior of PL at high temperatures is a strong argument for inter band transitions. At these temperatures the excitons and the shallow impurity states are completely ionized and consequently the transitions should be predominantly plasma like.

To fit the experimental data we used the approach, developed above, for the case of valence band-to-conduction band direct transitions, taking into account well-known relation between the rate of spontaneous recombination $r_{sp}(\omega)$ and the absorption coefficient $\alpha(\omega)$:

$$r_{sp}(\omega)=\frac{n^2(\hbar\omega)^2}{\pi^2c^2\hbar^3}\alpha(\omega)\left[\exp(E_c-E_v-\Delta F_{cv}/kT)-1\right]^{-1}, \tag{6}$$

where $\Delta F_{cv}$ is the difference of Fermi energies for non equilibrium (excess) carriers in both the conduction and the valence band. We suppose the possible contribution of crystal heating inevitable at comparatively intense crystal pumping. PL signal was usually detected at the excitation level of the order of (3-5)-$10^4$ KJ/pulse, providing the excitation density of a few tenths $kW/cm^{-2}$. Under this condition the area of crystal actual for luminescence becomes "overheated" at about 35 K. The dashed curves, shown in Fig.3, were obtained by the fitting procedure, carried out at corrected temperatures. The experimental points in Fig.2, plotted at the highest temperatures are just those one measured from the IR PL. They are well agreed with the experimental dependence of $E_g(x,y,T)$ for MCMS. This result shows also that the energy states involved both into absorption and high temperature PL are the same band-like nature. Another question of interest is the nature of low-energy tail both for absorption and IR PL. We have used the fitting of these data by the Urbach law

$$\alpha(\omega)=\alpha_0\exp\left(\frac{\hbar\omega-E_0}{\sigma_0}\right) \tag{7}$$

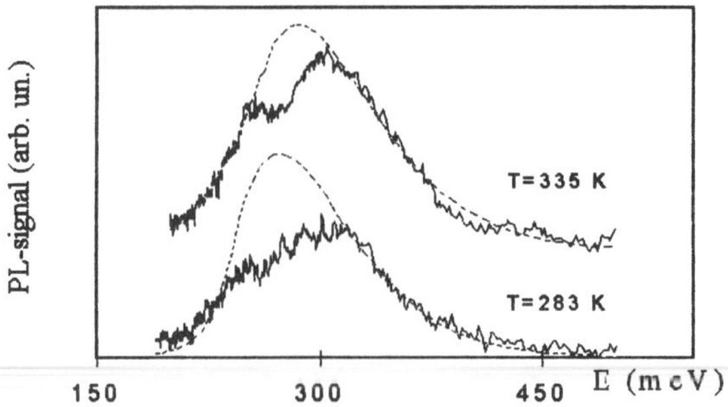


Fig.3 PL spectra of MCMS (x=0.093, y=0.023, $n_e$=1.05*$10^{17}$ $cm^{-3}$).
Dashed lines - the theoretical fit.

According to [13] it is possible to describe self-consistently the normal optical absorption as well as non equilibrium phenomena such as luminescence or pumped absorption. The fitting is shown in Fig.3 by dotted lines. We have established that the Urbach tails both for absorption and IR PL of MCMS are determined by the same set of fitting parameters $\alpha_0, E_0, \sigma_0$. It gives a good ground to declare that the physical processes responsible for the formation of Urbach tail as for the transmission and the emission of light in MCMS involve the same set of energy states of band-like type.

## 4. Temperature dependences of the refractive index and dielectric constant for MCMS

Spectral transmission data allow to calculate also the refractive index n. Fig.4 shows the temperature dependence of refractive index of MCMS alloy with x=0.093, y=0.024 between 5K and 300K. It can be seen the strong temperature dependence of n. The refractive index n decreases from n=4.6 to n=3.8 as the temperature is increased from 5 K to 300 K. The extrapolation of the experimental data to the conventional content of Cd in the $Hg_{1-x}Cd_xSe$ alloys is in a reasonable agreement with the refractive index values reported in[12] and the dependencies of n on temperature and composition obtained experimentally. The behavior observed can be analyzed by means of Kramers-Kronig transformation. The complex dielectric function ε(ω) of mixed crystal includes independent contributions from inter band transitions ( $\Delta\varepsilon_{inter}$ ), intra band ones ($\Delta\varepsilon_{intra}$ ) and phonons ($\Delta\varepsilon_{ph}$ ). Therefore ε(ω) can be written as follows

$$\varepsilon(\omega) = \varepsilon_\infty + \Delta\varepsilon_{inter} + \Delta\varepsilon_{intra} + \Delta\varepsilon_{ph} \qquad (8)$$

General expressions for $\Delta\varepsilon_{inter}$, $\Delta\varepsilon_{intra}$ and $\Delta\varepsilon_{ph}$ are given in [14] in form admitted the numerical calculation. In the case of mixed MCMS crystal all contributions are dependent on the x, y content. It has been shown also in [7],that this crystal has three-mode behavior and plasmon-phonon interaction has to be taken into

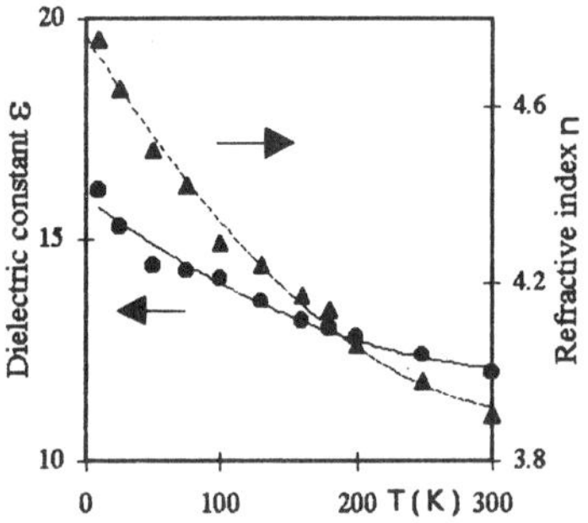


Fig.4 Temperature dependencies of dielectric constant $\varepsilon_\infty$ and refractive index n for MCMS (x=0.095,.y=0.023, $n_e$=8.8*$10^{16}$ $cm^{-3}$).

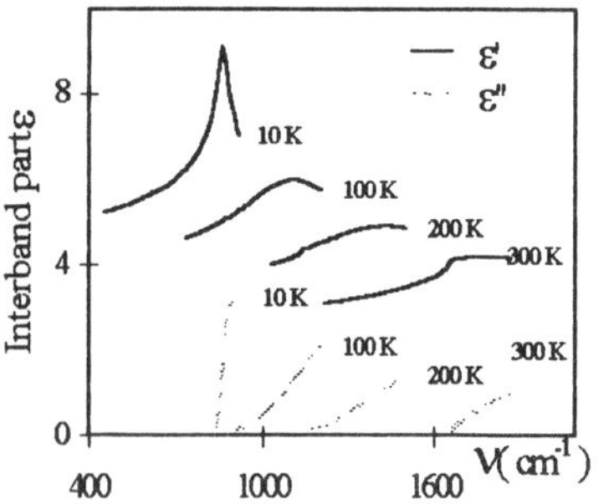


Fig.5 Contributions of interband transitions into dielectric function of MCMS at different temperatures. Crystal parameters as at Fig.4.

account at sufficiently high carrier concentration. Possessing the data for $E_g$ and $E_F$ dependencies on the temperature we can calculate independently each of the contributions. Fig.5 shows the contributions caused by inter band transitions into the real and the imaginary parts of total complex dielectric function at various temperatures. Analogous curves were calculated for the $\Delta\varepsilon_{intra}$ and $\Delta\varepsilon_{ph}$ contributions. For the spectral range of interference fringes in transmission of MCMS (x=0.1, y=0.02) $\Delta\varepsilon_{ph}$ occurs of negligible magnitude. Then we transformed Eq.(8) into dependence

$$\varepsilon'_\infty = n^2 - \varepsilon'_{inter} - \varepsilon'_{intra} \qquad (9)$$

which is justified at $n^2 >> k^2$. The optical dielectric constant $\varepsilon'_\infty$, determined in such manner, is then calculated from the experimental data for $n^2$. Temperature dependence of $\varepsilon'_\infty$ is plotted in Fig.4 being an intrinsic optical characteristic of semiconductors.

## 5. Conclusion

Manganese ions doping the $Hg_{1-x}Cd_xSe$ modify the physical properties of the latter one. Our previous data on Hall measurements have shown that the carrier mobilities do not decrease, at least, at small amount of Mn and are of the same value as those one for the best of $Hg_{1-x}Cd_xSe$ samples [5]. We believe that this effect is the manifestation of stabilizing properties of manganese ions leading to increase of structural perfectness of ternary compounds. Optical absorption near the fundamental edge is well described by the model of direct band-to-band electron transitions. Based on the Kane's band theory the modified scheme of absorption coefficient calculation is suggested. This technique allows to determine the exact energies and the wave functions of the states which are allowed by Kane's secular equation and involved into the transition by the photon of energy $\hbar\omega$. The fitting procedure for the absorption coefficient gives the temperature dependences of $E_g$ and $E_F$ for MCMS from 5K to 300K. Absorption tail is good fitted by Urbach law. At higher density of pumping the IR photoluminescence was detected for the first time in the mercury selenides. The signal of photoluminescence grows in magnitude at temperature increase from the noticeable one at T=150K to well-pronounced peak at room temperature. Investigation of the physical mechanisms of photoluminescence shows that direct band-to-band transitions are responsible for emission. The data, obtained from transmission and IR RL , are then used for determination of principal physical characteristic of semiconductor, $\varepsilon'_\infty$. Summarizing we believe that new data presented herein on the optical properties of semimagnetic semiconducting MCMS makes it an intriguing object for theoretical and experimental investigations as well as for possible applications.

This work was supported by the Volkswagen-Stiftung.

## References

[1] Balcerak R and Brown L 1992 *J. Vac. Sci. Technol. B* **10** 1353-8
[2] Galazka R R 1985 *Journal of Crystal Growth* **72** 364-70
[3] Wall A, Caprile C, Franciosi A, Reifenberger R and Debska U 1986 *J.Vac.Sci. Technol.* A **4** 818-22
[4] Lehoczky S L, Broerman J G, Nelson D A and Whitsett C R 1974 *Phys. Rev.* B **9** 1598-620
[5] Nelson D A , Broerman J G, Summers C J and Whitsett C R 1978 *Phys. Rev.* B **18** 1658-72
[6] Gavaleshko N N , Kriven S I, Mazur Yu I, Paranchich S Yu and Sizov F F 1990 *Acta Physica Polonica* A **77** 287-90
[7] Gavaleshko N N, Kriven S I, Litvinchuk A P, Mazur Yu I and Paranchich S Yu 1990 *Semicond. Sci. Technol.* **5** 307-10
[8] Anderson W W 1980 *Infrared Phys.* **20** 363-72
[9] Kane E O 1957 *Physics Chem.Solids* **1** 249-56
[10] Tomm J W , Herrmann K H and Yunovich A E 1990 *Phys.Stat.Sol.*(a) **122** 11-42
[11] Mazur Yu I, Kriven S I, Tarasov G G and Shevchenko N V 1993 *Semicond. Sci. Technol.* **8** 1187-90
[12] Summers C J and Broerman J G 1980 *Phys. Rev.* B **21** 559-73
[13] Herrmann K H, Happ M, Kissel H, Mollmann K-P, Tomm J W, Becker C R, Kraus M M, Yuan S and Landwehr G 1993 *J. Appl. Phys.* **73** 3486-92
[14] Polian A., Toullec R and Balkanski M 1976 *Phys. Rev.* B **13** 3558-3565

*Inst. Phys. Conf. Ser. No 144: Section 2*
*Paper presented at 7th Int. Conf. on Narrow Gap Semicond. Santa Fe 9–12 January*



# New photoluminescence peculiarities in Mn-enriched $Hg_{1-x}Cd_xTe$ single crystals

Yu I Mazur[1], G G Tarasov[1] and J W Tomm[2]

[1] Institute of Semiconductors Physics, National Academy of Sciences, prospect Nauki 45, 252650 Kiev, Ukraine
[2] Institut fur Physik, Humboldt-Universitat zu Berlin, Sitz: Invalidenstr. 110, Unter den Linden 6, D-10099 Berlin, Germany

**Abstract.** Infrared photoluminescence has been investigated for the quaternary compounds $Hg_{1-x-y}Cd_xMn_yTe$ (MCMT) with mole fractions of x=0.3-0.4 and y=0.01-0.015. Narrow features in the photoluminescence spectra demonstrate the improved structural properties of MCMT as compared with ternary $Hg_{1-x}Cd_xTe$ (MCT), related to manganese influence. Non-monotonic behaviour of the photoluminescence peaks is observed if a magnetic field is applied. The maximum of the PL signal magnitude and the half-width minimum occurs at an intermediate magnetic field. Some evidence in favour of magneto-polaron manifestation is presented.

## 1. Introduction

The problem with stabilising the Hg-Te bonds and producing material more flexible for applications stimulates the transition to quaternary compounds by doping $Hg_{1-x}Cd_xTe$ with rare earth or transition element ions. The quaternary system $Hg_{1-x-y}Cd_xMn_yTe$ offers the possibility to tune the energy gap ($E_g$) with both the Cd or Mn content and to control its magnetic properties by the Mn content. Thus $E_g$ and magnetic properties can be designed independently. The basic technological challenges seem to be the homogeneity of the crystals, the mole fraction uniformity and the optical and transport properties [1].

The physical aspect of the problem is very interesting. The xchange contribution caused by the interaction of the free carriers with 3d electrons localised at manganese ions reconstructs dramatically the energy spectrum of the semimagnetic semiconductors in a magnetic field. Effects as luminescence flaring up [2] or “hole boiling” out of the acceptors [3] have been observed in magnetic fields as well. The energy pinning for optical transitions related to the anti-crossing of the ground state Zeeman component of shallow acceptors with its excited state has been reported for p-MCMT ($0.02<x<0.22$, $0.04<y<0.12$) [4]. Electrical measurements typically indicate a higher crystal homogeneity of quaternary compounds for a given interval of dopant content [4,5]. The photoconductivity spectra of MCMT reveal more peaks for the case of less Mn content, whereas a strong peak broadening is caused by composition fluctuation for larger Mn contents. Our previous data on far infrared transmission [6], reflection [7] and photoluminescence [8] in MCMT single crystals with low Mn content show a trend toward improvement of dynamic lattice characteristics and a decrease of the structural disordering degree. These novel effects and data stimulate new efforts to study the properties of the MCMT system.

Here we present results of photoluminescence (PL) investigations with temperature and magnetic field variation for MCMT crystals with $E_g$ < 400 meV. New peculiarities of the PL behaviour are observed in comparison with those for $Hg_{1-x}Cd_xTe$ of the same $E_g$.

## 2. Experiment

MCMT single crystals with zinc-blend structure were grown by a modified travelling heater technique. All of them were n-type. The wafers had a diameter about 10 mm and a thickness of about 300 μm. Their surfaces were treated by mechanical polishing and subsequent etching in a Br methanol solution. The homogeneity of the Mn and Cd distribution was determined by X-ray microprobe and was found to be constant over the whole sample area within the limits of $\Delta x$ and $\Delta y = \pm 0.003$ units. The MCMT crystals had electron concentrations

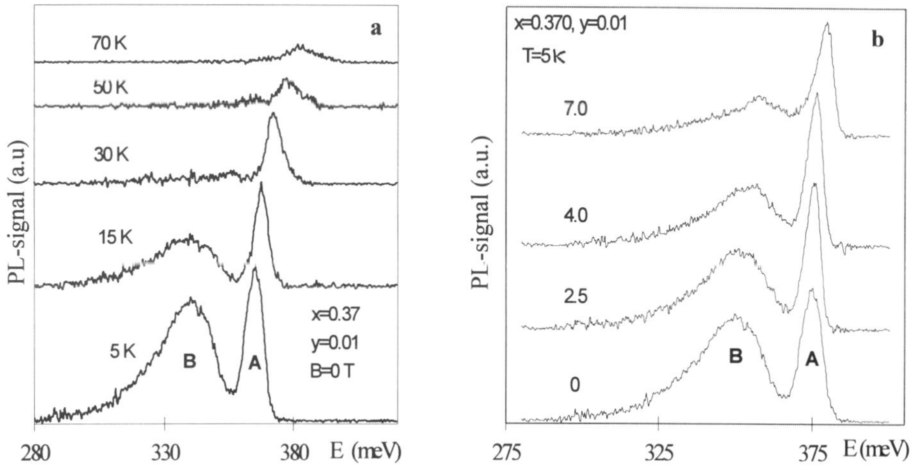


Fig.1. PL spectra of MCMT ( Sample 1) for various a) temperatures and b) magnetic fields.

of about of $n = (1\text{-}5)\times10^{15}$ $cm^{-3}$ and Mn contents in the range $y = 0.010\text{-}0.015$. Varying the cadmium content ($0.3<x<0.4$) $E_g$ was changed from 280 to 380 meV.

The PL was excited by a diode laser emitting 40 mW (cw) at 1.58 eV. A modified double modulation technique was applied. The investigations were performed using a IFS-88 Fourier spectrometer. The spectral resolution was about 0.3 meV. The excitation level was typically about 1-10 W/cm$^2$ (cw). An Oxford Spectromag 4000 system allowed to carry out measurements in the temperature range between 2 to 70 K and at magnetic fields in the 0-7 T range. The Voigt configuration was chosen for the studies in the magnetic field.

Fig.1 shows the temperature and magnetic field induced evolution of the PL spectrum observed for sample 1 having an $E_g$ value of about 375 meV. Note, that the PL spectra observed from all of our samples have two prominent features. The high energy peak (A) occurs at the position close to the energy of band-to-band transition. It is a narrow band with a half-width (FWHM) of 6-8 meV. This is less than that one usually observes in $Hg_{1-x}Cd_xTe$ of corresponding $E_g$. The much broader low energy feature (peak B) follows the A line and is of impurity nature. For the mole fraction of x=0.370 and y=0.011 (sample 1, depicted in Fig.1) we found the following behaviour with magnetic field and temperature variation:

i) If the temperature increases peak B decreases in magnitude and vanishes at T=50 K whereas peak A is detected even at T=70 K.
ii) With increasing magnetic fields both A and B peaks exhibit a blue shift.
iii) The magnitude ratio for A and B peaks changes in a non-monotonic way.
iv) A significant narrowing of the A peak takes place at an intermediate magnetic field.
v) Peak A becomes strongly asymmetrical at the high energy side.
vi) The very weak fine structure in the B line will not be discussed here, since this paper is rather devoted to the edge emission.

Similar tendencies of the temperature and magnetic field behaviour were observed also for the other mole fractions in the $Hg_{1-x-y}Cd_xMn_yTe$ system (x=0.33, y=0.015, $E_g$=336 meV (sample 2); x=0.317, y=0.009, $E_g$=280 meV (sample 3 )).

## 3. Theory

The comparison of MCMT PL data with those for MCT and $Hg_{1-x}Mn_xTe$ (MMT) single crystals show common features as well as some differences introduced by the Mn component in MCT. The exchange interaction of conduction electron spins with localized moments at the $Mn^{2+}$ sites can be taken into account by a Heisenberg-like Hamiltonian

$$H_{sd} = \sum_{\vec{R}_k} J\left(\vec{r} - \vec{R}_k\right)\left(\hat{S}_k \hat{\sigma}\right) \tag{1}$$

where $\hat{S}_k$ and $\hat{\sigma}$ are spin operators of the *k*th manganese ion and the band electron at $\vec{R}_k$ and $\vec{r}$ positions, respectively. Then the 8×8 effective-mass Hamiltonian using the Luttinger-Kohn basis functions [9] with account on Eq.(1) reduces the problem of the calculation of the energy spectrum of semimagnetic semiconductors (SMSC) to the problem of eigenvalues and eigenfunctions of 8×8 Hermitian matrix. In this case the interaction between $\Gamma_6$, $\Gamma_7$ and $\Gamma_8$ states, which are in close energetic proximity is treated exactly while more remote bands are included as perturbation up to $k^2$ by means of effective-mass parameters $\gamma_1$, $\gamma_2 = \gamma_3$, and k. The solution to be sought as a simple vector-column of harmonic oscillator eigenfunctions the approximation of mean field should be applied and Eq.(1) has to be replaced by [10]

$$H_{sd} = y\sigma_z \langle S_z \rangle \sum_{\vec{R}} J(\vec{r} - \vec{R} \qquad (2)$$

where $\langle S_z \rangle$ gives an averaged $S_z$ component of the localized spin and $\sum_k \langle S_{zk} \rangle = N_0 y \langle S_z \rangle$. Being periodical by $\vec{r}$ Eq.(2) allows to modify the Pidgeon-Brown model, introducing s-d and p-d exchange integrals $\alpha = \langle S|J(r)|S \rangle$ and $\beta = \langle X_i|J(r)|X_i \rangle$ (i=1,2,3) as adjustable parameters [9]. The magnetic subband structure for MCMT can be then calculated numerically as a function of $k_z$, the wave-vector component in the direction of the magnetic field B. It is substantially determined by the mean value $\langle S_z \rangle$, related immediately to the magnetization M of SMSC crystal

$$M = -N_0 y g_{Mn} \mu_B \langle S_z \rangle \qquad (3)$$

where $g_{Mn}$ is the gyromagnetic factor of the Mn ion and $\mu_B$ denotes the Bohr magneton. The $\langle S_z \rangle$ value is found to be

$$\langle S_z \rangle = -S_0 B_S(q) \qquad (4)$$

where the Brillouin function $B_S(q)$ for a spin *S*=5/2 is determined by $q = g_{Mn}\mu_B S B / k(T + T_0)$ with $S_0$ and $T_0$ are parameters taking account the cluster existence. In these notations the renormalized effective g-factors can be introduced for the conductivity band ($\Gamma_6$) to be $g_{eff} = g_0 + N_0 y\alpha \langle S_z \rangle / \mu_B B$ and for the valence band ($\Gamma_8$) to be $g_{eff} = g_0 + N_0 y\beta \langle S_z \rangle / 3\mu_B B$ [11]. For the case of the MCMT investigated here $g_{eff}$ exceeds zero and the usual sequence of Landau sublevels has to be observed, $E_a(n) > E_b(n)$, the *a* set corresponds to $\sigma_z = 1/2$ and the *b* set $\sigma_z = -1/2$ spin sublevels. In the Voigt geometry the interband transitions $a_{\Gamma_8}(n) \rightarrow b_{\Gamma_6}(n+1)$ are allowed by symmetry [11].

## 4. Experimental results and discussion

The range of MCMT mole fractions investigated in this study is also interesting for considering the competition of different emission mechanisms. They can be direct band-to-band transitions, free exciton, bound exciton, donor-acceptor pairs and so on. The observation of excitonic contributions in narrow-gap semiconductors is reduced by free carrier screening, impurity-produced random field, strain, etc. Therefore the recombination tends towards intrinsic and extrinsic non-radiative processes. Nevertheless, the PL spectra of narrow-gap MCMT also give some evidence for the presence of excitonic effects. Fig.2 depicts the dependence of the PL peak position for the main line (A) and for transition including defect levels (peak B) for sample 2 on temperature. Two regions with different slopes can be picked out. The high temperature region shows a linear behaviour with a slope near to the value $dE_g/dT = 3.86 \cdot 10^{-4}$ eV/K, calculated in the approximation of virtual crystal. The low temperature part is steeper, $d(\hbar\omega_{max})/dT \rangle dE_g/dT$, indicating a change of radiative recombination mechanism. It could be connected with the carriers or carrier pairs localized in the density-of-states tails, caused by potential fluctuations. However, the relatively small FWHM of the corresponding PL peak (in comparison with MCT crystals of the same $E_g$) gives evidence that these fluctuations are substantially relaxed due to the presence of Mn. A new proof of the manganese spectroscopic manifestation is the $\hbar\omega_{max}$ and FWHM (peak A) behaviour in the magnetic field (Fig.3,4). The position of the A line is well fitted by the calculations within the scheme, presented in Section 3, when the following

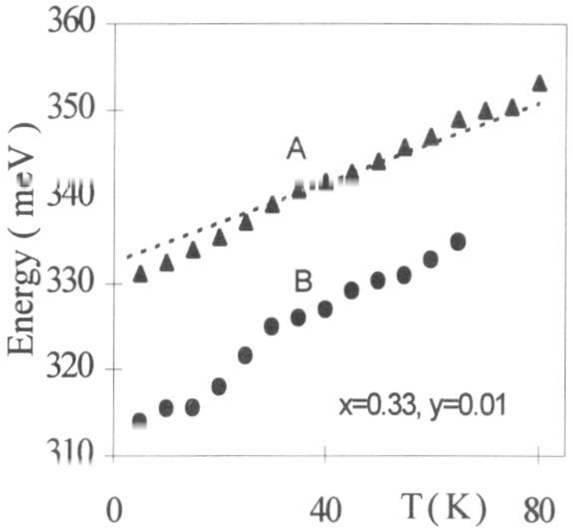


Fig. 2 PL peak positions (A and B) vs. the temperature (sample 2). The dashed line shows $E_g(T)$ in the approximation of virtual crystal.

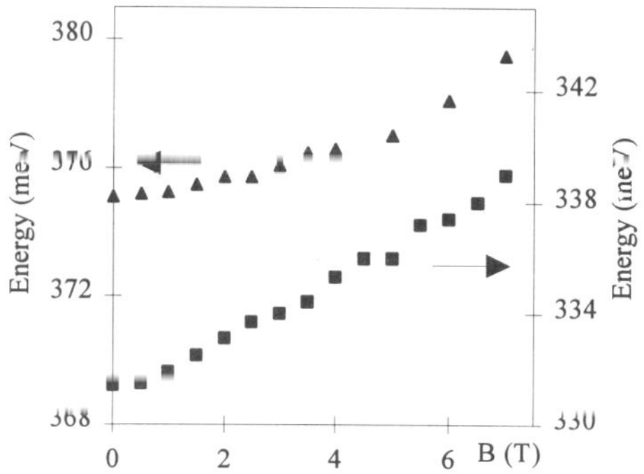


Fig. 3 A peak positions vs. magnetic field for sample 1 (left curve) and sample 2 (right curve).

set of parameters is chosen for the $b_{\Gamma_6}(0)\rightarrow a_{\Gamma_8}(-1)$ transition: $\gamma_1$ =3, $\gamma_2$ =$\gamma_3$ =0.05, $k$ =-1.65, $N_0\beta$=0.9, $N_0\alpha$=-0.45, $S_0$=2.0, $T_0$=0.6 K. With increasing temperature or decreasing $E_g$ (by reducing the cadmium content) the quality of the PL line shape fit within the model of direct band-to-band transitions [12] with account on Urbach tail grows giving the arguments in favour of interband transitions under this conditions. When the temperature decreases or the energy gap opens wider the role of excitonic effects grows, changing the $\hbar\omega_{max}$ (B) dependencies. It should not be excluded that a flare-up of excitonic photoluminescence [13] in the magnetic field happens since the binding energy triples in the 0-7 T range. Calculations as presented in reference [3] show the possibility of magneto-polaron manifestation. The probability of this effect can be found enlarged if one admits the new features of HgTe lattice dynamics, explained by the model of the off-center localization of Hg atom within the unit cell [14] and the strong anharmonity of Hg-Te vibrations. The temperature behaviour of the B line (Fig.3) is non-monotonic revealing different impurity states actual at every temperature range.

Fig. 4 shows a significant narrowing of the A peak from 8 meV to 5.7 meV at B=2 T with subsequent slow growth up to 6.5 at B=7 T (sample 1). This trend is hold for all our samples (cf. curve 2 in Fig.4). The PL feature becomes strongly asymmetrical due to the carrier cooling in magnetic field. The physical source of the PL narrowing in low magnetic fields could be ascribed to the magnetic component of quaternary solid solution. It is well known, that for disordered materials an interconnection of correlation and disorder effects takes place. A magnetic field diminishes the kinetic energy of carriers, promoting their localization. The localization can be the reason of the band narrowing observed.

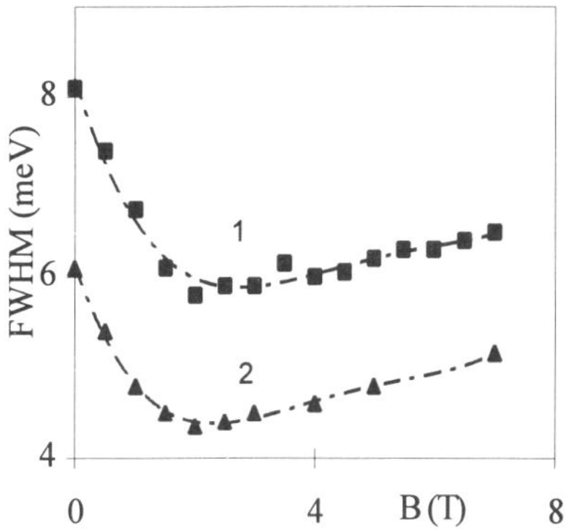


Fig. 4 FWHM versus magnetic field for two samples (1-sample 1, 2-sample 2).

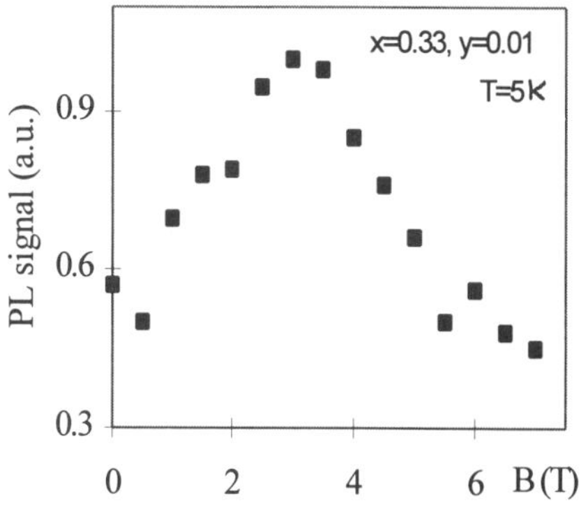


Fig. 5 Dependence of PL strength (peak A) on magnetic field for sample 2.

Additionally two other experimental results have to be mentioned:

(i) The temperature dependence of the spectral position of the PL peak for higher temperatures (linear region, cf. above) was observed to be dependent on the magnetic field. For example sample 1 has $\lambda = d(\hbar\omega_{max})/dT = 3.95 \cdot 10^{-4} eV/K$ for B=7 T versus $\lambda = 3.17 \cdot 10^{-4} eV/K$ for B=0 T. This means that the gap opens with increasing temperature stronger if a magnetic field is present. An alternative explanation of this effect without assuming a magnetic field dependent $E_g(T)$ relation could be given assuming a remarkable excitonic contribution to line A: Since the binding energy ($R_X$) without field amounts about 1.5 meV and at 7 T about 4.5 meV the thermal dissociation of the excitons should produce a stronger shift of the emission line if a magnetic field is applied. Additionally the exciton radius reduction caused by the magnetic field could allow a stronger localization (larger localization energy) e.g. in potential fluctuations. The thermal delocalization of localized excitons of course also increases the value of $\lambda$. The latter effect is expected to occur for lower temperatures (< 35 K), whereas the thermal exciton dissociation is really expected for the linear region mentioned where the thermal energy kT equals $R_X$. A 3 meV increase of the peak shift in the 40-80 K range would deliver $\Delta\lambda=0.75 \cdot 10^{-4}$ eV/K. This number agrees with the observed difference of the slopes (cf. above).

(ii) The PL amplitude exhibits a non-monotonic behaviour in the magnetic field (cf. Fig. 5). The structure observed resembles that for $Hg_{1-x}Mn_xTe$ at high excitation level reported in reference [15]. There the authors assign it to a crossing the magnetic subbands by the Fermi level.

## 5. Conclusions

The PL data presented in this paper show that the quaternary compound MCMT displays many features common for MMT crystals for low concentrations of Mn ions. We observed a non-monotonic behaviour of the PL lines in a magnetic field, a line narrowing for intermediate B fields (B about 2 T) and a stronger temperature dependence of the line positions in the presence of a magnetic field. A comparatively small FWHM of the PL feature for all mole fractions allows to conclude about the relaxing of MCT disordering in the presence of the manganese component.

This work was supported by Volkswagen-Stiftung

## References

[1] Piotrowski T 1985 *J. Crystal Growth* **73** 117-9

[2] Gelmont B L, Golonska R R , Vakhabova E M, Ivanov-Omskii V I, Postolaki I T and Smirnov V L 1986 *Soviet Semicond. Phys.* **20** 73-9

[3] Mycielski A and Mycielski J 1980 *Proc. XIth Int. Conf. Phys. Semicond.* 807-10

[4] Ivanov-Omskii V I, Kharchenko V A and Tsipishka D I 1992 *Sov. Semicond. Phys.* **26** 1728-36

[5] Wall A, Caprile C, Franciosi A, Reifenberger R and Debska U 1986 *J. Vac. Sci. Technol.* A **4** 789-804

[6] Mazur Yu I , Kriven S I , Tarasov G G and Shevchenko N V 1993 *Semicond. Sci. Technol.* **8** 1187-90

[7] Kriven S I, Mazur Yu I, Tarasov G G and Shevchenko N V 1992 *Soviet Phys. Solid State* **34** 10-3

[8] Tarasov G G, Mazur Yu I, Tomm J W , Kriven S I , Lavoric S R and Shevchenko N V 1994 *Proc. SPIE* **2113** 17-26

[9] Leung W and Liu L 1973 *Phys. Rev.* B **8** 3811-6

[10] Galazka R R and Kossut J 1980 In: *Narrow Gap Semiconductors Physics and Applications* **133** 245, Springer-Verlag

[11] Bastard G, Rigaux C, Guldner Y, Mycielski A, Furdyna J K and Mullin D P 1981 *Phys. Rev.* B **24** 1961-70

[12] Anderson W W 1980 *Infrared Phys.* **20** 363-72

[13] Ivanov-Omskii V I, Kokhanovskii S I, Seysyan R P, Smirnov V A, Yukish V A, Yuldashev S U and Efros A L 1983 *Solid State Commun.* **46** 25-8

[14] Kozyrev S P 1993 *Soviet Solid State Physics* **35** 1729-40

[15] Gelmont B L, Ivanov-Omskii V I, Postolaki I.T and Smirnov V A 1986 *Soviet Semicond. Phys.* **20** 508-10

*Paper presented at 7th Int. Conf. on Narrow Gap Semicond. Santa Fe 9–12 January*


# Measurement of surface recombination velocity in HgCdTe by using Photoconductive decay method

**Y C Chung† J K Hong† S B Lee† B I Chang† S U Kim† M J Park† and J M Kim‡**

† Department of Physics, Korea University, AnamDong 5-1, Seongbukgu, Seoul, 136-701, Korea

‡ Agency for Defence Development, Korea

**Abstract.** This paper reports a method of measuring the surface recombination velocity by the photoconductive decay method. The surface recombination velocity measurements on n-type $Hg_{1-x}Cd_xTe$ specimen for composition between 0.21 and 0.25 have been carried out with three different pulse width of applied light. The values for the surface recombination velocity are obtained by fitting the curves to the one dimensional transport equation for chemically etched samples and KOH passivated samples.

## 1. Introduction

The minority carrier lifetime in HgCdTe is an important parameter which determines the detectivity of photodetector[1]. There are several techniques to determine the minority carrier lifetime in semiconductor materials[2]. Among those techniques, photoconductive decay method is most widely used to measure the minority carrier lifetime in HgCdTe[3-5]. These measured values, often, are not bulk lifetime but effective lifetime, because recombination of excess carriers occur not only in bulk but also at surface[6]. Both surface and bulk recombinations are important recombination process. Some of authors have obtained surface recombination velocity by fitting the decay curve to the one dimensional transport equation[7][8]. These decay curves, however, are calculated only when a very short light pulse is applied to the specimen. Usually photoconductive decay signal is hard to be detected without any fluctuation. So It is not easy to get unique value for surface recombination velocity by fitting only one decay curve. Therefore the calculation is expanded to the general case in this study for the various length of pulse width. It has been observed that the photoconductive decay curves changes in their shape by the applied light pulse width. It is after all demonstrated in this paper to extract the parameters for the surface recombination velocity S and the life time $\tau$ which make the best fits to the various decay curves simultaneously.

## 2. Theoretical background

When the light pulse is applied to the n-type HgCdTe, time dependent spartial distribution of minority carriers (holes) are described by the one dimensional transport equation[9]

$$(\frac{\partial}{\partial t} - D_p\frac{\partial^2}{\partial x^2} + \frac{1}{\tau_p})p(x,t) = g(x,t), \tag{1}$$

where $D_p$ is the diffusion coefficient of hole, $\tau_p$ is the minority carrier lifetime, and $p$ is the excess hole concentration. The surface recombination velocity is taken as the boundary condition at both surface ($x$=0,$d$).

$$S = \pm D_p\frac{\partial p}{\partial x}\frac{1}{p}|_{x=0,d} \tag{2}$$

Previous authors[7][8] assume the external hole generation rate as follows,

$$g(x,t) = g_0\delta(t)e^{-\alpha x} \tag{3}$$

where $g_0$ is constant and $\alpha$ is absorption coefficient which is determined by the composition of HgCdTe and wavelength of applied light source. This equation gives the solution only for the case, when the applied light pulse approximated to the delta function. In this case the solution does not give any pulse width dependence. Therefore we revise an assumption for the hole generation rate $g(x,t)$ which is a function of time as follows,

$$g(x,t) = g_0e^{-\alpha x}\zeta(t) \tag{4}$$

where $\zeta(t) = 1$ for $-t_p < t < 0$ and $\zeta(t) = 0$ for $t \geq 0$ and $t_p$ is the pulse width. Solving equation for $-t_p < t < 0$ about several different pulse width $t_p$, we find the initial excess hole distributions when $t = 0$. These distributions means that the longer the pulse width is, the wider the distribution is. The laser light with 780nm of wavelength(the absorption coefficient is taken to be 10000$cm^{-1}$) is applied to the specimen. For $t_p$ = 5$\mu$s and $\alpha$ = 10000$cm^{-1}$, the initial excess hole distribution looks like that of $\alpha \simeq$ 300$cm^{-1}$, when $t_p$ reaches 0. It means that the photoconductive decay curves which are measured with same wavelength but different pulse widths is the same as the curves measured from same pulse width, as far as the pulse width is kept close to be zero, but different wavelength. The initial excess hole distributions with different pulse width are shown in figure 1.

## 3. Experiment

HgCdTe samples are grown by the Travelling Heater Method(THM) in our laboratory[10]. The as-grown samples are p-type, these samples are annealed in Hg vapour for type conversion. The samples used for this demonstration are n-type x=0.21 $\sim$ 0.25 HgCdTe, with carrier concentration n $\simeq 5\times10^{14}$ $cm^{-3}$, $\mu \simeq 1 \times 10^5 cm^2/\text{V}\cdot\text{s}$, at 77K. For the measurements, all samples were etched with 0.5% bromine methanol solution. Some samples are anodically oxidized in KOH solution. The light pulse is applied by GaAs laser ($\lambda$ = 780nm) having rise and fall time less than 10ns. In order to prevent minority carrier sweep out, the bias current less than 5mA is applied. The bias dependence of

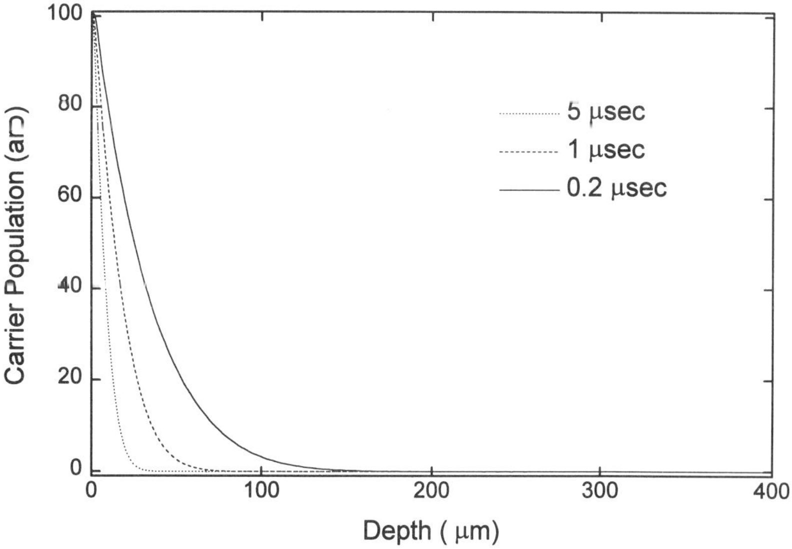


**Figure 1.** Initial excess Hole distribution with different pulse width. The sample thickness is 400 $\mu$m, x=0.23, $\alpha=10^4 cm^{-1}$, $\tau_p=5\mu s$ and S=500cm/s

decay curves is observed to know whether sweep out occurs or not. If it does, the decay does not occur normally. All measurements are taken at 77K. The signals of photoconductive curves are clear enough to measure directly, but the signals from 10 to 100 times are averaged to get more reliable curves. The photoconductive decay curves have been measured for three different types of sample which has different surface condition. One is the freshly etched sample, another is the sample left in air for a week after etched, the other is KOH passivated sample and these photoconductive decay curves are measured for three different pulse width (0.2, 1, 5$\mu$s). In all cases, they appeared to be good pulse width dependence. To insure the accuracy of measurements, the measurements have been repeated, resulting in good agreement within 2%.

## 4. Results and Discussion

For freshly etched samples, the measured and fitted curves are shown in figure 2(a). The shorter the pulse width is, the faster the curve decays as it is predicted. When the specimen is illuminated for 200ns which is too short a pulse width in comparison, the excess holes are created only near surface. So the recombination at surface occurs dominantly, even if excess holes move toward inside the bulk by diffusion during decay. When the 5$\mu$s pulse is applied to the specimen, many holes diffuse to bulk, resulting that the recombination in bulk becomes dominant. In any case, the decay does occur exponentially, because the decay occurs simultaneously both at surface and in bulk. It is found that longer pulse is applied to the specimen, the more decay curve approaches

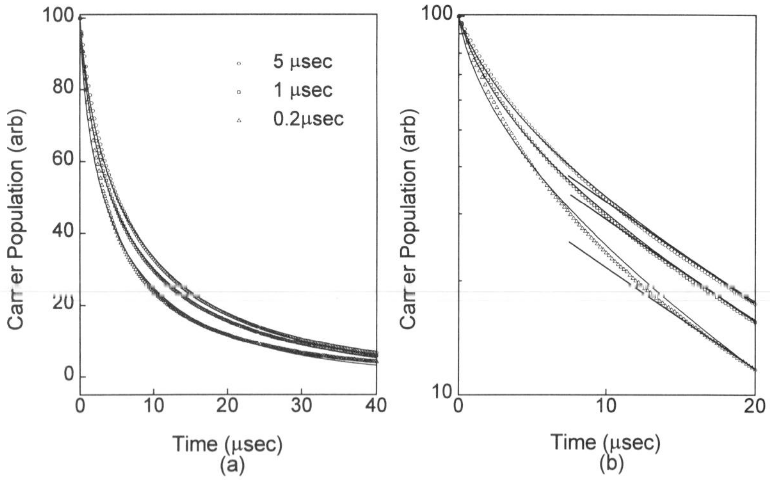


**Figure 2.** The photoconductive decay curves measured on chemically etched sample with different pulse widths(0.2, 1, 5$\mu$s). The solid lines are the best fit by using one dimensional transport equation with S = 800cm/s. (a) The plot in linear scale (b) The plot in log scale

to exponential curve. When the excess hole is generated throughout the sample in the case of the small absorption coefficient, the decay also occurs exponentially. The solid line in Figure 2(a) is the fit of the decay curves. For s=800cm/s, fitted curves are in good agreement with experimental data points. The same curves are shown in log scale at figure 2(b). It is considered that the curves could be divided in two region,a straight region and the bended region, as shown in figure 2(b). In straight region, there are gaps between straight parallel lines. It is found, however, that the faster the surface recombination is, the wider the gap tend to be. Surface recombination velocity could be extracted from the measurements of gap size more accurately. For other freshly etched samples, measured values for recombination velocity range from 500 to 1000cm/s, these values are slightly higher than the values reported by Graft et al.[8], but are similar with the result of Wright et al.[9] Another measurement has been done for the sample left to be exposed in air at room temperature for about a week. Measured curves decay a little faster than that of the previous measurement. Calculated surface recombination velocity is 1200cm/s. It can be concluded that the increase of the surface recombination velocity is due to the native oxide which can be formed by air exposure. The recombination velocities for the KOH passivated samples are also measured. These values range from 200 to 1500cm/s,which are a little higher than the values which was reported by Lopes et al.[7]

By fitting three different curves for different pulse widths, the surface recombination velocities for the specimens can be measured more accurately than the previous reported technique.

## 5. Summary

A more accurate method of obtaining the surface recombination velocity by photoconductive decay technique has been suggested. The surface recombination velocities of thick n-type samples which has been grown by THM have been measured for different pulse widths. For freshly etched samples, the surface recombination velocities have been measured to the range from 500 to 1000cm/s, and for anodically oxidized samples the values ranged from 200 to 1500cm/s.

## Acknowledgments

We wish to thank S. H. Shin and J. S. Woo at Department of Korea University, Seoul, Korea for their helpful disscussion and for their corporation in measuring the electrical properties of the specimens. This work is supported partially by the Basic Science Research Program, Ministry of Education in 1994.

## References

[1] Pratt R G, Hewett J, Capper P, Jones C L and Quelch M J 1983 *J. Appl. Phys.* **54** 5152–7

[2] Lopes V C, Syllaios A J and Chen M C *Semicond. Sci. Technol.* **8** 824–41

[3] Bajai J, Shin S H, Pasko J G and Khoshnevisan 1983 *J. Vac. Sci. Technol.* **A1(3)** 1749–51

[4] Zucca R, Edwall D D, Chen J S and Johnston S L 1991 *J. Vac. Sci. Technol.* **B9(3)** 1823–8

[5] Fastow R and Nemirovsky Y 1989 *Appl. Phys. Lett.* **55** 1989

[6] Kinch M A, Brau M J and Simmons A 1973 *J. Appl. Phys.* **44** 1649–73

[7] Lopes V C, Wright W H and Syllaios A J 1990 *J. Vac. Sci. Technol.* **A1(8)** 1167–70

[8] Graft R D, Carson F F, Dinan J H, Boyd P R and Longshore R E 1983 *J. Vac. Sci. Technol.* **A1** 1696

[9] Wright W H, Lopes V C, and Syllaios A J 1989 *SPIE Proc.* **1107** 212

[10] Lee H Y, Kim Y S, Chung Y C, Oh J W, Kim B J, Kim S U and Park M J 1991 *J. Kor. Appl. Phys.* **4** 45–50

*Inst. Phys. Conf. Ser. No 144: Section 2*
*Paper presented at 7th Int. Conf. on Narrow Gap Semicond. Santa Fe 9–12 January*



# EFFECT OF THE DISTRIBUTION OF PHOTOGENERATED CARRIER ON HgCdTe PHOTOCONDUCTIVITY

Yongsheng Gui, Junhao Chu,
National laboratory for Infrared Physics, Shanghai Institute of Technical Physics, Chinese Academy of Sciences, Shanghai 200083, China

Limin Gui
Department of electronic technique and science, East China Normal University, Shanghai 200062, China

**Abstract**: According to the one-dimension longitudinal model and some latest empirical formulae of intrinsic absorption coefficient in $Hg_{1-x}Cd_xTe$, the distribution of photogenerated carriers has been calculated by taking into account the surface recombination velocity, the lifetime of the minority carrier in bulk material, sample thickness and the wavelength of incident light. It has been found that the distribution of minority carrier depends sensitively on the wavelength, thickness, as well as the surface recombination velocity. The obtained results are used to analyze the performance of $Hg_{1-x}Cd_xTe$ photoconductive detectors for 8-14μm at 77K. The optimum parameters of the device chips have been obtained by studying the relationship of low frequency response voltage versus sample thickness and light wavelength. The theoretical predictions agree with the experimental data of detectors.

## 1. INTRODUCTION

The theory of responsivity in $Hg_{1-x}Cd_xTe$ photoconductive detector has been given in the literature, based mainly on one-dimensional modes [1-9]. It appears that the actual devices are not one-dimensional and may have contacts that deviate from ohmicity. Two-dimensional model and three-dimensional model for $Hg_{1-x}Cd_xTe$ photoconductor responsivity have been developed by Kolodny [10] and Oda [11].

In this paper the distributions of photogenerated carriers have been calculated by taking account of the surface recombination velocity, bulk lifetime, sample thickness and the wavelength of incident light according to the one-dimension longitudinal mode and some latest empirical formulae of intrinsic absorption coefficient in $Hg_{1-x}Cd_xTe$. The obtained results are used to analyze the performance of $Hg_{1-x}Cd_xTe$ photoconductive detectors for 8-14μm and 3-5μm at 77K. The optimum thickness of the device chips is about 4-6μm for the samples with composition from x=0.2 to x=0.3 when the surface recombination velocity is 200cm/s.

## 2. THEORETICAL MODEL FOR THE PHOTOCONDUCTIVE RESPONSE

A typical photoconductor fabricated by thin bulk film, rectangular in shape with length l, width w, and thickness d in Fig. 1, is considered. The device is biased by applying a constant electric field E between terminals a and b. Assuming that a uniform signal photon flux Φ is incident on the photoconductor area $A=w\times l$. To derive an expression for voltage responsive, one must first find the spatial distribution of photogenerated electrons and holes in the semiconductor.

For simplicity we take a one-dimensional approach neglecting the effect of recombination at the metal-semiconductor interface. The distribution of photogenerated carriers is also assumed to be uniform along the x and z directions. The photogenerated excess carriers' concentration $\Delta p(y) = p(y) - p_0$ within the semiconductor is governed by ambipolar transport. The ambipolar continuity equation in one-dimensional case can be written as[12]

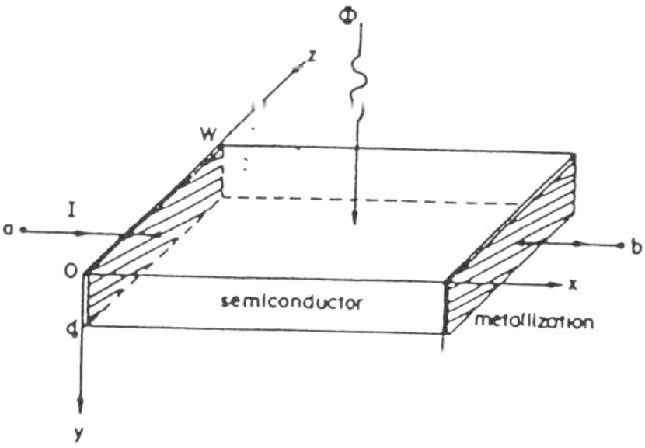


Fig. 1 Photoconductor with surface area w×l exposed to a uniform photon flux density Φ

$$D_0 \frac{\partial^2 \Delta p}{\partial y^2} - \frac{\Delta p}{\tau} + g(y) = 0 \tag{1}$$

where

$$D_0 = \frac{(n_0 + p_0) D_e D_h}{p_0 D_e + n_0 D_h} \tag{1A}$$

$$\tau = \tau_0 \{1 - \frac{\tau_0}{\tau_a}[1 - \exp(\frac{\tau_0}{\tau_a})]\} \tag{1B}$$

$$\tau_a = L/(\mu_0 \mathrm{E}) \tag{1C}$$

$$\mu_0 = \frac{(n_0 - p_0)\mu_e \mu_h}{n_0 \mu_e + p_0 \mu_p} \tag{1D}$$

$$g(y) = \frac{\alpha \Phi (1-R)(e^{-\alpha y} + R\, e^{-2\alpha d} e^{\alpha y})}{1 - R^2 e^{-2\alpha d}} \tag{!E}$$

In the above expressions g(y) stands for the volume generation rate of electron-hole pairs. The multi-reflection in the photoconductor has been considered. $R$ is the reflection coefficient. The thermal equilibrium electron and hole concentrations are $n_0$ and $p_0$, $\mu_e$ and $\mu_h$ denote carrier drift mobilities, $D_e$ and $D_h$ denote the respective diffusion coefficients. The carrier lifetime of photogenerated carriers $\tau_0$ was calculated from the Auger limited lifetime expression. Considering the sweepout when the detectors are operated at high electric field, lifetime $\tau_0$ should be replaced by efficient lifetime $\tau$ shown in equation (1B).

The intrinsic absorption coefficient in $Hg_{1-x}Cd_xTe$ is complicated, here the latest empirical formula derived by Chu et al[16] is used.

$$\alpha = \alpha_0 (\alpha_g / \alpha_0)^{[(E-E_0)/(E_g - E_0)]}, \quad \text{when } E < E_g \tag{1F}$$

$$\alpha = \alpha_g \exp[\sqrt{A(E - E_g)}], \quad \text{when } E \geq E_g \tag{1G}$$

where the meanings of E, $E_0$, $E_g$, $\alpha$, $\alpha_g$ and A are explained in reference (16).

The boundary conditions are

$$D_0 \frac{\partial \Delta p(0)}{\partial y} = S_1 \Delta p(0) \tag{2}$$

$$D_0 \frac{\partial \Delta p(d)}{\partial y} = -S_2 \Delta p(d) \tag{3}$$

where $S_1$ and $S_2$ are the front and rear surface recombination velocity, respectively.

The solution of equation (1) is

$$\Delta p(y) = Ash(y/L) + Bch(y/L) + \tau g(y)/(1-\alpha^2 L^2) \tag{4}$$

where

$$A = [-k_1 n_3 + k_2 n_2 ch(d/L) + n_2 sh(d/L)]/n_1 \tag{4A}$$

$$B = [-n_3 - n_2 ch(d/L) - k_2 n_2 sh(d/L)]/n_1 \tag{4B}$$

$$n_1 = (k_1 + k_2)ch(d/L) + (1 + k_1 k_2)sh(d/L) \tag{4C}$$

$$n_2 = (\alpha L + k_1)C + (k_1 - \alpha L)D \tag{4D}$$

$$n_3 = (-\alpha L + k_2)C\exp(-\alpha d) + (k_2 + \alpha L)D\exp(\alpha d) \tag{4E}$$

$$C = \frac{\tau}{1-\alpha^2 L^2}\frac{\alpha(1-R)\Phi}{1-R^2\exp(-2\alpha d)} \tag{4F}$$

$$D = \frac{\tau}{1-\alpha^2 L^2}\frac{\alpha(1-R)R\Phi\exp(-2\alpha d)}{1-R^2\exp(-2\alpha d)} \tag{4G}$$

$$k_{1(2)} = S_{1(2)}\tau/L \tag{4H}$$

In the above $L = \sqrt{D_0 \tau}$ is the diffusion length.

The average number of excess photogenerated carriers' concentration is given by

$$\Delta\bar{p} = \frac{1}{d}\int_0^d \Delta p(y)dy \tag{5}$$

At low frequency modulation, the voltage responsivity is directly proportional to the average excess photogenerated carriers' concentration given by equation (5), and can be written as

$$R_v = \frac{\lambda}{hc}\frac{\mathrm{E}}{\Phi}\frac{\Delta\bar{p}(\mu_e + \mu_h)}{(n\mu_e + p\mu_h)W} \tag{6}$$

## 3. RESULTS AND DISCUSSION

The base material(n-type) concentration is $5\times10^{14}\,cm^{-3}$ . The width and length of the detector have been taken as 50 and 100μm, respectively. The mobilities of electrons and holes were calculated from the expression given by Rosbeck[13]. The values of $E_g$ and $n_i$ were obtained from the semi-empirical expression of Chu et al[14,15].

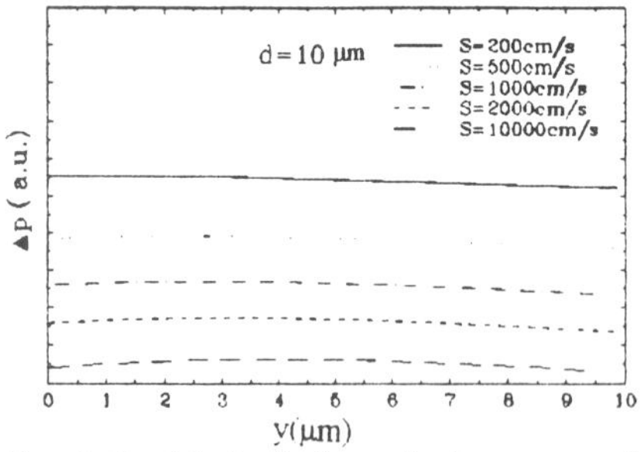


Fig. 2 Spatial distribution of photogenerated carriers for various surface recombination velocities. The thickness of sample is 10μm.

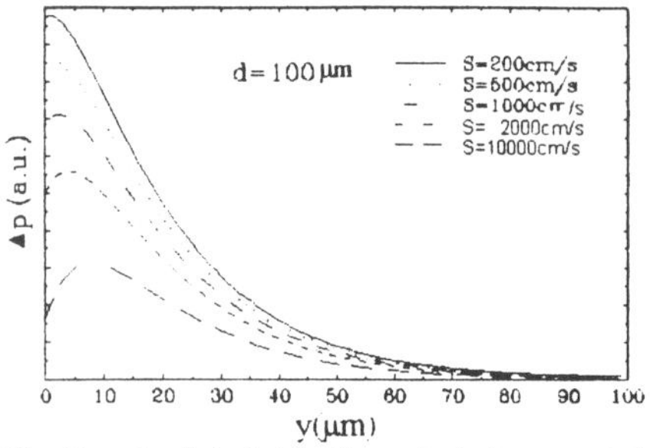


Fig. 3 Spatial distribution of photogenerated carriers for various surface recombination velocities. The thickness of sample is 100μm.

The distribution of photogenerated carriers is calculated and shown in Fig. 2 and Fig. 3 for the samples with the thicknesses of 10μm and 100μm, respectively, under the incident radiation fluxes of wavelength $\lambda_g$ corresponding to the photon energy equal to $E_g$(when x=0.2, $\lambda_g$=13.1μm). It is seen that the excess photogenerated carriers distribution is uniform if the thickness is smaller than the diffusion length L and photogenerated carriers' concentration reduce rapidly if the thickness is larger than L. Because the light is absorbed within a few micrometer under the surface, the photogenerated carriers diffuse from the surface according to the exponential declining, the average excess photogenerated carriers' concentration decreases when the thickness is larger than L. From equation (6), the voltage response is proportional to the average excess photogenerated carriers' concentration, so the responsivity also decreases while the thickness increases. The dependence of responsivity on the carrier recombination velocity at detector front and rear surfaces (here assuming $S_1$ and $S_2$ are equal to S) is also shown in Fig. 2 and Fig. 3. Because those photogenerated carriers recombined in the surface do not contribute to the voltage responsivity. The greater the surface recombination velocity, the worse the detector performance is.

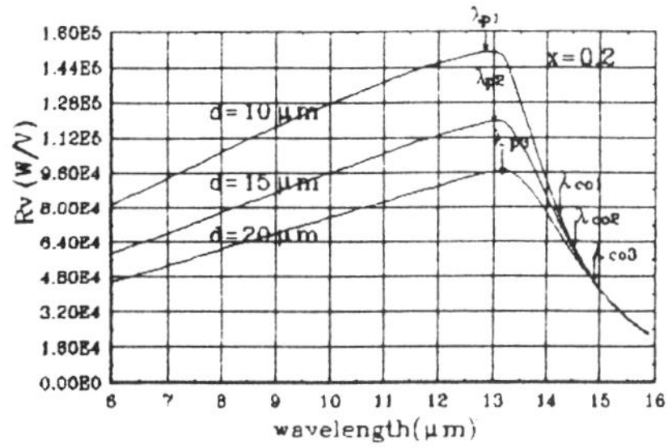


Fig. 4 Frequency response for $Hg_{0.8}Cd_{0.2}Te$ with various thicknesses

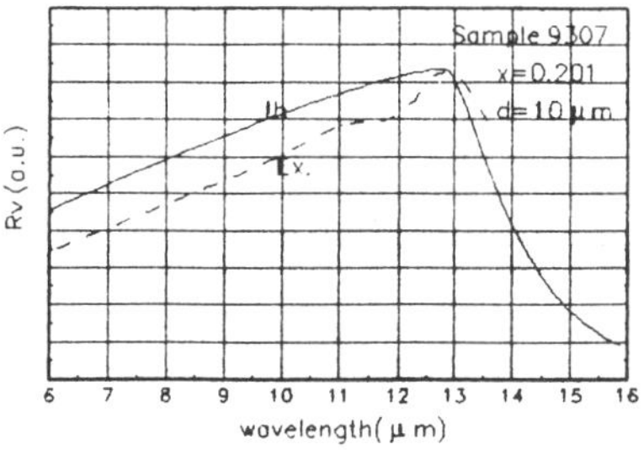


Fig. 5 Photocurrent spectra of sample 9307. Solid line refers to calculated result while dashed line to experimental data.

Fig. 4 shows the voltage responsivity for the photoconductors with different thicknesses. Fig. 4 also shows the variation of voltage responsivity as a function of photon wavelength for a HgCdTe photoconductor with composition (x=0.2). The peak positions of voltage responsivity (also the cutoff wavelengths) are shifted to longer wavelength, that is, from 12.9μm (when d=10μm) to 13.2μm(when d=20μm) when the thickness increases. Because the absorption coefficient for long wavelength is smaller than that of short, the photon has to travel farther before being absorbed. The number of long wavelength photons is more than that of short in a uniform Φ. The theoretical results agree with the experimental data of sample 9307 (x=0.201, $N=7.9\times10^{14}\,cm^{-3}$), as shown in Fig. 5, the peak position of is also near $\lambda_g$ .

The thickness of the detector should be optimized for practical operating condition, especially for spectral response and operating temperature. Fig. 6 shows the relationship between the thickness of photoconductive detector and the voltage responsivity at 77K for the samples with the mole fraction x=0.2, x=0.25 and x=0.3, respectively. In the calculation, the optimum thickness $d_{op}$ is about 4-6μm for the three different mole fractions. The voltage

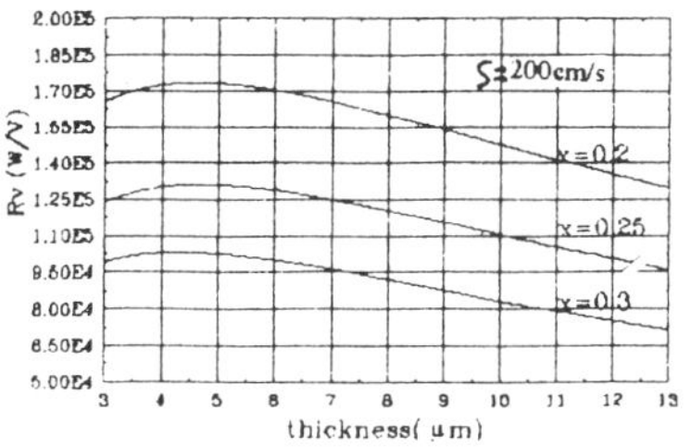


Fig. 6 Relationship of response with sample thickness for HgCdTe.

responsivity increases more than 20% if the thickness is chosen as 4-6μm not the typical thickness 10μm for application.

The calculated results show that the optimum thickness $d_{op}$ will increase as the surface recombation velocity S increase. The optimum thickness $d_{op}$ is 5.5-7.5μm for S=400cm/s and 7-9μm for S=800cm/s.

On the basis of the above discussion it is concluded that the performance of HgCdTe photoconductive detector can be deduced by using the spatial distribution of photogenerated carriers. The thickness of photoconductive detectors affects the peak position and the intensity of spectral response. The optimum thickness is about 4-6μm for a wide range of composition.

## 4. REFERENCES

[1] R.M. Broudy and V.J. Mazurcyk 1981 *Semicinductors and Semimetals* (edited by R.k. Willardson and A.C. Beer), Vol.18 Academic Press, New York (1981) p 157
[2] A.Rogalski and J. Piotriwski 1988 *Prog. Quant. Electron.* **12** 87
[3] R.L. Willams 1968 *Infrared Phys.* **8** 337
[4] M.A. Kinch and S.R. Borrello 1975 *Infrared Phys.* **15** 111
[5] S.R. Borrello, M.A. kinch and D. Lamont 1977 *Infrared Phys.* **17** 121
[6] M.A. Kinch, S.R. Borrello and A. Simmons 1977 *Infrared Phys.* **17** 127
[7] Rachana Kumar, Sudha Gupta, Vishnu Gopal and K.C. Chhabra 1991 *Infrared Phys.* **31** 101
[8] Vishnu Gopal, Rachana Kumar and K.C. Chhabra 1991 *Infrared Phys.* **31** 435
[9] A. Jozwikowska, K. Jozwikowski and A. Rogalski 1991 *Infrared Phys.* **31** 543
[10] A.Kolodny and I. Kidron 1982 *Infrared Phys.* **22** 9
[11] N.Oda 1988 *Proc.SPIE* **915** 20
[12] Mckelvey T.P. 1966 *Solid State and Semidconductor Physics* (Harper & Row) p 320
[13] J.P. Rosbeck, R.E. Starr, S.L. Price and R.J. Riley 1982 *J. Appl. Phys.* **53** 6430
[14] J.H. Chu et al 1987 *Chinese Progress in Physics* **7(3)** 280
[15] J.H. Chu et al 1983 *Chinese Journal of Infrared Research* **2** 24 1
[16] J.H. Chu et al 1994 *J. Appl. Phys.* **74(2)** 1234

*Inst. Phys. Conf. Ser. No 144: Section 3*
*Paper presented at 7th Int. Conf. on Narrow Gap Semicond. Santa Fe 9–12 January*



# The study of a resonant Cr donor in PbTe, PbSe, $PbTe_{1-y}Se_y$ and $Pb_{1-x}Sn_xTe$

**E Grodzicka, W Dobrowolski, T Story, Z Wilamowski and B Witkowska**

Institute of Physics, Polish Academy of Sciences, Al. Lotników 32/46, 02–668 Warszawa, Poland

**Abstract.**
We summarize transport and magnetic properties of lead based IV-VI semiconductors containing resonant Cr donor, with concentrations of dopants corresponding to the Fermi level being stabilized at the resonant state.

## 1. Introduction

Stoichiometrically grown IV-VI crystals are known of the presence of metal vacancies, that act as acceptor centers, and therefore demonstrate *p*-type conductivity. The Cr dopant substitutes for a host cation and takes the form of $3d^3/3d^4$ $Cr^{3+/2+}$ ion. As a result of an interplay between the quantity of $Cr^{3+}$ donor centers and the concentration of metal vacancies with an increasing Cr concentration takes place the change of the conductivity type. The Fermi level shifts towards the conduction band. Provided that the concentration of the impurities is high enough it reaches the Cr resonant level superimposed on the conduction band continuum and is being pinned there.

## 2. Experimental results and discussion

We have studied following sets of samples: $Pb_{1-x}Cr_xTe$ for $0 \leq x \leq 0.01$; $Pb_{1-x}Cr_xSe$ for $0 \leq x \leq 0.01$; $Pb_{1-x}Cr_xSe_{0.1}Te_{0.9}$ for $0 \leq x \leq 0.03$; $Pb_{1-x-y}Cr_xSn_yTe$ for $x \approx 0.005$ and $0 \leq y \leq 0.3$; $Sn_{1-x}Cr_xTe$ for $x = 0.005$ and $x = 0.02$; All the samples studied were grown by the Bridgman method. The chemical composition of samples was checked by X-ray energy dispersive fluorescence analysis revealing a relatively strong gradient of chromium along an ingot. Transport properties (Hall effect and conductivity) in the temperature range $4 - 300$ K, the low temperature ($1.6 - 4.2$ K) ac magnetic susceptibility, and EPR measurements were carried out. We studied both as grown and annealed samples.

### 2.1. $Pb_{1-x}Cr_xSe$ and $Pb_{1-x}Cr_xTe$

The results of our transport measurements of $Pb_{1-x}Cr_xSe$ are fairly similar to our earlier investigations of $Pb_{1-x}Cr_xTe$ [1], that agree well with available literature data [2]. Both can be summarized as follows. As grown samples with the small concentration of Cr turn out to be $p$-type. The $n$-type conductivity appears as a result of doping with Cr (for PbTe in amount $N^*_{Cr} \geq 1.5 \cdot 10^{19}$ cm$^{-3}$), as shown in Fig.1. The increase of the Cr content leads at first to an increase in the electron concentration proportional to the amount of Cr introduced. From $x^* = 0.004$, which is equivalent to $N^*_{Cr} \approx 7 \cdot 10^{19}$ cm$^{-3}$ in $Pb_{1-x}Cr_xSe$ and $x^* = 0.002$, corresponding to $N^*_{Cr} \approx 3 \cdot 10^{19}$ cm$^{-3}$ in $Pb_{1-x}Cr_xTe$, the electron concentration remains constant at the level of $n = 1.8 \cdot 10^{19}$ cm$^{-3}$ or $n = 1.25 \cdot 10^{19}$cm$^{-3}$ respectively. Introducing more Cr does not change the concentration of electrons. What's more, the isothermal annealing in the atmosphere of Pb or in the atmosphere of Se or Te is not a way to influence the concentration value. The conductivity of $n$-type samples has a typical metallic character. The carrier concentration is only weakly temperature dependent. In samples from the saturation region (with the Fermi level trapped by the resonant Cr state), a decrease of $n$ with an increase in $T$ is mostly related to a downward temperature shift of the impurity level. No thermal activation of electron concentration is observed. The crystals of $Pb_{1-x}Cr_xSe$ are characterized by rather high electron mobilities of the order of $10^5$ cm$^2$/Vs at $T = 4.2$ K. The mobility of our $Pb_{1-x}Cr_xTe$ samples (especially after annealing) shows even greater values, characteristic of PbTe crystals, in which the high dielectric constant ($\varepsilon \approx 1300$) represents screening out the charged impurity Coulomb potentials that normally limit the mobility in semiconductors at low temperatures. The highest mobility value observed in our samples was $2 \cdot 10^6$cm$^2$/Vs. That was seen in the sample only very slightly doped with Cr ($x = 0.00028$). Samples in the mixed valence regime (i.e., with concentrations of Cr greater than $1.3 \cdot 10^{19}$ cm$^{-3}$) have lower mobilities $\mu \simeq 7 \cdot 10^4$cm$^2$/Vs, however, they still correspond to the best mobilities observed in the heavily doped PbTe (c.f.[3]). The lack of any dependence of the electron concentration upon the level of chromium doping in the mixed valence regime (when the Fermi level is pinned to the chromium state) enabled us to determine the position of the Cr resonant energy. For PbSe the saturation value of $n$ corresponds to the Fermi level position 125 meV above the bottom of the conduction band, which we have calculated taking into account the band structure parameters after [4]. A similar value was obtained for PbTe. Because these values are very close each other it is apparent that the whole band-offset between PbSe and PbTe results from the difference in the values of the energy gaps and the valence bands positions and can be estimated, according to [5], to be equal to 43 meV.

The origin of the peak of the density of localised states, which limits the electron concentration, can be determined from magnetic measurements. The $Cr^{2+}$ ion in an octahedral coordination is characterized by a singlet ground state and the Van Vleck magnetism, i.e. by a temperature independent magnetic susceptibility and a lack of EPR signal. Differently, $Cr^{3+}$ and $Cr^{+}$ that are characterized by the Curie magnetism, i.e. by a magnetic susceptibility which obeys the Curie-Weiss law and a presence of EPR signal.

For each of our $n$-type $Pb_{1-x}Cr_xSe$ and $Pb_{1-x}Cr_xTe$ samples a single EPR line has been detected. The position of the line is isotropic which indicates that the investigated impurity center is of the cubic symmetry. The $g$-factor value is substantially lower than 2.

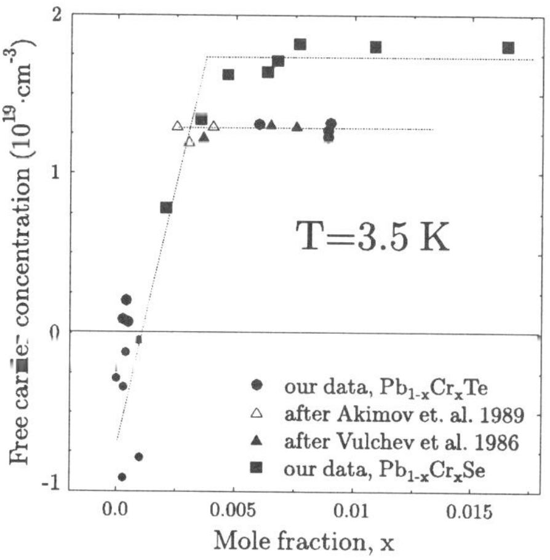


**Figure 1.** Conduction electron concentration in $Pb_{1-x}Cr_xSe$ and $Pb_{1-x}Cr_xTe$ as a function of Cr molar fraction. The lines are only a guide to the eye.

For the lowest impurity concentrations $g = 1.96$ and its value decreases proportionally with the increasing electron concentration to $g = 1.93$ [6]. Since, for $Cr^+$ with singlet orbital ground state $^6S$, the expected value of $g$-factor is very close to $g = 2.00$, the observed value of $g$-factor indicates that EPR signal originates from $Cr^{3+}$ ($3d^3$, $S = 3/2$) in the lattice site. Such conclusion is confirmed by measurements of the ac susceptibility, which shows the Curie magnetism. An example of the temperature dependence of the magnetic susceptibility is shown in Fig. 2.

The experimental results presented above indicate that, Cr in PbTe and PbSe creates a resonant donor state. Doping with Cr leads at first to a proportional increase of the electron concentration (see Fig. 1). The increase of the chromium content above some critical value does not change the electron concentration which stabilizes at a constant level. This behavior results from the selfionization process ($Cr^{2+} \Rightarrow Cr^{3+} + e_{band}$). At relatively small quantities the chromium impurity forms $Cr^{3+}$ state and the conduction electron concentration is proportional to the amount of Cr dopant. Above the characteristic chromium concentration $N^*_{Cr}$, doping with Cr introduces only $Cr^{2+}$ ions. These ions are electrically neutral, the conduction electron concentration becomes constant, independent of further doping since the Fermi level reaches the Cr level and is trapped there. The selfionization process stops. In the high doping regime only a part of Cr impurities are ionized, the chromium ions coexist in two charge states: $Cr^{2+}$ and $Cr^{3+}$ creating this way a mixed valence regime. The observed paramagnetic resonance is due to $Cr^{3+}$ions, while the electrically neutral $Cr^{2+}$ ions contribute to the magnetization as Van Vleck ions. The lack of the response of the electron concentration to the annealing procedure is also the consequence of the resonant character of Cr donor. Changing the number of electrically active native defects one changes only the number of $Cr^{2+}$ ions that have to be autoionized to shift the Fermi level to the energy level of $Cr^{2+/3+}$. For Cr concentration being large enough the Fermi level will remain pinned to the energy level of $Cr^{2+/3+}$ and carrier concentration will be constant.

Finally it is important to underline that despite the fact that the impurity level

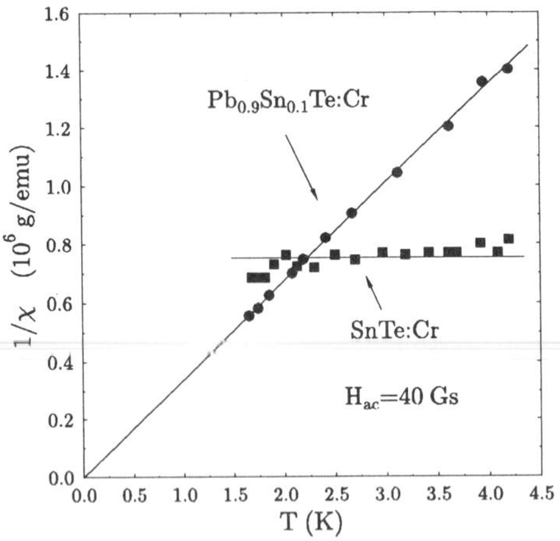


**Figure 2.** Inverse magnetic susceptibility in $Pb_{1-x-y}Cr_xSn_yTe$ and $Sn_{1-x}Cr_xTe$ vs temperature

$Cr^{3+/2+}$ is degenerate with the conduction band, the $Cr^{3+}$ state is very weakly hybridized with the conduction band states [7]. The EPR linewidth of 60 G, which is observed in the mixed valence regime [8], means that the lifetime, in particular the lifetime determined by the finite lifetime of the impurity in $Cr^{3+}$ state, is longer than $10^{-9}$s. In other words, it means that the rate of the self-ionization process $Cr^{2+} \Rightarrow Cr^{3+} + e_{band}$ is smaller than $10^9$ $s^{-1}$ and the level broadening caused by the hybridization with the conduction band states is smaller than 1 $\mu$eV.

## *2.2. $Pb_{1-x-y}Cr_xSn_yTe$ and $Sn_{1-x}Cr_xTe$*

Starting from PbTe and increasing tin content in $Pb_{1-0.005-y}Cr_{0.005}Sn_yTe$ alloy one observes a decrease of the electron concentration. For approximately $y \geq 0.1$ the as grown crystals are $p-$type. It is caused probably by the lack of Cr to fill all the empty states in the valence band. The number of these states are directly measured by the concentration of the conducting holes in the $Pb_{1-y}Sn_yTe$ matrix and are known to increase rather fast with an increasing tin content. SnTe crystals present $p-$type conductivity with very high concentrations of carriers $p \simeq 10^{21}$ $cm^{-3}$ and very low mobilities $\mu \simeq 100$ $cm^2/Vs$. Magnetic properties of $Pb_{1-x-y}Cr_xSn_yTe$, $y = 0.3$, are typical for Curie paramagnets and are, in our opinion, due to the spin magnetic moments of $Cr^{3+}$ ions. A different behavior is observed in $Sn_{1-x}Cr_xTe$. There is practically no temperature dependence of the magnetic susceptibility. It indicates the van Vleck type of the magnetic behavior which may be attributed to the $3d^4$ states of $Cr^{2+}$ ions that are expected to have non-magnetic ground state in the octahedral crystal field. One may attempt to interpret the experimental findings presented above in a model assuming the shift of the $Cr^{2+}$ level from the conduction band in the case of $Pb_{1-x}Cr_xTe$ and $Pb_{1-x-y}Cr_xSn_yTe$ to the valence band in the case of $Sn_{1-x}Cr_xTe$. It will induce the decrease of the electron concentration and changes in the relative concentrations of $Cr^{2+}$ and $Cr^{3+}$ ions. In $Sn_{1-x}Cr_xTe$ the level of $Cr^{2+}$ is expected to be located below the Fermi level of the crystal. Only magnetically and electrically "neutral" (with respect to the host cations) $Cr^{2+}$ ions are expected to be present in this case.

In conclusion, we have observed that Cr acts as a resonant donor both in $Pb_{1-x}Cr_xTe$ and in $Pb_{1-x-y}Cr_xSn_yTe$ with small tin content $y < 0.2$.

**Acknowledgments**

Work supported in part by U.S.-Poland Maria Curie-Skłodowska Joint Fund II through grant No.PAN/NSF-92-113.

**References**

[1] Grodzicka E, Dobrowolski W, Kossut J, Story T and Witkowska B 1994 *J. Crystal Growth* **138** 1034–9
Story T, Grodzicka E, Witkowska B, Górecka and Dobrowolski W 1992 *Acta Phys. Pol. A* **82** 879–81

[2] Akimov B A, Dmitriev A V, Khokhlov D R and Ryabova L I 1993 *phys. stat. sol. (a)* **137** 9–55
Akimov B A, Verteletskiĭ P V, Zlomanov V P, Ryabova L I, Tananaeva O I and Shirokova N A 1989 *Fiz. Tekh. Poluprov.* **23** 244–9 (*Sov. Phys. Semicond.* **23** 151)
Vulchev V D and Borisova L D 1987 *phys. stat. solidi (a)* **99** K53–6

[3] Nimtz G and Schlicht B 1985 *Narrow Gap Semiconductors, Springer Tracts in Modern Physics, Vol.98* (Editor Höhler G, Berlin: Springer Verlag) 1–98

[4] Bauer G, Pascher H and Zawadzki W 1992 *Semicond. Sci. Technol.* **7** 703–23

[5] Langer J M and Heinrich H 1985 *Phys. Rev. Lett.* **55** 1414–6
Caldas M J, Fazzio A and Zunger A 1984 *Appl. Phys. Lett.* **45** 671–3
Zunger A 1985 *Phys. Rev. Lett.* **54** 849–50

[6] Wilamowski Z and Story T 1995 *J. Magn. Magn. Mater.* **140-144** in press

[7] Wilamowski Z, Świątek K, Dietl T and Kossut J 1990 *Solid State Commun.* **74** 833–7

[8] Wilamowski Z, Story T and Jantsch W 1995 *Acta Phys. Pol. A* **87** 189–92

*Inst. Phys. Conf. Ser. No 144: Section 3*
*Paper presented at 7th Int. Conf. on Narrow Gap Semicond. Santa Fe 9–12 January*



# Anomalous transport properties of $Sn_{1-x}Gd_xTe$

**W Dobrowolski, E Grodzicka, T Story, Z Gołacki and R R Gałązka**

Institute of Physics, Polish Academy of Sciences, Al. Lotników 32/46, 02–668 Warszawa, Poland

**Abstract.**
The low temperature Hall mobility in $Sn_{0.955}Gd_{0.045}Te$ with $p^* \approx 3 \cdot 10^{20}$ cm$^{-3}$ achieves considerably higher values than in samples with different composition $x$ or hole concentration $p^*$. This result may be understood assuming that the presence of Gd creates a resonant donor state degenerate with the valence band continuum.

## 1. Introduction

SnTe forms with GdTe semimagnetic substitutional solid solutions $Sn_{1-x}Gd_xTe$ up to $x = 0.1$ [1,2]. The X-ray analysis has shown that $Sn_{1-x}Gd_xTe$, in the investigated composition range, crystallizes in the rock-salt structure. There were no second phase inclusions detected. The mixed crystals, similarly to the host material SnTe, are always $p$–type with a metallic type of conductivity. This property is attributed to a large concentration of Sn vacancies, that behave as acceptor centers with zero activation energy. The number of these vacancies (and thereby the concentration of carriers) can be, within one order of magnitude, controlled by an isothermal annealing.

The investigation of the magnetic susceptibility, the electron paramagnetic resonance (EPR), and the Hall effect led to a conclusion that Gd substitutes Sn as an electrically active $Gd^{3+}$ ion possessing a free ion-like permanent magnetic moment of $7\mu_B$ [2, 3]. The study of magnetic properties has shown a weak antiferromagnetic Gd-Gd interaction which was ascribed to the superexchange via Te [2]. Recently a new, unforeseen attribute of $Sn_{1-x}Gd_xTe$ mixed crystals was reported [4,5]. Their magnetic properties critically depend on a carrier concentration and a composition. A rapid increase of the paramagnetic Curie temperature $\Theta/x$ for samples with hole concentrations $p = p^* \approx 3 \cdot 10^{20}$cm$^{-3}$ and Gd contents $x < x^* \approx 0.045$ as compared to crystals with $p \neq p^*$ was revealed.

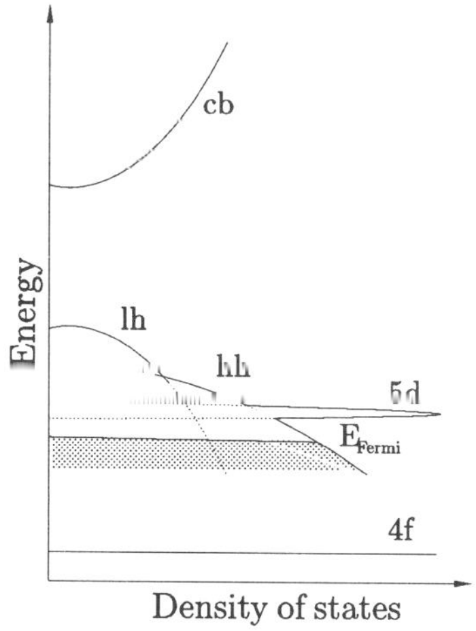


**Figure 1.** Schematic diagram of energy levels and densities in $Sn_{1-x}Gd_xTe$

## 2. Model

These magnetic properties of $Sn_{1-x}Gd_xTe$ were explained on the basis of the model assuming that the Gd $5d$ energy state is degenerate with the valence bands. In such circumstances, depending on a relative energetic position of the Fermi level and the Gd donor resonant state, Gd ions may exist in one of the two charge states: as singly ionized $Gd^{3+}$ $(4f^7)$ donors or as electrically neutral $Gd^{2+}$ $(4f^7 5d^1)$ ions. Figure 1 presents a schematic representation of a density of states function end energy levels for the case when the Fermi energy is below the resonant state energy. In such a situation at low temperatures all the ions will be in $Gd^{3+}$ $(4f^7)$ state.

According to Ref. [4] and [5] the exchange interaction between $Gd^{3+}$ $(4f^7)$ ions may be considered as an intra ion $4f - 5d$ exchange followed by an interatomic $5d - 5d$ exchange coupling via conducting holes. The interatomic part of this interaction is proportional to the density of states at the Fermi level and inversely proportional to the energy difference $E_{Fermi} - E_{5d}$ [6]. Therefore, $\Theta/x$ is resonantly enhanced when $E_{Fermi}$ approaches $E_{5d}$. Thus, in samples with $x < 0.05$, the $5d$ energy state is degenerate both with the heavy hole $\Sigma$-band and with the light hole $L$-band, as shown in Fig. 1. In samples with $x > 0.05$ the $5d$ state lies higher in the valence band and is degenerate only with the light hole $L$-band. The Fermi level may then raise higher in the valence band, but will be still too far from the Gd $5d$ level to cause the resonant enhancement of the exchange coupling.

The aim of the presented paper is to verify the model described above by the investigation of transport effects.

## 3. Experimental results

The $Sn_{1-x}Gd_xTe$ samples used in our experiments were grown by the Bridgman technique. The chemical composition and homogeneity were checked by an electron microprobe. The measurements of the electric conductivity and the Hall effect were performed

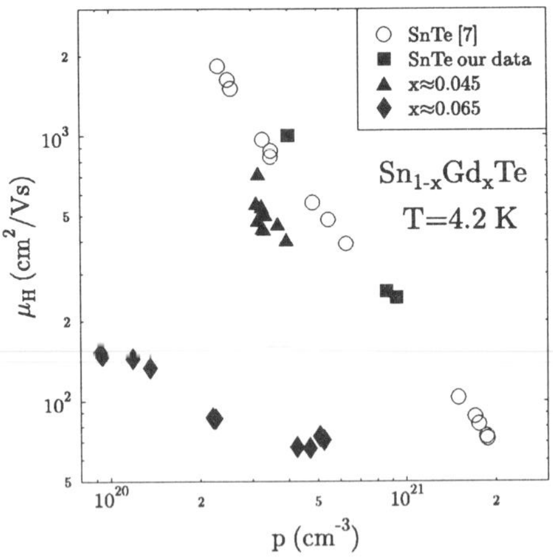


**Figure 2.** The Hall mobility vs apparent carrier concentration ($p = 1/eR_H$) for SnTe and $Sn_{1-x}Gd_xTe$

in the crystals of $Sn_{1-x}Gd_xTe$ ($x \leq 0.065$) over the temperature range $T = (1.5 - 300)$K. The hole concentration varied in the range $p = (7 \cdot 10^{19} - 1.5 \cdot 10^{21})$ cm$^{-3}$. In order to reduce or to increase their carrier concentration the samples were isothermally annealed at about 700 C, in Sn or Te atmosphere, respectively.

The dependence of the Hall mobility data as a function of an apparent carrier concentration in SnTe after Ref. [7], and our data for two sets of $Sn_{1-x}Gd_xTe$ samples ($x \approx 0.045$ and $x \approx 0.065$), is shown in Fig. 2.

The mobilities observed for the samples with Gd content $x \approx 0.065$ are considerably lower than those for pure SnTe with analogous holes concentrations. This is an evident indication of an additional scattering channel which emerge with an admixture of the Gd to the host material. Taking into account that Gd was established to be a substitutional donor, the likely candidates seem to be: scattering by the screened Coulomb potential and scattering by the short range potential of the ionized Gd. The first case may be, however, excluded because of the large value of the dielectric constant in SnTe [8].

Qualitatively and quantitatively different results were obtained for samples with the lower portion of Gd, $x = 0.045$. A sharp increase in the carrier mobility value takes place when the concentration of carriers decreases. The highest mobilities found in samples with $p = 3.3 \cdot 10^{20}$cm$^{-3}$ are only slightly smaller than in the case of SnTe samples with comparable carrier concentrations, and 6 times higher than the mobility values found in the $Sn_{0.935}Gd_{0.065}Te$ samples. The measurements of the Hall effect revealed also another interesting feature of $Sn_{1-x}Gd_xTe$ system. Namely, the concentration of carriers in $Sn_{0.945}Gd_{0.045}Te$ could not be reduced below $p \approx 3 \cdot 10^{20}$cm$^{-3}$ regardless of the fact that it was possible for the samples with $x = 0.065$ as well as for the crystals of SnTe or $Sn_{1-x}Mn_xTe$ [9].

Figures 3a and 3b present Hall mobility and apparent carrier concentration data as a function of the temperature between 3.5 and 295 K.

The mobility of $Sn_{0.935}Gd_{0.065}Te$ samples is practically temperature independent. A

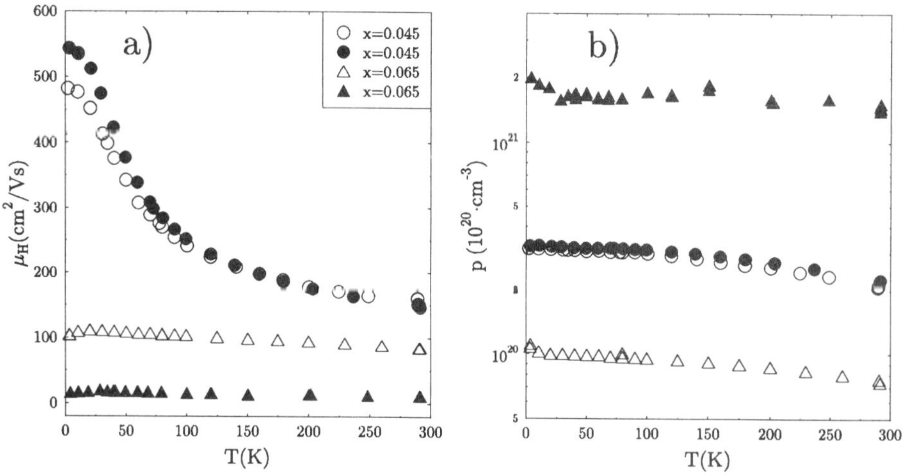


**Figure 3.** The Hall mobility (a) and apparent carrier concentration (b) vs temperature for $Sn_{1-x}Gd_xTe$ samples

different behavior exhibits the material with the lower Gd content. The curves obtained for the samples with $x \approx 0.045$ resemble qualitatively those for SnTe and its alloys with PbTe or MnTe. The rapid increase in the mobility with decrease in the temperature is due to the acoustic phonon and polar phonon scattering. The flatness of the $\mu$ vs temperature curve below 30 K is generally ascribed to the predominance of the defect scattering in this temperature region. The apparent carrier concentration in both sets of samples only slightly depends on the temperature.

Therefore, in respect of transport properties, and in accordance with their chemical composition, we can distinguish two groups of samples.

## 4. Discussion

The experimental data described above can be qualitatively explained within the frames of the model proposed in Ref. [4] and [5]. The analysis is to be limited to the low temperature range (i.e. $T < 30$ K). Moreover we assume that in the investigated composition range Gd has a negligible influence on the band structure of SnTe, except for creating the resonant state which is degenerate with the valence bands continuum. In the alloys studied the active imperfections are metal (i.e. tin) vacancies and substitutional Gd atoms. While the metal vacancies are always doubly negatively charged scattering centers, the Gd atoms, depending on the position of the Fermi level relative to the Gd resonant level, may appear or in neutral, or in the singly ionized state.

For $x = 0.045$ we evaluate that the energy level of Gd resonant $5d$ state is located about 0.2 eV below the top of the valence band (cf Fig.1). Thus it is degenerate with both $\Sigma$ and $L$ valence bands. This energetic position leads to a situation that the concentration of $Gd^{3+}$ ions (the most effective charged scattering centers) decreases down to zero when the Fermi energy approaches the $5d$ resonant level (in other words carrier concentration approaches $p^*$). The observed increase of the carrier mobility is due to the fact that a relatively strong scattering on the core potential of ionized Gd donors is during that process exchanged by the weaker alloy scattering (neutral impurity) potential.

For samples with $x \approx 0.065$ the $5d$ energy state is located closer to the top of the valence band, being degenerate only with light hole $L$-band. In the whole carrier concentration range obtainable by the isothermal annealing substitutional Gd atoms are ionized to the $Gd^{3+}$ state. Thus, additionally to the scattering on the core potential of tin vacancies, holes may scatter on the core potential of Gd ions. This explains the low mobilities encountered in these crystals. The relaxation time for the ion core potential scattering has been calculated by Ravich et. al [10]. Their model accounts for the effects of non-parabolicity of the matrix-elements as well as for the density of states. Using formula from Ref. [10] it was possible to evaluate the core potential as $U_v(core) \approx 3 \cdot 10^{-34}\mathrm{erg} \times \mathrm{cm}^{-3}$.

In conclusion, we found that the model successfully explaining the magnetic properties of $Sn_{1-x}Gd_xTe$ (see Ref. [4] i [5]) is also sufficient to account for all the observed transport effects.

## Acknowledgments

Work supported in part by U.S.-Poland Maria Curie-Skłodowska Joint Found II through Grant No. PAN/NSF-92-113 and by Committee for Scientific Research under Grant 2 2314 91 02.

## References

[1] Gołacki Z and Jasiołek G 1989 *J. Mat. Sci.* **24** 2999–3001

[2] Górska M, Anderson J R, Kido G, Green S M and Gołacki Z 1992 *Phys. Rev.* **B 45** 11702–8

[3] Urban P and Sperlich G 1975 *Solid State Commun.* **16** 927–30

[4] Górska M, Story T, Arciszewska M, Grodzicka E, Gołacki Z and Łusakowski A 1995 *Acta Phys. Pol. A* **87** 197–200

[5] Story T, Górska M, Arciszewska M, Grodzicka E, Gołacki Z and Gałązka R R 1995 *J. Magn. Magn. Mater.* **140-144** in press

[6] Łusakowski A, to be published

[7] Sagar A and Miller R C 1962 *Proc. 6-th Int. Conf. on Phys. of Semicond., Exeter, 1962* (London: Institute of Physics and Physical Society) 653–8

[8] Nimtz G and Schlicht B 1985 *Narrow Gap Semiconductors, Springer Tracts in Modern Physics, Vol.98* (Editor Höhler G, Berlin: Springer Verlag) 1–117

[9] Story T, Karczewski G, Świerkowski L and Galazka R R 1990 *Phys. Rev.* **B 42** 10477–87

[10] Ravich Yu I, Efimova B A and Tamarchenko V I 1971 *phys. status solidi (b)* **43** 11–33

*Inst. Phys. Conf. Ser. No 144: Section 3*
*Paper presented at 7th Int. Conf. on Narrow Gap Semicond. Santa Fe 9–12 January*



# Stimulated emission in multi-valley lead salts with star degeneracy lifted by strain and magnetic fields

**J W Tomm, K H Herrmann, M Mocker, T Kelz and T Elsässer**
Humboldt-Universität zu Berlin, Institut für Physik, Unter den Linden 6,
D-10099 Berlin, Germany
**R Klann**
Paul-Drude-Institut, Hausvoigteiplatz 5-7, D-10117 Berlin, Germany
**B V Novikov**
NIIF, Dept. Sol. State Phys.,Ulyanovskaya 1, 198904 St. Petersburg, Russia,
**V G Talalaev and V E Tudorovskii**
NIIKI GOI, 188537 Sosnovy Bor, Russia
**H Böttner**
Fraunhofer-Institut für Physikalische Meßtechnik, Heidenhofstr. 8
D-79110 Freiburg / Br., Germany

**Abstract.** The results of a study of the threshold of stimulated emission in optically pumped PbSe layers are reported. Direct time resolved measurements and data from quasi steady state luminescence measurements were used to define this parameter 'threshold'. A qualitative explanation of the observed threshold shifts due to both external parameter strain and magnetic field is given.

## 1. Introduction

Binary PbSe and PbTe are common basic materials for the preparation of infrared (IR) diode lasers, e.g. as substrates and as wells in laser structures employing quantum confinement effects. Furthermore bandstructure and material parameters as energy gap $E_g$ and effective masses m* are sufficiently well-known for the binary materials and thus make them good model substances. This paper discusses the relation between time resolved absorption measurements on a ps timescale and quasi-steady-state IR photoluminescence (PL) experiments carried out with the same sample set. Stimulated emission thresholds after optical pumping are compared with the onset of a fast recombination channel after ps excitation. We further demonstrate that the multi-valley bandstructure of the samples causes stimulated emission threshold shifts caused by the break-up of star degeneracy due to the action of strain or a magnetic field.

## 2. Experimental

A sample set consisting of 9 PbSe layers was investigated. The layers were grown on freshly cleaved (111) oriented $BaF_2$ substrates (about 1 $cm^2$) by molecular beam epitaxy. At 77 K typical carrier concentrations (p and n-type) were in the order of low $10^{17}$ $cm^{-3}$, whereas the mobility was between 5000 $cm^2/(Vs)$and 30000 $cm^2/(Vs)$. The layer thicknesses ranged between 0.85 μm and 6.3 μm.

For time-resolved measurements a pump and probe technique ($\tau_{pulse}$=8 ps) was applied, allowing us to monitor temporally resolved absorption changes in the spectral range between 0.1 eV and 0.4 eV covering the spectral region around the absorption edge. Further information on this experimental technique is provided in [1,2].

IR PL experiments in backscattering geometry were performed employing a Q-switched Nd:YAG laser excitation ($\hbar\omega$=1.17 eV, $\tau_{FWHM}$=80 ns, repetition rate 0.1 to 1 kHz) source and a

$Hg_{1-x}Cd_xTe$ photodiode as detector. A grating monochromator delivered the spectral resolution. The samples were mounted in a Spectromag 4000 system, allowing measurement in magnetic fields up to 7 T.

## 3. Results and discussion

### *3.1 Fast recombination mechanism at high excitation densities*

IR PL measurements in the spectral range of the band edge emission in PbTe and PbSe as discussed in this paper are usually performed using pulsed excitation by Q-switched CO- or Nd:YAG-lasers. The relatively high excitation levels in this excitation regime deliver spectra of *stimulated emission.* The commonly observed appearance of stimulated emission after non-equilibrium carrier creation is a special feature of lead salts and is caused by both the small $E_g$ of these materials and the absence of excitonic recombination as efficient non-equilibrium carrier annihilation mechanism. Thus the absolute value of the rate of spontaneous emission from these materials is relatively small when a separation of the quasi-Fermi-levels by $E_g$ is reached and stimulated emission starts to display significant contributions.

This peculiarity was already discussed in very early papers on PL in lead salts e.g. by Gureev et a. [3] and Yunovich et al. [4]. Later the gain decay time $\tau_g$ ('instantaneous lifetime of stimulated emission') was estimated from quasi-steady state measurements using stationary rate equations. 30-200 ps are given by Tomm et al. [5] for $Pb_{0.91}Sn_{0.09}Te$ layers and Valeiko et al. [6,7] report on 10-100 ps for $Pb_{0.89}Sn_{0.11}Te$/PbTe superlattices. Schwarz et al. [8] reported the first direct measurement of $\tau_g$ by time resolved measurements and found a value of $\tau_g$ = 60 ps. Later systematic direct studies of fast recombination channels for PbTe and PbSe for 70 and 300 K and various doping levels of the samples were reported in a series of papers by Klann et al. [1,2,9]. Fig. 1 a displays the temporal change of the absorption

$$\Delta A = -\log (T/T_0) \sim (\alpha(n_0,p_0)-\alpha(n_0+\delta n,p_0+\delta p)) \qquad (1)$$

(full circles corresponding to the right scale) of a PbSe sample having a carrier concentration $p_0 = 1.6 \times 10^{17}$ cm$^{-3}$ for an excitation photon energy of $\hbar\omega$ = 365 meV and a test photon energy of $\hbar\omega_{test}$ = 282 meV at 300 K.

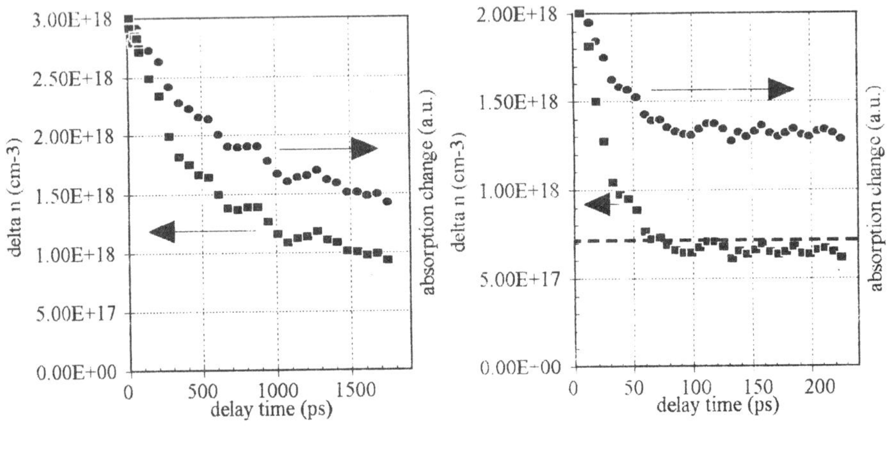


Fig. 1 Absorption change (full circles, right scales) and calculated nonequilibrium carrier density (full squares, left scales) for PbSe/$BaF_2$ at a) 300 K (left) and b) 70 K (right).

Table 1: Instantaneous lifetime data ($\tau_g$) obtained from $\delta n(\tau)$ in Fig. 1a and Fig 1 b

| temperature (K) | $\delta n$ ($cm^{-3}$) | instantaneous lifetime ($\tau_g$ ) |
|---|---|---|
| 70 | $1.5\times10^{18}$ | 40 ps |
| 70 | $7\times10^{17}$ | about 10 ns |
| 300 | $2\times10^{18}$ | 0.8 ns |
| 300 | $1\times10^{18}$ | 2 ns |

Note, that the PbSe room-temperature gap is $E_g$= 277 meV. Consequently the test beam monitors the temporal evolution of the population of the bands with non-equilibrium carriers ($\delta n$), i.e. the dynamic Burstein-Moss-effect. Note, that the initially excited non-equilibrium carrier concentration was $\delta n = \delta p = 3 \times 10^{18}$ $cm^{-3}$. From the change of the absorption the change of $\delta n$ can be directly obtained from (1). For the calculation of $\alpha$ we used the set of formulas given by Anderson [10]. With the knowledge of $p_0$, $\delta p$, $E_g$, $\hbar\omega_{test}$, the temperature, the assumption of bipolar generation ($\delta n = \delta p$) and $n_i^2 = n_0 \times p_0$ we calculated the temporal evolution of $\delta n(\Delta A)$. These data are represented by the full squares in Fig. 1 a. From this data directly instantaneous lifetimes can be obtained by the differentiation of the curves. The same procedure was made for the 70 K experimental data. Since the energy gap at T=70 K is reduced to about 165 meV excitation and test quantum energies were reduced. Instantaneous lifetimes $\tau(\delta n)$ are collected in Table 1. For 300 K the instantaneous lifetimes may be understood in the framework of Auger recombination (cf. reference [9]) whereas for 70 K obviously stimulated emission is the main recombination mechanism.

### *3.2. Infrared photoluminescence*

The onset of the fast recombination with $\tau_g$ = 40 ps is considered as threshold of stimulated emission. From Fig. 1 b the threshold excitation density may be directly determined to be $\delta n = 7 \times 10^{17}$ $cm^{-3}$ (cf. dashed line). Fig. 2 depicts the PL signal-versus-excitation power dependence for the same sample (full triangles) along with data for other samples. From a linear plot of the data points the threshold of stimulated emission is found to be at about 10 $kWcm^{-2}$ (cf. arrow in Fig. 2). The corresponding generation rate of $G = 9\times10^{25}$ $cm^{-3}$ $s^{-1}$ and $\delta n = 7\times10^{17}$ $cm^{-3}$ give

$$\tau = \delta n/G = 8 \text{ ns} \quad (2)$$

as effective lifetime *below* the threshold. This number well agrees with the 10 ns for the instantaneous lifetime below the threshold estimated from picosecond spectroscopy (cf. Table 1) and is confirmed by PL decay and thus consistency is demonstrated.

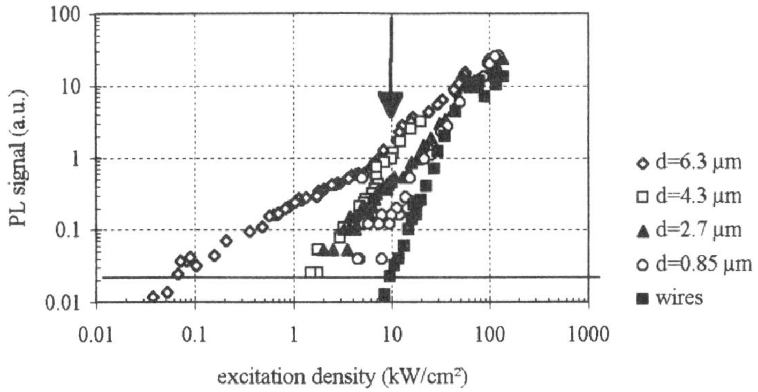


Fig. 2 PL signal versus excitation density for $PbSe/BaF_2$ at T = 77 K with the layer thickness as parameter. The full line marks the PL signal where the signal-to-noise ratio equals one.

However, Fig. 2 also indicates, that the layer thickness and the surface-to-bulk ratio (see also data from wires; cf. reference [11]) are important parameters which influence (i) the threshold and (ii) the quantum efficiency of stimulated emission (slope in Fig. 2). In the following we will limit the discussion to (i). The obvious layer thickness dependence of the *threshold of stimulated emission* in Fig. 2 is influenced by a number of parameters such as surface recombination, confinement effects and strain. In order to exclude degradation effects due to repeated cooling cycles the data in Fig. 2 were obtained from samples which had not been cycled before. External parameters as strain or a magnetic field can influence the threshold via various mechanisms: Both influence directly the threshold since Eg (or the 'effective gap' $Eg^{eff}$) is shifted and thus $\delta n$ necessary to separate the quasi Fermi levels by $Eg^{eff}$ changes. Furthermore an additional break up of star degeneracy influences the lifetime [12] and thus also the threshold. In order to discriminate between both mechanisms we performed additional magnetoluminescence measurements. Fig. 3 shows the PL signal-versus-excitation power dependence for an nearly unstrained sample ($Eg^{eff}$ = 144.5 meV after more than 15 cooling cycles) with and without magnetic field influence. The general decrease of the threshold in the magnetic field is caused by a magnetic field induced lifetime enhancement (cf. [12,13]). The difference between Faraday (B parallel to $\langle 111 \rangle$ direction since the PL was realized in backscattering geometry) and Voigt geometry (B in the layer plane) one has to consider the multi valley bandstructure of the lead salts: An estimation of the strain induced shifts as well as the shift of the Landau levels can be made on the basis of the model given in reference [14]. The appropriate direction dependent masses $m^{eff}$ as a measure for the expected magnetic field shifts can be estimated from the main axis of the ellipses with the main axes $m_a$ and $m_b$ from

$$m^{eff} = (m_a \times m_b)^{1/2} \quad . \tag{3}$$

These cross-sectional ellipses are given from the intersection of the four constant energy ellipsoids with the plane in which the cyclotron orbit lies. For Faraday geometry only two different cross sections (two valley groups) are found and consequently only two masses are obtained. In reference [15] we reported magnetoluminescence spectra obtained under these conditions and have experimentally shown, that only the energetically lower group (3 valleys), i.e. that where the heavier mass is to be taken into account, significantly contributes to the PL and defines $Eg^{eff}$. This statement is even more valid at the threshold of stimulated emission which is discussed here. The relevant 'heavier' mass amounts $m^{eff} = 1.29 \times m_\perp$, where $m_\perp$ is given by the two similar small main axes of the constant energy ellipsoids. For Voigt geometry usually all 4 different mass values are found to be different, however, the heaviest and consequently the relevant one (only one valley) of them can be determined and amounts $m^{eff} = 1.34 \times m_\perp$. Thus the corresponding $Eg^{eff}$-

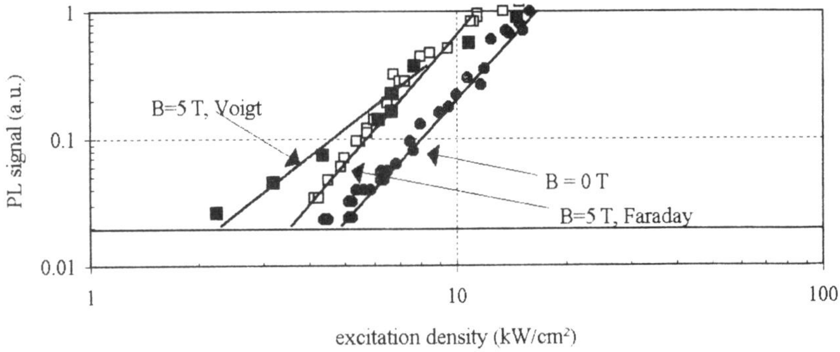


Fig. 3 PL signal versus excitation density for a 6 μm thick $PbSe/BaF_2$ sample at T = 10 K with a magnetic field of B = 5 T in Voigt and Faraday configuration and at B = 0 T.

difference for Voigt and Faraday geometry at B=5 T is marginal. Thus in this experiment it is not the $E_g^{eff}$ shift but the redistribution of the carriers mainly in three valleys (Faraday geometry) or mainly in one valley (Voigt geometry) which causes the remarkable threshold shift between both geometries. Consequently we observed in Voigt geometry the lowest and for B=0 (4 equivalent valleys) the largest threshold of stimulated emission.

## 4. Conclusions

We present a study of the threshold of stimulated emission in optically pumped PbSe layers. We find consistent experimental data from picosecond gain decay measurements and thresholds of quasi steady state PL spectra obtained from the same samples.

Thresholds of stimulated emission for different samples have been measured and were found to depend systematically on the layer thickness. One of the parameters influencing the threshold - the break up of the star degeneracy if a magnetic field or strain is applied - was analyzed with magnetoluminescence and found to be able to cause significant (factor of about 2) threshold shifts.

## References

[1] Klann R, Buhleier R, and Elsässer T 1991 *Appl. Phys. Lett.* **59** 885-7

[2] Klann R, Höfer T, Buhleier R, Elsässer T and Lambrecht A 1992 *Appl. Phys. Lett.* **61** 2866-8

[3] Gureev D M, Davarashvili O I, Zasavitzky I I, Matsonashvili B N and Shotov A P 1975 *Fiz. Tekh. Poluprov.* **9** 1902-8

[4] Yunovich A E, Ten V P, Fedorov M S and Chramzov A P 1975 *Fiz. Tekh. Poluprov.* **9** 904-6

[5] Tomm J W, Herrmann K, Yunovich A E and Solotov S I 1985 *phys. stat. sol. (a)* **88** 277-81

[6] Valeiko M V, Zasavitzky I I, Matveenko A V and Matsonashvili B N 1986 *Pisma v. Zh. Eks. Teor. Fiz.* **43** 140-2

[7] Valeiko M V, Zasavitzky I I, Matveenko A V, Matsonashvili B N and Sakseev D A 1987 *Fiz. Tekh. Poluprov.* **21** 57-62

[8] Schwarz B D, Huber C A and Nurmikko A V 1982 *Lecture Notes in Physics* **152** (Berlin, Heidelberg, New York: Springer) p 163-6

[9] Klann R, Höfer T, Buhleier R, Elsässer T and Lambrecht A 1993 *Semicond Sci. Technol.* **8** S305-8

[10] Anderson W W 1987 *IEEE Journ. Quant. El.* **QE-13** 532-543

[11] Herrmann K H, Tomm J W, Haertel S, Hoerstel W, Möllmann K-P, Heimbrodt W, Tacke M and Böttner H 1994 *Journ. Luminescence* **60&61** 371-5

[12] Genzow D, Mocker M and Normantas E 1986 *phys. stat. sol. (b)* **135** 261-70

[13] Rupprecht R, Hofmann W, Geist F and Pascher H 1994 *Verhandl. DPG (VI)* **29** 1049 (cf. also this proceedings)

[14] Singleton J, Kress-Rogers E, Lewis A V, Nicholas R J, Fantner E J, Bauer G and Otero A 1986 *J. Phys. C: Solid. State Phys.* **19** 77-92

[15] Tomm J W, Möllmann K-P, Peuker F, Herrmann K H, Böttner H and Tacke M 1994 *Semicond Sci. Technol.* **9** 1033-41

*Inst. Phys. Conf. Ser. No 144: Section 3*
*Paper presented at 7th Int. Conf. on Narrow Gap Semicond. Santa Fe 9–12 January*



# Faraday effect in epitaxial films of IV-VI semiconductors

W. Herbst[1], H. Pascher[1], G. Bauer[2]

[1]Experimentalphysik I, Universität Bayreuth, D-95440 Bayreuth,
[2]Institut für Halbleiterphysik, Johannes Kepler Universität Linz, A-4040 Linz

**Abstract.** The Faraday effect is a powerful method to obtain information on the spin splittings of semiconductors. We measured the Faraday rotation of the narrow gap semimagnetic lead salt IV-VI semiconductors $Pb_{1-X}Eu_XTe$, $Pb_{1-X}Eu_XSe$ ($x < 5\%$) with energy gaps in the mid infrared.

An optical bridge technique recently published for Faraday effect measurements in the visible spectral region, was modified for use in the mid infrared. This allowed for the first time to observe Faraday rotation in epitaxial grown narrow gap semiconductor films with thicknesses down to 2 μm. The experiments were performed in external magnetic fields up to 7 T and at various temperatures down to 1.7 K.

The Faraday effect can be calculated in the framework of a $\mathbf{k}\cdot\mathbf{p}$ - theory. In the case of diluted magnetic semiconductors the exchange interaction between the localised paramagnetic ions and the band carriers is added to the original $\mathbf{k}\cdot\mathbf{p}$ - Hamiltonian. Each allowed transition between an occupied and an unoccupied state contributes to the total amount of rotation. The experimental results can be quantitatively reproduced by such a calculation, acting as verification for the used band parameters. Thus in future the method will allow to determine band and exchange parameters in thin films and superlattices.

## 1. Introduction

Faraday effect is called the rotation of the plane of linearly polarised light passing a sample along an external magnetic field. This effect has turned out to be a powerful method examining the bandstructure of narrow or wide gap semiconductors [1,2]. Since the Faraday effect is proportional to the sample thickness, one is usually limited to bulk material. In epitaxial films with thicknesses of some μm the expected rotations are very small. In this paper we have modified an optical bridge technique, which was established by Kohl *et al.* [3] for the visible spectral region, for use in the mid infrared. In the visible it was even possible to observe the Faraday rotation of superlattices of CdTe/CdMnTe [3] with a total thickness of a couple of hundred nm.

Narrow gap IV-VI semiconductors, like PbTe or PbSe, and their related semimagnetic compounds, e.g. $Pb_{1-X}Eu_XSe$, $Pb_{1-X}Mn_XTe$, are very interesting materials for use as detectors or laser diodes. For any application a detailed knowledge of the bandstructure is necessary.

In semimagnetic IV-VI compounds one of the most interesting features is the strong modification of the band structure in a magnetic field compared to the diamagnetic host materials, due to the exchange interaction between the localised magnetic moments and the free band carriers. As a consequence the effective g-factors in conduction and valence band get strongly dependent on magnetic field and on temperature.

Faraday rotation is affected by the sum of the spin splittings in conduction and valence band. If one band is characterised by other experimental methods like coherent Raman scattering [5], it is possible to get quantitative information on the other band and it is thus possible to determine a complete set of bandparameters regardless of the doping of the sample. In this work we provide a theoretical model for the calculation of the Faraday rotation in a narrow gap IV-VI semiconductor in the framework of the well established **k·p** theory according to Mitchell and Wallis [4] and its extension for semimagnetic compounds, described in Ref. [5].

## 2. Theoretical model

### 2.1 Principal treatment of the Faraday rotation

When linearly polarised light propagates along an external magnetic field B through a sample the plane of polarisation is rotated (Faraday effect). The amount of rotation is proportional to the sample length L. For degenerate and also for semimagnetic semiconductors the Faraday rotation is not a linear function of the external magnetic field, but Verdet's constant V and the specific rotation, respectively, get strongly dependent on the magnetic field and also on temperature. The correct treatment of the Faraday rotation in a quantum mechanical approach is given by several groups [6,7] and yields for the angle of rotation $\Theta_F$:

$$\Theta_F = -\frac{\pi\hbar e^2 \cdot L}{n_0 c}\sum_i\sum_j \frac{\left|(v_+)_{ij}\right|^2 - \left|(v_-)_{ij}\right|^2}{\hbar^2(\omega_{ij}^2 - \omega^2)} \cdot f_i \cdot (1-f_j) = V \cdot L \cdot B \tag{1}$$

where $n_0$ is the refractive index of the material without an external magnetic field, c denotes the velocity of light in the vacuum, $(v_\pm)$ are the matrix elements of the velocity operator for each transition between an occupied state i and an unoccupied one j allowed for $\sigma^-$ respectively $\sigma^+$ polarised light, $\hbar\omega_{ij}$ are the corresponding transition energies, $\hbar\omega$ is the photon energy and $f_i$, $f_j$ are the occupation probabilities for initial and intermediate state i and j, respectively.

To calculate the Faraday rotation for a special semiconductor, $(v_\pm)$, $\hbar\omega_{ij}$, $f_i$,and $f_j$ have to be taken from a band model.

### 2.2 Band structure of IV-VI semiconductors

Narrow gap IV-VI semiconductors like PbTe or PbSe and their related semimagnetic compounds are direct semiconductors, with the fundamental energy gap located at the L-point of the Brillouin zone. Since there are four equivalent L-points, the IV-VI materials belong to the many valley semiconductors. The surfaces of constant energy are nearly ellipsoids of revolution with their symmetry axis parallel to the <111> directions.

It is well established to describe the band structure by a **k·p** - theory according to Ref. [4]. Thereby the uppermost valence band and the lowest conduction band are treated exactly and remote bands are considered in perturbation up to the order $k_B^2$, $k_B$ being the electron wave vector parallel to the external magnetic field. With that procedure the energies of the Landau levels depend on 11 band parameters: 3 two band parameters (energy gap $E_g$, and matrix elements $P_{||}$, $P_\perp$) and 8 far band parameters which describe the contributions to the effective masses and g-factors of conduction and valence band. The eigenvalues and eigenvectors are calculated by diagonalising a 4x4 matrix for each Landau quantum number n.

If one deals with diluted magnetic IV-VI compounds, it is necessary to extend the original **k·p** Hamiltonian by an exchange term, which describes the interaction between the localised magnetic moments of the paramagnetic ions and the mobile band carriers [5]. In the basis of the band edge eigenfunctions of the non magnetic material there are 4 non zero exchange integrals (2 for the conduction and valence band, respectively). Depending on the angle between the <111> direction and the magnetic field the exchange interaction couples Landau levels with different spin and different Landau quantum numbers [5].

The basis of the semimagnetic problem is obtained as a linear combination of the non magnetic band edge functions. In principle this results in an infinite system of equations which has to be truncated for practical reasons. This procedure is valid because the strength of the coupling is reduced with increasing difference of Landau quantum numbers.

The exchange interaction is treated in the framework of a meanfield approximation. The magnetic field and temperature dependence follows a modified Brillouin function, where an effective spin $S_0$ and an effective temperature $T_0$ is introduced to account for a weak coupling among the magnetic ions themselves [8]. Within these approximations the exchange interaction can be described for IV-VI compounds by 7 parameters: the content of the paramagnetic ions x, $T_0$, $S_0$, and the four exchange integrals.

2.3 Calculation of the Faraday rotation for IV-VI semiconductors

There are two main contributions to the Faraday effect in a doped narrow gap semiconductor (see e.g. Ref. [6,7]): All allowed interband transitions which are not blocked due to Pauli's principle, and the plasma contribution, where all allowed intraband transitions are taken into account.

The summation over all allowed transitions in Equ.(1) can be separated in a sum over the Landau quantum number n and an integration over $k_B$. The selection rules in Faraday geometry for $\sigma^-$ - and $\sigma^+$ polarised transitions are given in table 1. The surface normal of the samples examined in this work is parallel to the <111> direction. Therefore in Faraday configuration one of the surfaces of constant energy is oriented parallel to the magnetic field and three of them with an angle $\Phi$=70.53° with respect to **B**. The contribution of the 0°- and the 70.53°-valleys to the total Faraday rotation is weighted by the number of equivalent valleys. The calculations were performed numerically according to the scheme outlined in Fig. 1:

**Table 1:** *Selection rules for inter- and intraband transitions in Faraday geometry according to Ref.[5]. (**B** || **k** || <111>) for left and right handed polarised light.*

| | |
|---|---|
| Interband: | $\Delta n = 0$, $\Delta S = \pm 1$ |
| Intraband: | $\Delta n = \pm 1$, $\Delta S = 0$ |

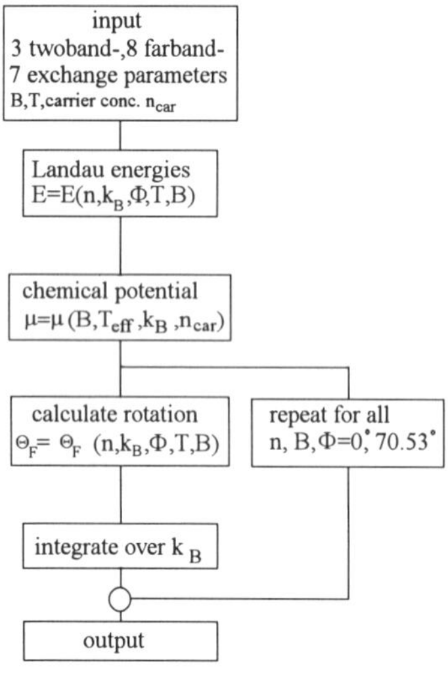


**Fig.1:** *Schematic calculation of the Faraday rotation of a IV-VI semiconductor.*

As a first step the Landau levels of conduction and valence bands are calculated for given magnetic field B, temperature T and carrier concentration $n_{car}$ for the two relevant values of $\Phi$. To account for population effects the chemical potential is determined from the given carrier density by integrating over the density of states multiplied by the corresponding Fermi distribution function. The next step is the calculation of the Faraday rotation for each transition according to Equ.(1). The dependence of the transition energies on $k_B$ is considered in parabolic approximation. For the transition matrix elements the same procedure is used as outlined in Ref. [7]:

$$\mathbf{v}_{\pm} = \mathbf{2} \cdot \frac{\partial \mathbf{H}}{\partial \mathbf{k}^{\mp}} \qquad (2)$$

where $k^{\pm}$ is proportional to the raising and lowering operators and H denotes the complete Hamiltonian.

In non magnetic narrow gap IV-VI semiconductors, like PbTe or PbSe, the effective g-factors in conduction and valence band are more or less equal in their absolute value. Therefore the population effects are dominating the total Faraday rotation. This is not true in semimagnetic compounds due to the exchange interaction.

## 3. Experimental and theoretical results

In this section the comparison between experiment and theory is shown exemplary for two semimagnetic systems. The band- and the exchange parameters for these two samples which are used for the calculation have been carefully determined by magneto transmission and coherent Raman experiments.

To measure the Faraday effect of thin epitaxial films we used a difference method as described in Ref.[3]. Linearly polarised light with photon energy $E_L$ propagates through the sample, which is mounted in a split-coil superconducting magnet, and passes a polarising beam splitter. The two beams are detected by means of identical liquid nitrogen cooled HgCdTe detectors. The signals served as input for a difference amplifier. Without magnetic field the initial polarisation in front of the sample is chosen such, that the difference signal is zero. At finite magnetic field an easily detectable signal arises, which if normalised is a function of the angle of rotation due to the Faraday effect, only.

### 3.1 Results for $Pb_{1-x}Eu_xSe$

In Fig.3a Verdet's constant V of a sample of $Pb_{1-x}Eu_xSe$ (X = 4.1%) is shown for several photon energies $E_L$ at a fixed temperature of 1.8 K. The specific rotation is strongly enhanced if the photon energy approaches the fundamental gap of the sample. V decreases with increasing magnetic field, because the exchange interaction saturates and the spinsplittings show more or less the behaviour of the non magnetic host crystal. In Fig.3b the temperature dependence of the total Faraday rotation is plotted, where the Brillouin function like bending is clearly seen. With increasing temperature the initial slope becomes smaller. This again reflects the influence of the exchange interaction on the Faraday effect.

The calculated curves are marked with circles in Figs.3a and 3b. There is an excellent agreement of the calculated and experimental results. The deviation for small magnetic fields arises because not enough Landau levels are taken into account in the calculation. We have performed the calculation with Landau quantum numbers up to 70, due to limitations of the computer hardware. The calculations were done without any fitting parameter.

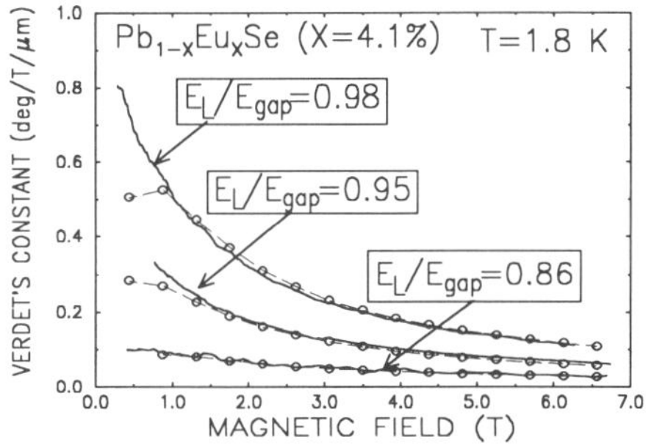


**Fig.3a:** *specific Faraday rotation of $Pb_{1-x}Eu_xSe$ (x=4.1%),T=1.8 K; $p_{car}=1\cdot10^{17}cm^{-3}$, L = 10 μm, $E_g$=273.4 meV,*
*<u>experiment:</u> solid lines;*
*<u>calculation:</u> dashed lines with circles*

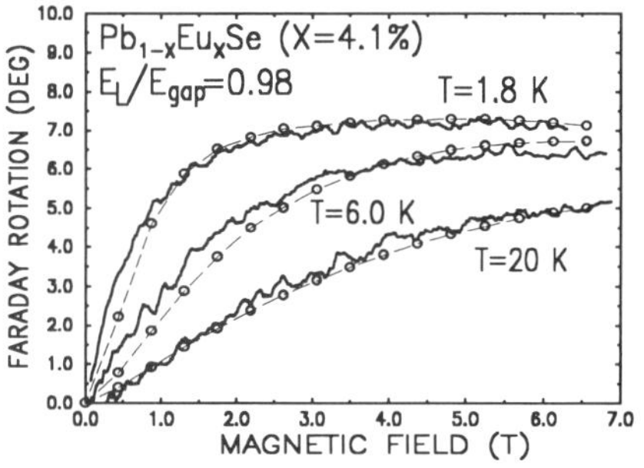


**Fig.3b:** *total Faraday rotation of $Pb_{1-x}Eu_xSe$ (x=4.1%), $E_L$= 269.6 meV; $p_{car}=1\cdot10^{17}cm^{-3}$, L = 10 μm, $E_g$=273.4 meV,*
*<u>experiment:</u> solid lines;*
*<u>calculation:</u> dashed lines with circles*

### 3.2 Results for $Pb_{1-x}Eu_xTe$

In contrast to $Pb_{1-x}Eu_xSe$ the data obtained with $Pb_{1-x}Eu_xTe$ show a much weaker exchange interaction (Figs.4a,b). The exchange parameters of PbEuTe are a little bit smaller then for PbEuSe [9], mainly the concentration x is lower in this sample than in the PbEuSe sample shown in Figs. 3a,b. The band parameters of $Pb_{1-x}Eu_xTe$ samples with x < 2% have again been determined by magneto transmission and by CARS experiments [9]. Faraday effect of a sample with 1.4 % Europium was measured at 1.8 K and 20 K. The specific rotation of the sample is more or less identical for both measured temperatures. The result is shown in Figs.4a,b for T=1.8 K. There is no Brillouin function like dependence as occurred in the $Pb_{1-X}Eu_XSe$ sample. The magnetic field dependence is dominated by a rather weak maximum of the specific rotation at a magnetic field of approximately 4 T, due to population effects in this p-doped sample. These experimental results can also be reproduced by the theory where the band and exchange parameters of Ref.[9] have been used. This fact demonstrates the validity of the band and exchange parameters given for $Pb_{1-X}Eu_XTe$ in Refs.[9,10].

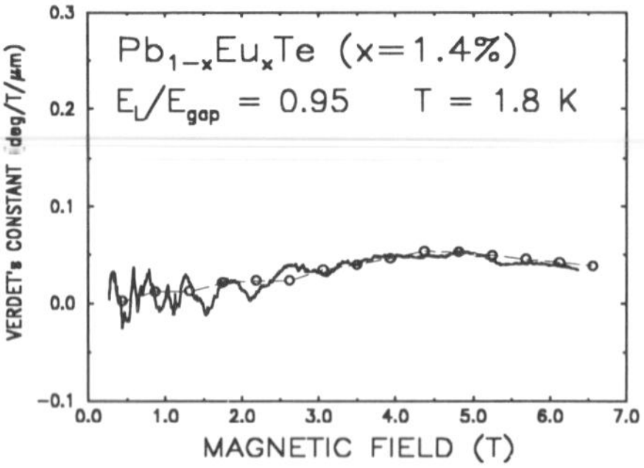


**Fig.4a:** *specific Faraday rotation of* $Pb_{1-X}Eu_XTe$ *(x=1.4%),T=1.8K;* $p_{car}=3\cdot10^{17}cm^{-3}$, *L = 3.0 μm,* $E_g$*=252.6 meV,*
*experiment: solid line;*
*calculation: dashed lines with circles*

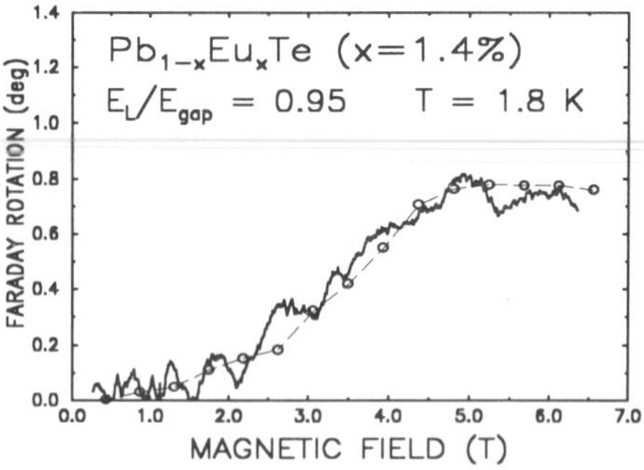


**Fig.4b:** *total Faraday rotation of* $Pb_{1-X}Eu_XTe$ *(x=1.4%),T=1.8 K;* $p_{car}=3\cdot10^{17}cm^{-3}$, *L = 3.0 μm,* $E_g$*=252.6 meV,*
*experiment: solid line;*
*calculation: dashed lines with circles*

## 4. Summary

The results described in Section 3 showed that it is possible to achieve a detailed understanding of the Faraday rotation of semimagnetic IV-VI semiconductors within the established band model and the quantum mechanical theory for the Faraday effect. A quantitatively excellent agreement between theory and experiment is observed in a material with strong exchange interaction of the free carriers and localised magnetic moments and in a material where these effects are rather small due to the low content of magnetic ions. This fact was proofed with sets of band- and exchange parameters, which have been determined by independent experimental methods, and calculating the Faraday rotation without adjustable parameters. Particularly for the $Pb_{1-X}Eu_XTe$ system with a higher content of Europium there are still problems to understand magneto transmission and CARS results. For this system the Faraday rotation will be a very useful method.

## References


[1] Pidgeon C R, Brown R N 1966 *Phys. Rev.* **146**, 575
[2] Bartholomew D U, Furdyna J K, Ramdas A K 1986 *Phys. Rev. B* **34**, 6943
[3] Kohl M, Awshalom D D 1991 *Phys. Rev.* B, **43**, 2431
[4] Mitchell D L, Wallis R F 1966 *Phys. Rev.*, **151**, 581
[5] Bauer G, Pascher H, Zawadzki W 1992 *Semicond. Sci. Technol.* **7**, 703
[6] Halpern J, Lax B, Nishina Y 1964 *Phys. Rev.*,**134**,A140
[7] Jiménez-Gonzáles H J, Aggarwal R L, Favrot G 1994 *Phys.Rev.B,* **49**,4571
[8] Gaj J A, Planel R and Fishman G 1979 *Solid State Commun.***29**, 435
[9] Geist F, Mejia-Garcia C, Pascher H, Bauer G, Springholz G, *Proceedings of the Int. Workshop on Semimagnetic Semiconductors,*Linz 1994 (in press)
[10] Geist F, Pascher H, Springholz G, Bauer G, *Proceedings of the 7th Int. Conf. on Narrow Gap Semiconductors*, Santa Fe 1995 (to be published)

*Inst. Phys. Conf. Ser. No 144: Section 3*
*Paper presented at 7th Int. Conf. on Narrow Gap Semicond. Santa Fe 9–12 January*



# Band- and Exchange Parameters of $Pb_{1-x}Eu_xTe$

F. Geist[1], H. Pascher[1], G. Springholz[2], G. Bauer[2]

[1]Experimentalphysik I, Universität Bayreuth, D-95440 Bayreuth,
[2]Institut für Halbleiterphysik, Johannes Kepler Universität Linz, A-4040 Linz

**Abstract.** MBE grown PbEuTe samples with Eu-contents between 0 and 1.7% were investigated by interband magnetoabsorption as well as by coherent anti-Stokes Raman scattering in the mid infrared. For samples with energy gaps larger than 220meV it is preferable to perform the CARS measurements with frequency doubled $CO_2$-lasers (resonance CARS).

The data were interpreted in the framework of the Mitchell & Wallis bandstructure model with the exchange interaction taken into account in mean field approximation. The magnetotransmission minima of samples with higher Eu-contents were rather broad superpositions of unresolved transitions, particularly in Voigt configuration. So CARS turned out to be important not only for the determination of the spin splittings of valence and conduction band, but also yielded more precise information on interband matrixelements.

With increasing Eu content a pronounced decrease of $P_\perp$ is observed, whereas $P_{||}$ remains constant. The exchange interaction between free carriers and localised magnetic moments of the Eu-ions is by far weaker than in Mn-based semimagnetic IV-VI compounds but of the same order of magnitude as in PbEuSe.

## 1. Introduction

The ternary IV-VI semiconductor alloys are of technological importance particularly as materials for laser diodes emitting in the mid infrared [1]. For a systematic control of the device parameters a detailed knowledge of the band structure is important. In previous investigations either interband- [1] or intraband transitions [2,3] have been analysed and a dramatic change of band parameters even with low Eu-contents was found. In this material a complication arises from the possible modification of the valence band due to an energy level crossing with the localised Eu 4f levels [4,5].

Moreover the diluted magnetic compounds are interesting due to the exchange interaction between free carriers and localised magnetic moments of the paramagnetic ions. A complete understanding of the band structure needs the observation of interband as well as intraband transitions in a magnetic field. In the present work PbEuTe epitaxial films with energy gaps up to 340meV were investigated by interband magnetoabsorption in Faraday and Voigt configuration and by CARS in magnetic fields up to 7T.

The data are analysed using the **k·p** model by Mitchell & Wallis [6]. The exchange interaction between free carriers and localised magnetic moments of the $Eu^{++}$-ions is considered in mean field approximation with the magnetisation proportional to a modified Brillouin function according to Gaj [7]. Finally the model contains 3 two-band parameters ($E_g$, $P_\perp$ and $P_{||}$), 8 far band contributions to effective masses and g-factors and the exchange parameters A, $a_1$ for the valence and B, $b_1$ for the conduction band. From those the 4 exchange integrals can be calculated [8].

## 2. Experimental results

The PbEuTe samples with Eu contents up to 3% were prepared by MBE on $BaF_2$-substrates with [111] growth direction. Most of the samples are n-type with carrier concentrations of about $2x10^{17}cm^{-3}$ at 77K. One of them with x=0.014 is p-type.

### 2.1. Interband magnetoabsorption

The magnetooptical interband transmission experiments were performed in magnetic fields up to 7T and for temperatures between 1.7 and 20K. According to the selection rules in Faraday configuration with **k** ‖ **B** ‖ [111] $\sigma^+$ and $\sigma^-$ circularly polarised light ($\Delta n=0$, $\Delta s=\pm 1$) and in Voigt configuration with **k** ⊥ **B** ‖ [1 -10] linearly, parallel **B** polarised light ($\Delta n=0$, $\Delta s=0$) was used. The measurements of the transmission as a function of the magnetic field were performed either with CO-lasers, frequency doubled $CO_2$-lasers or especially for wavelength shorter than 4.6μm with a globar.

Fig. 1 shows a comparison of the $\sigma^+$ and $\sigma^-$ spectra of 3 samples with increasing Eu content in Faraday configuration. The respective photon energies are about 33meV higher than the energy gaps, so that similar spectra should occur. With increasing concentration a significant broadening of the transitions is observed, so that especially at lower fields less transitions can be identified. Moreover, some of the transitions at higher magnetic fields are no longer observable for more than 1.5% Eu (see minimum at about 4.5T in Fig.1).

The resonant magnetic fields which are taken from the transmission minima for the two polarizations for non magnetic PbTe occur at the same position because of the approximately identical g-factors of valence and conduction band. A difference of the minima indicates an exchange interaction affecting valence- and conduction band in a different manner. Even for the sample with 3.4% Eu only very small differences are observed.

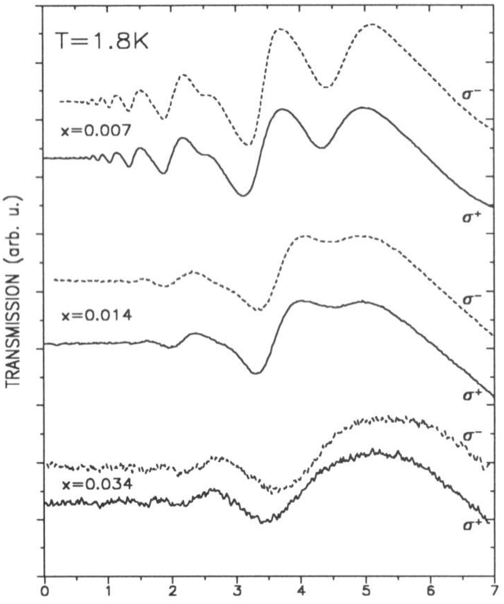


*Fig. 1: Interband transmission vs. magnetic field*
*$x = 0.007$: $E_g = 222.0$ meV; $\hbar\omega = 254.2$ meV*
*$x = 0.014$: $E_g = 252.6$ meV; $\hbar\omega = 285.2$ meV*
*$x = 0.034$: $E_g = 343.1$ meV; $\hbar\omega = 375.1$ meV*

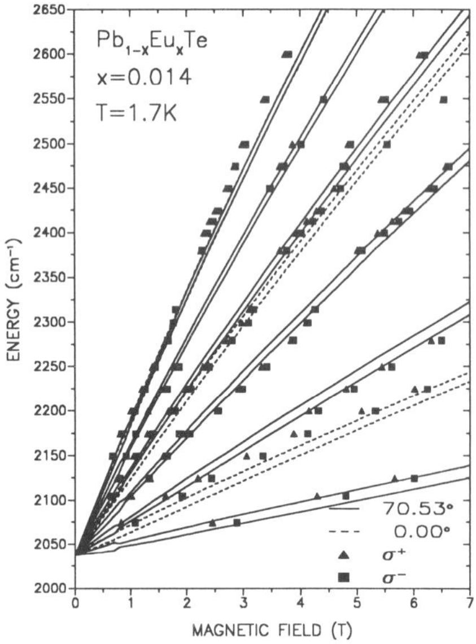


*Fig. 2: Fan chart for interband transition energies in Faraday configuration*
*p-PbEuTe; x=0.014; T=1.7K*

Fig. 2 shows the dependence of the interband transition energies between Landau levels on the magnetic field for Faraday configuration. The full and dotted lines are calculated with band and exchange parameters as given in tables 1 and 2 for the two series according to transitions in valleys with an angle of 0° and 70.53° between valley axis and magnetic field,

respectively. From the magnetoabsorption mainly the energy gap and the interband matrix elements are obtained. The Eu content is determined according to the published dependence of $E_g$ on x: $E_g$=189.7meV+4500meV∗x [9] which agrees well with our determination from Eu- flux during MBE growth of the films.
With increasing x some deviations between the measured and calculated transition energies are arising. Due to the relatively poor resolution obtained with the interband transmission experiments it was not possible to fit far band parameters. They are taken from PbTe [10], an approximation justified by the small Eu-contents of the material. On the other hand the energy gap is increased up to 50% with respect to PbTe.

## 2.2. Coherent anti-Stokes Raman scattering (CARS)

In our infrared CARS experiments two simultaneously Q-switched $CO_2$-lasers with frequencies $\omega_L$ and $\omega_S$ are collinearly focused onto the sample. By non-linear optical interactions radiation with a new frequency $2\omega_L - \omega_S$ is generated. The intensity at $2\omega_L - \omega_S$ has maxima if $\omega_L - \omega_S$ corresponds to the excitation energy of a Raman allowed transition. For a semiconductor held in Voigt configuration in a magnetic field the spin resonances are the strongest of these transitions. CARS is the most precise method to determine the spin splittings.

Fig 3 shows experimental recordings of a n-type and a p-type sample. In many cases the radiation of the strong lasers generates enough minority carriers to observe their spin resonances in addition to those of the majority carriers. Particularly in p-type material photoexcited electrons give strong signals due to the fact that their mobility is higher than that of the holes. The spectra of the n-type PbEuTe exhibit no temperature dependence indicating a rather weak exchange interaction in the conduction band. The resonant magnetic fields corresponding to the valence band are shifted with increasing temperature to lower magnetic fields. So the spin splitting in the valence band is reduced due to the exchange interaction as it is also the case in PbEuSe.

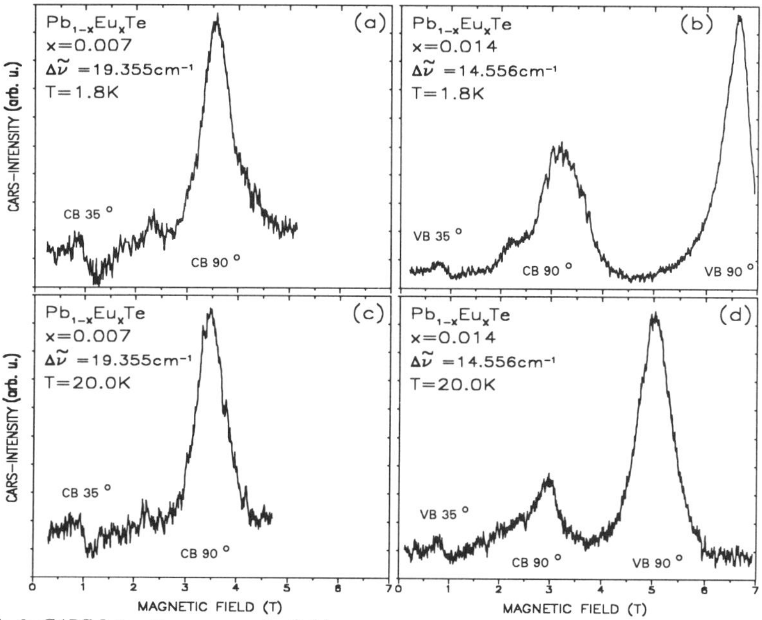


***Fig 3:*** *CARS-Intensity vs. magnetic field*
*(a) n-PbEuTe; x=0.007; T=1.8K* *(b) p-PbEuTe; x=0.014; T=1.8K*
*(c) n-PbEuTe; x=0.007; T=20.0K* *(d) p-PbEuTe; x=0.014; T=20.0K*

In samples with Eu-contents higher than 0.5% only resonance CARS with frequencies nearly corresponding to the energy gap could be observed. For this the radiation of the $CO_2$-lasers was frequency doubled in phasematched $AgGaSe_2$-crystals. If the energy gap is smaller than the used photon energies additional one-photon resonances occur. The spin resonances are identified by analysing the temperature dependence or by varying the laser lines but with a constant frequency difference.

Such spectra were taken for a lot of frequency differences and in Fig. 4 the resulting fan chart of a n-type sample with 1.7% Eu is shown.

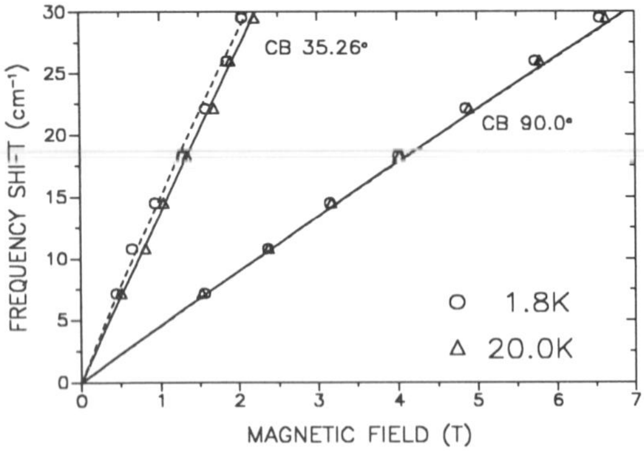


***Fig. 4:*** *Spin splitting vs. magnetic field*
*n-PbEuTe x=0.017*

*dotted lines: calculated for T=1.8K*
*full lines: calculated for T=20K*

Because of the very weak exchange interaction of the conduction band and due to the narrow linewidth besides the exchange parameters the interband matrix elements can be fitted to the CARS data more precisely than to interband transitions.

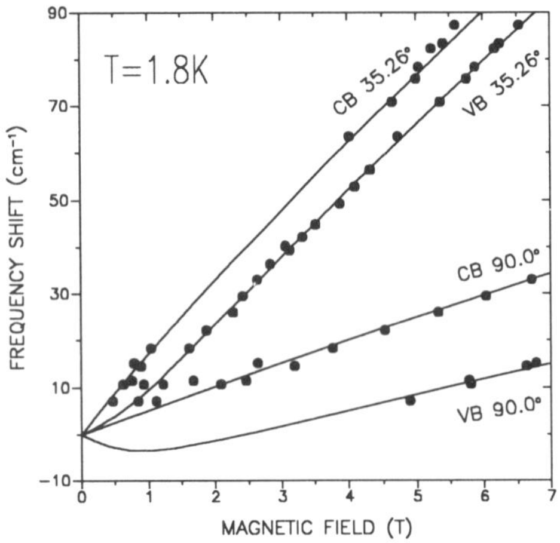


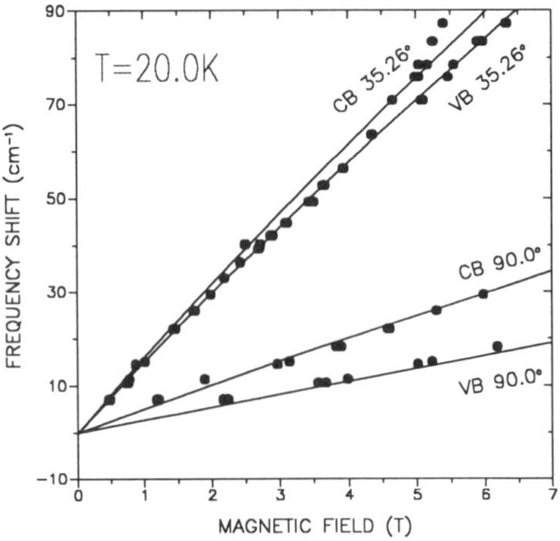


***Fig. 5a:*** *Spin splitting vs. magnetic field*
*p-PbEuTe x=0.014; T=1.8K*

***Fig. 5b:*** *Spin splitting vs. magnetic field*
*p-PbEuTe x=0.014; T=20.0K*

Figs. 5a and 5b show the corresponding fan chart of the p-type sample with 1.4% Eu. Here transitions of electrons and holes are observed. The perfect agreement of the calculated transition energies (full lines) for 1.8 K as well as 20 K confirm the reliability of the obtained band and exchange parameters. It should be noted that the interband transition energies in Fig.2 are calculated with the same set of parameters and also agree well with the experimental results.

## 3. Band- and exchange parameters

Table 1 lists the band parameters of PbEuTe. The experimental error of the energy gaps is less than 1%. From the energy gaps the Eu content can be calculated with the relation mentioned above. The interband matrixelements were determined by a fit to interband transmission and to CARS data, because especially the spin resonances of the electrons yielded precise values of $P_\perp/P_{||}$. For Eu contents higher than 1.5% they become more important for the calculation. The uncertainty of the matrixelements is estimated to be 5%. All experimental results could be described very well using PbTe [10] far band parameters.

***Table 1:*** *Band parameters of $Pb_{1-x}Eu_xTe$ (T=1.8K)*

| Parameter | x=0.0 | x=0.004 | x=0.007 | x=0.014 | x=0.017 |
|---|---|---|---|---|---|
| $E_g$ (meV) | 189.7 | 207.7 | 222.0 | 252.6 | 267.8 |
| $2P_\perp^2/m_o$ (eV) | 6.02 | 5.6 | 5.3 | 5.1 | 5.0 |
| $2P_{\|\|}^2/m_o$ (eV) | 0.51 | 0.48 | 0.47 | 0.50 | 0.50 |

The significant decrease of $P_\perp$ with increasing x agrees well with FIR magnetoreflectivity measurements reported by Yuan [3]. The value of the matrixelement $2P_\perp^2/m_o$=8.23eV, obtained by Karczewski et al. [2] for a sample with 2% Eu is in contrast to the results obtained in this work.

The exchange parameters given in table 2 are almost an order of magnitude smaller than in the Mn-based compounds. The exchange interaction of the conduction band is weaker than in the valence band. Compared to PbEuSe the exchange parameters have the same signs and the values are of the same order of magnitude but slightly smaller. The effective temperature $T_o$ and the effective spin $S_o$ [7] are not fitted to our data but taken as $S_o$=7/2 and $T_o$= 0K, because due to the small exchange parameters their effect on the magnetisation is not seen in the CARS data. The published values by Gorska et al. [11] are obtained with samples with higher x-values.

***Table 2:*** *Exchange parameters of $Pb_{1-x}Eu_xTe$*

| A (meV) | $a_1$ (meV) | B (meV) | $b_1$ (meV) |
|---|---|---|---|
| $26 \pm 5$ | $24 \pm 5$ | $5 \pm 3$ | $0 \pm 3$ |

Dietl et al. [12] calculated the exchange parameters, using a microscopic tight-binding model of the spin dependent coupling between free carriers and localised spins in the diluted IV-VI semiconductors. The resulting parameters of PbEuTe are about a factor of three larger than observed in the present work. But the theoretical calculation confirms the signs and the relation between the exchange interaction in the valence and conduction band.

## References

[1] Partin D L 1988 *IEEE J. Quant. El.* **24** 1716-1724

[2] Karczewski G, Furdyna J K, Partin D L, Thrush C N and Heremans J P 1992 *Phys. Rev.* **B46** 13331-13338

[3] Yuan S 1994 Thesis, University Linz

[4] Iida M, Shimizu T, Enomoto H and Ozaki H 1993 *Jpn. J. Appl. Phys.* **32** 4449-4453

[5] Geist F, Mejia-Garcia C, Pascher H, Bauer G and Springholz G 1994 *Proc. of the Int. Workshop on Semimagnetic Semiconductors (Sept. 28-30, Linz)*, to be published

[6] Mitchell D L and Wallis R F 1965 Phys. Rev. 151 581

[7] Gaj J A, Planel R and Fishman G 1979 *Solid State Commun.* **29** 435-595

[8] Bauer G, Pascher H and Zawadzki W 1992 *Semicond. Sci. Technol.* **7** 703-723
Pascher H, Röthlein P, Bauer G and Palmetshofer L 1987 *Phys. Rev.* **B36** 9395-9398

[9] Ishida A, Matsuura S, Mizuno M and Fujiyasu H 1987 *Appl. Phys. Lett.* **51** 478-480

[10] Pascher H 1984 *Appl. Phys. B* **34** 107-122

[11] Gorska M, Anderson J R, Kido G and Golacki Z 1990 *Sol. State Comm.* **75** 363-367

[12] Dietl T, Sliwa C, Bauer G and Pascher H 1994 *Phys. Rev.* **B49** 2230-2233

*Inst. Phys. Conf. Ser. No 144: Section 3*
*Paper presented at 7th Int. Conf. on Narrow Gap Semicond. Santa Fe 9–12 January*



# FTIR Characterization of IV-VI Semiconductors Grown by LPE on (100) $BaF_2$

Patrick J. McCann, Lin Li, Shu Yuan, and John E. Furneaux*
School of Electrical Engineering, *Department of Physics and Astronomy
Laboratory for Electronic Properties of Materials
University of Oklahoma, Norman, OK 73019

**Abstract**

The recent development of a new liquid phase epitaxy (LPE) process for growth of narrow bandgap IV-VI semiconductor alloys on low cost $BaF_2$ substrates offers new opportunities to advance IV-VI semiconductor device technology. Fourier transform infrared (FTIR) absorption spectroscopic characterization of lattice-matched $PbSe_{0.78}Te_{0.22}$ layers grown by LPE on (100) $BaF_2$ substrates shows that the absorption edge varies from 213 meV at 4 K to 275 meV at room temperature. Characterization of quaternary $Pb_{0.95}Sn_{0.05}Se_{0.80}Te_{0.20}$ layers shows that the absorption edge varies from 118 meV at 10 K to 200 meV at room temperature. Strong Fabry-Perot interference fringes are also observed in the FTIR spectra which allow calculation of epilayer thicknesses. The FTIR data show that a double heterostructure (DH) laser fabricated from these materials, with the ternary alloy as cladding layers and the quaternary alloy as the active layer, will have band edge discontinuities of approximately 40 meV and a tuning range from 10.3 μm at 77 K to 6.4 μm at 300 K.

Because the bandgaps of IV-VI semiconductors depend strongly on temperature they are useful for fabricating lasers that can be tuned over relatively large spectral ranges. Such devices are uniquely suited for high resolution molecular spectroscopy applications such as air pollution monitoring. Efforts to improve IV-VI semiconductor laser performance characteristics such as mode stability and maximum operating temperatures have been hindered by the high cost and limited availability of IV-VI semiconductor substrates. The recent development of a new liquid phase epitaxy (LPE) process [1] for growth of narrow bandgap IV-VI semiconductor alloys on low cost $BaF_2$ substrates has changed this situation offering new opportunities to advance IV-VI semiconductor device technology. In this article, we present Fourier transform infrared (FTIR) absorption spectroscopic data for two IV-VI semiconductor alloys, $PbSe_{0.78}Te_{0.22}$ and $Pb_{0.95}Sn_{0.05}Se_{0.80}Te_{0.20}$, grown by LPE on (100) $BaF_2$ substrates. These materials can comprise the cladding and active layers, respectively, of a double heterostructure laser. We show that such a laser will have a tuning range from 10.3 μm at 77 K to 6.4 μm at 300 K.

IV-VI semiconductor epilayers were grown on polished 4 mm x 4 mm x 380 μm thick (100) $BaF_2$ substrates cut from ingots purchased from Bicron, Inc., Solon, OH. Two growth solution compositions were used, $Pb_{0.97}(Se_{0.40}Te_{0.60})_{0.03}$ and $(Pb_{0.95}Sn_{0.05})_{0.965}(Se_{0.46}Te_{0.54})_{0.035}$ with liquidus temperatures of 618°C and 642°C, respectively. Growth within the 610°C to

660°C temperature range is necessary in order to obtain continuous epitaxial layers because a temperature dependent substrate surface reaction [2] catalyzes nucleation and thus promotes two-dimensional layer growth [3]. After growth, the layers were inspected using Nomarski microscopy to confirm that they were free of inclusions and voids and that good growth solution wipe-off occurred. Measured lattice parameters, as determined by x-ray diffraction, were 6.200 Å for the ternary layer and 6.185 Å for the quaternary layer. The $PbSe_{0.78}Te_{0.22}$ ternary layer, the composition of which is calculated using Vegard's Law, is lattice matched with the $BaF_2$ substrate ($a_o$ = 6.200 Å) while the quaternary layer has a lattice parameter mismatch of 0.25%. The composition of the $Pb_{0.95}Sn_{0.05}Se_{0.80}Te_{0.20}$ quaternary layer is also calculated using Vegard's law assuming that there is uniform liquid-to-solid segregation between the group IV components. This assumption is valid based upon Auger electron spectroscopy (AES) compositional analysis of IV-VI semiconductor quaternary alloys [4].

Infrared absorption measurements were performed with a BioRad Model FTS-60A Fourier transform infrared (FTIR) spectrometer using a variable temperature sample chamber equipped with KRS-5 windows. A $Hg_{1-x}Cd_xTe$ detector with a cutoff wavelength of 14.3 μm (87 meV) was used for all measurements. Sample transmission through the substrate was used because the $BaF_2$ substrate is transparent to infrared radiation from the visible to about 13 μm (95 meV). Spectra obtained with a bare $BaF_2$ substrate were used as the background and 1024 scans were co-added to eliminate noise.

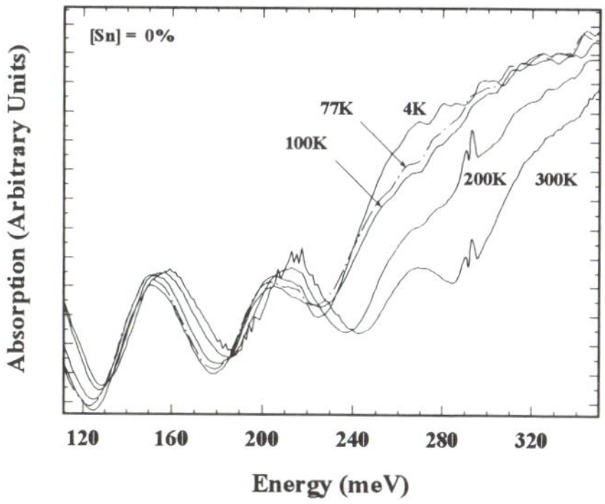


Figure 1: FTIR absorption spectra obtained at various temperatures for a $PbSe_{0.78}Te_{0.22}$ layer grown by LPE on a (100) $BaF_2$ substrate. The broad peaks at low energies are Fabry-Perot interference fringes. Absorption peaks near 290 meV for the 300 K and 200 K spectra are due to residual $CO_2$ in the spectrometer.

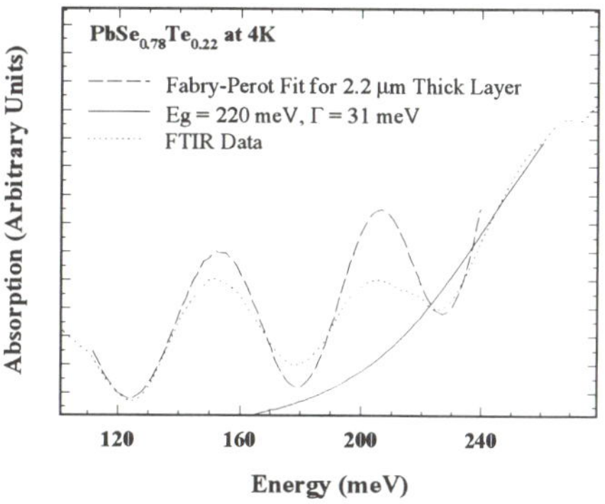


Figure 2: A theoretical fit to the 4K FTIR data in Figure 1. Two fits are shown, one for the Fabry-Perot interference fringes corresponding to a $PbSe_{0.78}Te_{0.22}$ layer thickness of 2.2 μm and another for the absorption edge. This latter fit corresponds to a bandgap at 4 K of 220 meV and a broadening parameter, Γ, of 31 meV.

Figure 1 is a set of FTIR spectra showing absorbed light as a function of photon energy for a $PbSe_{0.78}Te_{0.22}$ layer at various temperatures. Distinct absorption edges at each temperature are clearly evident. The absorption edge energies decrease at a rate of 0.46 meV/K as the temperature is lowered from 300 K to 100 K; further decrease from 100 K to 4 K is not significant. The absorption peaks near 290 meV for the 300 K and 200 K spectra are the P and R branches of the 2350 $cm^{-1}$ $\nu_3$ vibrational mode of $CO_2$, and the small absorption peaks between 180 meV and 230 meV for the 300 K spectrum are associated with the 1450 $cm^{-1}$ to 1850 $cm^{-1}$ $H_2O$ absorption band. These peaks are not evident at lower temperatures as the spectrometer was sufficiently purged during these measurements.

Strong Fabry-Perot interference fringes are apparent at energies below the absorption edges in Figure 1. Epitaxial layer thickness can be calculated from this fringe spacing using the equation

$$t = \frac{\lambda_1 \lambda_2 (m_1 - m_2)}{2(n_1 \lambda_2 - n_2 \lambda_1)}$$

where $\lambda_1$ and $\lambda_2$ are the wavelengths at the fringe maxima (or minima), $m_1$ and $m_2$ are integers corresponding to the orders of the fringes, and $n_1$ and $n_2$ are the refractive indices of the epilayer at the two different wavelengths. Using linearly extrapolated refractive index values calculated from published PbSe and PbTe data [5] an epilayer thickness of 2.18 μm is obtained for the $PbSe_{0.78}Te_{0.22}$ layer. This thickness is in very good agreement with physically measured thicknesses of ternary alloys having the same growth time of 6 minutes [1]. Figure 2 compares the FTIR spectrum for the $PbSe_{0.78}Te_{0.22}$ layer at 4 K with separate theoretical fits [6] for the Fabry-Perot interference effect and the absorption edge. The absorption edge fit corresponds to a bandgap of 220 meV and a broadening parameter, Γ, of 31 meV. This broadening parameter, which is a measure of the steepness of the absorption edge, is considerably larger than those obtained for other IV-VI semiconductors. PbTe layers grown by MBE [7], for example, have Γ values of less than 2 meV. An explanation for this difference is that subtle compositional variations can occur during LPE growth of these ternary alloys, possibly due to the low liquid-to-solid segregation coefficient for tellurium, thus producing a more "fuzzy" band structure. Free carrier absorption in these heavily n-type layers [1] may also contribute to the broadening.

Figure 3 shows absorption spectra at various temperatures for the quaternary alloy. Compared to the ternary alloy, the rate of absorption edge decrease is nearly the same but the absorption edges are shifted to lower energies. Fabry-Perot interference fringes are also observed and their spacing along with extrapolated refractive indices from published $Pb_{0.95}Sn_{0.05}Se$ and $Pb_{0.95}Sn_{0.05}Te$ data [5] yield a calculated epilayer thickness of 2.24 μm. Though this thickness is comparable to that of the ternary layer, the quaternary layer growth time was only 4 minutes. This apparent contradiction can be understood by considering the higher growth temperature for the quaternary alloy. According to the phase diagram and the lever rule more solid material will grow in a given growth period at higher temperatures thus increasing the effective growth rate. In addition, because epilayer growth is limited by chalcogen atom diffusion through the metal-rich liquid growth solution, it is possible that a mixture of lead and tin may also allow faster chalcogen atom diffusion thus increasing the growth rate.

In addition to absorption edges and Fabry-Perot interference fringes, some LPE samples have exhibited strong absorption features around 175 meV (7.1 μm, 1400 $cm^{-1}$). These features do not change with temperature and do not appear to be associated with any trace gases in the spectrometer. It is believed that this absorption correlates with an unusually thick interfacial reaction layer between the $BaF_2$ substrate and the epilayer. This reaction layer, which is necessary for continuous epitaxial layer growth, as discussed above, should be kept as thin as possible but can be as thick as 3000 Å [2]. Its composition in the case of $PbSe_{0.78}Te_{0.22}$ layer growth is predominantly BaSe. FTIR analysis of BaSe powder [8] indeed shows a strong absorption band in the same 1300 $cm^{-1}$ to 1500 $cm^{-1}$ spectral range as the LPE samples. Furthermore, FTIR analysis of a $BaF_2$ substrate pulled through a $Pb_{0.962}(Se_{0.40}Te_{0.60})_{0.038}$ growth solution at 645°C, a temperature sufficient for reaction product formation but above the growth solution liquidus temperature of 634°C so that no layer growth occurred, showed strong absorption peaks at 1375 $cm^{-1}$ and 1450 $cm^{-1}$. These peaks correspond exactly to the two strongest peaks observed in some of the $PbSe_{0.78}Te_{0.22}$ samples. We therefore conclude that the 175 meV absorption feature observed in some LPE samples is unambiguously associated with the epitaxy-enabling barium chalcogenide reaction product that forms on the substrate surface prior to LPE growth.

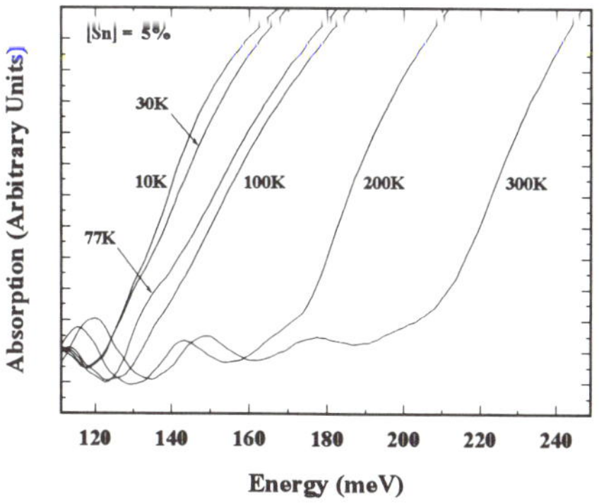


Figure 3: FTIR absorption spectra obtained at various temperatures for a $Pb_{0.95}Sn_{0.05}Se_{0.80}Te_{0.20}$ layer grown by LPE on a (100) $BaF_2$ substrate. Note the shift to lower energies as compared to the ternary sample shown in Figure 1. Fabry-Perot interference fringes are also apparent in this sample.

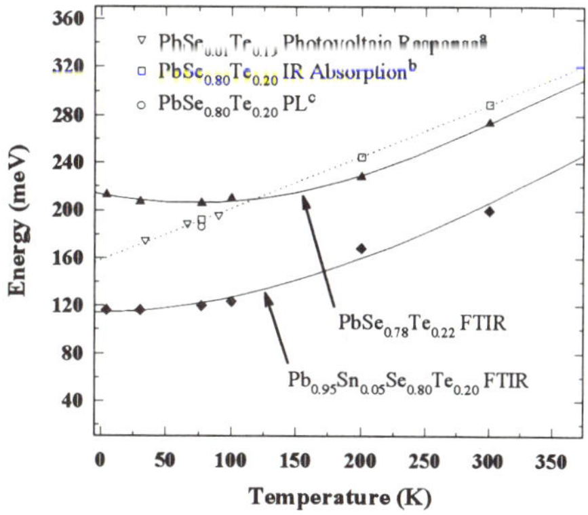


Figure 4: FTIR absorption edge energies versus temperature for undoped $PbSe_{0.78}Te_{0.22}$ and $Pb_{0.95}Sn_{0.05}Se_{0.80}Te_{0.20}$ layers grown by LPE. Bandgap data for similar ternary alloys are also shown. Data "a", "b", and "c" are taken from refs. [10], [11], and [12], respectively.

Figure 4 summarizes the FTIR data by plotting absorption edge energies as a function of temperature. Absorption edges were obtained from the intersection of the square of the absorption slope with the lowest measured absorption between the interference fringes. This method, which assumes a parabolic band structure, has previously been used to obtain the bandgaps of various IV-VI semiconductor compounds [9]. The obtained bandgaps also agree well with those obtained from the fitting procedure [6] discussed above, which does take into account non-parabolic band effects. FTIR absorption edge energy dependence on temperature above 100 K for the ternary alloy is similar to the bandgap dependence on temperature for $PbSe_{0.80}Te_{0.20}$ alloys grown by HWE based upon photovoltaic [10], optical absorption [11], and photoluminescence[12] studies. Agreement below 100 K, however, is not good. Our FTIR data for the ternary layer show that the absorption edge remains constant around 210 meV. We believe that the high electron concentration in these LPE-grown layers pins the Fermi energy in the conduction band at low temperatures thus shifting the absorption edge to higher energies even though the bandgap continues to shrink as the temperature is lowered [13]. Recently performed photoluminescence experiments [14] support the conclusion that such a Burstein-Moss effect [15] is occurring in these LPE-grown layers.

Our data show that the quaternary layer at 77 K has a bandgap of 121 meV, or an emission wavelength of 10.3 μm if used as the active region of a laser. Heating to 300 K causes a nearly linear increase of the bandgap to 200 meV. Since pulsed operation of IV-VI semiconductor lasers has been demonstrated at temperatures has high as 290 K [16], a laser made from this material can thus be expected to have a tuning range from about 10.3 μm (970 $cm^{-1}$) to about 6.4 μm (1560 $cm^{-1}$). Although the $Pb_{0.95}Sn_{0.05}Se_{0.80}Te_{0.20}$ quaternary alloy studied here is not lattice matched with $BaF_2$ or the $PbSe_{0.78}Te_{0.22}$ ternary alloy, its optical properties should be similar to the lattice-matched $Pb_{0.95}Sn_{0.05}Se_{0.76}Te_{0.24}$ alloy, which has recently been grown by LPE [17]. These ternary and quaternary alloys can comprise the cladding and active layers, respectively, of a lattice-matched double heterostructure laser. Our data show that the conduction and valence band edge discontinuities, assuming equal offsets, are approximately 40 meV, and that these offsets remain constant over a wide range of temperatures.

In summary, FTIR absorption spectra for ternary $PbSe_{0.78}Te_{0.22}$ and quaternary $Pb_{0.95}Sn_{0.05}Se_{0.80}Te_{0.20}$ layers grown by LPE on (100) $BaF_2$ substrates were presented. The FTIR data above 100 K show that the absorption edges of these alloys increase with temperature at a rate of 0.46 meV/K, in good agreement with measurements of other IV-VI semiconductor alloys. The absorption edges for the ternary alloy are not as steep as observed for other IV-VI semiconductor alloys suggesting that there may be subtle compositional variations in these LPE-grown layers, and lack of absorption edge change at low temperatures is believed to be a result of the Burstein-Moss effect. Observed Fabry-Perot interference fringe spacing also allowed accurate calculation of the epilayer thicknesses. FTIR characterization revealed that some samples had unusually thick barium chalcogenide interfacial reaction layers thus providing important information for improving the LPE growth procedure. Most importantly, though, the data presented here show that these IV-VI semiconductor materials can be used to design double heterostructure tunable lasers covering the 6 to 10 μm spectral range.

## Acknowledgments

This work was supported by grants from the National Science Foundation and the Oklahoma Center for Advancement of Science and Technology. We also gratefully acknowledge contributions to this work by Dongfan Zhong, Srinath Aanegola, Dwayne Appling, Jason Smith, and Robert Wright.

## References

[1] P. J. McCann and C. G. Fonstad, *Thin Solid Films* **227**, 185 (1993).
[2] P. J. McCann and C. G. Fonstad, *J. Electron. Mater.* **20**, 915 (1991).
[3] P. J. McCann, *Mat. Res. Soc. Symp. Proc.* **221**, 289 (1991).
[4] P. J. McCann, J. Fuchs, Z. Feit, and C. G. Fonstad, *J. Appl. Phys.* **62**, 2994 (1987).
[5] P. K. Cheo, *Handbook of Solid State Lasers*, Marcel Dekker, New York (1989).
[6] S. Yuan, N. Frank, G. Bauer, M. Kriechbaum, *Phys. Rev.B* **50**, 5286 (1994).
[7] S. Yuan, H. Krenn, G. Springholtz, G. Bauer, and M. Kriechbaum, *Appl. Phys. Lett.* **62**, 885 (1993).
[8] P. J. McCann, J. B. Smith, S. Aanegola, and J. E. Furneaux, (to be published).
[9] W. W. Scanlon, *J. Phys. Chem. Solids* **8**, 423 (1959).
[10] A. Jedrzejczak, D. Guillot, and G. Martinez, *Phys. Rev. B* **17**, 829 (1978).
[11] L. I. Bytenskii, V. I. Kaidanov, R. B. Melnik, S. A. Nemov, Yu. I. Ravich, *Sov. Phys. Semicond.* **14**, 40 (1980).
[12] M. V. Valeiko, I. I. Zasavitskii, V. L. Kuznetsov, A. V. Kurganskii, and B. N. Matsonashvili, *Sov. Phys. Semicond.* **19**, 388 (1985).
[13] P. J. McCann, S. Aanegola, and J. E. Furneaux, *Appl. Phys. Lett.* **65**, 2185, 1994.
[14] P. J. McCann, L. Li, J. E. Furneaux, and R. Wright, *Appl. Phys. Lett.* (March 13, 1995).
[15] E. Burstein, *Phys. Rev.* **93**, 632 (1954), T. S. Moss, *Proc. Phys. Soc. B* **67**, 775 (1954).
[16] B. Spanger, U. Schiessl, A. Lambrecht, H. Böttner, and M. Tacke, *Appl. Phys. Lett.* **53**, 2582 (1988).
[17] P. J. McCann and D. Zhong, *J. Appl. Physics* **75**, 1145 (1994).

*Inst. Phys. Conf. Ser. No 144: Section 3*
*Paper presented at 7th Int. Conf. on Narrow Gap Semicond. Santa Fe 9–12 January*



# Magnetotransport in PbTe nipi structures

J. Oswald, M. Pippan, G. Heigl, G. Span, T. Stellberger

Institut für Physik, Montanuniversität Leoben, Franz Josef Str. 18
A-8700 Leoben, Austria

**Abstract:** In this paper the 3D- and 2D- behaviour of wide single quantum wells which consist of one period of a PbTe nipi-structure is studied theoretically and experimentally. A simple model combines the 2D- subband levels and the 3D-Landau levels in order to calculate the density of states in a magnetic field perpendicular to the 2D plane. It is shown that at a channel width of about 500 nm one can expect to observe 3D- and 2D-behaviour at the same time. Finally the general design aspects for PbTe samples are discussed.

## 1. Introduction

PbTe is a narrow gap semiconductor with both the conduction and the valence band extrema located at the L points of the Brillouin zone [1,2]. The effective mass ellipsoids are oriented along the four <111> directions (Fig.1). The PbTe layer systems are grown on $BaF_2$ substrates by HWE (Hot Wall Epitaxy).

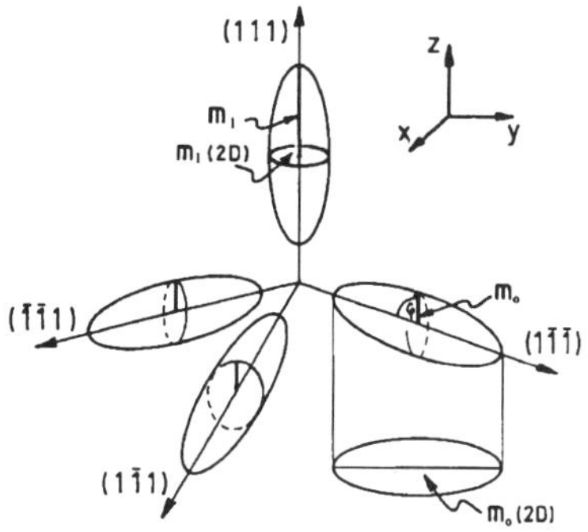


Fig.1. Effective mass ellipsoids of PbTe. With respect to the (111) growth direction there are two different effective masses for both the subband quantization: $m_l$, $m_o$ and the Landau level splitting: $m_l$(2D), $m_o$(2D).

The basic idea of doping superlattices is to create a periodic band-edge modulation in real space without changes in the chemical composition [3,4]. By using a typical doping level of $1x10^{17}...3x10^{17}cm^{-3}$ any potential modulation exceeding 10 meV leads to completely de-coupled potential wells in PbTe nipi-structures. The typical behaviour of such structures therefore can be investigated sufficiently by use of only a single period which in our case is a simple pnp-structure. In this pnp-structures the p-layers are designed to be non-depleted in order to screen the embedded n-channel from any influence from surface and interface states [5]. Selective contacts to the embedded n-channel are realised [6] in order to perform magneto transport experiments. The mobility in this selectively contacted structures is in the order of $3x10^5$ ... $1x10^6$ $cm^2V^{-1}s^{-1}$ at a temperature of T = 2.2K. The extremely high dielectric constant ($\varepsilon$ = 500 - 1000) leads to wide quasi parabolic potential wells with a typical subband separation ranging from 1- 3 meV. A magnetic field perpendicular to the layers leads to a Landau level splitting which is large as compared to the subband splitting even at magnetic fields below B = 11 Tesla.

## 2. Model

For the modelling of the electronic structure in a magnetic field B||[111] it is a good approximation to assume the subband states and the Landau states to be separable. In this case the energy levels of the system can be simply written as:

$$E_{ij}^{l,o}(B) = E_i^{l,o} + E_j^{l,o}(B) \tag{1}$$

l denotes the longitudinal and o the oblique valleys, i denotes the subband level index and j the Landau level index.

The subband energies are calculated selfconsistently by a numerical solution of the Schrödinger equation within the effective mass approximation. The energy level system consists of all Landau levels (LL's) sitting on each of the subband levels. Since in wide quantum wells the Landau level separation is much higher than the subband splitting, a picture which considers the Landau levels to be split by the subband quantization is more convenient. Therefore simply the Landau levels resulting from a 3D-calculation according to [7] are used in Eqn. 1 (for more details on 3D-Landau levels of PbTe see also e.g. [8] or [9]). The modelling of the density of states (DOS) then is easily done by a superposition of individual Landau peaks with a broadening factor b. The broadening function in this approximation is assumed to be Gaussian. The total DOS as a function of energy E and magnetic field B then reads as follows:

$$DOS(E,B) = \frac{q_e \cdot B}{h} \sum_{i,j} \frac{1}{b \cdot \sqrt{\pi}} \cdot \left( e^{-\frac{(E - E_{ij}^{l}(B))^2}{b^2}} + 3 \cdot e^{-\frac{(E - E_{ij}^{o}(B))^2}{b^2}} \right) \tag{2}$$

The numerical factor 3 in the second term in brackets accounts for the threefold degeneracy of the oblique valleys. For a basic understanding one can consider two different idealized cases which will mix up in reality. One case is an ideal square potential and the second case is an ideal parabolic potential. If we have an ideal square potential the subband energies depend on the square of the subband index. Therefore the lower subband levels are very close to each other and their associated Landau levels (of same Landau index) cause an overall enhancement of the density of states (DOS) near the subband ground state. The energy of the subband ground state is nearly identical to the position of the associated 3D - LL (LL of a three dimensional sample). At higher subband energies the subband levels are separated more and more and therefore a modulation of the DOS by the subband structure is more and more pronounced. In a parabolic potential well equally spaced subband levels are present. Therefore an ideal parabolic potential would not lead to an overall enhancement of the DOS at the position of the 3D - LL's. However, this idealized cases result from bare potentials and in real structures the situation will be different from both cases. The self consistent feedback of free electrons will change the potential as discussed in the following:

The effect of subband filling by adding electrons to an n-channel of a wide pnp-structure is limited by a Fermi energy which corresponds to the Fermi energy of bulk material at the same doping level. A further increase of the carrier sheet density therefore results in an increasing channel width and consequently a flat region in the middle of the electron channel is created. This behaviour starts in PbTe pnp structures which have a doping level of about $2x10^{17}cm^{-3}$ at carrier sheet densities of about $6x10^{12}cm^{-2}$. In this regime the subsequent filling of subbands by increasing the sheet carrier density is realised by a lowering of the subband energies while keeping the Fermi energy constant. The decrease of the subband energies happens by the increase of the channel width. In this way the subband structure turns more and more into the subband scheme of a square potential well. Consequently the lower part of the subband ladder looks like that of a square potential while the higher part will be dominated by the parabolic branches of the potential wells. For magneto transport experiments this means that there is an overall enhancement of the DOS at the energy position of the 3D-LL's and therefore also bulk like Shubnikov - de Haas (SdH) oscillations of the magneto resistance can be expected. At higher subband energies, which means the energy range between the 3D-LL's, a modulation of the DOS due to the subband structure can survive.

Fig. 2 shows the result of a model calculation of a pnp-structure with parameters according to the sample used in the experimental section. For this sample the carrier sheet density is $n^{2D} = 1.74x10^{13}cm^{-2}$ and the background doping level in the electron channel is $N_D = 3x10^{17}cm^{-3}$. The resulting shape of the potential near the bottom in the electron channel can be seen from the insert in Fig.2b. In Fig.2a the Fermi energy versus magnetic field is strongly influenced by the Landau levels of the subband ground states. This behaviour is the same like in 3D samples.

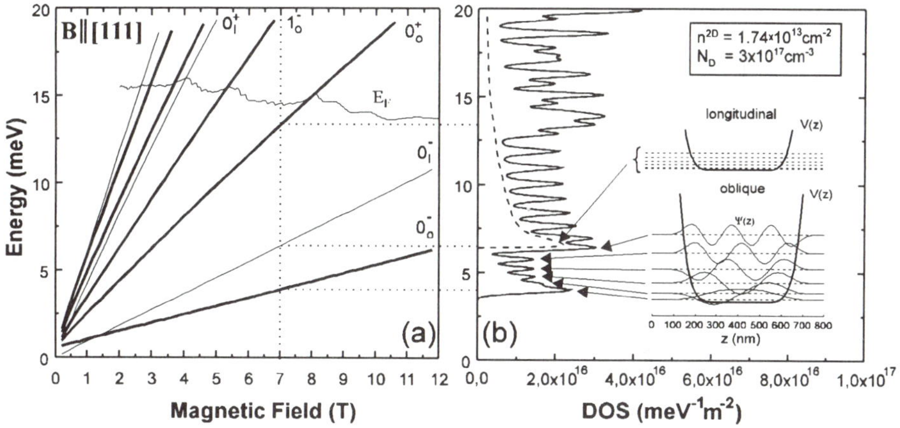


Fig 2. a) Landau levels of the subband ground states and Fermi energy versus magnetic field. b) density of states at B = 7 Tesla. Upper insert in b) potential and energy levels according to the longitudinal valley. Lower insert in b) potential, energy levels and wave functions according to the oblique valleys. The dashed curve in b) represents schematically the contribution of the longitudinal valleys.

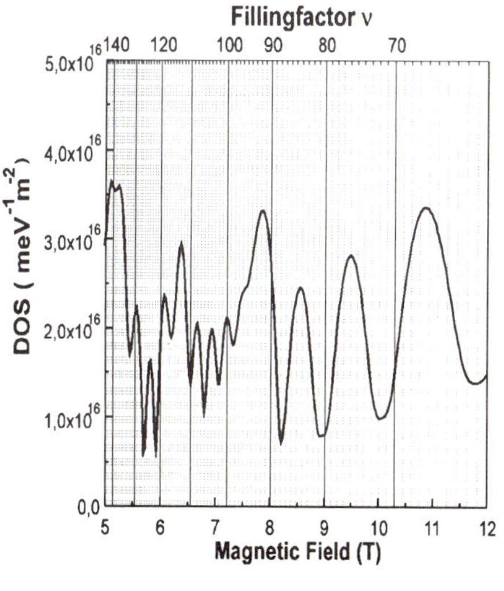


Fig.3: Density of states at the Fermi level versus magnetic field and filling factor according to Fig.2

The reason can be seen from Fig.2b. In this figure the DOS is plotted for a constant magnetic field of B = 7 Tesla. An overall enhancement of the DOS near the position of the 3D-LL's levels can be understood from the overlap of the DOS- peaks of the lowest Landau-subband states. The LL's of the subband ground states at B = 7 Tesla are marked by the dotted lines guiding from Fig.2a to Fig.2b. Because of the small subband separation in the longitudinal valley their contribution to the DOS is more or less a uniform background which is indicated schematically by the dashed line in Fig.2b.

### 3. Experiments

The sample parameters in this experiment are those used for the model calculation in the pervious section. The shape of the sample is a conventional Hall bar geometry. The experiments where performed in an 11 Tesla superconducting magnet at pumped $He^4$ . Using a conventional lock-in technique, 3D like SdH oscillations dominate the magneto transport (Fig.4a) without any additional structures in the data. In order to enhance the sensitivity of the experiment the set-up was extended to a double modulation arrangement:

An additional constant ac-current at a different (lower) frequency is superimposed to the ac-current used for the normal magnetotransport measurement. In selectively contacted pnp-structures this additional current causes also an additional (small) voltage drop which occurs only in the n-layer. This voltage drop has the same effect as a gate voltage applied by isolated metal gates which cannot be used for our samples. The voltage drop within the n-channel causes a change of the pnp-potential and therefore causes also a change of the electron channel width. A modulation of the channel width, in turn, results in a modulation of the subband energies. The signal resulting from the conventional lock-in technique is used as the input signal for a second lock-in amplifier which is tuned to the modulation current. In this way the experiment is getting sensitive to structures in the DOS which are associated with the subband spectrum. As shown in Fig.4b a structure in the experimental data is found in the upper magnetic field range where no structure is observed in the direct signal. The signal is well reproducible by an up- and down- sweep of the magnetic field.

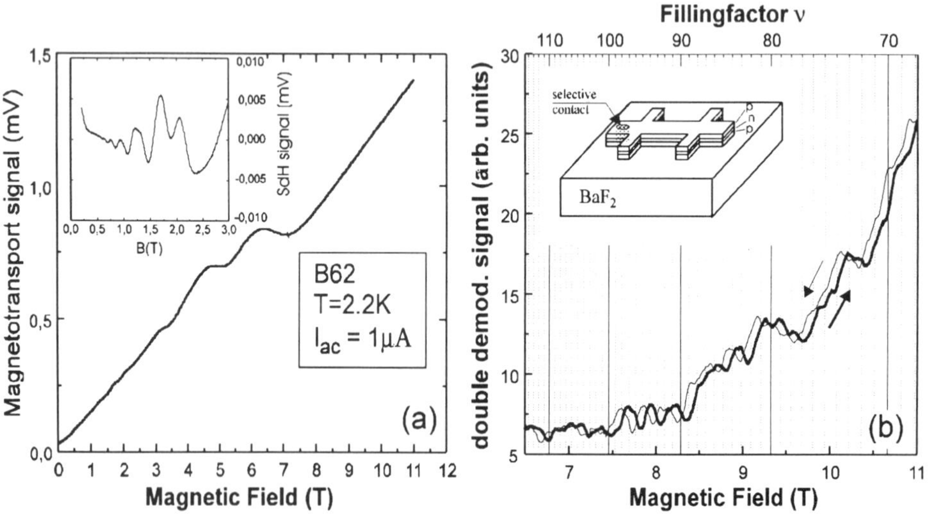


Fig.4 a) Magneto resistance data. Insert: magnification of the lower field range after subtraction of a linear slope. b) signal resulting from double modulation technique. Insert: Scheme of the sample configuration with one representative selective contact.

## 5. Discussion and concluding remarks

The modulation of the DOS at the Fermi level (Fig.3) results from changes of the individual filling factors of the different involved subband systems. The carrier distribution between the different subband systems changes with the magnetic field. Since only the total filling factor is periodic in 1/B a plot of the data over 1/B cannot be used to check the periodicity. It is not possible to expect a particular period length to exist over the entire field range. In the investigated magnetic field range the model calculation shows that two or three subband schemes overlap and the filling factors in that range are extremely high. Such a system is, of course, extremely sensitive to inhomogeneities or other influences and therefore it is not realistic to expect a one to one coincidence of the structures in the experimental and the simulated data. But typical changes $\Delta\nu$ of the filling factor between the dominating peak like features of the signal are found to be in agreement.. A typical period $\Delta\nu \geq 3$ indicates that the three oblique valleys dominate the subband structure. The model calculation includes also a strain induced energy shift between the longitudinal and the oblique valleys of the order of 1 meV [9]. Due to the sample parameters, in particular the high Fermi energy, this parameter is not very critical in this case. Concerning the general design aspects of PbTe pnp-samples one can say the following:

In order to get gaps in the DOS between the LL's all carriers have to be transferred to the oblique valleys. This is in principle always possible at sufficiently high magnetic fields (see LL's in Fig. 2). However, the required magnetic fields increase with increasing Fermi energy, which in turn depends on the doping level. Using a low Fermi energy the required magnetic field would shift to lower values (e.g. $B_{min}$ = 6 Tesla at $N_D$=1x$10^{17}$cm$^{-3}$). But the disadvantage of a low Fermi energy is that if the quantum limit is reached (only lowest 3D-LL of the oblique valleys is occupied), the Fermi level enters the oblique subband scheme also at low subband energies and the subband separation might be not yet high enough to form gaps in the DOS (see Fig.2 b). From this point of view it is not good to aim at low filling factors and/or low carrier densities. The optimum situation is reached if the sample enters the quantum limit in the upper most part of the available magnetic field range so that the Fermi level is as far as possible above the subband ground state. That means that also the filling factor is kept at high numbers. This is opposite to 2D systems in hetero structures where one aims at low filling factors. But the reason for the different behaviour is simple: In ideal 2D systems made by heterostructures the subband splitting is much higher than the LL-splitting in the entire field range. Therefore the filling factor indicates the number of occupied LL's within the lowest subband. Going to low filling factors with increasing magnetic fields opens the gaps in the DOS more and more. In wide PbTe pnp-structures the gaps in the DOS are determined by the subband spectrum, which does not change with the magnetic field. In this case the gaps between subband states are larger for higher subband indices ( = filling factors) provided that the sample has entered the quantum limit according to it's 3D-properties. The presented sample would enter this regime at magnetic fields between 15 and 20 Tesla.

## 6. References


[1] Mitchel D L and Wallis R F 1966 Phys. Rev. **151** 581

[2] Dimmock J O 1971 *The Physics of Semimetals and Narrow Band Semiconductors* edited by Carter D L and Bate R T 405 (Pergamon, Oxford, 1971)

[3] Esaki L and Tsu R 1970 IBM J. Res. Dev. **14** 61

[4] Döhler G H 1986 IEEE J. Quantum Electron. **22** 1682

[5] Oswald J, Tranta B, Pippan M and Bauer G 1990 J. Opt. Quantum Electron. **22** 243

[6] Oswald J and Pippan M 1993 Semicond. Sci. Technol. **8** 435

[7] Burkhard H, Bauer G and Zawadzki W 1979 Phys. Rev. B **19** 5149

[8] Oswald J, Goldberg B B, Bauer G and Stiles P J 1989 Phys. Rev. B **40** 3032

[9] Oswald J, Pichler P, Goldberg B B and Bauer G 1994 Phys. Rev. B **49** 17029

*Inst. Phys. Conf. Ser. No 144: Section 3*
*Paper presented at 7th Int. Conf. on Narrow Gap Semicond. Santa Fe 9–12 January*



# Epitaxy of IV-VI materials on Si with fluoride buffers and fabrication of IR-sensor arrays

H. Zogg, A. Fach, J. John, J. Masek, P. Müller, C. Paglino

AFIF at Swiss Federal Institute of Technology,
Technopark ETH-Teil, Pfingstweidstr. 30, CH-8005 Zürich, Switzerland

**Abstract.** A review and new results are presented on MBE growth and infrared sensor array fabrication with $Pb_{1-x}Sn_xSe$ layers on Si including a very thin $CaF_2$ buffer needed for compatibility reasons. Densities of visible defects below $10^3$ $cm^{-2}$ at the surfaces and HRXRD line widths of 105 arc sec (for 3 μm thick layers) indicate a good crystalline quality of the layers which are grown on 3" Si(111)-substrates.
The induced mechanical strain due to the thermal expansion mismatch relaxes down to cryogenic temperatures even after many temperature cycles. This is due to dislocation glide in the main <110>{100}-glide system even at low temperatures. The layers maintain their quality on temperature cycling, no significant HRXRD line broadening was observed for layers cycled as much as 1400 times between room temperature and 77 K. This despite these layers suffered a cumulative plastic deformation as high as 400%!
IR-sensor arrays with up to 256 elements were fabricated using a new delineation technique which is suited for whole wafers and for growth and fabrication on *active* Si-read-out chips. Preliminary 128x128 pixel LWIR images are presented.

## 1. Introduction

Highest sensitivity for thermal imaging is achieved with infrared (IR) sensor arrays fabricated in narrow gap semiconductors (NGS) [1]. Preferably, the NGS should be grown as an epitaxial layer on a Si-substrate because of size, costs, and the possibility to integrate the read-out electronics of large arrays directly into the Si-substrate.

In the following, we review our present state of growth and fabrication of photovoltaic NGS IR-sensors in epitaxial IV-VI (lead salt) layers on Si-substrates. We use IV-VI materials, mainly $Pb_{1-x}Sn_xSe$ for the 8-12 μm range, rather than the well known $Hg_{1-x}Cd_xTe$. This is because IV-VIs are much easier to grow and handle, because high quality epitaxial layers are obtained on lattice mismatched Si-substrates, and because IV-VIs are rather forgiving materials as regards charged defects (the very large permittivities of IV-VIs efficiently shield the electric field of these impurities) [2-4]. The optical absorption is as high that 2 μm thick layers, which are easy to pattern, suffice, whereas about 10 μm thick layers are needed with $Hg_{1-x}Cd_xTe$. The sensitivities of photovoltaic IR-sensors are determined by the $R_0A$ (resistance-area) product, the ultimate sensitivities achievable are similar both for $Pb_{1-x}Sn_xSe$ and $Hg_{1-x}Cd_xTe$ under similar conditions. Good uniformity in cut-off wavelength is much easier to achieve with $Pb_{1-x}Sn_xSe$, since its bandgap depends much less on the chemical composition x as compared to $Hg_{1-x}Cd_xTe$. This is since PbSe (x=0) is already narrow gap, whereas $E_g$ is as high as 1.6 eV in CdTe.

| Material | lattice constant Å | thermal expansion coefficient at 300 K $10^{-6}$ /K | band gap energy eV | cut-off wavelength µm | low freq. permittivity |
|---|---|---|---|---|---|
| Si | 5.431 | 2.6 | 1.1 | 1.1 | |
| $CaF_2$ | 5.46 | 19.1 | >>1 | | |
| $SrF_2$ | 5.80 | 18.4 | >>1 | | |
| $BaF_2$ | 6.20 | 19.8 | >>1 | | |
| PbS | 5.94 | 20.3 | 0.42 | 3.0 (300K) | 172 |
| | | | 0.31 | 4.0 (77K) | 184 |
| PbTe | 6.46 | 19.8 | 0.31 | 4.0 (300K) | 380 |
| | | | 0.22 | 5.6 (77K) | 428 |
| PbSe | 6.12 | 19.4 | 0.27 | 4.6 (300K) | 206 |
| | | | 0.18 | 7.0 (77K) | 227 |
| $Pb_{1-x}Eu_xSe$ ($x=0-\approx.02$) | 6.12 | | 0.18 - 0.42 | 7 - 3 | |
| $Pb_{1-x}Sn_xSe$ ($x=0-\approx.2$) | 6.12-6.06 | | 0.18 - 0 | 7 - ∞ (77K) | |
| $Pb_{1-x}Sn_xTe$ ($x=0-\approx.4$) | 6.46-6.40 | | 0.22 - 0 | 5.6 - ∞ (77K) | |
| $Cd_{1-x}Hg_xTe$ ($x=0-\approx.87$) | 6.48-6.46 | 5.0 | 1.6 - 0 | 0.8 - ∞ | 18 |

Table 1. *Some properties of the relevant materials*

However, at present, most effort for IR focal plane array applications for the 8-12 µmrange is concentrated on $Hg_{1-x}Cd_xTe$. The technical development stage of $Pb_{1-x}Sn_xSe$-on-Si IR-sensor arrays is therefore far behind that of $Hg_{1-x}Cd_xTe$ on lattice matched CdZnTe or other substrates, but significant improvements have been achieved in the last year.

## 2. Epitaxial growth of $Pb_{1-x}Sn_xSe$ onto Si(111) with $CaF_2$ buffer

In order to get high quality IV-VI layers on Si-substrates, a IIa-fluoride buffer is needed. Without a buffer layer, poor epitaxy only is achieved [5]. In most of our earlier work [2-4], we used a stacked $CaF_2/BaF_2$ buffer layer of about 200 nm thickness. This was because $CaF_2$ is near lattice matched to Si, and untwinned $CaF_2$ layers are obtained. On the other hand, high quality IV-VI layers suitable for infrared device fabrication have been grown by Holloway et al. since 1970 on bulk $BaF_2$ single crystals [6]. We grow the layers in a 2-chamber MBE system onto 3 inch Si(111) wafers. A Shiraki or room temperature cleaning is applied prior to the deposition of $CaF_2$ and $BaF_2$ at temperatures up to 750°C or 450°C, respectively. The lower temperatures are used if growth is performed onto active Si-chips which contain an Al-metallization [7,4]. A 2-3 µm thick $Pb_{1-x}Sn_xSe$ layer is then grown in the second chamber at 300-400°C by using a solid PbSe, SnSe and Se source. The Se source is needed for adjustment of the carrier concentration. About 1 µm binary PbSe is grown first, then the SnSe flux is switched on. The SnSe/PbSe ratio determines x and therefore the desired cutoff wavelength. P-conduction with concentrations in the low $10^{17}$ $cm^{-3}$ range is needed for the fabrication of photovoltaic devices with the blocking Pb-contact technique [6].

Until recently, we fabricated all layers using these stacked $CaF_2/BaF_2$ buffers. However, when processing devices with photolithographic techniques, it turned out that epitaxial $BaF_2$ layers pose problems. This is due to the solubility of $BaF_2$ in water, which is 100 times stronger than of $CaF_2$. We found that a very thin $CaF_2$ layer (≈20 Å) is suitable, too, and greatly facilitates device processing. This despite the lattice mismatch with PbSe is much larger, and it was claimed earlier that no epitaxy is obtained on $CaF_2$ substrates [6]. We obtain best results with layers with below 30 Å thickness. This approximately corresponds to the critical thickness for misfit dislocation creation at the typical growth temperature of 750 °C (with corresponding mismatch at this temperature of 2.6%), and one may speculate that dislocation free $CaF_2$ buffer layers give best results. The quality of the overgrown IV-VI layers is indistinguishable from that grown with buffers containing $BaF_2$.

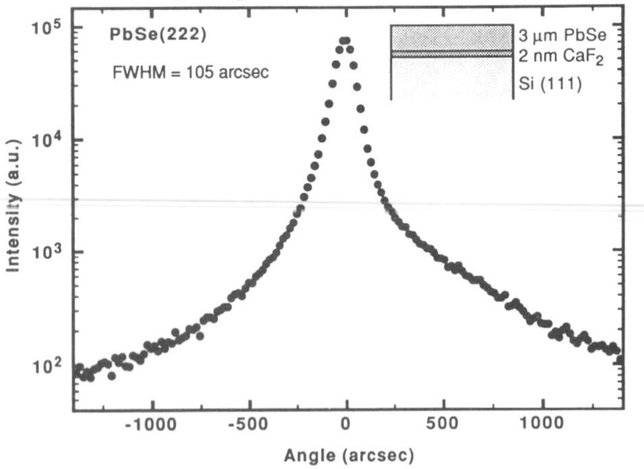


Fig. 1. *HRXRD rocking curve of PbSe on $CaF_2$/Si(111) with 12% lattice mismatch. FWHM is 105 arcsec.*

Mirror smooth surfaces result with the new $CaF_2$ as well as the stacked $CaF_2/BaF_2$ buffers. Densities of microscopic defects observable in the light microscope (>0.5 µm) are down to $10^3$ $cm^{-2}$. X-ray rocking curve line widths (FWHM) of 3 µm thick PbSe layers were 120-180 arcsec for $CaF_2/BaF_2$ buffer layers. The linewidths for the most optimized growth conditions with thin $CaF_2$ buffer layers are now as low as 105 arc sec for 3 µm thick layers (Fig. 1), and 60 arc sec for 10 µm thick layers grown for comparison. The only difference in the shape of the PbSe rocking curves grown with or without a $BaF_2$ part in the buffer is a slight broadening in the shoulder towards larger angles, as shown in Fig. 1. This can be interpreted as some higher tensile strain in the lowest part of the layer due to an increased dislocation density near the Si-interface; these dislocations would otherwise lie in the $BaF_2$ part of the buffer. These FWHM correspond to threading dislocation densities of $10^8$ $cm^{-2}$ and below. Consistent results were obtained by Nguyen-Van-Dau and Mathet et al., they obtained x-ray rocking curve widths in the range 100-150 arcsec with similar Si/$CaF_2$/PbSe stacks [8]. They had grown the layer sequence in the same MBE-chamber. From TEM studies, they found dislocation densities at the surface of 2 µm thick layers below $10^7$ $cm^{-2}$ [9].

Note that these IV-VI layers are heavily lattice mismatched with regard to the Si-substrate (table 1). In addition, the thermal expansion mismatch is as severe that mechanical strains as high as 0.6% are expected in the layers when cooling down from growth- to room temperature (RT) if the layer strains elastically only. However, for (111)-oriented layers, thermal strain relaxation occurs by dislocation glide. This is reflected with near atomic resolution with STM images (performed in air) which show slip-step lines (Fig. 2). They run along the three equivalent <110>-directions in the (111)-surface and reflect dislocation glide in the main {100}<110>-glide system with the {100}-glide planes inclined 54.7° with respect to the surface [4,10]. As soon as the thermal mismatch strain is large enough, dislocations start to glide. A

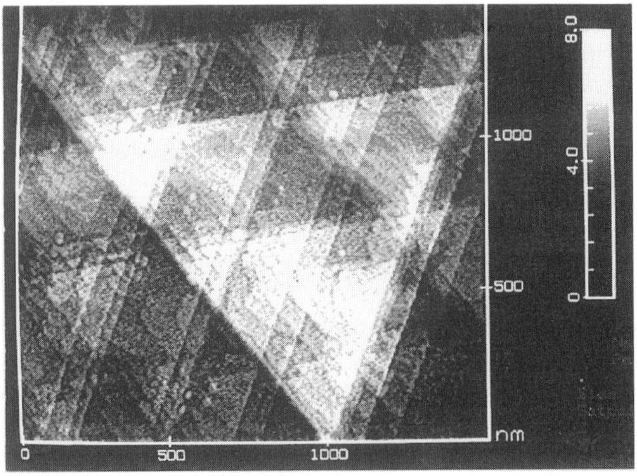


Fig. 2. *STM image of the surface of $Pb_{1-x}Sn_xSe$-on-Si(111) with intermediate $CaF_2$ buffer showing steps due to dislocation glide to relieve the thermal mismatch strain.*

misfit dislocation which has crossed the entire layer leaves no defect inside the material, but has created a surface step only. Due to the 3-fold symmetry, this heavy glide does not result in an angular tilt of the lattice, but rather in surface plateaus with slight height differences of multiples of the projection of the corresponding Burgers-vectors.

When cycling the layers between RT and 77 K (the operating temperature for the devices), the thermal mismatch strain relaxes near completely on each temperature change even at 77 K, and hysteresis effects occur in each cycle. Near no strain hardening is observed up to about 30 such cycles. Strain hardening occurs when continuing cycling, but all layers tested withstand such cycling without problems up to the highest number (1400 cycles) performed up to now. The FWHM of x-ray rocking curves do not appreciably broaden with cycling. This is shown in Fig. 3 where the FWHM at RT are plotted after different numbers of cycles. As discussed in Ref. 10, after some number of temperature cycles, near no new dislocation lines are generated, but the existing ones move back and forth on each cycle in order to relieve the thermal mismatch strain at low or higher temperature. Even after 1400 cycles the layers relax plastically on each cycle. The amount of plastic deformation after these many cycles is still 25% of the expected elastic strain dictated by the thermal mismatch. The total cumulative plastic deformation the layers have withstood corresponds to more than 400%. This is a surprisingly high value in view of the retained high structural quality. It reflects the ease of dislocation glide in IV-VI materials on the main glide system, but not on the second or higher systems [10]. To the authors knowledge, these materials seem to be the only ones where devices still work after such large plastic deformation.

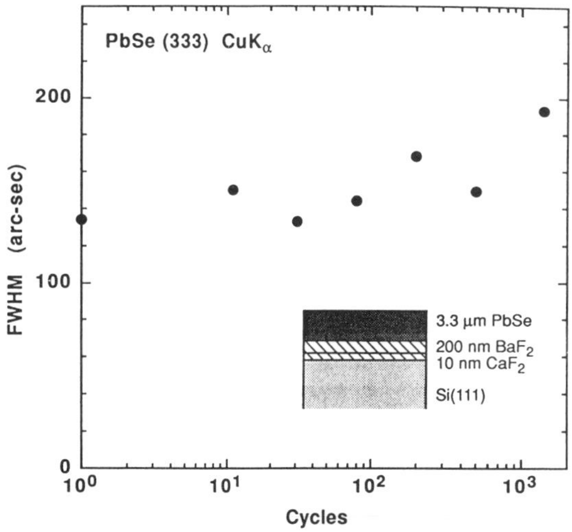

Fig. 3 *HRXRD rocking curve widths vs the number of temperature cycles between RT and 77 K.*

Due to the hysteresis, the point of zero strain has moved to a temperature midway between RT and 77 K. The layers are under a tensile strain below, and compressively stressed above this temperature. Due to this "strain symmetrization", we may conclude that when cycling is still further continued until no more plastic relaxation occurs (i.e. elastic deformation only, expected after >2000 thermal cycles for PbSe), and because only a further increase of 25% of maximal strain occurs as compared to the results after 1400 cycles, the layers withstand continued plastic deformation regardless if they strain elastically only or relax to some extent.

## 3. Photovoltaic IV-VI-on-Si infrared device fabrication

We fabricate the sensor arrays by vacuum deposition of Pb-blocking contacts, as first described by Holloway et al. [6]. Pb inverts the surface region of p-type material, thereby creating an induced p-n junction. The deposition procedure of this Pb layer is critical in order to obtain high quality sensors. We either deposit it in a separate high vacuum system, or directly in the MBE growth chamber after the IV-VI layer has cooled down to RT. A common ohmic contact fabricated with Pt or Au completes the arrays. We fabricated staggered linear arrays with 100 µm pitch containing up to 256 sensors, and also small area arrays with up to 16 x 16 pixels. They were fabricated in PbS, PbTe, $Pb_{1-x}Eu_xSe$, $PbS_{1-x}Se_x$, PbSe and $Pb_{1-x}Sn_xSe$ on Si(111) substrates with corresponding cut-off wavelengths ranging from 4 to 14 μm at 77 K. These sensors have been fabricated using a stacked $CaF_2/BaF_2$ buffer and already described in Refs. 2-4. They operate near the reflection loss limit with typical external quantum efficiencies above 50%. Illumination is from the backside through the IR-transparent buffer layer and substrate. No surface passivation and no anti-reflection coating was used up to now. Sensitivities are considerably above the background noise limit (BNL), however, they are still significantly below the theoretical limit given by Auger recombination. Some arrays were even grown on active Si-substrates which already contained the read-out electronic circuitries [7].

With the new thin $CaF_2$ buffer layer, we developed a new delineation technique which relies on simple wet-etching steps only and is suitable for full wafer processing [11]. The schematic cross section of these sensors is shown in Fig. 4. Pb, Ti and Ni are vacuum deposited, while Au is electroplated. Besides some test mask sets, we fabricated bilinear arrays with up to 2x128 sensors as well as 16 x 16 arrays with 50 μm pitch. Examples are shown in Figs. 5 and 6. The fan-out of the bilinear arrays is spread-out to 150 μm pitch for easier bonding to an existing special read-out chip.

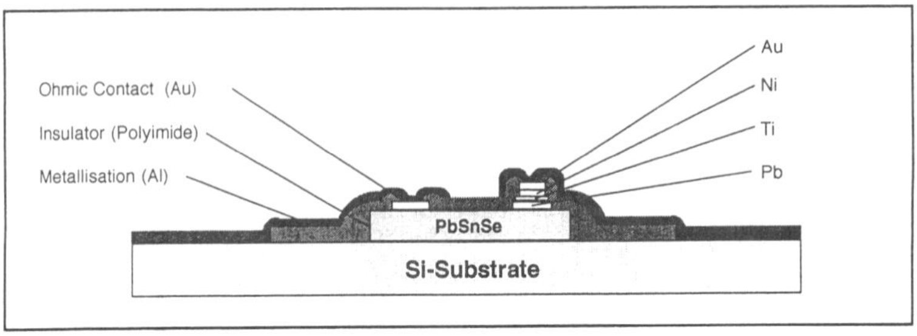


Fig. 4. *Cross section of the device lay-out as used for the new batch process.*

Fig. 5 *Partial and enlarged view of a 128x2 $Pb_{1-x}Sn_xSe$-on-Si array fabricated with the new batch technique. Pitch between the sensors is 50 μm.*

Typical quantum efficiencies of such $Pb_{1-x}Sn_xSe$ sensors are again above 50% without antireflection coating. The sensors are diffusion limited down to around 100 K, while generation recombination noise determines the sensitivities below. Fig. 7 shows the distribution of quantum efficiencies (photo currents) of a typical $Pb_{1-x}Sn_xSe$ array at 85 K. About 50 sensors out of 256 have been arbitrarily chosen for this purpose. The standard deviation is 2.7%, all sensors are working.

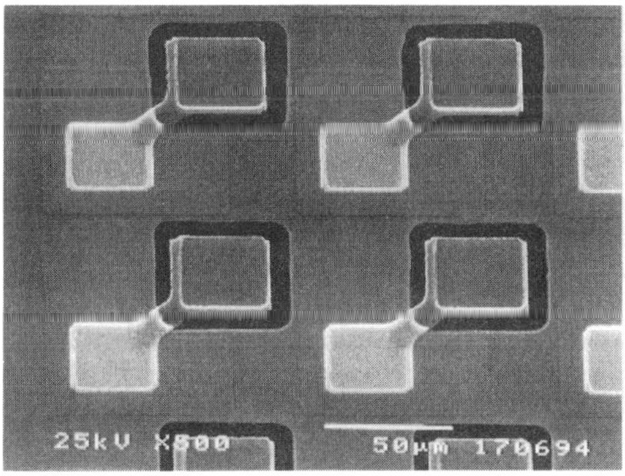


Fig. 6 *Partial view of a 16x16 $Pb_{1-x}Sn_xSe$-on-Si test array fabricated with the new batch technique. Pitch is 50 µm.*

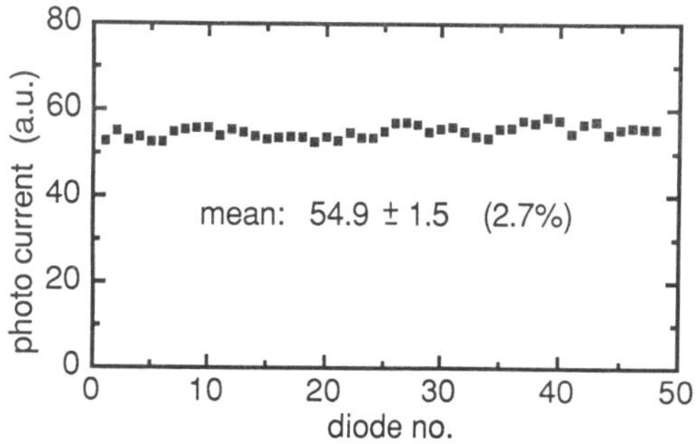


Fig. 7 *Distribution of quantum efficiencies (photocurrents) for 48 arbitrarily chosen sensors from a linear $Pb_{1-x}Sn_xSe$-on-Si IR-sensor array with 10 µm cut-off wavelength.*

## 4. Demonstrations of thermal imaging with $Pb_{1-x}Sn_xSe$-on-Si photovoltaic infrared sensor arrays

A simple thermal imaging camera was assembled. The FPA consists of 128 individual pixels fabricated with the new batch process. It is hybridly connected to a read-out chip placed at the side of the sensor chip. A line of the image consists of 128 elements, corresponding to the 128 pixels of the FPA, while the second direction is obtained by mechanical scanning of a mirror. The data are read-out serially, corrected for sensor nonuniformities, and displayed. Fig. 8 shows two typical images obtained with this unoptimized set-up. The temperature resolution is presently limited by the A/D conversion card and the nonuniformity correction procedure we use. A considerable improvement would result if the noise of the leakage currents of the sensors only would be considered.

## 5. Conclusions

We have successfully grown epitaxial IV-VI narrow gap semiconductor layers on Si(111) and fabricated photovoltaic IR-sensor arrays in the layers with cut-off wavelengths ranging from 3 µm to 12 µm. Advantages are the simple fabrication procedure, good homogeneity, and that layers of 2-3 µm thickness only can be used. Processing compatibility on standard Si-VLSI chips is demonstrated with a low temperature growth procedure with all

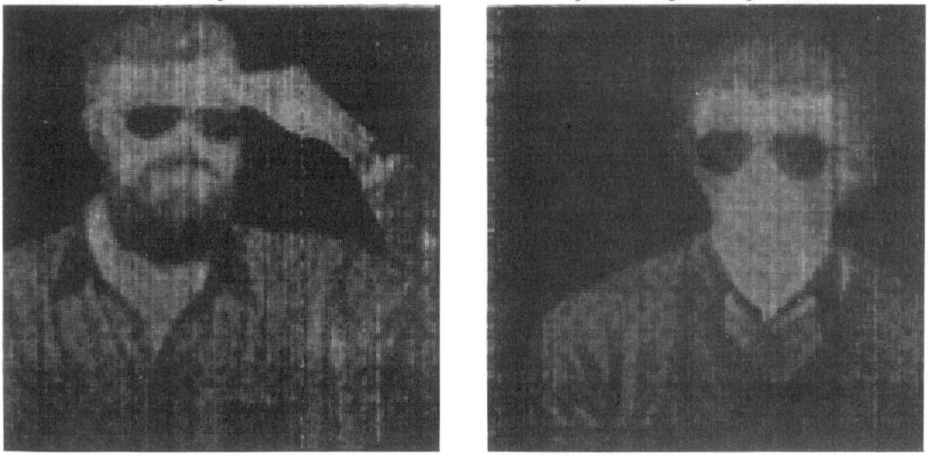

Fig. 8 *Two thermal images obtained with a 128 element $Pb_{1-x}Sn_xSe$-on-Si line sensor.*

growth steps performed below 450°C. This allows layer epitaxy and IR-sensor fabrication to be performed on completely processed and metallized active Si-wafers (with (111)-orientation). Wet etching is suitable for the delineation of the arrays, and a full wafer batch technique has been developed. This new technique is relied on a 2 nm thin $CaF_2$ buffer layer. The thermal expansion mismatch between the IV-VI layer and Si relaxes through glide of dislocations. Temperature cycling between RT and 77 K was performed more than 1400 times without adversely affecting the layer quality, this corresponds to a cumulative plastic deformation of more than 400%. After some thermal cycles, essentially no new thermal mismatch relieving dislocations are formed on temperature changes, but the threading ends of existing ones move back and forth on each cycle. Therefore, very few new defects are created.

Although we have already obtained sensors with high sensitivities, we are still below the theoretical limit. Considerable improvements are still possible by optimizing the device fabrication technique, especially replacing the blocking Pb-contacts used so far by buried p-n junctions with a thin wider bandgap cap layer to further reduce noise currents.

**ACKNOWLEDGMENT**. The financial Support from the Swiss National Science Foundation, the Swiss Defence Technology and Procurement Agency, the European Space Technology Centre, and the Priority Program "optics" of the board of the Swiss Federal Institutes of Technology is gratefully acknowledged.

## References

[1] Norton P R, Opt. Eng. **30**, 1645-1848, 1991.

[2] Zogg H, Blunier S, Hoshino T, Maissen C, Masek J, Tiwari A N, IEEE Trans. Electron Devices **38**, 1110-1117, 1991.

[3] Zogg H, Maissen C, Masek J, Hoshino T, Blunier S, Tiwari A N, Semicond. Sci. Technol. **6**, C36-C41, 1991.

[4] Zogg H, Fach A, Masek J, Blunier S, Opt. Eng. **33**, 1440-1449 (1994).

[5] Müller P, Zogg H, et al, to be published.

[6] Holloway H, "Thin film IV-VI semiconductor photodiodes", in "Physics of Thin Films", Academic Press, Vol. 11, 1980, pp. 105-203.

[7] Masek J, Hoshino T, Maissen C, Zogg H, Blunier S, Vermeiren J, Claeys C, Proc. **SPIE 1735**, 54-61, 1992.

[8] Nguyen-Van-Dau F, Mathet V, Galtier P, Padeletti G, Olivier J, Crété D G, Collot P, Mat. Sci. Eng. **B21**, 317-320, 1993.

[9] Mathet V, Galtier P, Nguyen-Van-Dau F, Padeletti G, Olivier J, J. Crystal growth **132**, 241-249, 1993.

[10] Zogg H, Blunier S, Fach A, Maissen C, Müller P, Teodoropol S, Meyer V, Kostorz G, Dommann A, Richmond T, Phys. Rev. **B50**, 10801-10810 (1994).

[11] Zogg H, Fach A, John J, Masek J, Müller P, Paglino C, Extended Abstracts of the 1994 int. conf. on solid state materials and devices, SSDM 94, Aug. 23-26, 1994, Yokohama, Japan, pp. 963-964.

*Inst. Phys. Conf. Ser. No 144: Section 3*
*Paper presented at 7th Int. Conf. on Narrow Gap Semicond. Santa Fe 9–12 January*



# Monolayer short period superlattices of narrow gap PbTe and antiferromagnetic wide band gap EuTe

G. Springholz[a], J. J. Chen[b], N. Frank[a], Y. Ueta[a], G. Marschner[a], C. Pichler[a], G. Bauer[a], M. S. Dresselhaus[b], G. Dresselhaus[b], T. M. Giebultowicz[c], V. Nunez[d], L. Salamanca-Riba[f]

[a] Institut für Halbleiterphysik, Johannes Kepler Universität Linz, A-4040 Linz, Austria
[b] Department of Physics, Massachusetts Institute of Technology, Cambridge, MA 02139, USA
[c] George Mason University, Fairfax, VA 22030, USA
[d] National Institute of Standards and Technology, Gaithersburg, MD 20899, USA
[f] University of Maryland, College Park, MD 20742-2115, USA

**Abstract.** Monolayer short period $(EuTe)_n/(PbTe)_{3n}$ superlattices (SL) with alternating magnetic EuTe and non-magnetic PbTe layers were grown by molecular beam epitaxy. For 2D layer-by-layer heteroepitaxial growth, only a narrow regime of growth conditions can be used in order to prevent surface roughening due to strain-induced coherent islanding. For such growth conditions, scanning tunneling microscopy reveals a very smooth surface morphology. The structural properties of the SL samples were studied by x-ray diffraction and transmission electron microscopy. Magnetization measurements show an antiferromagnetic-like phase transition for an even number $n$ of EuTe monolayers and a ferrimagnetic-like transition for odd n. In addition, neutron diffraction data indicate the existence of a long-range interlayer coupling between the magnetic layers across the intervening non-magnetic PbTe layers.

## 1. Introduction

Molecular beam epitaxy (MBE) of $PbTe/Pb_{1-x}Eu_xTe$ quantum well (QW) heterostructures has been studied for several years [1]. Exploiting the very strong increase of the energy band gap of the ternary $Pb_{1-x}Eu_xTe$ with Eu content, efficient infrared QW diode lasers have been produced [2]. The binary EuTe is a wide band gap (2eV) semiconductor, yielding very large confinement energies in PbTe/EuTe heterostructures. Since EuTe is usually semi-insulating, it can be used to separated the active layers of lead salt-based heterostructures. In the present paper, however, we focus more on the *magnetic* properties of EuTe/PbTe short period superlattices, since results on the electronic properties have been presented elsewhere[3].

## 2. Molecular beam epitaxy

Molecular beam epitaxy of PbTe and EuTe on $BaF_2$ (111) substrates was carried out in a Riber MBE growth chamber. The absolute beam flux rates from the effusion cells were measured with a quartz crystal thickness monitor, and the substrate temperature was calibrated using the $Te_2$ condensation points. For EuTe growth, *separate* beam flux sources for Eu and Te are used. Because of the high Te vapor pressure at the MBE substrate temperatures (~250°C), excess Te has to be supplied during EuTe growth, and therefore, the growth is controlled by the Eu flux impinging on the surface, whereas all excess Te reevaporates from the surface [4]. In contrast, since PbTe sublimes in the form of PbTe *molecules*, it is directly evaporated from a PbTe effusion cell. Since the PbTe molecules adsorbed on the layer surface remain bound in a molecular state until incorporation in the crystal lattice [5], MBE growth of PbTe is similar to MBE growth of elemental semiconductors, whereas EuTe MBE is comparable to III-V MBE .

Depending on the substrate temperature and the $Te_2$ to Eu beam flux ratio, a Te- or Eu-stabilized EuTe growth mode exists, characterized by different surface reconstructions [4]. However, due to the 2.1 % lattice mismatch between EuTe and PbTe, 2D layer-by-layer *heteroepitaxial* growth of EuTe on PbTe (111) exists only in a very narrow regime of growth conditions. This is a result of the fact that in thermodynamic equilibrium, the preferred surface configuration of a fully-strained EuTe overlayer on PbTe (111) is not that of a 2D layer, but that of a strongly corrugated layer, consisting of 3D islands on a thin wetting layer. In such a case, a significant part of the strain energy can be reduced by a lateral elastic deformation of the coherent 3D islands [6]. However, this effect can be suppressed when the mobility of the surface adatoms during growth is drastically reduced, so that the surface diffusion length is smaller than the critical wavelength for strain-induced islanding [7]. Thus for low substrate temperatures and high $Te_2$-to-Eu beam flux ratios, 2D layer-by-layer growth can be maintained, giving rise to pronounced RHEED intensity oscillations [4,8].

The STM image of a 25 monolayer (ML) thick EuTe layer on PbTe (111) grown under such conditions is shown in Fig. 1. Whereas the surface of PbTe is atomically flat on a length scale of more than 2000 Å and exhibits only large growth spirals around the core of threading dislocations originating from the lattice-mismatched growth on the $BaF_2$ substrates [8], the EuTe surface is much more corrugated on a length scale of about 200 Å with small monolayer islands nucleated on the surface. This is a consequence of the much shorter surface diffusion lengths involved in EuTe MBE growth under such conditions. In addition, however, beyond a layer thickness of 20 ML, straight monolayer high step lines (see Fig. 1) suddenly appear on the surface parallel to the crystallographic $\langle\bar{1}10\rangle$ and $\langle\bar{2}11\rangle$ directions. The end points of these step lines can be identified as penetration points of threading dislocations, which apparently have moved by several *μm* parallel to the surface. As described by the Matthews-Blakeslee model for strain relaxation [9], this movement results in the formation of misfit dislocation segments at the interface, as illustrated in the inset of Fig. 1. Thus, the appearance of these surface step lines indicates the onset of strain relaxation and thus the critical layer thickness can be detected with high precision [10]. Initially, the density of the visible dislocation glide lines increases only gradually with layer thickness, but increases abruptly beyond a thickness of 40

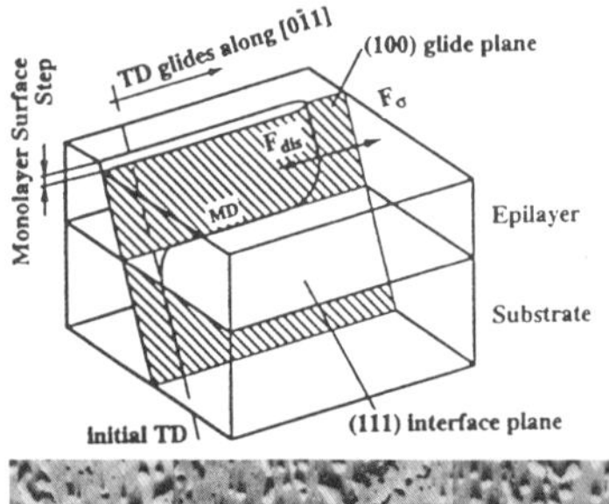


**Figure 1:** STM image (2 × 0.5 $\mu m^2$) of a 25 ML thick EuTe epitaxial layer on PbTe (111). The arrow indicates the end point of a threading dislocation. Inset: schematic illustration of the strain relaxation process due to misfit dislocation formation by glide of pre-existing threading dislocations.

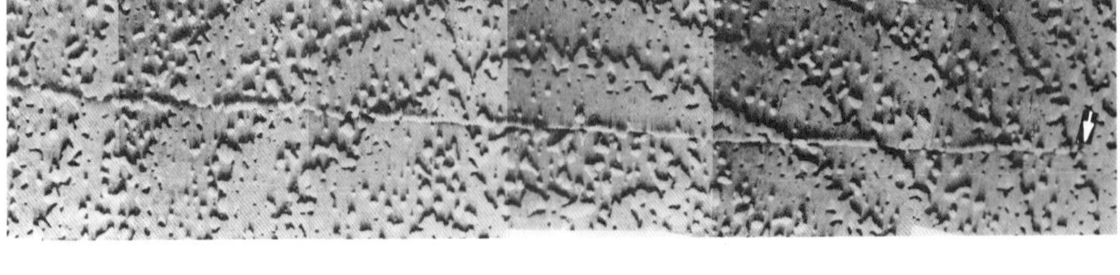

ML indicating a drastic acceleration of the strain relaxation process, which is accompanied by a distinct roughening of the surface [8].

## 3. Structural properties of short period $(EuTe)_n/(PbTe)_m$ superlattices

A series of short period $(EuTe)_n/(PbTe)_m$ superlattices was grown on relaxed PbTe buffers on $BaF_2$ (111) substrates using optimized growth conditions, i.e. substrate temperature of 260°C and a $Te_2$-to-Eu flux ratio of 1.2. The nominal number of monolayers contained in each layer ranged from $n= 2$ to 7 for EuTe and from $m = 6$ to 21 for PbTe with a constant ratio of $m/n = 3$ and a total number of 200 to 600 periods. Because of the constant ratio between the EuTe and PbTe layer thicknesses, the lattice-mismatch of 0.6 % of the SL stack as a whole with respect to the PbTe buffer is the same for all samples. Since strain induced islanding is suppressed, the epitaxial surface remained very smooth throughout the whole growth sequence, as indicated by the sharp RHEED streaks observed up to the end of growth. On top of the SL stack, an 800 Å PbTe cap layer was deposited to prevent oxidation of the EuTe layers.

The structural properties of the samples were investigated by Cu $K\alpha_1$ x-ray diffraction using a high-resolution Philips MRD diffractometer with a Bartels monochromator in the primary beam and a channel cut Ge crystal in the secondary beam. Figure 2 shows the triple axis (TA) diffractogram around the (222) reflection for a representative SL sample with 300 x (6 ML EuTe/18 ML PbTe). Besides the $0^{th}$ order superlattice peak SL0, several superlattice satellite peaks are observed. On the high angle side, two smaller peaks from the PbTe buffer and cap layer and the sharp $BaF_2$ substrate peak appear. For all samples, the determined superlattice period agrees quite well with nominal values. Since the thicknesses of the individual layers in the superlattices are well below the critical layer thicknesses, the SL layers are fully strained with respect to each other, with an identical in-plane lattice constant. However, the peak splitting between the buffer and cap layer indicates a significant degree of strain relaxation of the superlattice stack as a whole, since the thin PbTe cap layer is tensilely strained to the in-plane lattice constant of the SL stack which is larger than that of the relaxed PbTe buffer.

In order to determine the structural parameters of the sample, the measured diffractogram was fitted using dynamical diff-raction theory (dashed line in Fig. 2). From the best fit, the individual layer thicknesses were determined to be 6.4 ± 0.2 ML for EuTe and 17.6 ± 0.2 ML for PbTe, which agrees quite well with the nominal growth values, with a in-plane SL lattice constant of 6.489Å which was verified by a strain analysis using reciprocal space mapping of the asymmetric (422) Bragg reflection, yielding an in-plane SL lattice constant of 6.488Å.From the known individual layer thicknesses, the in-plane lattice constant of a corresponding free standing SL stack was calculated to be 6.501Å, i.e., the lattice-mismatch of the SL with respect to the PbTe buffer layer is 0.58 %. Since the actual

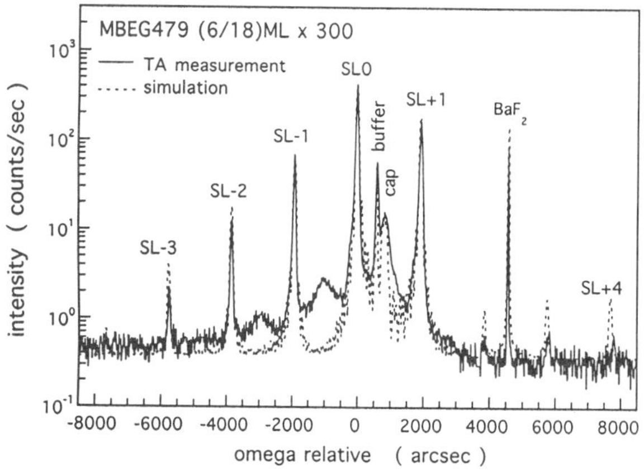


**Figure 2:** Triple axis Ω/2θ X-ray diffraction curves ($CuK\alpha_1$ radiation) for a 300 period $(EuTe)_6/(PbTe)_{18}$ superlattice: solid line: measured data and dashed line: simulated diffractogram.

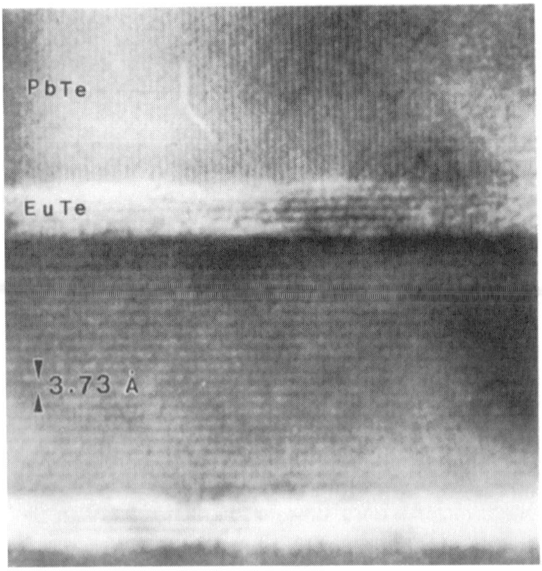


**Figure 3:** Cross-sectional TEM image of a $(EuTe)_5/(PbTe)_{22}$ superlattice sample.

in-plane lattice constant is smaller, only a partial 68 % strain relaxation of the SL stack has occurred in spite of the large total SL thickness of 2.7 μm. This indicates that even for such large thicknesses, a complete relaxation cannot a priori be assumed.

As an additional feature, for all samples additional diffraction peaks appear at angular positions which correspond exactly to "1/2" order superlattice satellite peaks, indicating the existence of an additional double period modulation along the [111] superlattice direction The intensity of these peaks is much lower than for the corresponding integer satellite peaks, and, in addition, the peak *shape*, i.e., the full width at half maximum (FWHM) is much larger. The additional modulation is related to the EuTe-PbTe thickness ratio and the strain in the SL stack and is not observed for SL samples with smaller *n*/*m* thickness ratio than about 1/5. For further investigations, a (5/22) ML SL sample was thinned for high resolution transmission microscopy (TEM) studies. The cross sectional TEM image depicted in Fig. 3 shows that the EuTe/PbTe interfaces are very abrupt and atomically flat on a length scale of at least 100 Å. This is in good agreement with the STM results which shows a lateral EuTe surface corrugation of about 200 Å (Fig. 1).

## 4. Magnetic properties

EuTe is a well known isotropic Heisenberg antiferromagnet with a Neel temperature of 9.6 K [11]. Its magnetic structure (MnO type) consists of ferromagnetically ordered $Eu^{2+}$ spins in the (111) planes but antiparallel or antiferromagnetic (AF) spin orientations in the adjacent (111) planes (type II AF). In short period (111) oriented $(EuTe)_n/(PbTe)_m$ superlattices, part of the ferromagnetic EuTe (111) planes are replaced by non-magnetic PbTe planes, and thus the influence of reduced dimensionality on the magnetic properties can be investigated.

Elastic neutron diffraction measurements [12] show the formation of a strong magnetic peak at the (1/2 1/2 1/2) position below the transition temperature, which shows that the EuTe layers in the superlattices retain their type II AF structure. However, in scans parallel to the [111] direction in Q space, additional SL satellite peaks appear around the magnetic (1/2 1/2 1/2) peak (see Fig. 4), indicating a coherence of the spin ordering along the superlattice direction between consecutive EuTe layers due to an *interlayer coupling* between the magnetic moments across the non-magnetic intervening PbTe layers. As the thickness of intermediate PbTe layers in the SLs increases from $m$ = 15 ML to 18 ML, the superlattice satellites of the magnetic peak disappear (see Fig. 4) and only a broad (1/2 1/2 1/2) peak remains, corresponding to a scattering from incoherent (uncoupled) EuTe layers in the superlattice stack. The small carrier concentration in the semiconductor samples (~ $10^{17}$ $cm^{-2}$) excludes the RKKY interaction as interlayer coupling mechanism, allthough no alternative long range coupling

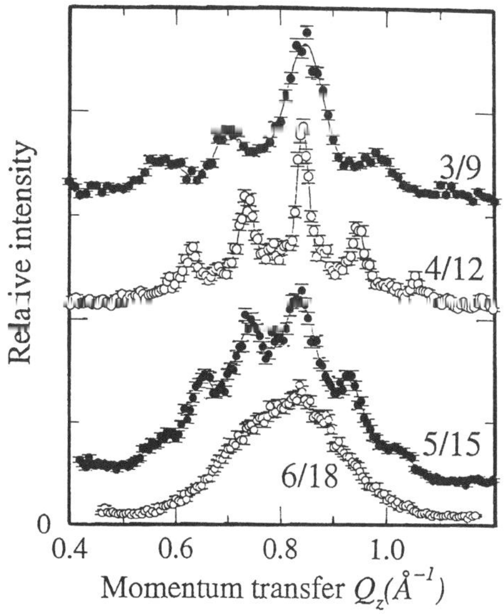


**Figure 4:** Neutron diffraction scans at 1.8 K along the [111] direction through the AF (1/2 1/2 1/2) reflection at $Q_z = 0.84Å^{-1}$ for 4 different $(EuTe)_n/(PbTe)_m$ superlattice samples with (*n*/*m*) given above.

mechanism (coupling extending to 55 Å) has yet been found, except for a recent model proposed by Bruno [13].

For the SL samples described above, temperature-dependent in-plane magnetization M(T) measurements were carried out by a SQUID magnetometer applying a weak external field in the (111) plane [14]. Whereas the M(T) curve of a reference 3 *μm* EuTe epilayer shows the typical behavior of bulk EuTe with $T_N$ of 9.6K, the M(T) of the SLs is quite different in several respects: (1) there is a large anisotropy in M(T) for external fields applied either parallel or perpendicular to the (111) growth plane, (2) with decreasing thickness of the EuTe layers, the critical transition temperature decreases from 9.6 K for *n* = 5 EuTe ML to 4.2 K for *n* = 2 ML and (3) the shape of the M(T) curves depends strongly on the number of monolayers contained in the EuTe layers in the superlattices. As illustrated in Fig. 4, for $(EuTe)_3/(PbTe)_9$ and $(EuTe)_5/(PbTe)_{15}$ (odd number of EuTe monolayers) a *ferrimagnetic-like* phase transition is observed, whereas for $(EuTe)_4/(PbTe)_{12}$ the phase transition is antiferromagnetic-like. In a simple model, the ferrimagnetic behavior is attributed to the uncompensated spin configuration in the SLs with odd number of

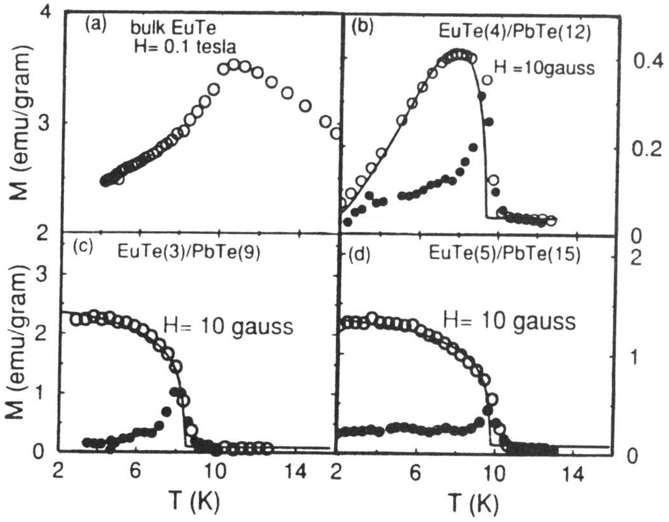


**Figure 5:** In-plane SQUID magnetization measurements after field cooling (full circles) and after zero field cooling (open circles) for (a) a bulk 3 μm EuTe film and (b) to (d) for $(EuTe)_n/(PbTe)_{3n}$ superlattice samples with $n_{EuTe}$ = 3, 4 and 5 ML. The solid line represents a mean field fit to the data.

monolayers in the EuTe layers. Taking into account the finite number of EuTe (111) planes contained in each layer, a mean field analysis including biquadratic coupling is able to reproduce the observed transition temperatures as well as the M(T) behavior of the SL samples (see solid lines in Fig. 4). More details are given in Ref. [14].

In our model, the striking difference between the SL samples with even or odd number of monolayers contained in the EuTe layers is based on an only partial compensation of the staggered magnetic moments in the EuTe AF II structure modulated along the [111] direction. Therefore, knowledge of the exact individual layer thicknesses in the real SL samples is of crucial importance for the interpretation of the magnetization data. Up to now, we have only considered an ideal structure in which an integer number of monolayers is always contained in each layer. However, not only do our x-ray diffraction data indicate that the EuTe layer thicknesses in the SL samples are 5-10 % smaller than the nominal values, but also the inherent monolayer surface fluctuations during layer-by-layer epitaxial growth lead to an additional variation in the layer thickness within each layer. In the (4/12) SL sample, for example, due to the incompletion of the $4^{th}$ monolayer, the EuTe layer consists in part of 3 ML and of 4 ML and, additionally, due to thickness fluctuations, small parts may even include 2 or 5 ML. Finally, intermixing or interdiffusion at the heterointerfaces cannot be totally excluded. To quantify these effects, more work has to be done to characterize the interface properties of the SLs and the theoretical description of the magnetization has to be refined in order to include thes layer thickness fluctuations.

## 5. Conclusions

Short period $(EuTe)_n/(PbTe)_{3n}$ superlattices were grown by MBE and characterized by STM, x-ray diffraction and TEM indicating a high structural perfection of the samples. Neutron diffraction and SQUID magnetization studies on the SL samples reveal very interesting magnetic properties due to the low dimensionality of the magnetic system and to an interlayer coupling between magnetic layers separated by non-magnetic spacer layers.

This work was supported by the Deutsche Forschungsgemeinschaft # SP 466/1-1 and by NSF # DMR-92-01878.

## References

[1] Partin D L 1988 *IEEE J. Quantum Electron.* QE-24 1716-1726.
[2] Feit Z, Kostyk D, Woods R J and Mak P 1991 *Appl. Phys. Lett.* 58 343-345.
[3] Springholz G, Yuan S, Bauer G and Krenn H 1993 *Mat. Res. Soc. Symp. Proc.* 301 353-358.
[4] Springholz G and Bauer G 1993 *Phys. Rev.* B48 10998-11009.
[5] Springholz G and Bauer G 1995 *J. Appl. Phys.* in print.
[6] Snyder C W, Orr B G, Kessler D and Sander L M 1991 *Phys. Rev. Lett.* 66 3032-6.
[7] Srolowitz D J 1989 *Acta Metall.* 37 621-5 Tersoff J and LeGoues F K 1994 *Phys. Rev. Lett.* 72 3570-3.
[8] Springholz G, Frank N and Bauer G 1994 *Appl. Phys. Lett.* 64 2970-2.
[9] Matthews J W and Blakeslee A E 1974 *J. Cryst. Growth* 27 118-125.
[10] Frank N, Springholz G and Bauer G 1994 *Phys. Rev. Lett.* 73 2236-9.
[11] Wachter P 1979 *Handbook of the Physics and Chemistry of the Rare Earths* Vol I (North Holland) 507-74.
[12] Giebultowicz T M, Nunez V, Springholz G, Bauer G, Chen J J, 1995 *J. Magn. Magn Mat.* 140-144.
[13] Bruno P 1994 *Phys. Rev.* B49 13231-64.
[14] Chen J J, Wang Z H, Dresselhaus M S, Dresselhaus G, Springholz G and Bauer G 1994 *Solid State Electronics* 37 1073.

*Inst. Phys. Conf. Ser. No 144: Section 3*
*Paper presented at 7th Int. Conf. on Narrow Gap Semicond. Santa Fe 9–12 January*



# Magnetoreflectivity study of electron-LO phonon interaction in $Pb_{1-x}Eu_xTe$ and $PbTe/Pb_{1-x}Eu_xTe$ Multi-Quantum Wells

Shu Yuan*[a], H. Krenn[a], G. Springholz[a], G. Bauer[a], and P. J. McCann[b]
a) Institut fuer Halbleiterphysik, Universitaet Linz, A-4040 Linz, Austria
b) University of Oklahoma, School of Electrical Engineering, Norman, OK 73019

**Abstract.** Molecular beam epitaxy(MBE) grown n-type $Pb_{1-x}Eu_xTe$ epilayers and $PbTe/Pb_{1-x}Eu_xTe$ multi-quantum well (MQW) samples were studied at 4 K by magnetoreflectivity in Faraday configuration (B//[111]) in magnetic fields up to 6 Tesla. A strong electron-LO phonon interaction due to the large Froehlich coupling constants is observed in the cyclotron masses for the $Pb_{1-x}Eu_xTe$ epilayers. The cyclotron masses of quasi two dimensional (2D) carriers in PbTe wells (n= 1.5 - 3x$10^{11}$ $cm^{-2}$ ) of $PbTe/Pb_{1-x}Eu_xTe$ MQW samples do not exhibit a significant resonant electron-polar phonon interaction.

## I. Introduction

During the past years a large number of studies were devoted to the investigation of modifications of the conduction and valence band structure due to nonparabolicity and electron-phonon interaction in polar semiconductors (polaron coupling). Magneto-transmission and reflectivity measurements were used to study bulk semiconductors, as well as quasi-two dimensional (2D) systems like heterostructures and quantum wells [1]. These investigations were focused on III-V and II-VI compounds. However, in the IV-VI compounds, which have comparatively high values of Froehlich coupling constants, no detailed investigations were performed. Early experimental observations of cyclotron resonance (CR) pinning phenomena in 3D systems were reported by Burkhard et al [2] for bulk-like PbTe epilayers, and theoretical investigations were carried out by Vogl et al. [3] who considered occupation effects due to the comparatively high carrier concentration in the IV-VI compounds. Recently, magnetotransmission studies on p-$Pb_{1-x}Eu_xTe$ (x=0.02) were reported by Karczewski et al. [4], however, no hole-phonon coupling could be observed. In 2D structures of IV-VI compounds, CR measurements were performed only in a rather limited magnetic field range [5-6], and thus no attention was paid to the presence of polaron pinning phenomena.

In this paper, we report on systematic magnetoreflectivity measurements carried out on $Pb_{1-x}Eu_xTe$ epilayers (n=2.2x$10^{16}$ to 5.0x$10^{16}$ $cm^{-3}$, x< 0.03) and on $PbTe/Pb_{1-x}Eu_xTe$ MQW samples in the Faraday geometry (magnetic field B // [111], up to 6 T) at 4.2 K using a Fourier transform infrared (FTIR) spectrometer. The samples were grown by MBE on (111) $BaF_2$ substrates [7]. Optical transmission and photoconductivity data for these samples have been previously reported [8-9]. In the $Pb_{1-x}Eu_xTe$ epilayers the electron mobility decreases from about $10^5$ $cm^2/Vs$ for x = 0.01 to 2x$10^4$ $cm^2/Vs$ for x = 0.03 (T< 10K). The $PbTe/Pb_{1-x}Eu_xTe$ MQW samples with $d_{QW}$= 90 - 120 Å were grown on about 5μm thick $Pb_{1-x}Eu_xTe$ buffers. The electron mobility in the PbTe wells of the four $PbTe/Pb_{1-x}Eu_xTe$ MQW samples studied, is more than $10^5 cm^2/Vs$, and the 2D carrier concentration varied from about n=1.5x$10^{11}$ $cm^{-2}$ to 3x$10^{11} cm^{-2}$. As reported previously [7], parallel conduction in the $Pb_{1-x}Eu_xTe$ buffer layer cannot be avoided in such samples.

## II. Magnetoreflectivity of PbTe and $Pb_{1-x}Eu_xTe$.

The energy eigenstates of bulk $Pb_{1-x}Eu_xTe$ in magnetic field are strongly influenced by the LO-phonons through Froehlich coupling[1,10], which is described by the Froelich constant α [10] :

$$\alpha = \frac{e^2}{4\sqrt{2}\pi\varepsilon_o}\left(\frac{1}{\varepsilon_\infty} - \frac{1}{\varepsilon_o}\right)\left(\frac{M_\perp^-}{\omega_{LO}\hbar^3}\right)^{1/2}$$

where $\varepsilon_\infty$ and $\varepsilon_o$ are the high frequency and static dielectric constants, respectively, and $M_\perp^-$ is the conduction band transverse effective mass [11]. The n=1 Landau level becomes double-valued for magnetic fields above a certain value where the CR frequency ($\omega_{CR}$) is close to the LO-phonon frequency $\omega_{LO}$. In Fig.1, Landau levels for PbTe are calculated taking into account band nonparabolicity (band parameters: Ref. 12) and electron-phonon interaction according to Ref. 10. For the epilayers grown along [111] direction, one of the four conduction band valleys (longitudinal valley) is oriented along the growth direction and three valleys (oblique valleys) are along $[\bar{1}11]$ directions in **k**-space. The n=1 level is split for B > 2.9 T for the $\ell$-valley and for B > 6 T for the o-valleys. Because of the many valley band structure two Froehlich constants exist for the different valleys with α=0.12 ("$\ell$") and α=0.18 ("o"). For $Pb_{1-x}Eu_xTe$, α increases with Eu content due to the increase of effective mass and decrease of $\varepsilon_\infty$ [8].

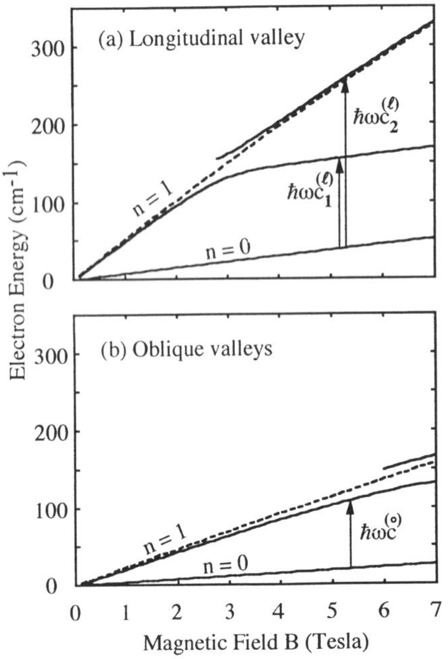


Fig. 1. Coupled LO-phonon electron energy levels for PbTe. The lowest Landau levels are obtained from Ref. 6. Dashed lines are uncoupled levels.

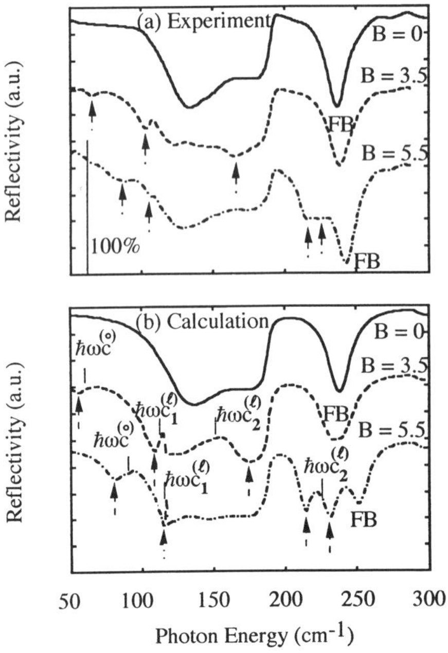


Fig.2. Experimental and calculated reflectivity spectra of PbTe#127 for B=0, 3.5, and 5.5 T. B=0 and B=3.5 T data were shifted up by 1 and 0.5, respectively. CR and DA are indicated by bars and arrows, respectively. FB denotes Fabry-Perot fringes.

Experimental magneto-reflectivity spectra for a PbTe epilayer (sample #127, n=6x10$^{15}$cm$^{-3}$, $\mu$=465000 cm$^2$/Vs at 20K, d=4.5$\mu$m) are shown in Fig. 2a for magnetic fields of B= 0, 3.5 and 5.5 Tesla. For B=0, the effective plasma edge is close to the LO-phonon energy (117 cm$^{-1}$). The sharp increase in the reflectivity at about 190 cm$^{-1}$ indicates the TO-phonon frequency ($\omega_{TO}$) of the $BaF_2$ substrate [6] and the dip at 240 cm$^{-1}$ is due to Fabry-Perot interference effects. In magnetic fields, CRs cause dielectric anomalies (DAs), which are evidenced by dips in the reflectivity spectra (indicated by dashed arrows). For a numerical simulation we use the proper dielectric function for the lead compounds [2] modified by Harper's expressions [10] for the electron-LO phonon interaction [11]. The magnetoreflectivity spectra shown in Fig.2b were calculated considering also the multilayer reflectivity by the transfer matrix method. Since in the FTIR experiments no circularly polarized radiation was used, the calculated reflectivity is an average of reflectivities for left and right hand circularly polarized light. All parameters for PbTe were taken from Ref. 12; for the $BaF_2$ substrate, $\hbar\omega_{TO}$ = 189 cm$^{-1}$, $\hbar\omega_{LO}$ = 290 cm$^{-1}$, $\varepsilon_o$ = 1.93, and broadening $\gamma$ = 3 cm$^{-1}$ were used [6]. The results are shown in Fig.2b with CR and DA positions indicated by full and dashed arrows, respectively. Figure 3 shows a fan chart for the PbTe sample of Fig.2a, where also the upper CR ($\hbar\omega_{c2}^{(\ell)}$) and associated DA branches are clearly observed. The lower branches for transitions involving the longitudinal ($\hbar\omega_{c1}^{(\ell)}$) and oblique masses ($\hbar\omega_c^{(o)}$) pin to the phonon energy as expected from the calculated Landau levels shown in Fig. 1. For o-valleys, the CR mass is much larger than for the $\ell$-valley (see Fig. 1b), and the splitting of n=1 Landau level occurs at B$\approx$ 6 Tesla, close to the upper limit of our experimental magnetic field range.

From similar results for $Pb_{1-x}Eu_xTe$ epilayers, band parameters for different Eu-contents were obtained (interband momentum matrix elements $2P_\perp^2 / m_o$=5.3eV, 5.2 eV, 4.8 eV, and 4.0 eV for x=0.99%,1.30%, 1.85%, 2.5%, respectively; $2P_{//}^2 / m_o$=2.5eV for x=0.99%). Our values are somewhat different from those obtained by Karczewski et al [4]. In our experiments, due to larger effective masses in the wider gap $Pb_{1-x}Eu_xTe$ samples [8], we clearly see pinning in the lower branch, whereas the upper branches for the $\ell$-valley are not observed in the experiments below B = 6T.

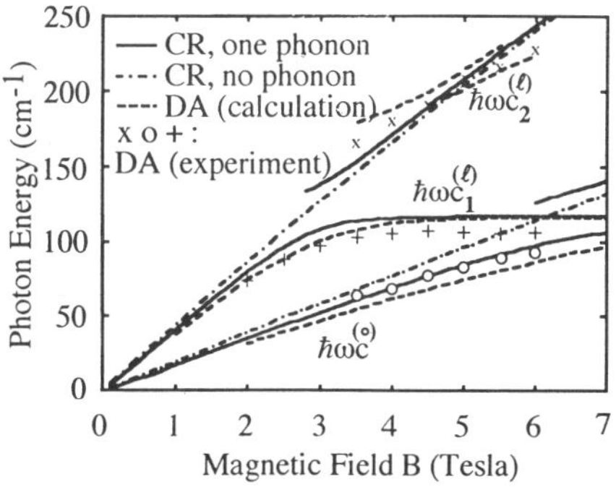


Fig. 3. Fan chart for PbTe#127 showing CR energies and DAs with and without taking into account the electron-LO phonon interaction.

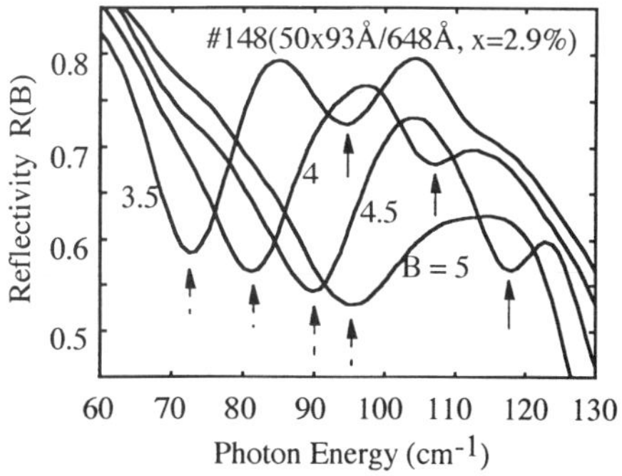


Fig. 4. Measured magnetoreflectivity spectra R(B) for PbTe/$Pb_{1-x}Eu_xTe$ MQW#148. DAs associated with CRs in the PbTe QWs and in the $Pb_{1-x}Eu_xTe$ buffer are indicated by solid and dashed arrows, respectively.

III. Magnetoreflectivity of $PbTe/Pb_{1-x}Eu_xTe$ MQW's

The $PbTe/Pb_{1-x}Eu_xTe$ MQW samples were characterized by Shubnikov-de Haas (SdH) measurements and by mid-infrared photoconductivity spectroscopy. The latter experiments show clearly the (1-1) interband transitions involving electrons in the i=1 electronic subbands of the valence and conduction bands, both in the $\ell$-valley and o-valleys of PbTe QWs [9]. Envelope function approximation calculations for these samples show that the $(1\text{-}1)^o$ interband transition energy is considerably larger than the $(1\text{-}1)^\ell$ one [9], in agreement with photoconductivity measurements. The observation of the $(1\text{-}1)^\ell$ interband transition indicates that the $1^o$ subband is not populated by electrons and hence only the $i=1^\ell$ subband of the $\ell$-valley of the PbTe QWs is populated by the electrons. For the $Pb_{1-x}Eu_xTe$ buffer layers of these MQW samples ($x \approx 3\%$), only the $\ell$-valley is populated by electrons, since strain pushes the oblique valleys by a few meV up with respect to the longitudinal valley [6].

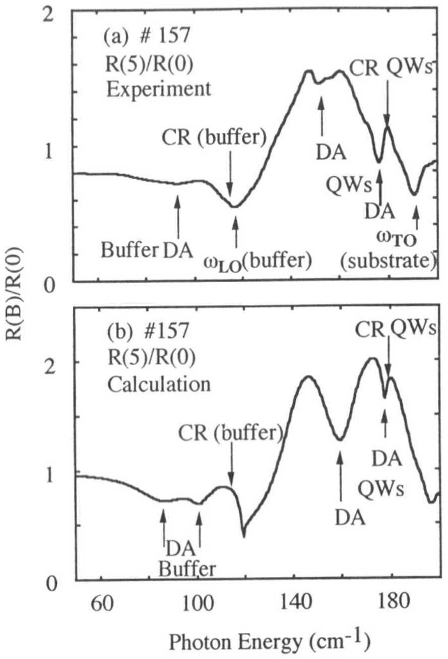


Fig. 5. Measured (a) and simulated (b) R(B)/R(0) spectra for $PbTe/Pb_{1-x}Eu_xTe$ #157 (30x118Å/486Å, $Pb_{1-x}Eu_xTe$ buffer: 4.6μm, x=1.5%). Arrows indicate CR and DA positions.

In Fig.4, as an example, magnetoreflectivity spectra for $PbTe/Pb_{1-x}Eu_xTe$ #148 are plotted for different magnetic fields. Dips due to DAs associated with CR of electrons both in the PbTe QWs as well as in the $Pb_{1-x}Eu_xTe$ buffer layers are clearly observed as indicated by arrows. In Fig. 5, the measured and calculated R(B=5T)/R(0) spectra for $PbTe/Pb_{1-x}Eu_xTe$ #157 are compared. For the 2D dielectric function in the simulations we use the expressions given in Ref. 13. We find that no polaron correction is necessary for the explanation of the 2D CRs close to the LO phonon energy. In our multilayer samples the CRs in PbTe QWs are commonly manifested by two adjacent dips in the reflectivity. However, in contrast to the bulk CRs, the higher frequency dip is very close to the 2D CR energy. From the fit it turns out that the 2D CR damping parameter $\omega_\tau$ is about 1 $cm^{-1}$, while the 3D CR damping parameter γ' below LO-phonon frequency is 5 $cm^{-1}$ in qualitative agreement with the lower mobility data in $Pb_{1-x}Eu_xTe$ [7]. In Fig.6 we plot the experimental position of the reflectivity dip associated with the 2D CR of the $i = 1^\ell$ level in the PbTe wells as a function of magnetic field for all four MQW samples. The 2D effective mass of PbTe QWs is inversely proportional to the (1-1) interband transition energy, i.e., the QW band gap. The calculated $E_{11}$ QW gaps at B=0 for these samples are given in Fig. 6. In a magnetic field, the corresponding 2D effective masses increase with magnetic field at more or less the same rate for all the samples, due to the same band nonparabolicity [11]. The larger the QW gap, the smaller the CR mass, and consequently the higher the CR energy. $PbTe/Pb_{1-x}Eu_xTe$ samples #148 and #156 have the same (1-1) QW gap, and thus in Fig. 6 the measured reflectivity dip positions for these two samples are very close to each other.

Landau-level fan charts for the MQW sample #156 are plotted in Fig.7. The experimental reflectivity dip positions were determined from the R(B)/R(0) spectra. From Figs. 6 and 7 it is clear that no LO-phonon pinning effect is observed for the upper branch of the reflectivity dips (dielectric anomalies).

In the MQW samples, the 2D CR is difficult to observe in magnetoreflectivity at low magnetic fields. The data shown in Fig. 7 for fields higher than 2 Tesla demonstrate that the comparatively broad features in magnetoreflectivity (Fig.5) result from the CR of the carriers in the longitudinal valley in the $Pb_{1-x}Eu_xTe$ buffer layer, i.e., the electrons responsible for the observed parallel conduction in the Hall measurements[7]. This buffer CR clearly exhibits an LO phonon pinning behavior. The lower branch associated with the CR in the PbTe QW wells seems to indicate a pinning behavior, too. However, it is not due to a change of the 2D cyclotron mass with magnetic field, but due to a superposition of the electromagnetic waves in the multilayer stack that includes the $Pb_{1-x}Eu_xTe$ buffer.

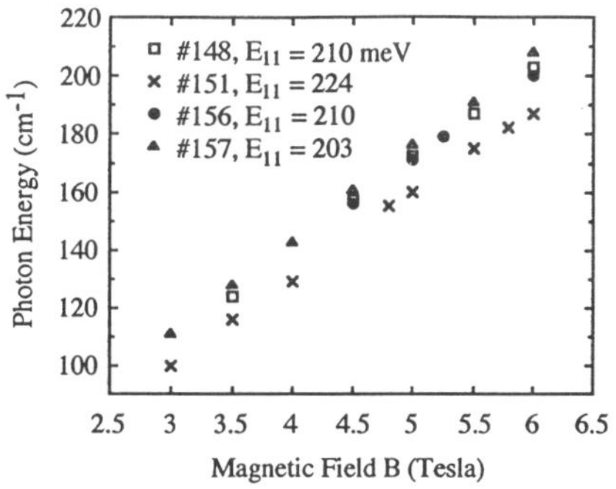


Fig. 6. Measured DAs associated with the CRs in the PbTe QWs for four samples. Sample parameters: #151: 46x62Å/621Å, $Pb_{1-x}Eu_xTe$ buffer: 5.2μm, x=2.6%; #156: 39x91Å/641Å, $Pb_{1-x}Eu_xTe$ buffer: 5.0μm, x=2.5%.

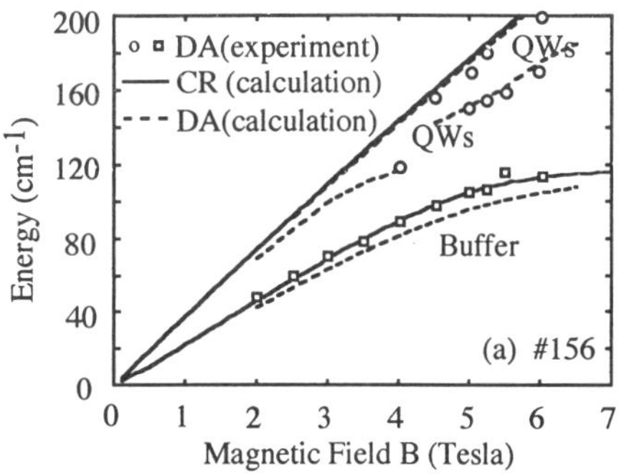


Fig. 7. Fan chart for $PbTe/Pb_{1-x}Eu_xTe$ MQW#156, showing experimental and calculated DAs and CR energies vs. magnetic field for both PbTe QW and $Pb_{1-x}Eu_xTe$ buffer layer.

IV. Discussion and Conclusions

Investigations on III-V compound heterostructures have shown that the polaron mass enhancement and the resonant coupling to the LO phonons depend sensitively on the carrier concentration [13], e.g., no apparent polaron effect was observed by Ziesmann et al. in an InAs QW sandwiched between GaSb layers with a 2D carrier concentration of 8.8 x $10^{11}$ $cm^{-2}$ [14]. Wu et al. [1] performed an extensive theoretical study of polaron-cyclotron resonance masses in QW´s as a function of electron density and quantum well width, including the finite width effect of the 2D layer and screening of the electron phonon interaction within the static random phase approximation. From a quantitative comparison with experimental data by Singleton et al. [15] on GaAs QW´s, it was concluded that polaron effects are quite weak in QW's. On the basis of the data for III-V QW´s and that the carrier concentration in the $PbTe/Pb_{1-x}Eu_xTe$ QW samples is in the range between $1.5x10^{11}$ $cm^{-2}$ and $3x10^{11}$ $cm^{-2}$, it is not astonishing that polaron effects are basically absent in the 2D CR of the $PbTe/Pb_{1-x}Eu_xTe$ MQW samples.

In summary, we have investigated cyclotron resonance in $Pb_{1-x}Eu_xTe$ single layers and $PbTe/Pb_{1-x}Eu_xTe$ QW samples using FTIR magnetoreflectivity spectroscopy, and obtained good agreement between measured and simulated magnetoreflectivity spectra. Furthermore, for PbTe wells in the $PbTe/Pb_{1-x}Eu_xTe$ MQW samples, the 2D cyclotron mass and its dependence on QW gap and magnetic field were obtained. Due to electron-LO-phonon interaction (Froehlich polaron), obvious polaron pinning effects are observed in 3D case both in the $Pb_{1-x}Eu_xTe$ epilayers and in the $Pb_{1-x}Eu_xTe$ buffer layers of the $PbTe/Pb_{1-x}Eu_xTe$ MQW samples. In contrast, for the 2D carriers in the PbTe QWs the resonant electron-LO phonon interaction is quite small.

Work supported by FWF, Vienna, Austria .

References.

[1] X. Wu, F. M. Peeters and J. T. Devreese Phys.Rev. B40, 4090 (1989) and references therein.
[2] H. Burkhard, G. Bauer, and W. Zawadzki, Phys. Rev. **B19**, 5149 (1979).
[3] P. Vogl, and P. Kocevor, 1978, in: Proc. Int. Conf. on Physics of Seiconductors (Institute of Physics, London)p.1317
[4] G. Karczewski, J. K. Furdyna, D. L.Partin, C. N.Thrush, J. P. Heremans, Phys.Rev. B**46**,13331 (1992).
[5] L. S. Kim, H. D. Drew, R. E. Doezema, J. P. Hermans, D. L. Partin ,Phys.Rev.B **35**, 2521 (1987).
[6] G. Bauer, in Landau Level Spectroscopy, edited by G. Landwehr and E. I. Rashba, (Elsevier Science Publishers, Amstdam, 1991), p. 277.
[7] G. Springholz, G. Ihninger, G. Bauer, M. M. Olver, J. Z. Pastalan, S. Romaine, and B. B.Goldberg, Appl. Phys.Lett. **63**, 2908 (1993).
[8] S. Yuan, H. Krenn, G. Springholz, and G. Bauer, Phys. Rev. **B47**, 7213 (1993).
[9] S. Yuan, G. Springholz, G. Bauer, and H. Kriechbaum, Phys. Rev. **B49**, 5476 (1994).
10] P. G. Harper, Proc. Phys. Soc. **192**, 793(1967).
[10] L. Swierkowski, W. Zawadzki, Y. Guldner and C. Rigaux, Solid State Commun. **27**, 1245 (1978).
[11] S. Yuan, Ph. D. dissertation, University of Linz, 1994, unpublished.
[12] G. Bauer, H. Pascher, and W. Zawadzki, Semicond. Sci. Technol., **7**, 703(1992).
[13] M. Ziesmann, D. Heitmann, L. L. Chang, Phys. Rev. **B35**, 4541(1987).
[14] R. J. Nicholas, J. C. Portal, C. Houlbert, P. Perrier, and T. P. Pearsall, Appl. Phy. Lett. **34**, 492(1993).
[15] J. Singleton, R. J. Nicholas, D. C. Rogers, C. T. B. Foxon, Surf. Sci. **196**, 429 (1988).

* present address: School of Electrical Engineering, University of Oklahoma, 202 W. Boyd, Room 219, Norman, OK73019. e-mail: lepmsy@mailhost.ecn.uoknor.edu. After September 1, 1995: Department of Electronic Materials Engineering, Research School of Physical Sciences and Engineering, Australian National University, Canberra, ACT 0200, Australia.

*Inst. Phys. Conf. Ser. No 144: Section 3*
*Paper presented at 7th Int. Conf. on Narrow Gap Semicond. Santa Fe 9–12 January*



# Many-Particle Effects in Magneto-Photoluminescence from Epitaxial Films of Narrow Gap IV-VI Semiconductors

R. Rupprecht, H. Pascher

Experimentalphysik I, Universität Bayreuth, D-95440 Bayreuth, Germany

**Abstract.** Existing models for radiative interband recombination, valid only for idealised band structures, are extended to the case of many-valley semiconductors with aspherical surfaces of constant energy. The theoretical lineshape calculations are based on the (combined) density of states of valence and conduction band in a magnetic field and on the occupation of the Landau levels at finite temperatures by free carriers due to doping as well as due to photoexcitation. The models explicitly take into account band gap renormalisation due to many-particle effects, carrier heating by the exciting laser radiation, possible violation of momentum conservation and onset of stimulated emission.

The lineshape models are used for the interpretation of photoluminescence spectra of a variety of IV-VI semiconductors. In this paper we exemplary present the results obtained on $Pb_{1-x}Eu_xSe$ epitaxial films (Eu contents x approximately 4 %) with well-known energy band structures. Photoluminescence was excited by the radiation of a Q-switched Nd:YAG laser at different pump intensities around and well above the threshold for the onset of stimulated emission. Experiments were done at a temperature of 1.7 K in magnetic fields up to 7 T.

Increasing excitation intensity shifts the spectra to lower energies, demonstrating that at any pump level a homogeneous nonequilibrium electron hole plasma is observed. Plasma densities up to $5\times10^{17}$ cm$^{-3}$ yielded a band gap reduction up to 7 meV. Keeping the excitation intensity constant, the band gap renormalisation is increased proportional to $B^{1/3}$ by 4 meV in a magnetic field of 4 T. These numerical values, surprisingly high for materials with high dielectric constants and small effective masses, clearly demonstrate that many-particle effects in IV-VI compounds must not be neglected.

## 1. Introduction

Photoluminescence (PL) is one of the most important experimental methods for the band gap determination of semiconductor materials. The detailed understanding of the luminescence of IV-VI narrow gap semiconductors is of particular importance due to their application as laser diodes operating in the mid infrared.

Up to now, the modifications of the band structure of these compounds by many-particle effects (exchange and correlation energy terms) have been neglected using the argument of high dielectric constants and small effective masses. It has to be checked whether this approximation is justified or not when performing PL experiments on the IV-VI'es, because there are possible effects of the high free carrier concentrations due to photoexcitation as well as due to doping of the samples. By means of a detailed analysis of PL lineshapes and -positions it is shown in the present work, that there are indeed experimentally observable effects of many-particle interactions. These effects have to be taken into account in order to determine an exact value for the *single*-particle band gap. Using simple methods for this aim, like evaluating the peak or the low-energy side of a PL spectrum, would cause an systematic error.

## 2. Effects of many-particle interactions

In general, many-particle interactions affect both PL lineshapes and -positions. One reason for their effect on PL line*shapes* calculated in the single-particle picture could be the renormalisation of matrix elements. However, the energy dependence of the corresponding enhancement factor [1], which describes the increase of the total rate of radiative interband transitions, shows negligible variation over the spectral range of a photoluminescence line. In addition, the small renormalisation of transverse and longitudinal effective masses $m_{t/l}^{CB}$ and $m_{t/l}^{VB}$ of conduction (CB) and valence band (VB) as well as of the mass anisotropies $\gamma_e = m_t^{CB} / m_l^{CB}$ and $\gamma_h = m_t^{VB} / m_l^{VB}$ [2][3] do not change the PL lineshapes noticeable. As a consequence, the main effect of many particle interactions is their influence on PL line*positions*. The CB and VB energy dispersion relations are nearly rigidly shifted on the energy scale due to exchange and correlation energy terms, resulting in a red shift of the PL spectra.

*Without external magnetic field*, the kinetic energy as well as the exchange energy of the interacting carriers in conduction and valence band of a many-valley semiconductor with ellipsoids of revolution as surfaces of constant energy can be obtained *quantitatively*.

Assuming quadratic energy dispersion laws - in the case of free carrier densities smaller than about $5\times10^{19}$ cm$^{-3}$ this approximation is valid for lead chalcogenides, if one is interested in the densities of states [4] - the Hartree-Fock calculation for $T=0$ [5] results in a average kinetic energy per electron which is *independent* on the shape of the Fermi surface:

$$\frac{E_{kin,e}}{N_e} = \frac{3}{5}\varepsilon_F^{CB} = \frac{3}{10}\frac{\hbar^2 (3\pi^2 n / \nu)^{2/3}}{m_d^{CB}} \quad \text{with } \varepsilon_F^{CB} = \hbar^2 k_{F,e}{}^2 / (2 m_d^{CB}), \tag{1}$$

where $\nu$=4 is the number of equivalent valleys, $n=N_e/V$ the density of free electrons and $m_d^{CB}=((m_t^{CB})^2 m_l^{CB})^{1/3}$ denotes the CB density-of-states effective mass. The meaning of the other symbols is as usual. The average exchange energy per electron can be obtained *exactly* by the following integration over the ellipsoidal Fermi volume $V_{F,e}(\gamma_e)$:

$$\frac{E_{exch,ee}}{N_e} = -\frac{1}{n}\frac{1}{4\pi\varepsilon_r\varepsilon_0} \iint\limits_{V_{F,e}(\gamma_e),\vec{k} \neq \vec{k}'} \frac{d\vec{k}\, d\vec{k}'}{(2\pi)^6} \frac{4\pi e^2}{|\vec{k} - \vec{k}'|^2} = -\frac{e^2}{4\pi\varepsilon_r\varepsilon_0}\frac{3}{4\pi} k_{F,e}\varphi(\gamma_e), \tag{2}$$

where $\varphi(\gamma_e)$ explicitly accounts for the anisotropy:

$$\varphi(\gamma_e) = \left[\gamma_e^{1/6} \arcsin\sqrt{1-\gamma_e}\right] \Big/ \sqrt{1-\gamma_e} \quad \text{for } \gamma_e < 1. \tag{3}$$

For the interacting holes in the valence band the Hartree-Fock energy terms $E_{kin,h}/N_h$ ($p=N_h/V$) and $E_{exch,hh}/N_h$ are calculated in complete analogy to Eqs. (1)-(3).

Considering a hole concentration due to doping of $p_{dop}=1.3\times10^{17}$ cm$^{-3}$ and using the high frequency dielectric constant $\varepsilon_{r,\infty}$=25.2 of PbSe, the CB and VB kinetic and exchange energy terms are plotted in Fig. 1a for $Pb_{1-x}Eu_xSe$, x=0.041. Note, as a consequence of the assumed p-type background doping, the VB Hartree-Fock energy terms do not extrapolate to zero in the limit $r_{s,e}\rightarrow\infty$.

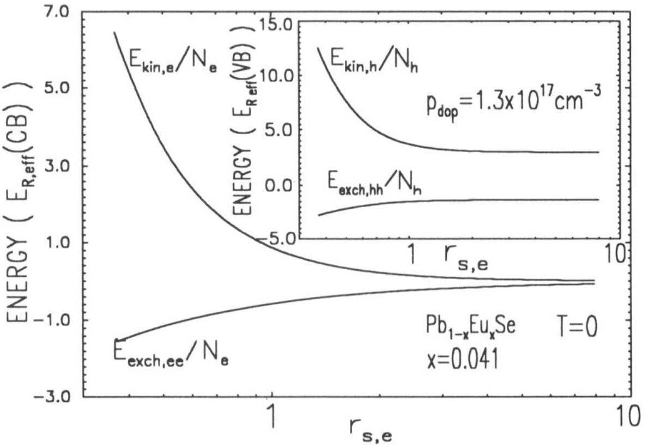


*Fig.1a: CB Hartree-Fock energy terms in units of the effective Rydberg $E_{R,eff}(CB)\equiv E_R m_d^{CB}/\varepsilon_r^2$=1.69meV as a function of the interparticle spacing $r_{s,e}\equiv[3/(4\pi n)]^{1/3}/[a_0\varepsilon_r/m_d^{CB}]$ (in units of the effective Bohr radius) for $Pb_{1-x}Eu_xSe$, x=0.041. Inset: VB energy terms in units of $E_{R,eff}(VB)$=1.60meV as a function of $r_{s,e}$.*

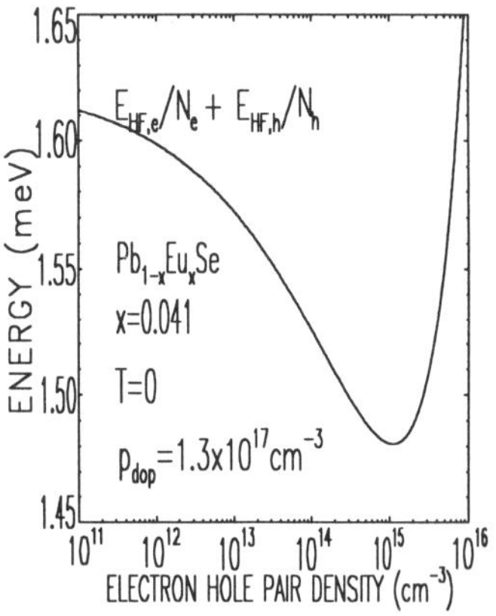


*Fig.1b: Sum of CB and VB Hartree-Fock energies as a function of the electron hole pair density for $Pb_{1-x}Eu_xSe$, x=0.041.*

As shown in Fig. 1b, the sum of the CB and VB Hartree-Fock energies remains a *positive* quantity in the whole range of the electron hole pair density. In addition, the resulting minimum at the small equilibrium electron hole pair density of about $1\times10^{15}$ cm$^{-3}$ with a depth of $\cong 0.14$ meV $\hat{=}$ 1.6K is *not* pronounced. For these reasons, a homogeneous nonequilibrium electron hole plasma is expected to be observed performing PL experiments on the IV-VI'es.

As another consequence of the high densities of photoexcited carriers ($10^{16}$ cm$^{-3}$ to $10^{18}$ cm$^{-3}$ in typical PL experiments) and of the comparatively short carrier lifetimes in the direct gap IV-VI'es, the effective temperature $T_{eff}$ of the carrier subsystem differs considerably from the lattice temperature $T_L$. This carrier heating is also caused by the quantum energy of the exciting Nd:YAG laser radiation, which is much larger than typical band gaps of the samples. So even at lattice temperatures of $T_L$=1.7 K the carrier densities as well as the integral occurring in the exchange energies (see Eq. (2)) are calculated numerically.

Correlation energies, that means interaction energy terms going beyond the Hartree-Fock approximation [6], are considered according to Eq. (4) and the analogous one for the holes.

$$E_{corr,ee}/N_e = \left[0.0622\ln(r_{s,e}\nu^{1/3}) - 0.096\right] E_{R,eff}(CB) \text{ for } r_{s,e} < 1 \tag{4}$$

In order to calculate the band gap reduction $\Delta\varepsilon_g$, in addition to the sum of the CB and VB exchange and correlation energy terms ($\rightarrow$ electron hole plasma shift $\Delta\varepsilon_{EHP}(\varepsilon_r)$) the effect of electron(hole)-LO-phonon coupling has to be considered in case of strongly polar semiconductors. The corresponding polaron shift $\Delta\varepsilon_{P,e}+\Delta\varepsilon_{P,h}$ [7] is:

$$\Delta\varepsilon_{P,e} + \Delta\varepsilon_{P,h} = -\alpha_e\varphi(\gamma_e)\omega_{LO} - \alpha_h\varphi(\gamma_h)\omega_{LO}, \tag{5}$$

where $\omega_{LO}$ denotes the frequency of the LO-phonon and $\alpha_{e/h}$ are polar coupling constants:

$$\alpha_{e/h} \equiv (e^2/4\pi\varepsilon_0)\left(1/\varepsilon_{r,\infty} - 1/\varepsilon_{r,0}\right)\sqrt{m_d^{CB/VB}/(2\hbar\omega_{LO})}. \tag{6}$$

In the limit of very small electron and hole densities (plasma frequency much smaller than LO-phonon frequency) the band gap reduction is given by the sum of the EHP-shift, calculated using the static limit $\varepsilon_{r,0}$ of the background dielectric constant, and of the polaron shift ("$\varepsilon_{r,0}$ -approximation" [7]). In contrast, in the high density case the polaron shift is negligible compared to the EHP-shift, the latter calculated using the high frequency dielectric constant ("$\varepsilon_{r,\infty}$-approximation" [7])).

To our knowledge, *in the case of external magnetic fields* a quantitative treatment of many-particle interactions is possible only if one assumes idealised band structures (single parabolic spherical bands). Calculations in the limit of high magnetic fields and in the high density limit of an equilibrium electron hole plasma yield power-law dependencies of the exchange energies on the magnetic field strength, like e. g. $|E_{exch,ee}/N_e+E_{exch,hh}/N_h|$ proportional to $B^{1/3}$ [8]. Despite these approximations concerning the band structure are not valid in case of the IV-VI'es, our experimental data also yield the $B^{1/3}$-law (see Fig. 4).

### 3. Magneto-photoluminescence lineshapes

A detailed analysis of (magneto-)photoluminescence lineshapes was performed in order to get information whether many-particle interactions cause experimentally observable effects in case of the VI-VI'es. In the following, the lineshape models are given explicitly for non-vanishing magnetic fields; B=0 is the easily found limes. However, only for B=0 the band gap reduction $\Delta\varepsilon_g$ occuring in the corresponding equations can be calculated quantitatively and therefore the *single*-particle band gap can be obtained for real band structures.

For interband transitions (quantum energy $\hbar\omega$) *conserving* the component $k_B$ of the electronic wave vector along the magnetic field, the microscopic net stimulated emission rate $dr_{stim}(\hbar\omega)$ is proportional to the absorption constant $\alpha(\hbar\omega)$ [9]:

$$dr_{stim}(\hbar\omega) \propto (\hbar\omega)^2\alpha(\hbar\omega), \tag{7}$$

The microscopic spontaneous emission rate $dr_{spon}(\hbar\omega)$ is obtained from $dr_{stim}(\hbar\omega)$ as [9]:

$$dr_{spon}(\hbar\omega) \propto (\hbar\omega)^2\alpha(\hbar\omega)\left\{1-\exp\left[(\hbar\omega - \mu^{CB} + \mu^{VB})/(k_B T_{eff})\right]\right\}^{-1}, \tag{8}$$

where $\mu^{CB}$ and $\mu^{VB}$ denote the CB and VB quasi-Fermi levels for T=$T_{eff}$ and $k_B$ Boltzmann's constant. As a consequence, the absorption constant is the fundamental quantity:

$$\alpha(\hbar\omega) \propto \sum_{\Phi} (B/\hbar\omega) Z(\Phi) \sqrt{m_B^{red}(\Phi)}\, |P_B(\Phi)|^2 \times \tag{9}$$

$$\times \sum_{\substack{\text{allowed}\\ \text{transitions}}} \Gamma_{\tau\alpha}(1/\sqrt{\hbar\omega - \hbar\Omega}) \left[f_e^{VB}(\varepsilon_a) - f_e^{CB}(\varepsilon_b)\right] \text{ with } \hbar\Omega \equiv \varepsilon_g + \Delta\varepsilon_g + \tilde{\varepsilon}_L^{CB} - \tilde{\varepsilon}_L^{VB}.$$

In Eq. (9), $f^{CB/VB}$ denote the CB and VB Fermi distribution functions and $\Gamma_{\tau\alpha}$ stands for the usual broadening function of Lorentzian type [10] with the argument including the energies of the Landau levels $\tilde{\varepsilon}_L^{CB/VB}$ [11] respective to their renormalised band edges. The one summation in Eq. (9) runs over all interband transitions allowed according to the selection rules valid for the IV-VI'es [11]. The other sum extends over the different types of valleys, characterised by the angle $\Phi$ between the main ellipsoidal axis and the magnetic field direction and by the number $Z(\Phi)$ of equivalent ellipsoids. The effective masses along the magnetic field [10] and therefore the reduced mass also, as well as the interband matrixelement [11] depend on $\Phi$:

$$m_B^{CB/VB}(\Phi) = m_l^{CB/VB}\cos^2(\Phi) + m_t^{CB/VB}\sin^2(\Phi), \tag{10}$$

$$|P_B(\Phi)|^2 = P_\perp^{\,2}\sqrt{\cos^2\Phi + (P_\perp / P_\parallel)^{-2}\sin^2\Phi}, \tag{11}$$

where $P_\perp$ and $P_\parallel$ are the interband matrixelements perpendicular and parallel to the [111] direction, respectively. The energies of the CB and VB levels fulfilling $\varepsilon_b(k_B)$-$\varepsilon_a(k_B)$=$\hbar\omega$ were calculated using the parabolic approximation concerning $k_B$:

$$\varepsilon_a(k_B) = (m_B^{red}/m_B^{CB})\tilde{\varepsilon}_L^{VB} + (m_B^{red}/m_B^{VB})(\varepsilon_g + \Delta\varepsilon_g + \tilde{\varepsilon}_L^{CB} - \hbar\omega) \text{ and} \tag{12}$$

$$\varepsilon_b(k_B) = (m_B^{red}/m_B^{VB})(\varepsilon_g + \Delta\varepsilon_g + \tilde{\varepsilon}_L^{CB}) + (m_B^{red}/m_B^{CB})(\tilde{\varepsilon}_L^{VB} + \hbar\omega). \tag{13}$$

In order to calculate the microscopic emission rates in case of interband transitions taking place under *violation* of $k_B$-conservation [9], one has to integrate over all occupied electron and hole states using the individual CB and VB densities of states corresponding to Eq. (9).

The macroscopic emission rates, that means the PL lineshapes, are proportional to the microscopic ones in case of a linear optical process like *spontaneous* emission. In contrast, the model of a unsaturated one-dimensional optical amplifier [12] was used in order to describe the exponential amplification of the recombination radiation in case of stimulated emission.

## 4. Experimental results on band gap renormalisation

The observed dependence of the red shift of the PL spectra measured for B=0 on the carrier densities is in agreement with the $\varepsilon_{r,\infty}$-approximation (see Fig. 2a). Despite yielding a still larger total self-energy shift, the $\varepsilon_{r,0}$-approximation (including polaron shift) results in exchange and correlation energy terms being too weakly dependent on the plasma density. The lineshape fits result in a single-particle band gap of 274.3meV, a value which was confirmed by interband magnetotransmission performed with very low radiation powers.

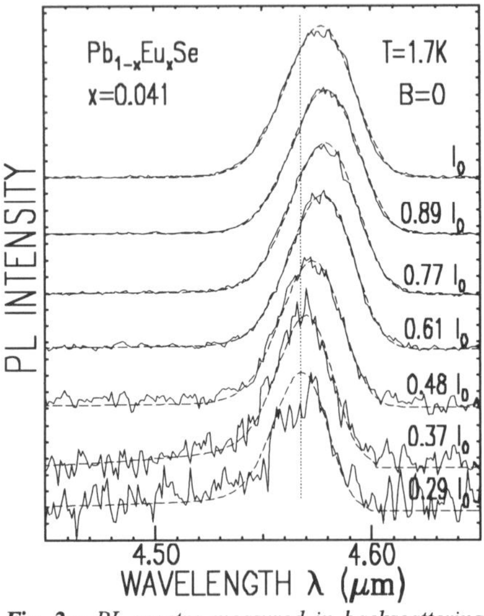


*Fig. 2a: PL spectra measured in backscattering geometry with different excitation intensities ($I_0$=1.3GW/m$^2$). Broken lines: calculated lineshapes (k-conserving interband transitions)*

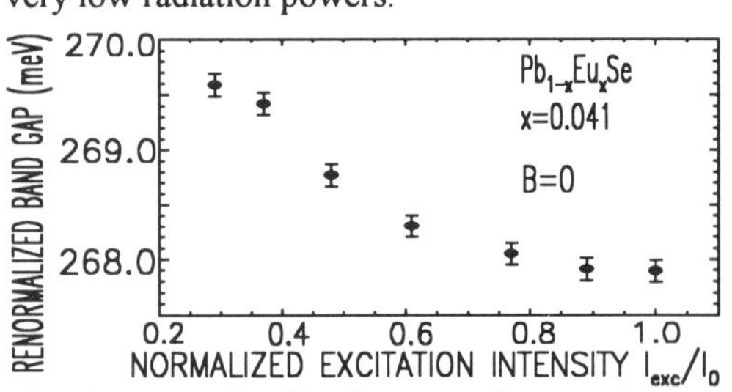


*Fig. 2b: Renormalised band gap as a function of excitation intensity for the spectra plotted in Fig. 2a ($\varepsilon_{r,\infty}$-approximation).*

A possible influence of an external magnetic field B on many-particle interactions has been investigated using $Pb_{1-x}Eu_xSe$ samples with very well-known band structures, determined by CARS (coherent anti-Stokes Raman scattering) and transmission magnetospectroscopy. Keeping the excitation intensity constant, the observed blue-shift with increasing B is *smaller* than expected according to the well-known B dependence of the Landau levels of the x=0.041 $Pb_{1-x}Eu_xSe$ sample (compare calculated and experimental spectra in Fig. 3). Thus there must be a magnetic field dependence of the many-particle energy terms. Indeed, in order to obtain agreement between measured spectra and calculated lineshapes, the value of the renormalised band gap for B=0, which is the energy reference for the calculation of the Landau levels, has to be lowered with increasing B. In case of the spectra shown in Fig. 3, this red shift (0.3 meV between B=4T and B=7T) is clearly observable, but not very large due to the *low* constant excitation intensity. In addition, only for $B \geq 3T$ the magnetic field dependence of the band gap reduction follows the $B^{1/3}$-law, valid for high plasma densities. Therefore, in order to make band gap renormalisation effects as large as possible, very ***high*** but still constant excitation levels were applied. In this case, the $B^{1/3}$-law fits the experimental data very well even at small magnetic fields, yielding a band gap reduction of 4meV at 4T (Fig. 4).

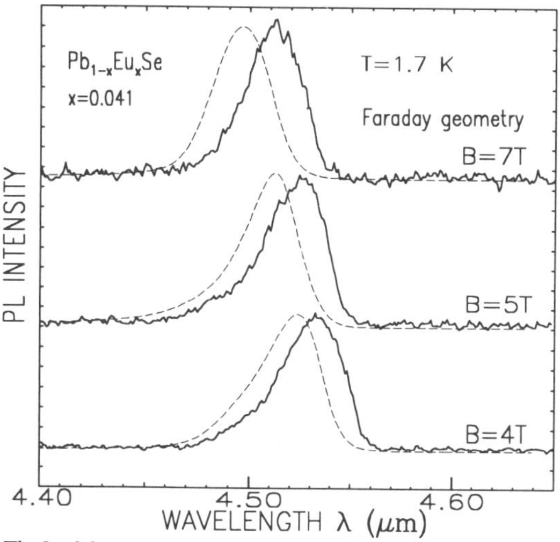


***Fig.3**: Magneto-PL spectra measured in backscattering geometry with low, but contant excitation intensity. Broken lines: calculated lineshapes (interband transitions violating $k_B$-conservation), taking the magnetic field dependence of the band gap renormalisation not into account.*

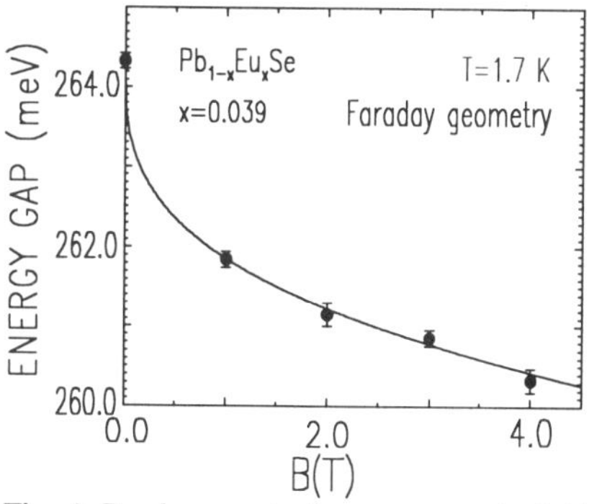


***Fig. 4**: Band gap reduction vs. magnetic field as obtained for constant excitation intensity. Full line: $B^{1/3}$-dependence.*

## Acknowledgement

Helpful discussions with G. Bauer are gratefully acknowledged. Samples were given by M. Tacke.

## References


[1] Rice T M 1977 in *Solid State Physics* ed. by Ehrenreich H, Seitz F and Turnbull D, vol 32 (New York: Academic) p 1

[2] Rice T M 1974 *Nuovo Cimento B* **23** 226

[3] Rösler M, Zimmermann R 1977 *Phys. Stat. Sol. (b)* **83** 85

[4] Dalven R 1973 in *Solid State Physics* ed. by Ehrenreich H, Seitz F and Turnbull D, vol 28 (New York: Academic) p 179

[5] Combescot M, Noziéres P 1972 *J. Phys. C* **5** 2396

[6] Gell-Mann M, Brueckner K A 1957 *Phys. Rev.* **106** 364

[7] Beni G, Rice T M 1978 *Phys. Rev. B* **18** 768

[8] Chui S T, 1974 *Phys. Rev. B* **8** 3438

[9] Lasher G, Stern F 1964 *Phys. Rev.* **133** 533

[10] Roth L M, Lax B, Zwerdling S 1959 *Phys. Rev.* **114** 90

[11] Bauer G, Pascher H, Zawadzki W 1992 *Semicond. Sci. Technol.* **7** 703

[12] Shaklee K L, Nahorny R E, Leheny R F 1973 *J. Lumin.* **7** 284

*Inst. Phys. Conf. Ser. No 144: Section 3*
*Paper presented at 7th Int. Conf. on Narrow Gap Semicond. Santa Fe 9–12 January*



# Upper and lower valence band energy spectra of narrow gap semiconductor solid solutions of $Bi_{2-x}Sn_xTe_3$ crystals

V.A. Kulbachinskii
Low Temperature Physics Department, Physics Faculty, Moscow State University, 119899, Moscow, Russia

M. Inoue, M. Sasaki, H. Negishi, W.X. Gao, and Y. Giman
Department of Materials Science, Faculty of Science, Hiroshima University, Higashi-Hiroshima 724, Japan

P. Lostak and J. Horak
Institute of Chemical Technology, 532 10 Pardubice, Czech Republic

Transport properties and pulsed laser induced 'transient thermoelectric effect' (TTE) along the $C_2$ axis have been measured in the temperature range 4.2-300 K for solid solutions of p-type $Bi_{2-x}Sn_xTe_3$ crystals ($0 \le x \le 0.035$). By doping Sn atoms, the hole Fermi energy of the upper valence band (UVB) is increased and the Sn-induced 'impurity' band is formed near the top of the lower valence band (LVB) lying by about 15 meV below the top of the UVB. The observed TTE voltages are characterized by a multiple relaxation process with relaxation times for thermal diffusions of photogenerated carriers. The TTE data show that there are four different relaxation times ($\tau_1$-$\tau_4$) due to holes and two $\tau_5$ and $\tau_6$ due to electrons. Combined with the static transport data, as well as existing band model, we have discussed the energy spectra of these crystals.

## 1. Introduction

Bismuth telluride $Bi_2Te_3$ is a layer semiconductor with the principal axes $C_2$ and $C_3$, whose structure consists of five atomic planes in the sequence Te(1)-Bi-Te(2)-Bi-Te(1). It has always a p-type conductivity due to high density of crystal defects, while practical applications of $Bi_2Te_3$-based compounds to thermoelectric devices require to produce both n- and p-type conductivity. The valence and conduction bands (band gap: $E_g \cong 0.15$ eV) both have many-valley structures consisting of six general ellipsoids in each case, with each ellipsoid centered on a mirror plane of the Brillouin zone.[1-4] According to Köhler,[4] the band edge density-of-states mass for a single ellipsoid at the upper valence band (UVB) top amounts to $0.106m_0$, the corresponding integral density-of-states mass for the six-valley model is $0.35m_0$, and their main axis in the mirror plane is tilted by $31.5°$; the energy separation $\Delta E$ between UVB and the lower valence band (LVB) is known to be 20.5 meV.[4] To our knowledge, much less is known about the band parameters for the LVB.

In the present work, we have carried out systematic measurements of electrical resistivities, Hall effect, Shubnikov-de Haas (SdH) effect, and pulsed laser induced 'transient thermoelectric effect' (TTE) along the $C_2$ axis over the temperature range 4.2-300 K and in magnetic fields up to 10.5 T for p-type $Bi_{2-x}Sn_xTe_3$ single crystals. In particular, the TTE method is a powerful dynamic technique for understanding transport properties of multicarrier systems of electrons and holes, as evidenced for silicon,[5] quasi-two-dimensional charge-density-wave material $\eta$-$Mo_4O_{11}$.[6] Our main concern is to study the effect of Sn doping on the electronic properties of the host $Bi_2Te_3$ and to get valuable information about its energy spectrum, in particular, the anisotropy in the lower valence and electron bands. Single crystals of $Bi_{2-x}Sn_xTe_3$ ($0 \le x \le 0.035$) were grown by a modified Bridgeman method. The experimental setup and measuring principle of the TTE method have been described earlier.[5] A pulsed laser (laser power: ~350 mJ) produced by a Nd:YAG laser source with the wavelength of 1060 nm (=1.17 eV) and pulsewidth of 8 ns was irradiated normal to one end of a crystal [see inset of Fig. 3(b)]. Laser-light induced TTE voltages were detected over the wide time range 50 ns to 500 ms by a digital storage oscilloscope through a home-made preamplifier, whose output signal was led to a computer for record and numerical analysis.

## 2. Results and discussions

### A. Transport properties

Figure 1 shows the temperature dependence of the resistivities $\rho_{C2}$ along the $C_2$ axis for $Bi_{2-x}Sn_xTe_3$ with different Sn contents. With decreasing temperature, the resistivity of undoped sample is decreased monotonically, showing a metallic behavior. For solid solutions, the $\rho_{C2}$-$T$ curves have a shoulder near 80 K ($x$=0.0025) or a maximum near 50 K ($x$=0.0125 and 0.025), but the resistivity with $x$=0.035 exhibits rather a increasing function of $T$ as the temperature is lowered. Such anomalous behaviors in the $\rho_{C2}$-$T$ curves appearing below 100 K only for doped samples are one of the characteristic features of $Bi_{2-x}Sn_xTe$ crystals, which may be due to intervalley scatterings between the UVB and Sn-induced 'impurity band' (see later). The Hall coefficient $R_H$ of undoped sample does not depend on an applied magnetic field $B$, while those of Sn-doped samples are also nearly field-independent at low fields below about 0.2 T, above which these are increased with $B$; the value of $R_H$ is found to decrease with increasing $x$.

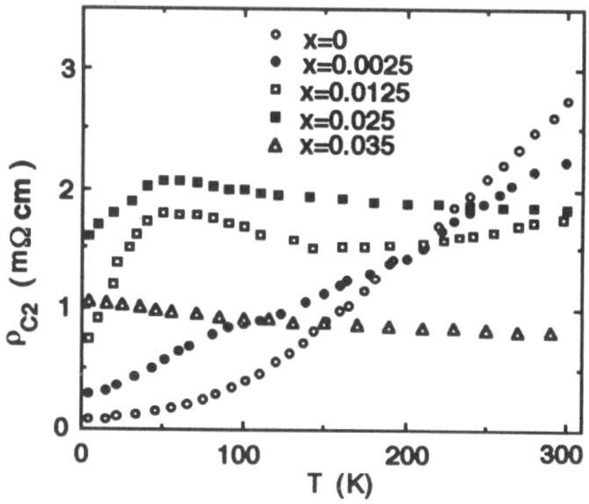


Fig. 1. Temperature dependence of resistivity.

The magnetoresistance is proportional to $B^2$ in low magnetic fields up to 0.5 T, while SdH oscillations are observable in higher $B$ for samples with $x$=0 and 0.005, those for $x$=0.015 have a extremely small amplitude, but not observable for $x$=0.035. According to the ellipsoidal nonparabolic model,[4] we have calculated the Fermi energy $E_F$ of the holes in the UVB of $Bi_{2-x}Sn_xTe_3$ with the same parameters obtained for $Bi_2Te_3$ by Köhler. According to him, the values of $(1/eR_H)^{2/3}$ evaluated from the Hall data are plotted against $E_F$ for $Bi_{2-x}Sn_xTe_3$ (solid circles) and $Bi_2Te_3$ (open circles) by Köhler for comparison in Fig. 2; the straight line is the calculated one for a constant value of the integral density-of-states mass $m(0)^* = 0.35m_0$.[4] For the case of $Bi_2Te_3$, up to $E_F$=20.5 meV the agreement between the experiments and the calculation is good, beyond which the deviation is appreciable, indicating that a second type of hole band is present at high carrier densities. On the other hand, for the case of $Bi_{2-x}Sn_xTe_3$ the deviation appears at a lower concentration or lower Fermi energy $E_F \cong 15$ meV, as marked by arrow. We have also found that $R_H$ for undoped sample is independent of a magnetic field, whereas those for Sn-doped ones are field dependent, as mentioned earlier.

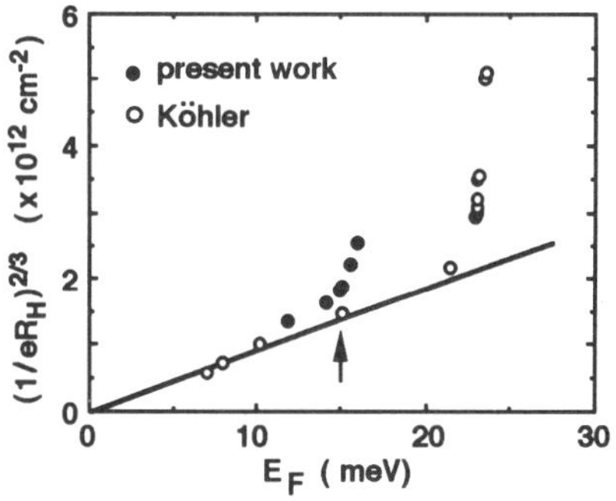


Fig. 2. $(1/eR_H)^{2/3}$ vs. $E_F$ plot.

Considering these results, we may infer that there present some Sn-induced *impurity (acceptor) band* (IB) located near the top of the LVB, which lies below the UVB top separated by 15 meV (see later). The existence of such an IB has been examined by the dynamic TTE measurements, as discussed below.

**B. TTE data**

Typical TTE signals $Bi_2Te_3$ at $T$=6 and 300 K in different time intervals are shown in Figs. 3(a) and (b). Immediately after the laser irradiation, the induced TTE voltage $V(t)$ at 300 K rises with time $t$ up to about 65 μV in the short time range 0-50 μs and then decreases in the long time scale spanning more than 200 ms; at low temperature, say 6 K, the induced voltage is enhanced to nearly 140 μV, followed by a decrease, showing a minimum, to zero at $t$ > 100 ms. These decay profiles in the time range 0-2 ms are mostly due to thermal diffusions by photogenerated 'holes' and those in the slow stage 10-200 ms by thermal phonons.[5] Similar TTE signals for $Bi_{2-x}Sn_xTe_3$ have been observed, which have additional contributions from 'electrons' in the intermediate temperature range (25 K<$T$<80 K), where a shoulder appears in the $\rho_{C2}$-$T$ curves (Fig. 1).

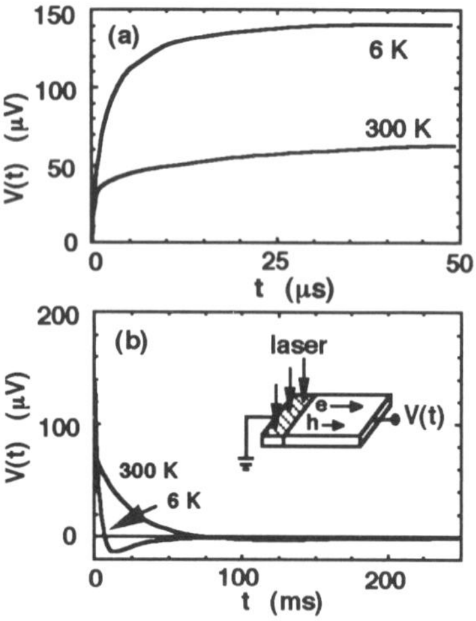


Fig. 3. Typical recorder traces of the TTE signals; (a) 0-50 μs, (b) 0-250 ms.

As for various solids,[5-7] the decay curves of TTE voltages are due to thermal diffusions of photoinduced electrons and holes, which can be expressed in the exponential form, $V(t) = V_0 + \Sigma a_i \exp(-t/\tau_i)$, where $V_0$ is a constant value at $t \to \infty$, $a_i$ a relaxation amplitude, and $\tau_i$ a relaxation time for the $i$th relaxation process or the type of carriers; $a_i$ >0 or $a_i$ <0 corresponds to a thermal diffusion of electrons or holes, respectively. For sample with $x$=0, there are four relaxation times $\tau_1$-$\tau_4$ (their relaxation amplitudes are all negative; $a_1$-$a_4$<0) that correspond to four kinds of holes; the numbering of the processes is set in the order from a fast to a slow one. The theoretical analysis of the diffusion equation[5] gives $\mu_i = eL_i^2/(2k_BT\tau_i)$, where $L_i$ is a diffusion length and $\mu_i$ a carrier mobility of the $i$th carrier. Assuming that $L_i$ is independent of the type of carriers ($L_i$ =$L$), we can evaluate a carrier mobility $\mu_i$ from the observed relaxation time $\tau_i$ for the $i$th carrier, provided that $L$ is known. However, we have no such data for $L$ and only get the ratio of $\mu_i = (\tau_j/\tau_i)\mu_j$ ($i, j$=1-4). Corresponding to the observed inequality $\tau_1<\tau_2<\tau_3<\tau_4$, we have $\mu_1>\mu_2>\mu_3>\mu_4$ for the corresponding mobilities. Therefore the carrier mobilities have been evaluated semiempirically using the observed Hall mobility $\mu_H$ (here obtained at low magnetic field 0.2 T).

Considering the possible band structure of our material system [Fig. 5(b)], we attribute the carrier transport along the $C_2$ direction to the holes with $\tau_1$ (or $\mu_1$) and $\tau_2$ ($\mu_2$) in the UVB (or light hole band) and those with $\tau_3$ (or $\mu_3$) and $\tau_4$ ($\mu_4$) in the LVB (heavy hole band). In the former case, we must consider the six equivalent valleys.[4] Thus we may assign $\tau_1$ (or $\mu_1$) to the holes of two $UVB_1$ and $\tau_2$ (or $\mu_2$) to those of four $UVB_2$. Since the relaxation times $\tau_3$ and $\tau_4$ for the LVB are much longer than those for UVB, we may neglect holes in the LVB, which may not contribute to the observed Hall mobility $\mu_H$. Thus we may write $\mu_H = (2/6)\mu_1 + (4/6)\mu_2 \cong (1/3)\mu_1$, where the contribution of $\mu_2$ is further neglected because of $\tau_1/\tau_2=\mu_2/\mu_1<<1$. Then we can evaluate semiempirically the carrier mobility $\mu_i$ ($i$=2-4) as, $\mu_i = (\tau_1/\tau_i)\ \mu_1 = 3(\tau_1/\tau_i)\mu_H$. The temperature dependence of these estimated hole mobilities is plotted in Figs. 4(a) for x=0 and 4(b) for x=0.0025, as well as the observed Hall mobility $\mu_H$ (solid squares). It is apparent that the hole mobility $\mu_i$ ($i$=1-4) obeys the power law $\mu_i \propto T^{2.0\text{-}2.5}$ at high temperatures, while these tend to saturate at lower temperatures, indicating that neutral impurity scatterings are predominant. The power law of the above form at high tempe-

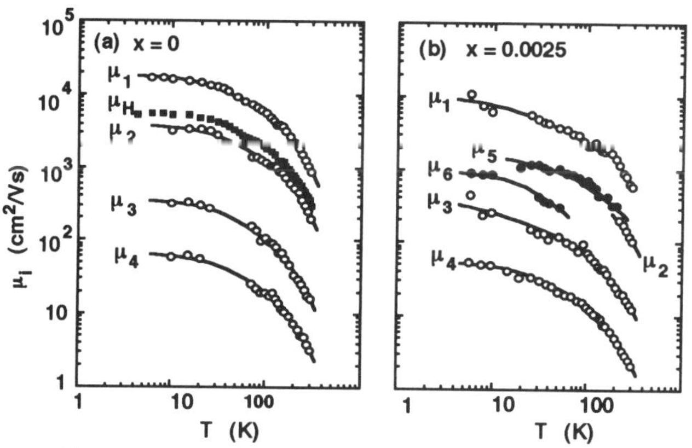


Fig. 4. Temperature dependence of the carrier mobilities $\mu_1$ - $\mu_6$ .

ratures may suggest that in this multivalley system intra- and inter-valley scatterings are effective. For solid solutions, we have observed two additional relaxation times $\tau_5$ and $\tau_6$ ($a_5$, $a_6$>0) in the intermediate temperature range, which are considered as due to two kinds of electrons. The evaluated mobilities for $x$=0.0025 are shown in Fig. 4(b); solid circles represent the components corresponding to $\tau_5$ and $\tau_6$. We see that $\mu_3$ and $\mu_4$ have nearly the same temperature dependence as those for the host material ($x$=0).

Taking into account the extensive studies for $Bi_2Te_3$,[1-4] together with the present static and dynamic data, we propose the possible band structure for $Bi_{2\text{-}x}Sn_xTe_3$, as shown schematically in Fig. 5(a). The narrow gap semiconductor $Bi_2Te_3$ ($E_g \cong 0.15$ eV) has six equivalent ellipsoidal light hole pockets (UVB) and the heavy hole pockets (LVB) lying by $\Delta E$ =20.5 meV below the top of the UVB. The lower electron band (LEB) has similarly six equivalent valleys,[8] but the upper electron band (UEB) is not well known. In addition, the Sn-induced 'impurity band' (IB) in $Bi_{2\text{-}x}Sn_xTe_3$ locates by about 15 meV below the top of the UVB. Figure 5(b) depicts the six equivalent valleys of the UVB, where two equivalent pockets and four ones that are responsible for the carrier transport along the $C_2$ direction are indicated by $UVB_1$ and $UVB_2$, which correspond to the observed relaxation times $\tau_1$ and $\tau_2$, respectively. Using the band edge parameters, the tilt angle 31.5° for $Bi_2Te_3$,[4] and the Fermi energy ($E_F \cong 15$ meV) for the UVB, we have calculated the effective masses along the $C_2$ axis; $m_1$*=0.033$m_0$ for $UVB_1$ and $m_2$*=0.091$m_0$ for $UVB_2$, as marked by arrows in Fig. 5(b). The ratio between these masses $m_2^*/m_1^*$ is 2.8 (see below).

Now the effective mass ratio for the $i$ and $j$th hole (electron) pockets can be written as, $m_i^*/m_j^* = (\tau_i/\tau_j)^{1/\gamma} = (\mu_j/\mu_i)^{1/\gamma}$, where $\gamma$ is a parameter characterizing a scattering mechanism. Assuming phonon scatterings for all carriers in each pocket ($\gamma$=5/2), we obtain the anisotropy between $UBV_1$ and

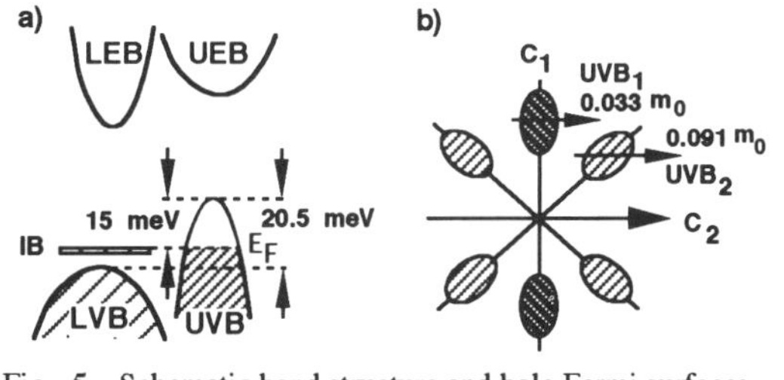


Fig. 5. Schematic band structure and hole Fermi surfaces.

$UVB_2$ along the $C_2$ direction $m_2^*/m_1^*$=1.8 for the host and 2.0 for doped samples ($x$=0.0025 and 0.0125), which are smaller than the calculated value (=2.8). These discrepancies may be due to deviations of actual crystallographic direction of samples used for TTE measurements from the true $C_2$ axis. Similarly we get $m_3^*/m_1^*$=4.9 and $m_4^*/m_1^*$=9.5 for LVB in undoped sample ($x$=0). If we cite the theoretical value of $m_1^*=0.033m_0$ obtained using the Köhler's parameters, we obtain the magnitude of each effective mass for undoped sample; $m_2^*=0.060m_0$, $m_3^*=0.16m_0$, and $m_4^*=0.31m_0$. From these data analyses, we have found that the anisotropies in the effective masses along the $C_2$ direction are 1.8-2.0 and 1.9-2.1 for UVB and LVB, respectively; the anisotropy for LVB seems to be slightly higher than that for UVB. On the other hand, our TTE signals in the intermediate temperature range show the existence of additional relaxation times $\tau_5$ and $\tau_6$ (or mobilities $\mu_5$ and $\mu_6$) in doped samples, which can be attributable to optically excited electrons from the IB into the LEB, since its location is near to the LVB compared to the UEB and the wavelength of the pulsed laser 1060 nm (=1.17 eV) is much larger than the band gap energy. The presence of such two relaxation times may suggest that the LEB for $Bi_{2-x}Sn_xTe_3$ is also anisotropic; similar evaluations of the effective masses $m_5^*$ and $m_6^*$ (>$m_5^*$) for the LEB along the $C_2$ direction yield $m_5^*=0.027m_0$ and $m_6^*=0.062m_0$, and their ratio $m_6^*/m_5^*$=2.3, while from the TTE data we obtain $m_6^*/m_5^*$=1.5. The observed anisotropy is also smaller than the theoretical one (=2.3) due to the same reason as that for UVB.

## 3. Conclusion

By doping Sn atoms into the host $Bi_2Te_3$, the hole Fermi energy of the upper valence band (UVB) is increased and a Sn-induced 'impurity band' (IB) is formed near the top of the lower valence band (LVB) lying by about 15 meV below the top of the UVB. The observed TTE voltages are characterized by a multiple relaxation process with relaxation time $\tau_i$ ($i$=1-6) for thermal diffusions of photogenerated carriers, where $\tau_1$-$\tau_4$ are due to holes in the UVB and LVB, and $\tau_5$ and $\tau_6$ (observable only in Sn doped samples) due to electrons in the LEB. The corresponding carrier mobilities $\mu_i$ are evaluated, from which we have estimated, based on the existing band model, the effective masses $m_1^* \cong 0.033m_0$ and $m_2^* \cong 0.060m_0$ for the hole Fermi surfaces of the UVB, as well as two effective masses $m_3^* \cong 0.16m_0$ and $m_4^* \cong 0.31m_0$ for the LVB along the $C_2$ direction, corresponding to $\tau_1$-$\tau_4$, respectively. Furthermore, we have obtained the electron effective masses $m_5^* \cong 0.027m_0$ and $m_6^* \cong 0.041m_0$ for the lower electron band (LEB). Our TTE data are the first direct measurements for the anisotropic LVB along the $C_2$ direction in $Bi_2Te_3$ based materials. It should also be emphasized that the Sn-induced impurity band plays an important role to transport properties, such as anomalous temperature dependence of electrical resistivity.

## References


1 J. R. Drabble, R. D. Groves, and R. Wolfe, Proc. Phys. Soc. **71**, 430 (1958).

2 R. Togel and G. R. Miller, *The Physics of Semimetals and Narrow Gap Semiconductors*, edited by D. L. Carter and R. T. Bate (Pergamon, Oxford, 1971), p. 349.

3 A. von Middendorff and G. Landwehr, Solid State Commun. **11**, 203 (1972).

4 H. Köhler, Phys. Status Solidi B **74**, 591 (1976).

5 M. Sasaki, H. Negishi, and M. Inoue, J. Appl. Phys. **59**, 796 (1986).

6 M. Sasaki, G. X. Tai, S. Tamura, and M. Inoue, Phys. Rev. B **47**, 6216 (1993).

7 M. Sasaki , G. X. Tai, M. Inoue, and H. Bidadi, J. Appl. Phys. **75**, 4520 (1994).

8 H. Köhler, Phys. Status Solidi B **73**, 95 (1976).

*Inst. Phys. Conf. Ser. No 144: Section 3*
*Paper presented at 7th Int. Conf. on Narrow Gap Semicond. Santa Fe 9–12 January*



# IV-VI misfit dislocation superlattices as zero-dimensional quantum boxes

**O A Mironov**
Institute of Radiophysics & Electronics, National Academy of Sciences of Ukraine, Kharkov 310085, UKRAINE
Department of Physics, University of Warwick, Coventry, CV4 7AL, UK

**A I Fedorenko, A Yu Sipatov**
Politechnical University, Kharkov 310002, Ukraine

**K H Herrmann, J W Tomm**
Humboldt-Universität Berlin, Institut für Festkörperphysik, D-10099 Berlin, Germany

**V I Litvinov**
Institute of Material Science Problems National Academy of Sciences of Ukraine, Chernovtsy 274001, UKRAINE

**Abstract.** We propose to generate quantum dots by a new method based on the formation of misfit dislocation in perfectly periodic arrays during the epitaxial growth of (001)IV-VI compound superlattices. Infrared photoluminescence spectra of non-uniform strained superlattices with misfit dislocations are measured at temperatures T=5K-90K. A large blue shift and temperature independent linewidth were observed in PbSe-PbS/(001)KCl superlattices with a three dimensional periodic array of misfit dislocations. We suppose that these features are due to quantum dot luminescence.

## 1. Introduction

Any new attempts to create quantum dots are of great importance because compared with 2D -structures it is more difficult to fabricate 0D and 1D objects. Recently a new way has been found to create quantum dots of InAs with the help of self-organising processes of Stranski-Krastanov island growth during MBE epitaxy on GaAs substrates [1]. Another example is the creation of 1D wires by dislocation slipping induced dimensionality reduction of a 2D-structure down to one dimension [2].

The present contribution is devoted to the investigation of the photoluminescence (PL) spectra of IV-VI superlattices (SLs) with a certain sign of lattice misfit. These SLs reveal the specific properties of 0D-systems such as a remarkable blue shift and temperature independent PL line width. We thus propose a new way to create quantum dots in IV-VI SLs via the generation of misfit dislocations during the epitaxial growth process.

It is known that in nonuniform strained IV-VI SLs there exists a nearly ideal periodic grid of interface misfit dislocations (MD) [3,4]. The lead and rare-earth chalcogenides are quite suitable for the preparation of such SLs. These materials have the NaCl-crystal structure, and provide a wide range of lattice misfit, f = 0.5 - 13 %, as well as misfit dislocation periods, D = 3 - 50 nm. Below we report the photoluminescence data and qualitatively discuss them in order to identify the sample as a quantum-dot array.

## 2. Sample characterisation

The multilayers were prepared by thermal or electron beam evaporation of lead and rare-earth chalcogenides in a vacuum chamber ( $P = 10^{-4} - 10^{-5}$ Pa ), with alternate deposition of the two compounds onto a (001)KCl substrate, in the temperature range 500 - 570 K. The growth conditions approached those of congruent sublimation, which made it possible to grow reproducibly films with a high structural quality and control the number of intrinsic point defects distributed homogeneously across the thickness and area of the film,even if no annealing took place after growth. The layer thicknesses and the growth rate were controlled by a quartz resonator. SLs with thicknesses 5 500 nm and number of layers 3-50 were grown. Structural investigation shows that growth in the <001> direction is according to a layer-by-layer Frank-van der Merwe mechanism and square grids of MDs are formed at the interface with a period D = 3 -50 nm. The MD grid periodicity is much better for a larger lattice misfit, $f > 2\%$ [4].

Such a SL with a MD grid may be considered a new kind of superlattice: *we call it a misfit dislocation superlattice* (MDSL).

The samples' structural properties were checked by X-ray diffraction (XRD) ( Fig.1, right) and transmission electron microscopy (TEM) in plan-view ( Fig.1, left ). XRD investigations supported the evidence of periodicity along the superlattice growth direction and TEM proved the existence of high quality in-plane MD grids and thus gave evidence for further dimension reduction.

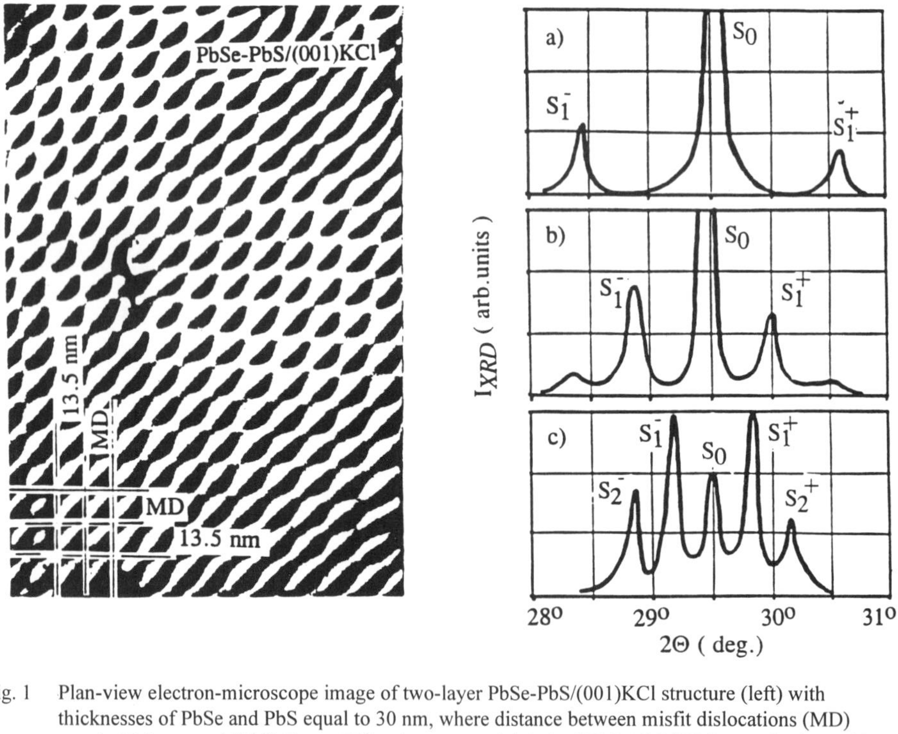


Fig. 1 Plan-view electron-microscope image of two-layer PbSe-PbS/(001)KCl structure (left) with thicknesses of PbSe and PbS equal to 30 nm, where distance between misfit dislocations (MD) equals 13.5 nm and (200) X-ray diffraction pattern ( right ) of PbSe-PbS/(001) superlattices with periods $D_{SL}$=8.1 nm (a), 15.8 nm (b) and 28 nm (c); where $S_{1,2}$- satellite reflections.

## 3. Results of photoluminescence measurements and discussion

Photoluminescence (PL) spectra of PbSe-PbS SLs were discussed earlier in [5,6] for T = 80 K - 260 K where two-dimensional size quantisation effects were observed (blue shift in PL spectra). In the above papers, only (111)PbSe-PbS SLs were investigated. In the case of <111> growth direction we have to take into account the Folmer-Weber mechanism and hence there are no misfit dislocations at all [3].We investigated the infrared PL spectra at temperatures 5K-90K in (001)-oriented PbSe-PbS SLs with MD grids and the results are shown in Fig.2. The specimen for PL measurements has 20 PbSe layers and 20 PbS ones, each 7.0 nm in thickness, and MD periodic grids with D = 13.5 nm. Quasi-stationary PL excited by a Q-switched Nd:YAG laser was measured (quantum energy 1.17eV, pulse width $\tau_{FWHM}$ = 80 ns, repetition rate 100 Hz - 1000 Hz). Due to the typical excitation density ~50 kW/cm$^2$ the emission should be considered as stimulated emission. The carriers are created with a considerable excess energy. There are four main lines with positions and widths which exhibit no temperature dependence, but show non-monotonic temperature dependence of amplitude. Compared to bulk PbSe (cf. reference [7]) and PbSe-PbS SL's without MD ( cf. references [5,6] ) where the PL lines exactly follow the $E_g(T)$ dependence of bulk material ( about 4.5×10$^{-4}$eV K$^{-1}$ )

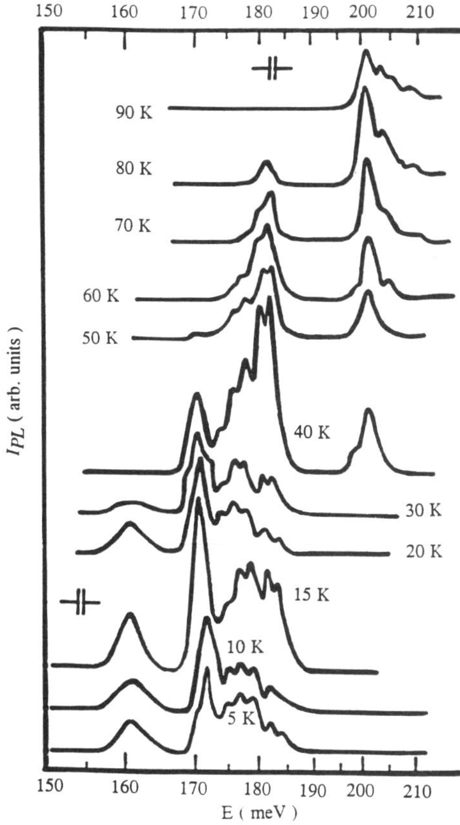


Fig.2 Photoluminescence emission spectra of PbSe-PbS/(001)KCl superlattice with misfit dislocation network (sample #543-1) at temperatures T = 5K - 90K. Superlattice period $D_{SL} = d_{PbSe} + d_{PbS}$ = = 14 nm ($d_{PbSe} = d_{PbS}$) , the number of periods - 20.

and the SL ground state ( for which the temperature dependence in this kind of SL is not much different from $E_g(T)$ of the well), respectively, we have found here *for lead salts a very unique* situation: *the PL lines change neither their positions nor width even at medium ( e.g. T = 50K ) temperatures.*

A MDSL has a compositional modulation along the normal to SL plane and additionally a deformation induced band gap modulation by periodic in plane MD arrays (fig.3). The MD stresses have been calculated and it appears that the amplitude of the oscillations of the MD stresses near interface are as high as $10^9$ Pa (Fig.3, left). The local position dependence of the PbSe energy gap at T=300K is schematically shown in Fig.3 (right). The co-ordinates (X,Y) denote the position in the interface plane and in the perpendicular direction, respectively. One can see from this picture that lattice deformation and gap modulation decrease with the distance from interface Y. The strain variation of the band gap width $\Delta E_g$ under the strain $\sigma_{11}$ could be obtained with the help of $\Delta E_g = (\delta E/\delta p)\times\sigma_{11}$ where, $(\delta E/\delta p)$ for PbSe equals $9\times10^{-11}$ eV/Pa. From the above estimation it is clear that, if we consider real values of normal strain $\sigma_{11}$ near the phase interface PbSe-PbS (fig.5), the variation of the band gap width $\Delta E_g(Y)$ for PbSe can be 30%-15% for distances from the interface of Y=1nm and Y=2nm respectively. Hence the modulation of the energy gap is deep enough to form electron-hole quantum confinement in only some parts of the interface plane and confines particles in 0D-quantum dots.

It is very important, that for a certain sign of the misfit in PbSe-PbS heterointerfaces the deformation changed PbSe energy gap $E_g(X)$ has minima in between two dislocation lines, which are separated by a distance of 13.5 nm. It means that each optically active PbSe region, bounded by four MD lines and two hetero-interface PbS-PbSe and PbSe-PbS planes, is free of traps which are a placed close to MD lines: thus the properties are sufficient for the PL signal observation. Compared to $E_g(T)$ [7], blue shifts of 15 - 30 meV were observed in the emission spectra in Fig.3 ( $E_g$=145 meV for bulk PbSe at T = 5 K-10 K ).

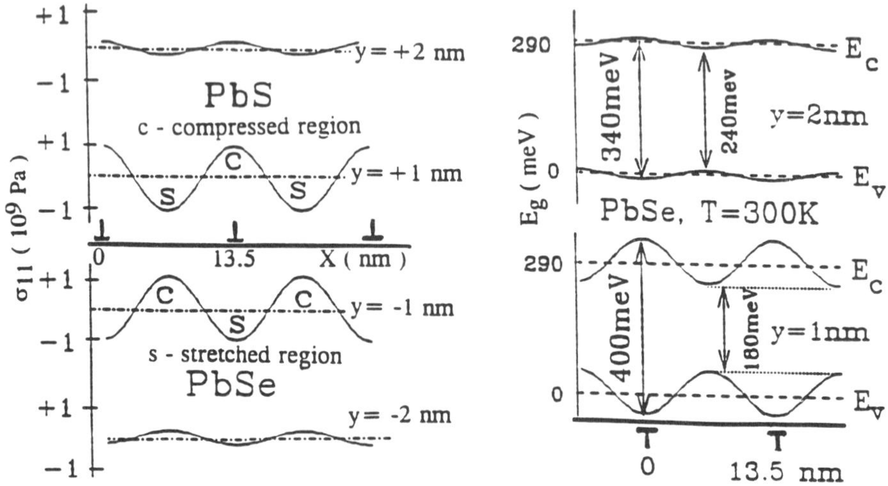


Fig. 3 Normal stress $\sigma_{11}$ distribution in two-layer PbSe-PbS/(001)KCl structure near the interface (left) and the PbSe energy band gap $E_g$ modulation by misfit dislocation stresses in abovestructure at T=300K (right), where : y - distance from the PbSe-PbS interface ;⊥- misfit dislocation position.

The positions as well as the width of the four lines we observed in PL spectra do not depend on temperature, which is due to the 0D size-quantization of the PbSe quantum dots. The same temperature independence of widths of the PL lines has been experimentally observed for Q-dots of III-V systems previously [1].

Using our data from Fig.1 [5] one can estimate size-quantization subband energies for PbSe-PbS SL (with 7.0 nm PbSe thickness) without MDs, e.g. $E_g^*$=195 meV at T=5K and $E_g^*$=225 meV at T=80 K. These estimations contradict present MDSL data (30 and 45 meV "red shift") for the low energy PL lines at 5K and 80K, respectively. Additionally, it is possible to show that in the two band isotropic approximation for the L-point of the PbSe Brillouin zone, in the case of full confinement (quantum box with sizes ~10 nm) it should be the same order of magnitude for increasing inter band energies $E_g^{**}$~20-35 meV for T =5K-40K. Unfortunately, it is again in contradiction with the present temperature independent PL line positions (Fig.2).

In the model of quantum dots [8] there exist four 'intraband' dipole transitions which change only the state of one particle in the electron-hole pair. The energy of such transitions should not depend on the band gap $E_g(T)$ and consequently on temperature. It is reasonable to suppose that the width of the lines is caused mainly by quantum dot size fluctuation, which is also temperature independent.

## 4. Conclusions

Thus we propose that epitaxially-grown dislocation-induced superlattices offer an interesting new object of extremely reduced dimensionality.

The MDLSs described above require not only further study of their optical properties but also investigation of the set of special properties such as the unusual high $T_c$-like superconductivity in SL (001) PbTe-PbS [9].

## References


[1] Bimberg D, Grundmann M, Ledentsov N N, Ruvimov S S, Werner P, Richter U, Gösele U, Heydenreich J, Ustinov V M, Egorov A Yu, Zhukov A E , Kop'ev P S , and Alferov Zh I 1994 *Abstracts Booklet. Workshop on MBE Growth Physics and Technology* (December 12-15, Institute of Vacuum Technology Warsaw, Poland) CI 2

[2] Guasch C, Voillot F, Goiran M, Peyrade J P, Bedel E, Fontaine C, Atmani H, and Rocher A 1994 *Solid State Electronics* **37** 567-569

[3] Palatnik L S and Fedorenko A I 1981 *J.Crystal Growth* **52** 917-924

[4] Michailov I F, Savitskii B A, Sipatov A Yu, Fedorenko A I and Shpakovskaya L P 1983 *Sov.Phys.Solid State* **25** 1166-1171

[5] Kolesnikov I V,Kovalev A N, Sipatov A Yu, Paramonov V I, Fedorenko A I, and Yunovich A E 1989 *Sov.Phys.Semicond.* **23** 601-603

[6] Herrmann K H, Auth J, Mollmann K P, Tomm J W, Bottner H, Lambreht A, Tacke M, Kolesnikov I V, Yunovich A E, Fedorenko A I, Mironov O A, and Sipatov A Yu 1993 *Semicond. Sci. Technol.* **8** S176-S179

[7] Preier H 1979 *Appl.Phys* **20** 189-206

[8] Haug H, Koch S W 1990 *Quantum Theory of the Optical and Electronic Properties of Semiconductors* (Singapore: World Scientific)

[9] Mironov O A, Zorchenko V V , Sipatov A Yu, Fedorenko A I, Nashchekina O N, and Chistyakov S V 1993 *Defects and Diffusion Forum* (Scitec Publication LTD, Switzerland) **103-105** 473-484; Mironov O A, Makarovskii O N, Fedorenko A I, Sipatov A Yu, Nashchekina O N, Zaritskii I M, and Konchits A A 1994 *Acta Physica Polonica* **A85** 603-606

*Inst. Phys. Conf. Ser. No 144: Section 4*
*Paper presented at 7th Int. Conf. on Narrow Gap Semicond. Santa Fe 9–12 January*



# RHEED and XPS Study of Oxide Desorption from InSb(100) Substrates

W.K. Liu and M.B. Santos

Laboratory for Electronic Properties of Materials and Department of Physics and Astronomy
University of Oklahoma, Norman, OK 73019

**Abstract.** Oxide desorption from InSb(100) substrates is studied using RHEED, XPS, AES and SEM. We observe a multi-stage desorption process for surfaces prepared using either of two common chemical etchants. A composite oxide containing $In_2O_3$ and $Sb_2O_5$ is evident in as-loaded substrates. During the first stage, most of the Sb oxide is reduced, leaving behind a thin oxide layer consisting predominantly of $In_2O_3$. As $T_{sub}$ is raised further in the presence of an Sb flux, desorption of $In_2O_3$ is evident. SEM images of substrates removed from the MBE chamber at various times during the desorption process indicate that In droplets condense on the surface before a smooth morphology ensues. We suggest that In reduced from $In_2O_3$ form droplets which then react with the supplied Sb flux to form InSb. Annealing at this temperature eventually produces a smooth, stoichiometric surface which results in a well-defined, Sb-stabilized pseudo-(1×3) RHEED pattern. InSb films grown on such substrates exhibit good structural and electrical properties.

The preparation of smooth, oxide-free substrate surfaces is a prerequisite for high quality film growth by molecular beam epitaxy (MBE). The growth and subsequent *in-situ* removal of native oxides has proven to be the most successful method for cleaning Si and GaAs substrates for MBE work [1]. Ion bombardment of InSb substrates followed by annealing is very effective in the removal of active contaminants such as oxygen and carbon, but has several disadvantages. Indium droplets can form due to preferential sputtering [2], crystallographic damage occurs [3], and the electrical properties of the surface layer may be modified [4,5]. The passivating oxide layer formed on InSb substrates during chemical etching is also not as readily desorbed thermally as that of other substrate materials. The main difficulty in thermal desorption arises from the low melting point of InSb (~527°C) and the low vapour pressure of one of its constituent oxides ($In_2O_3$) [6]. We present a detailed study of the oxide desorption mechanism in InSb(100) substrates using reflection high energy electron diffraction (RHEED), X-ray photoelectron spectroscopy (XPS), Auger electron spectroscopy (AES) and scanning electron microscopy (SEM). Complete oxide removal can be achieved through thermal desorption while preserving the stoichiometry and smooth morphology of the substrate surface.

Thermal desorption and RHEED measurements were carried out in the growth chamber of an Intevac Gen II MBE system loaded with solid elemental sources of In (7N,

Nikko Kyodo) and Sb (6.5N, Dowa). XPS and AES experiments were conducted at room temperature in an adjoining analysis chamber to eliminate complications caused by capping layers and exposure to atmosphere. All samples reported here were n-InSb(001) substrates (Johnson Matthey) cleaned *ex-situ* using a modified CP4A etchant ($HNO_3$ : $CH_3COOH$ : HF : DI $H_2O$ @ 2:1:1:10) [7]. Substrates prepared using a mixture of lactic acid : HCl : $HNO_3$ in the proportion of 25:4:1 [8-10] exhibit similar behavior. Substrate temperatures ($T_{sub}$) were deduced from readings of a thermocouple located at the center of the substrate heater and their reproducibility was ensured by monitoring the c(4×4)↔pseudo-(1×3) transition ($T_{tr} \approx 380°C$) [11] and the melting point of In.

The evolution of RHEED patterns observed along $[1\bar{1}0]$ during oxide desorption is shown in Fig. 1. The corresponding XPS and AES spectra are shown in Figs. 2 and 3, respectively. Shifts in the positions of the XPS peaks from the same element in different surroundings can be attributed to a rearrangement of the charge distribution of valence electrons [12]. In this study, chemical shifts were used to detect the presence of oxide layers on the substrate surface. Previous experiments have established that the predominant oxide species are the relatively volatile $Sb_2O_5$ and the ionic $In_2O_3$ which has a very high melting point and low vapor pressure [7].

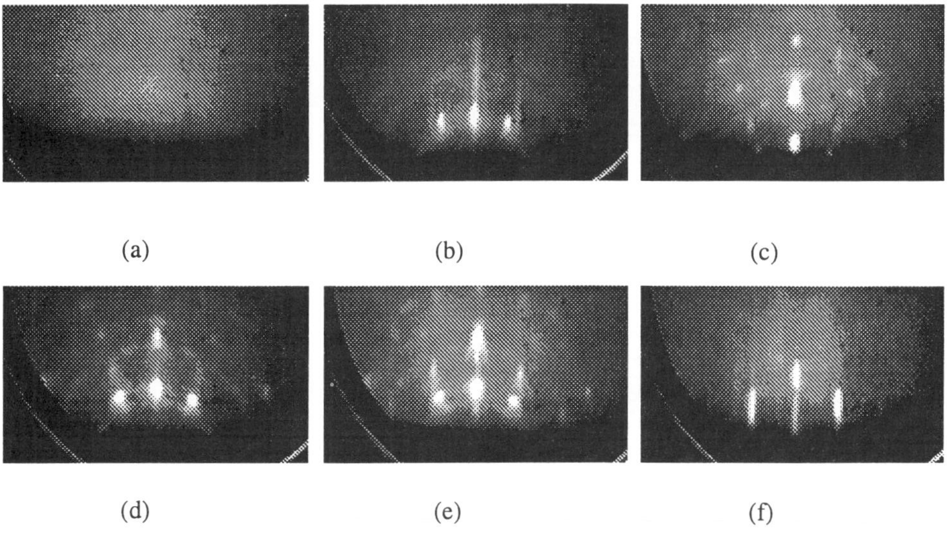


Fig. 1 RHEED patterns taken at different stages of the oxide desorption process along $[1\bar{1}0]$ at (a) $T_{htn}$ = 250°C, (b) $T_{tr}$-25°C, (c)-(e) $T_{tr}$+40°C-50°C w/$Sb_4$, (f) $T_{tr}$+60°C w/$Sb_4$.

Diffuse rings with a strong background are observed in the RHEED pattern after initial outgassing (Fig. 1a). Composite structures are clearly visible in the Sb3d XPS spectra where the low energy peaks (~537.5eV for $Sb3d_{3/2}$ and ~528.2eV for $Sb3d_{5/2}$ ) belong to Sb in bulk InSb and the larger peaks at higher BE correspond to oxidized Sb species. In both transitions, the observed chemical shifts of ~2.5eV indicate that the dominant Sb oxide is $Sb_2O_5$. A further complication occurs in the $Sb3d_{5/2}$ case due to the proximity of the Sb oxide peak to the O1s line at ~531eV. Since these peaks cannot be resolved in our XPS spectra, the high energy peak (@ ~530.5eV) must be treated as a superposition of two peaks. The In3d peaks appear singular but are slightly asymmetric with high energy tails. Since chemical shifts of the In3d levels are very small [13], overlapping bulk and oxide signals result in such a lineshape. These observations and the large oxygen peak in the corresponding AES spectra (Fig. 3a) are consistent with the presence of a mixed surface oxide layer ($In_2O_3$+$Sb_2O_5$).

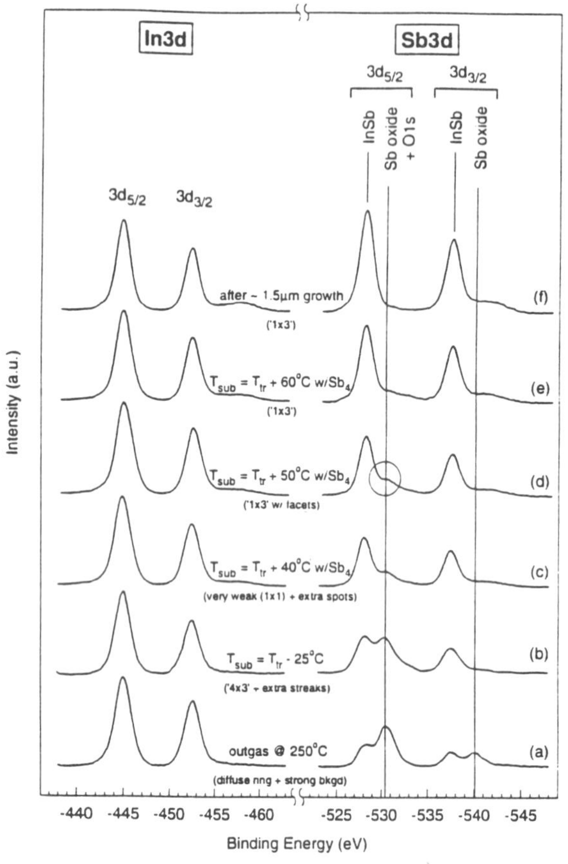


Fig. 2 XPS 3d spectra of InSb(100) substrate showing the progressive removal of oxides by thermal desorption (AlKα, $E_p$=15keV, HV=20eV).

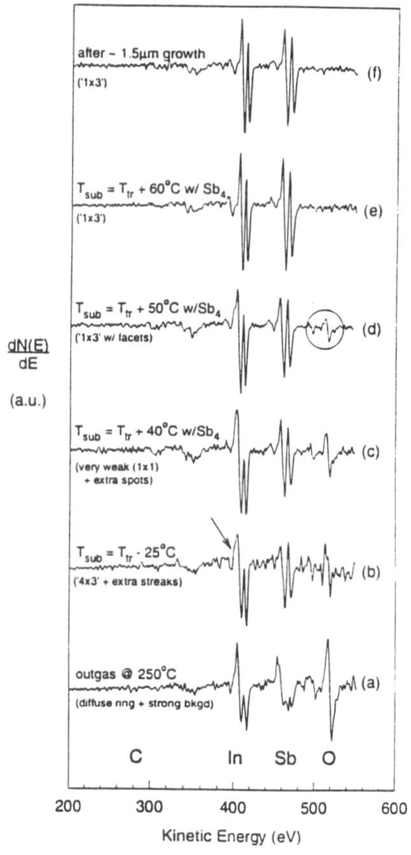


Fig. 3 AES spectra of InSb(100) substrate showing the progressive removal of oxides by thermal desorption ($E_p$=2keV, CRR=4).

Upon increasing $T_{sub}$, a weak InSb pseudo-(1×3)/(4×3) RHEED pattern emerges with additional streaks whose spacing does not pertain to InSb (possibly $In_2O_3$, $a_o$=10.12Å) (Fig. 1b). The intensity of the high energy peak in the $Sb3d_{3/2}$ spectrum diminishes while the In3d peaks remain asymmetric and the intensity of the O1s signal stays roughly the same (Fig. 2b). The oxygen peak in the AES spectra persists while the In MNN peak broadens with an asymmetry developing at the low KE (high BE) side (marked by arrow in Fig. 3b). During this first stage, most of the Sb oxide is reduced, but a thin oxide layer consisting predominantly of $In_2O_3$ remains. Judging from the change in the relative intensity of the O1s + $3d_{5/2}$ Sb oxide peak and the $3d_{3/2}$ Sb oxide peak, further thermal oxidation by oxygen released during the reduction of $Sb_2O_5$ might occur according to the following reactions:

$$2Sb_2O_5 \rightarrow 2Sb_2O_3 + 2O_2 \rightarrow 4Sb + 5O_2$$

$$4InSb + 3O_2 \rightarrow 2In_2O_3 + 4Sb$$

Similar reactions have been observed during GaAs oxide desorption [14]. SEM micrographs (Fig. 4a) taken at this stage reveal partial oxide desorption with small patches of bare InSb substrate. It is not surprising that epitaxial layers grown on such surfaces exhibit poor interfacial quality [10].

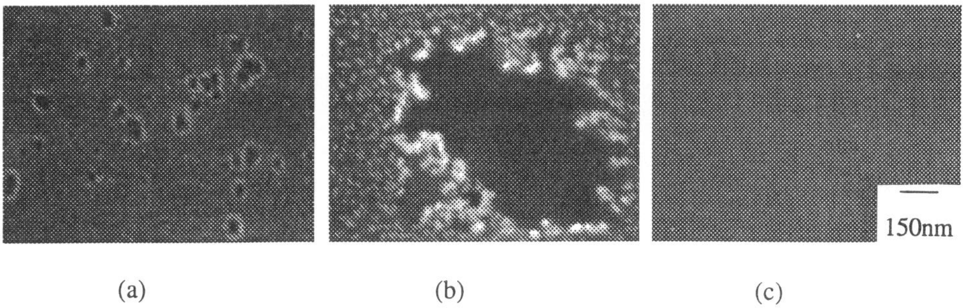


Fig. 4 SEM micrographs (BEI mode) taken at different stages of the oxide desorption process with RHEED patterns showing: (a) InSb pseudo-(1×3) + extra streaks (b) spotty pattern, (c) pseudo-(1×3).

As $T_{sub}$ is raised further in the presence of an Sb flux, desorption of $In_2O_3$ is evident through increasing symmetry in the In3d peak (decreasing FWHM) and a large decrease of the oxygen peak (Fig. 2c and 3c). The '3x' periodicity disappears completely from the RHEED pattern (Fig. 1c) while the integral-order streaks weaken and become broken. The appearance of more 3D spots aligned along <113> may be attributed to In islands, evident from SEM (Fig. 4b), which condense on the surface upon reduction of $In_2O_3$. The RHEED pattern becomes increasingly spotty (Fig. 1c and d), similar to the "anomalous" pattern reported by others [10,15]. Further increasing $T_{sub}$ weakens the intensity of the diffraction spots and the InSb(1x1) streaks re-emerge with arrowheads indicative of the presence of <113> facets (Fig. 1e). As $T_{sub}$ approaches ~$T_{tr}$+50°C, the InSb pseudo-(1x3) pattern with weakened arrowhead features is barely seen. Both XPS and AES reveal that a small amount of oxygen persist on the surface (Fig. 2d and 3d) until eventually disappearing at $T_{sub} \approx T_{tr}$ + 60°C (Fig. 2e and 3e). At this point the In:Sb signal approaches stoichiometric values and a well-defined InSb pseudo-(1×3) pattern is seen (Fig. 1f). SEM also reveals a virtually featureless surface (Fig. 4c). This suggests that the condensed In droplets has gained sufficient surface mobility at this temperature to combine with the Sb flux to form a smooth layer of InSb. Oxygen concentration level below the detection limits of XPS and AES is achieved and the lineshape and FWHM of spectra taken at this stage are similar to those obtained after the growth of ~1.5μm of InSb (c.f. graphs e and f in Figs. 2 and 3). Typically, the thermal desorption process takes about 2-3 hours to complete and InSb films grown on such substrates exhibit good structural and electrical properties [16,17].

In summary, a multi-stage desorption process is observed for surfaces prepared using either of two common chemical etchants. A composite oxide containing $In_2O_3$ and $Sb_2O_5$ is evident in the as-loaded substrate. During the first stage, most of the Sb oxide is reduced. A thin oxide layer consisting predominantly of $In_2O_3$ remains and is enriched by thermal oxidation occurring during the reduction of $Sb_2O_5$. As $T_{sub}$ is raised further in the presence of an Sb flux, desorption of $In_2O_3$ is evident and In droplets condense on the surface before a smooth morphology ensues. We suggest that In reduced from $In_2O_3$ form droplets which later react with the supplied Sb flux to form InSb. Annealing at this temperature produces a smooth, stoichiometric surface which gives rise to a well-defined, Sb-stabilized pseudo-(1×3) RHEED pattern.

The authors thank Bill Chissoe for the SEM studies and Joel Young for technical support. This work is partially supported by the National Science Foundation through grant no. OSR-9108771 and ECS-9410015.

## References


[1] Chang C C, Citrin P H and Schwartz B 1977 *J. Vac. Sci. Technol.* **14**(4) 943

[2] Farrow R F C, Cullis A G M , Grant A J and Pattison J E 1978 *J. Cryst. Gr.* **45** 292

[3] Bean J C, Becker G E, Petroff P M and Seidel T E 1977 *J. Appl. Phys.* **48**(3) 907

[4] Zheng Y-D, Chang Y H , McCombe B D, Farrow R F C, Temofomte T and Shirland F A 1986 *Appl. Phys. Lett.* **49**(18) 1187

[5] Yuen W T, Schweitzer M O, Jones T S, McConville C F, Johnson E A, Mackinnon A, Richardson N V and Stradling R A 1993 *Semicond. Sci. Technol.* **8** S396

[6] Remy H 1956 *Treatis on Iorganic Chemistry* vol 1 (Elsevier)

[7] Liu W K, Yuen W T and Stradling R A 1995 submitted for publication in *J. Vac. Sci. Technol.*

[8] Noreika A J, Francombe M H and Wood C E C 1981 *J. Appl. Phys.* **52** (12) 7416

[9] Ashley T, Dean A B, Elliott C T, McConville C F and Whitehouse C R 1988 *Electronics Lett.* **24**(20) 1270

[10] Klem J F , Tsao J Y, Reno J L, Datye A and Chadda S 1991 *J. Vac. Sci. Technol.* **A9**(6) 2996

[11] Whitehouse C R, Emeny M T, Davis L, Williams G M and Martin T 1986 *Proc. 4th Int. Conf. on Molecular Beam Epitaxy*, York.

[12] Werner H W and Garten R P H 1984 *Rep. Prog. Phys.* **47** 221

[13] Fadley C S 1978 *Electron Spectroscopy: Theory, Techniques and Applications* eds. Brundle C R and Baker A D (London: Academic Press)

[14] Vasquez R P, Lewis B F and Grunthaner F J 1983 *J. Vac. Sci. Technol.* **B1**(3) 791

[15] Wood C E C, Singer K, Ohashi T, Dawson L R and Noreika A J 1983 *J. Appl. Phys.* 54 2732

[16] Liu W K and Santos M B 1994 *Surf. Sci.* **319** 172

[17] Zhang Y, Su J, Winesett J, Furneaux J E, Liu W K, Santos M B and Doezema R E in this volume

*Inst. Phys. Conf. Ser. No 144: Section 4*
*Paper presented at 7th Int. Conf. on Narrow Gap Semicond. Santa Fe 9–12 January*



# RHEED Studies of the Surface Reconstructions of InSb(001) During Molecular Beam Epitaxy

M.B. Santos and W.K. Liu

Laboratory for Electronic Properties of Materials and Department of Physics and Astronomy
University of Oklahoma, Norman, Oklahoma 73019

**Abstract.** Surface reconstructions of InSb(001) during molecular beam epitaxy are studied using reflection high energy electron diffraction (RHEED). It is postulated that the only intrinsic structures in the growing InSb(001) surface are the (2x4)/c(2x8) Sb-dimers-terminated and the corresponding (4x2)/c(8x2) In-dimers-terminated structures. The pseudo-(1x3) pattern is believed to result from a mixture of different (2x4) phases, rather than a single surface structure.

Molecular beam epitaxy (MBE) has opened up many new areas of semiconductor and device physics through the vast range of low-dimensional structures fabricated using this technique. Of intrinsic importance to the MBE growth process, and ultimately to device properties, is the surface structure of the growing film. Reflection high energy electron diffraction (RHEED), with its grazing angle geometry and surface sensitivity, is the most widely employed *in situ* probe of surface structure during MBE. While RHEED studies of GaAs surfaces have provided a prototype for general understanding of III-V semiconductor surfaces, the details of InSb surface structure and MBE growth can be dramatically different. For example, the optimal growth conditions on GaAs(001) were found to be characterized by a (2x4) RHEED pattern. However, an equivalent surface structure on InSb(001) has not been observed. It is therefore important to study the reconstructions of the InSb(001) surface and their dependence on growth parameters such as substrate temperature ($T_{sub}$) and surface stoichiometry. Through these studies, InSb MBE can be better understood and, consequently, the growth of InSb-based heteroepitaxial films and strained layer superlattices can be optimized. In this paper, we present a RHEED study of InSb(001) homoepitaxy which further elucidates the dependence on growth parameters reported in earlier studies [1]. This discussion will be limited to the high-temperature, Sb-rich growth regime, but a discussion of the entire phase diagram can be found in Ref. 2.

Our experiments were carried out in an Intevac Gen II MBE system loaded with solid elemental sources of In (7N, Nikko Kyodo) and Sb (6.5N, Dowa). All specimens were grown on p-InSb(001) substrates (Cd-doped, Johnson Matthey) which were cleaned *ex situ* using a modified CP4A etchant [3]. The passivating oxide was thermally desorbed *in situ* under an Sb flux. Substrate temperatures ($T_{sub}$) were deduced from readings of a thermocouple located at the centre of the substrate heater and their reproducibility was ensured by monitoring the c(4x4)↔pseudo-(1x3) transition [4] and the melting point of In. The ratios of $Sb_4$ flux to In flux ($J_{Sb4} / J_{In}$) were inferred from beam equivalent pressure ratios.

Fig. 1 shows the phase diagram of the growth temperature and flux ratio dependence of our observed surface reconstructions. Examples of our RHEED patterns are presented in Fig. 2. We observe several surface reconstructions under Sb-rich growth conditions. The c(4x4) structure, seen at 270°C < $T_{sub}$ < 365°C for $J_{Sb4} / J_{In} \approx 1.5$, appears to be the most ordered, as evidenced by sharp, well-defined diffraction streaks and low background intensity (Fig. 2a). A structural model proposed [5] on the basis of earlier RHEED and photoemission measurements [5] of the c(4x4) surface has gained recent support from STM studies which revealed rectangular blocks of three Sb-dimers and a missing dimer with dimer bonds aligned along the [110] direction [6]. These images are similar to those observed for the GaAs(001)-c(4x4) surface [7]. Both surfaces are believed to have an overlayer of trigonally-bonded chemisorbed group-V atoms [8].

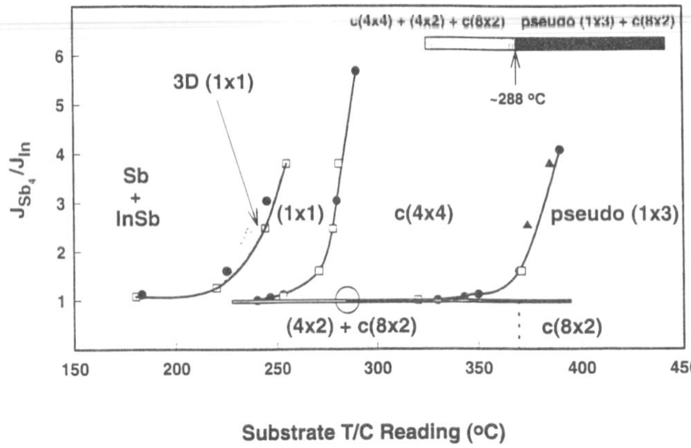


Fig. 1 Surface reconstruction phase diagram for InSb(001) during growth. Different symbols represent data taken from different samples.

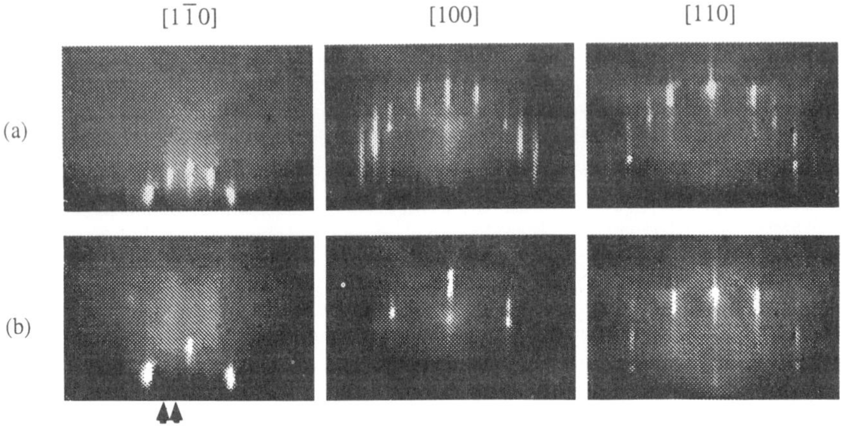


Fig. 2 RHEED patterns observed along the main azimuths showing [a] c(4x4) and [b] pseudo-'1x3' (the two unevenly spaced fractional-order streaks are marked by arrows) surface structures.

Although several surface reconstructions of InSb(001) and GaAs(001) (in particular, the co-existence of c(8x2) and (4x2), and the chemisorbed c(4x4)) are similar, the distinct (2x4) reconstruction observed over a wide range of surface stoichiometries in GaAs is not present in InSb. As $T_{sub}$ is increased in InSb(001), the 2x pattern along $[1\bar{1}0]$ gradually degrades with an apparent broadening of the 1/2 order streak. Careful inspection of the pattern reveals that, rather than a single broad 1/2-order streak, there are actually two fractional-order streaks (marked by arrows in Fig. 2b) with a strong intensity region in between. The splitting fraction, defined as the ratio of the distance between the two fractional-order streaks to that between the integral-order streaks, varies around the value of 0.25 as a function of $T_{sub}$. A change along [100] from 4x to 1x is nearly simultaneously observed. The temperature ($T_{tr}$) at which the 4x↔1x reversible transition occurs is widely used to calibrate substrate temperatures for InSb growth [4,9]. The poorer ordering of this pseudo-(1x3) pattern, also known as asymmetric-(1x3) because the fractional-order streaks are significantly closer together than in a standard (1x3) pattern, is evidenced by a strong background intensity and faint streaks in the RHEED pattern (Fig. 2b). The change in the [110] azimuth from 2x to 1x is considerably more gradual and, in some cases, a very weak 2x periodicity persists at $T_{sub}$ well above $T_{tr}$.

Continuous splitting of diffraction features has also been observed in LEED measurements of oxygen on nickel [10]. Those findings and similar results on other materials have been attributed to uniform mixing of at least two superstructures with different lengths [11]. The pseudo-(1x3) pattern that we observe may be due to similar effects. Oliveira *et al.* [12] suggested that this pattern is generated by the co-existence of (2x4) structures with different Sb coverages. The exact positions of the '3x' fractional-order streaks (i.e. the splitting fraction) are determined by the details of the mixing.

Several different (2x4) structures are thought to exist on GaAs(001). The so-called α, β and γ (2x4) structures have As coverages of 1/2, 3/4 and 1ML, respectively [13,14], and can be distinguished by the different relative intensities of their fractional-order diffraction streaks [13,15]. Although it remains controversial which of these phases is the most energetically favourable [16], the differences in energy among them were calculated to be <0.2eV/unit mesh [14,17]. Recent STM observations demonstrate that these different phases can co-exist [7,18]. *Random* mixtures of these phases will still result in a (2x4) RHEED pattern, but with less well-defined features and an intensity distribution that depends on the specific mix of phases on the surface [13].

However, the simultaneous presence of two or more of these phases can also give rise to a pseudo-(1x3) pattern if uniform mixing of phases occurs, as may be the case in InSb(001). The presence of superstructures generated by the *ordered* introduction of one (2x4) phase into a sequence of another phase can destroy the 4x periodicity [12]. In general, the three fractional-order streaks can split into six streaks, resulting in a 7x periodicity. However, destructive interference terms in the scattering amplitude pertaining to specific arrangements of dimers within the superstructures can result in zero intensity for some of these streaks. The observed pseudo-'3x' periodicity is believed to arise from splitting of the 2/4-order streak and the diminished intensity of the 1/4 and 3/4-order streaks. Note that while the splitting factor is determined by the average length of the superstructures, the dynamic nature of MBE growth necessitates a statistical spread in superstructure length. This is reflected in the RHEED pattern shown in Fig. 2b, where a continuous band of intensity is enveloped by the two fractional-order streaks. The persistence of a weak 2x periodicity along [110] is also consistent with uniformly mixed (2x4) phases. The upper limit of $T_{sub}$ imposed by the low melting point of InSb may

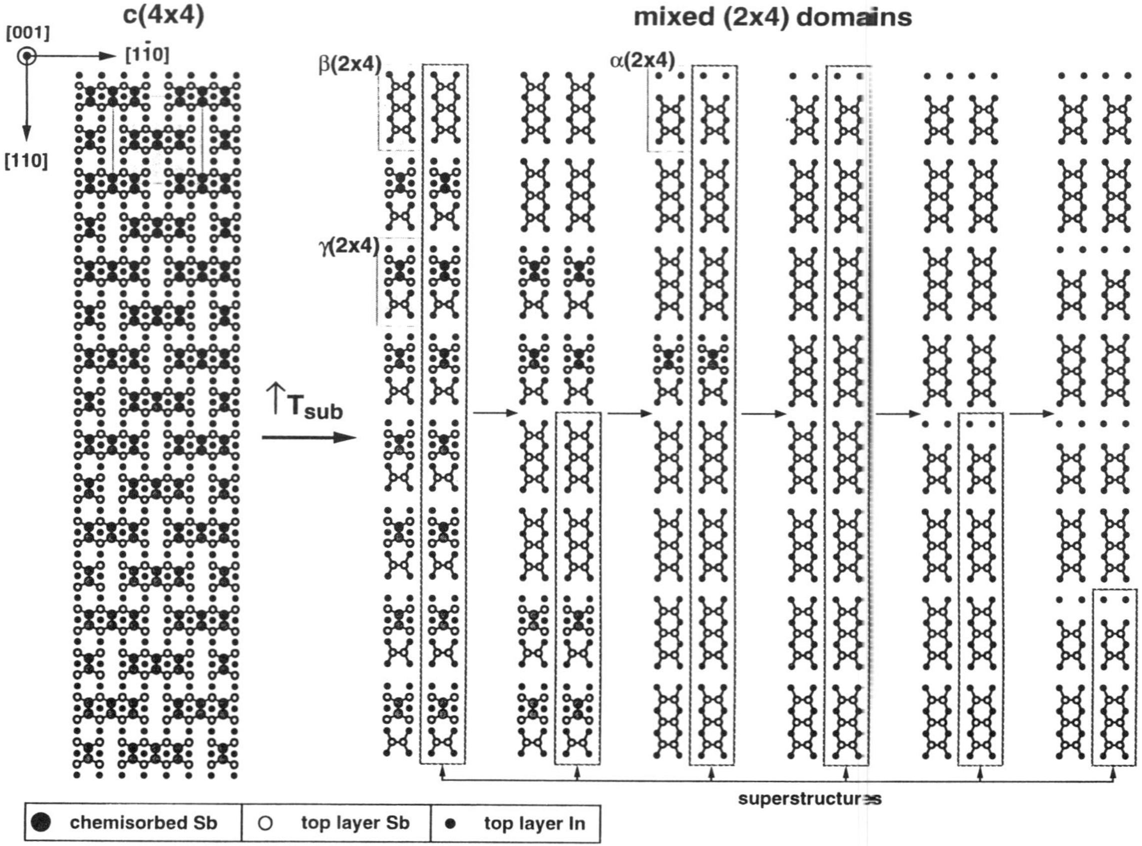


Fig. 3 Atomic model of the InSb(001) surface showing the evolution of a mixture of different (2x4) phases from c(4x4) structures as a function of growth temperature. Note the presence of superstructures with different lengths.

make it impossible to drive the InSb surface into equilibrium, where one particular (2x4) phase dominates and consequently produces a distinct (2x4) RHEED pattern.

Based on the above arguments and the assumption that the Sb coverage ($\theta_{Sb}$) decreases with increasing $T_{sub}$ for a fixed $J_{Sb4} / J_{In}$, we propose a simplistic atomic model for the evolution of a single domain c(4x4) surface to one composed of a mixture of (2x4) structures. As depicted in Fig. 3, the number of Sb dimers on the *top* layer and the surface stoichiometry change with $T_{sub}$, resulting in different superstructure lengths. In reality, the surface Sb-dimers have sufficient mobility to move in and out of phase in adjacent unit cells and can introduce a small number of closely-related c(2x8) domains [7,18,19]. This leads to degradation of the 2x ordering and further weakens the intensity of the 1/2-order diffraction streak in the [110] direction.

In summary, the pseudo-(1x3) pattern is believed to be due to a mixture of different (2x4) phases instead of representing a single surface structure. The '3x' periodicity is generated by differing amounts of each phase which vary with $T_{sub}$ and $\theta_{Sb}$. The c(4x4), (1x1) and symmetric-(1x3) patterns all arise from Sb-chemisorption induced structures ($\theta_{Sb}$>1ML) with Sb atoms trigonally-bonded to each other in the overlayers. These interpretations may prove useful when optimizing interface abruptness and growth conditions for InSb-based heteroepitaxial films and strained layer superlattices.

We thank Ryan Doezema, John Furneaux, Joel Young, and Bill Chissoe for helpful discussions and technical support. This work is partially supported by the National Science Foundation through grant no. OSR-9108771.

**REFERENCES**


[1] Oe K, Ando S and Sugiyama K 1980 *Jap. J. Appl. Phys.* **19** 417; Noreika A J, Francombe M H and Wood C E C 1981 *J. Appl. Phys.* **52** 7416; Williams G M, Whitehouse C R, Cullis A G, Chew N G and Blackmore G W 1988 *Appl. Phys. Lett.* **53** 1847

[2] Liu W K and Santos M B 1994 *Surf. Sci.* **319** 172

[3] Yuen W T, Liu W K, Joyce B A and Stradling R A 1990 *Semicond. Sci. Technol.* **5** 373.

[4] Whitehouse C R, Emeny M T, Davis L, Williams G M and Martin T 1986 *Proc. 4th Intl. Conf. on Molecular Beam Epitaxy*, York

[5] John P, Miller T and Chiang T-C 1989 *Phys. Rev. B* **39** 1730

[6] McConville C F, Jones T S, Leibsle F M and Richardson N V 1994 *Surf. Sci.* **303** L373

[7] Biegelsen D K, Bringans R D, Northrup J E and Swartz L-E 1990 *Phys. Rev. B* **41** 5701

[8] Larsen P K, Neave J H, van der Veen J F, Dobson P J and Joyce B A 1983 *Phys. Rev. B* **27** 4966

[9] Droopad R, Williams R L and Parker S D 1989 *Semicond. Sci. Technol.* **4** 111

[10] Germer L H, May J W and Szostak R J 1967 *Surf. Sci.* **8** 430

[11] Fujiwara K 1957 *J. Phys. Soc. Jap.* **12** 7; Houston J E and Park R L 1970 *Surf. Sci.* **21** 209

[12] de Oliveira A G, Parker S D, Droopad R and Joyce B A 1990 *Surf. Sci.* **227** 150

[13] Farrell H H and Palmstrøm C J 1990 *J. Vac. Sci. Technol. B* **8** 903

[14] Ohno T 1993 *Phys. Rev. Lett.* **70** 631

[15] The effect of surface ordering on fractional-order streak intensity is explore in Zhou J, Xue Q, Chaya H, Hashizume T, and Sakurai T 1994 *Appl. Phys. Lett.* **64** 583

[16] Northrup J E 1994 American Physical Society March Meeting, C24 6, Pittsburgh PA

[17] Chadi D J 1987 *J. Vac. Sci. Technol. A* **5** 834

[18] Pashley M D, Haberern K W, Friday W, Woodall J M and Kirchner P D 1988 *Phys. Rev. Lett.* **60** 2176

[19] Joyce B A, Neave J H, Dobson P J, and Larsen P K 1984 *Phys. Rev. B* **29** 814

*Inst. Phys. Conf. Ser. No 144: Section 4*
*Paper presented at 7th Int. Conf. on Narrow Gap Semicond. Santa Fe 9–12 January*



# Low Temperature, Damage-Free Surface Cleaning Of InSb(001) Prior To MBE Growth

A.D.Johnson, G.M.Williams, A.J.Pidduck, C.R.Whitehouse[1], T.Martin, C.T.Elliott & T.Ashley

Defence Research Agency, St Andrews Road, Great Malvern, U.K., WR14 3PS.
[1]Dept. of Electronic & Electrical Engineering, University of Sheffield, Mappin Street, Sheffield, S1 3JD, UK.

Atomically flat surface cleaning of InSb(001) has been achieved by exposure to an Electron Cyclotron Resonance (ECR) Hydrogen plasma prior to Molecular Beam Epitaxy (MBE). ECR plasma cleaned surfaces are shown by Scanning Electron Microscopy (SEM) and Atomic Force Microscopy (AFM) to be far superior to those achieved by other commonly used surface preparation techniques. It is also shown that in order to maintain surface stoichiometry, plasma cleaning must be performed with the substrate under an incident antimony flux and that this leads to the most atomically flat surfaces.

## 1. Introduction

The preparation of clean, atomically flat III-V semiconductor surfaces prior to epitaxy is of major importance in the manufacture of electronic devices. This is because the most important surface plane in device fabrication is the (001) surface rather than the (110) plane, which can be prepared by cleavage under vacuum. Consequently, the most widely used technique for removal of native oxide is thermal desorption under an excess group V overpressure to maintain surface stoichiometry. However, this technique is known to lead to rough surfaces [1] which have to be smoothed out by the growth of buffer layers before epitaxy of the device layers can begin.

Indium antimonide (InSb) has electronic properties which are well known for their applications to mid-infra-red emitters and detectors and ultra high speed electronic devices [2-4]. However, its oxide desorption temperature is very close to its bulk melting point (~525°C) and well in excess of its non-congruent evaporation temperature (~325°C). This makes surface preparation prior to epitaxy very difficult as it can take several hours to thermally remove the native oxide [5] and the resulting surface is rough and is characterized by large indium droplets at the substrate/epilayer interface.

A further technique that is widely used to remove the InSb native oxide is argon ion bombardment and annealing. This method uses argon ions of several hundred eV to sputter the oxide from the surface followed by low temperature anneals (~250°C) to reduce any structural damage induced by the ion beam. Like thermal cleaning, this technique is also very time consuming and usually takes 4-5 hours to fully remove the oxide layer. Argon ion cleaned surfaces are usually indium rich due to preferential removal of the group V species and also characterized by an n-type layer at the surface associated with damage induced by the ion beam. This n-type layer is known to extend 500-1000Å into the substrate and have an equivalent sheet doping level of ~$10^{13}$cm$^{-2}$ [6,7].

Therefore, we have been investigating new methods of removing the InSb native oxide so as to leave an atomically flat surface with as little structural and electrical damage as possible. Since it has been demonstrated that hydrogen plasmas are effective in removing native oxides from Si [8], GaAs [9], GaSb [10] and InP [11], we have investigated the effect of an electron cyclotron resonance (ECR) hydrogen plasma incident upon InSb(001). In addition we have examined the effect of substrate temperature and microwave power on oxide removal and the resulting surface morphology.

## 2. Experimental

Experiments were carried out in a VGV80H MBE system with a purpose built chamber for surface cleaning connected to its preparation chamber to allow for sample transfer under ultra-high vacuum (UHV) conditions. The cleaning chamber was equipped with reflection high energy electron diffraction (RHEED) and Auger electron spectroscopy (AES) for in-situ characterization, and also ports for optical monitoring techniques and Knudsen cells. InSb(001) substrates were prepared by degreasing in hot acetone, isopropyl alcohol and trichloroethylene before being etched in a hydrofluoric acid:acetic acid:nitric acid:water (1:1:2:10) mixture. After rinsing in deionized water, the wafers were In bonded to a molybdenum plate before loading into the MBE reactor. Once in the cleaning chamber, samples were held in a standard MBE sample manipulator to allow for sample heating and rotation during ECR cleaning. The plasma generator was an Astex AX4300 compact ECR source, operated in the range of 50-250W. Hydrogen gas (6N purity) was then fed into the system to a chamber pressure of $\sim 2 \times 10^{-4}$mbar, corresponding to a flow rate of ~ 10sccm, before striking the plasma.

The ECR source was mounted such that its angle of incidence with respect to the substrate surface was ~15°. This is because it has previously been shown that GaAs surfaces prepared by hydrogen plasmas give better RHEED patterns when the plasma strikes the surface at glancing rather than normal incidence [12]. All substrate temperatures refer to the values indicated by a thermocouple mounted behind the substrate mounting block.

## 3. Results

On exposing the InSb(001) samples to the ECR hydrogen plasma, it was indeed found that the surface oxide was removed. The dependence of the time taken to remove the oxide using the ECR hydrogen plasma as a function of the applied microwave power is shown in figure 1. During these measurements, the substrate temperature was kept constant at 200°C. The cleaning time is defined as the time taken until no oxygen peak could be detected using AES. The first point to notice is that the time required to remove the oxide is less than fifteen minutes even when the microwave power is as low as 50W. This compares very favourably with the several hours required by thermal desorption and argon ion bombardment. Secondly, it was found that the Sb/In Auger peak ratio becomes smaller for surfaces cleaned with microwave powers greater than 100W, indicating a more In rich surface. All subsequent experiments were therefore performed with the applied power at 100W to keep the surface as antimony rich as possible. Inspection of surfaces with RHEED indicated that in fact all ECR cleaned surfaces were In terminated as the RHEED pattern exhibited the c(8x2) In rich reconstruction [13]. The variation of cleaning time with substrate temperature at a constant applied microwave power (100W) is shown in figure 2. It can be seen that the time to remove the oxide is constant above 200°C and only rises to ~45mins at room temperature.

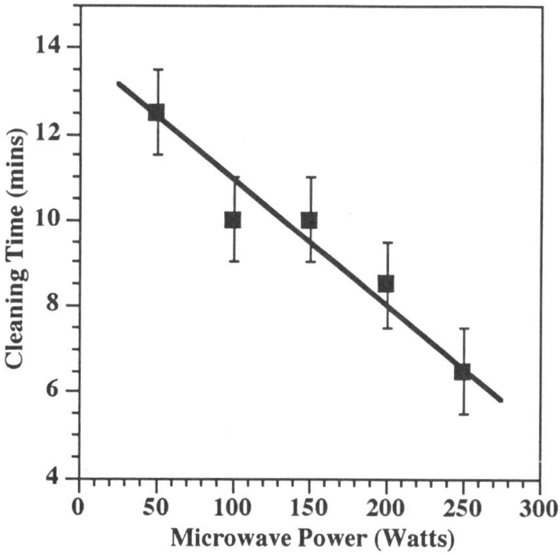


**Figure 1.** Cleaning time vs. applied microwave power for InSb(001) with substrate temperature constant at 200°C.

Surfaces cleaned by hydrogen ECR plasma were then compared with surfaces prepared by thermal cleaning and argon bombardment in a scanning electron microscope (SEM). Thermally cleaned and argon bombarded surfaces were characterized by a high density of indium droplets many hundreds of nanometres in diameter and a large number of surface pits with a depth of several tens of nanometres. This can be seen, on a smaller scale, in figure 3, which shows cross section transmission electron micrographs (TEM) of substrate/epilayer interfaces of layers with a) thermally cleaned and b) argon bombarded substrates. It can be seen that in both cases there is a large number of indium droplets at the interface, the thermally cleaned sample having a lower density of larger droplets.

SEM images of ECR cleaned surfaces showed no such features and were in fact extremely flat with no visible surface topography. Indeed, to ensure that the surface was in focus it was necessary to induce In islands on the surface with the incident electron beam and then to use them as features to focus on. However, SEM on samples exposed to the ECR plasma for longer periods (~1hour) showed very similar results to those obtained on thermally cleaned and argon bombarded surfaces, with many indium islands and deep pits. It was therefore necessary to fit an antimony Knudsen cell to the chamber so as to ECR clean the surface under an incident group V flux and thus retain surface stoichiometry.

The antimony flux was set so that it was equivalent to that for an InSb(001) growth rate of ~1$\mu$m/hour. In the presence of the incident antimony flux, the substrates had to be raised to a higher temperature to prevent condensation of antimony onto the surface, which occurs below ~300°C (actual substrate temperature). Samples were therefore cleaned at a substrate temperature of 500°C (~360°C actual substrate temperature). After exposure to the ECR plasma for ~10 minutes, RHEED patterns corresponding to the antimony rich c(4x4) surface reconstruction were obtained. Examples of such patterns are illustrated in figure 4. The sharpness and intensity of the patterns is equal to those normally obtained from epitaxial layers and better than those seen from any other type of cleaned InSb(001) surface. It was in fact possible to observe InSb RHEED oscillations directly from the ECR

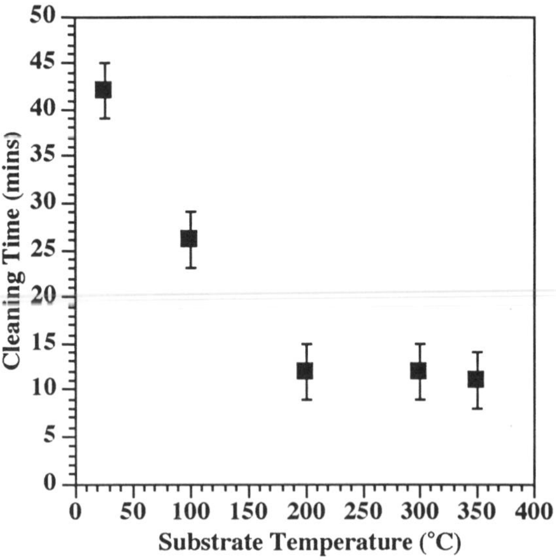


**Figure 2.** Cleaning time vs. substrate temperature at a constant applied microwave power (100W)

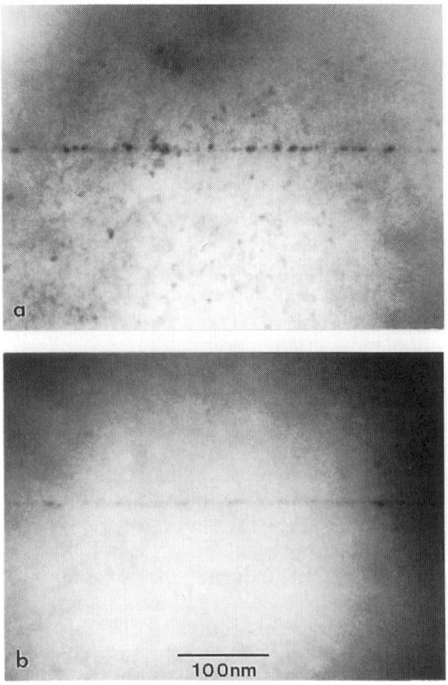


**Figure 3**. Cross section TEM images of substrate/ epilayer interfaces for a) thermally cleaned and b) Argon ion cleaned InSb(001) surfaces. Indium droplets can be seen in both cases.

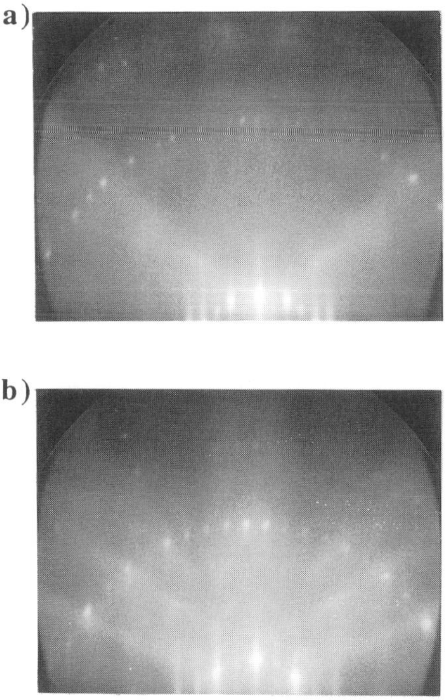


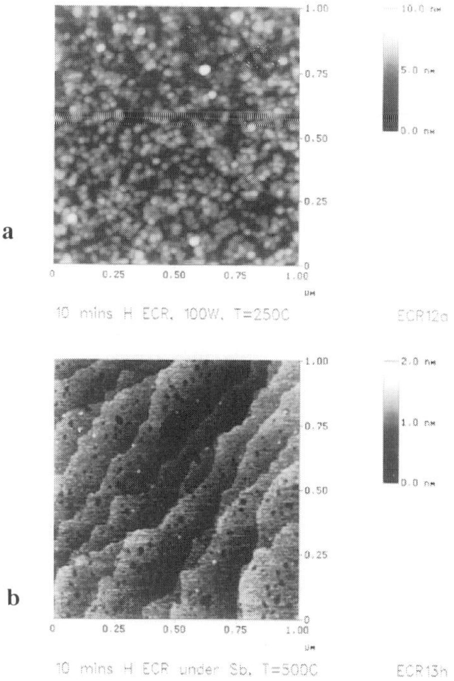


**Figure 4.** RHEED patterns of ECR plasma cleaned surfaces under an antimony flux. a) is the [010] azimuth 4x pattern and b) the [110] azimuth 2x pattern of the c(4x4) surface reconstruction.

**Figure 5.** AFM images of surfaces cleaned by ECR plasma, a) without and b) with an incident antimony flux. Note the difference in the vertical scales.

cleaned surface. On thermally cleaned or argon ion bombarded surfaces, it would normally be necessary to grow an InSb buffer layer of at least 1$\mu$m before it would be possible for RHEED oscillations to be seen, and such oscillations would be of a lower intensity than those seen on the ECR cleaned surfaces.

Finally, the surfaces cleaned by the various techniques were examined by AFM. Samples cleaned by thermal desorption and argon ion bombardment gave similar results to those obtained by electron microscopy. Both types of surface had a number of large (hundreds of nm) indium islands with a much higher density of small droplets, as seen by SEM and TEM. The argon ion bombarded surfaces were, however more heavily pitted. Figure 5 shows AFM images of ECR plasma cleaned surfaces for samples prepared both without (a) and with (b) an incident antimony flux. The most noticeable difference is the difference in vertical scales between the two images. The sample cleaned without the antimony overpressure has a height variation five times that of image b), and is thus much rougher.

Image b) on the other hand shows steps and terraces corresponding to the misorientation of the surface with respect to the (001) crystallographic plane. The steps appear to be single or bilayer in height and are regular across the surface. No such features are visible on a).

## Conclusions

It has been demonstrated that InSb oxide removal with an ECR hydrogen plasma produces atomically flat InSb(001) surfaces, superior to those achieved by any other method. Electron microscopy and AFM shows that, for samples prepared under a group V overpressure, the resulting surface is free of islands of indium, which are normally characteristic of InSb(001) surfaces prepared by other methods. RHEED patterns of samples prepared by ECR plasma in an antimony flux show cleaned surfaces to exhibit the antimony rich c(4x4) surface reconstruction, the patterns of a sharpness and quality equal to that of those seen on epitaxial layers. AFM images of such samples show the surface to be atomically flat with single or bilayer steps and terraces, as defined by the misorientation of the substrate crystal. Moreover, the ECR cleaning process typically takes only ~10 minutes compared with several hours for other methods. It is therefore proposed that ECR hydrogen plasma cleaning is the optimum method of cleaning InSb(001) prior to epitaxial growth.

**Acknowledgements**

We would like to thank R.Jefferies, D. Soley, V. Stimson, P. Moores and L. Smith for their valuable technical assistance in the course of making these measurements.

**References**

[1] G.W. Smith, A.J. Pidduck, C.R. Whitehouse, J.L. Glasper, A.M. Keir & C. Pickering, 1991 Appl. Phys. Lett. **59** (25) 3282.

[2] T.Ashley, A.B. Dean, C.T. Elliott, C.F. McConville, G.J. Pryce & C.R. Whitehouse, 1991 Appl. Phys. Lett. **59** (14) 1761.

[3] T. Ashley, C.T. Elliott, N.T. Gordon, R.S. Hall, A.D. Johnson & G.J. Pryce, 1994 Appl. Phys. Lett. **64** (18) 2433.

[4] T. Ashley, A.B. Dean, C.T. Elliott, G.J. Pryce, A.D. Johnson & H. Wills, 1995 Appl. Phys. Lett. **66** (4) 1.

[5] J.F. Klem, J.Y. Tsao, J.L. Reno, A. Datye & S. Chadda, 1991 J. Vac. Sci. Technol. A **9** (6) 2996.

[6] T.S. Jones, M.Q. Ding, N.V. Richardson & C.F. McConville, 1991 Surface Science **247** 1.

[7] W.T. Yuen, M.O. Schweitzer, T.S. Jones, C.F. McConville, E.A. Johnson, A. Mackinnon, N.V. Richardson & R.A. Stradling, 1993 Semicond. Sci. Technol. **8** S396.

[8] Tomohiro Shibata, Yasushi Nanishi & Masatomo Fujimoto, 1990 Jap. Journ. Appl. Phys. **29** (7) L1181.

[9] Akira Takamori, Sumio Sugata, Kiyoshi Asakawa, Eizo Miyauchi & Hisao Hashimoto 1987 Jap. Journ. Appl. Phys. **26** (2) L142.

[10] Z. Lu, Y. Jiang, W.I. Wang, M.C. Teich & R.M. Osgood Jr., 1992 J. Vac. Sci. Technol. B **10** (4) 1856.

[11] Yong Jin Chun, Takeyoshi Sugaya, Yoshitaka Okada & Mitsuo Kawabe, 1993 Jap. Journ. Appl. Phys. **32** (2) L287.

[12] I. Suemune, Y. Kunitsugu, Y. Kan & M. Yamanishi, 1989 Appl. Phys. Lett. **55** (8) 760.

[13] Kunishige Oe, Seigo Ando & Koichi Sugiyama, 1980 Jap. Journ. Appl. Phys. **19** (7) L417.

*Inst. Phys. Conf. Ser. No 144: Section 4*
*Paper presented at 7th Int. Conf. on Narrow Gap Semicond. Santa Fe 9–12 January*



# The Growth of Ternary Substrates of Indium Gallium Antimonide by the Double Crucible Czochralski Technique

T. Ashley, J. A. Beswick, B. Cockayne, C. T. Elliott.

Defence Research Agency, St. Andrews Road, Malvern, Worcestershire, WR14 3PS, UK.

**Abstract.** In order to further our research into uncooled Indium Antimonide based detectors and emitters, we require to reduce the cut-off wavelength below the 7$\mu$m room temperature cut-off wavelength of indium antimonide, into the 3-5$\mu$m atmospheric window. The approach we have chosen is to use the ternary alloy semiconductor indium aluminium antimonide, because of its rapid change of energy gap as a function of lattice parameter, grown onto a lattice matched substrate. The substrate we have chosen to use is indium gallium antimonide because it has a narrower liquidus-solidus separation than indium aluminium antimonide.

The change of melt composition during growth, due to the liquidus-solidus separation, is minimised by the use of a double crucible technique. We have developed a novel design of double crucible which has overcome difficulties associated with the liquidus-solidus temperature difference, friction between the inner and outer crucible, location of the inner crucible, back diffusion between the two crucibles, and freezing of the channel between the two crucibles. Simple means of fabricating the channel between the two crucibles and filling it with melt have been devised. Loss of crucibles due to expansion on freezing after growth has been eliminated. Use of the double crucible technique reported eliminates the need for ternary seeds to achieve lattice matching at seed-on and also the need for precise knowledge of the distribution coefficient.

[111] parallel-sided, single crystals of $In_{1-x}Ga_xSb$ with $0<x\leq0.11$, up to 24 mm in diameter and 5cm long, with dislocation densities of $2.10^4/cm^2$ and composition variation of better than 10% over the majority of the crystal, have been grown and assessed for lattice-matched epitaxy of indium aluminium antimonide alloys. Epitaxy on such crystals is expected to broaden the available wavelength and lead to improved leakage currents and other parameters in diodes, lasers, FETs, and detectors.

## 1. Introduction

Narrow gap semiconductors, such as Indium Antimonide (InSb) have many interesting properties arising from their band structure. The electronic properties include a very low electron effective mass, very high electron mobility, even at room temperature, and a high saturation velocity. The optical properties of interest include the small direct energy gap and the high ratio of stimulated to spontaneous transition probabilities. These properties are used in infrared detectors and emitters [1]. In order to further our research into uncooled Indium Antimonide based detectors and emitters, we require to reduce the cut-off wavelength below the 7$\mu$m room temperature cut-off wavelength of indium antimonide, into the 3-5$\mu$m atmospheric window. The approach we have chosen is to use the ternary alloy semiconductor indium aluminium antimonide, because of its rapid change of energy gap as a function of lattice parameter, grown onto a lattice matched substrate. The substrate we have chosen to use is indium gallium antimonide because it has a narrower liquidus-solidus separation than indium aluminium antimonide.

For the devices presently being studied we need to provide substrates which are lattice matched to the indium aluminium antimonide layers, in the case of $In_{(0.88)}Al_{(0.12)}Sb$ for example, this requires a substrate with the composition $In_{(0.892)}Ga_{(0.108)}Sb$. In the case of normal freezing such as occurs with a single crucible arrangement composition of the crystal is given by the normal freeze equation [2]:

$$C = kC_o(1-g)^{k-1}$$

in which C is the concentration of gallium in the grown ingot at the interface, $C_o$ is the initial concentration of gallium in the melt, k is the distribution coefficient and g is the fraction grown. This shows that, assuming a distribution coefficient of 2.4 [3] and an initial melt composition based on this, the composition of the crystal will change from $In_{(0.892)}Ga_{(0.108)}Sb$ at the start of growth to $In_{(0.959)}Ga_{(0.041)}Sb$ at the point where 50% of the melt has been crystallized. Such a change in composition of the crystal is clearly unacceptable, but unavoidable with a single crucible arrangement unless very large melt volumes are used.

One answer to this problem is the double crucible technique described by Leverton [4] for control of doping in germanium and developed in this work for control of segregation during the growth of ternary iii-v semiconductor crystals.

## 2. The double crucible design and procedure

An important difference between the application of the double crucible technique for low level doping of an elemental semiconductor crystal and for control of the composition of a ternary iii-v alloy arises from the wide separation of the liquidus and solidus lines, and hence difference in melting point of inner and outer melts in the case of the ternary compared with the negligible separation arising in the case of low level doping of an elemental simple semiconductor.

In the basic design of a double crucible the inner crucible floats on the surface of the melt contained in the outer crucible but must be located so that it remains central, melt is able to flow from the outer to the inner crucible melt through a connecting channel as growth proceeds thereby maintaining a constant melt volume in the inner crucible. Under ideal equilibrium conditions the composition of the crystal grown is equal to the composition of the outer melt, regardless of the distribution coefficient. A number of problems exist with the simple design of double crucible when applied to incongruently melting compositions such as the ternary iii-v semiconductors. The problems and their solution as incorporated in our design of the crucible [5] and our procedure are described below.

### *2.1 Freezing due to the liquidus solidus separation*

The separation of the liquidus and solidus lines results in the melting point of the inner melt being lower than that of the outer melt and consequently, under the low temperature gradients necessary for strain free crystal growth, the outer melt tends to freeze as it enters the inner crucible. It may also freeze onto the outside of the inner crucible and inhibit free movement. Freezing will also result in an undesirable change of composition.

Freezing of the melt in the outer crucible was cured by incorporating an insulated collar above the melt surface, and by using thick walled quartz for the inner crucible in order to lower the thermal conductance.

The use of thicker quartz and the insulated collar along with a novel design of crucible base also assisted in eliminating the freezing of the outer melt as it joined the inner melt [5]

### *2.2 Diffusion within the channel*

Diffusion can take place within the channel and result in undesirable changes of composition of the outer melt, thereby negating the principle of the double crucible and resulting in poor composition control [6] and [7]. This is particularly the case if the growth rates are low and the channel is wide, such as may be the case to facilitate easy filling.

The novel design of crucible base also facilitated use of a small hole for the communicating channel. Calculations based on the work of Matare [7] and Blackwell [8] showed that a 300$\mu$m diameter and 4mm long hole as employed in this work is equivalent to a 4mm diameter hole which is 700mm long and that it should be effective in preventing back diffusion at growth rates down to 6mm/hour or less for a 4mm diameter seed crystal and down to less than 1mm/hour for a 20mm diameter crystal.

### *2.3 Filling the channel*

The channel between the inner and outer crucible needs to be narrow to overcome back diffusion as described above. This is particularly the case with the ternary semiconductors where low pulling speeds are necessary to avoid constitutional supercooling. The narrower the channel the more difficult it is to fill with melt due to surface tension effects [6].

Difficulty in filling the channel was overcome in the present work by the procedure described in reference [5].

### *2.4 Lattice matching the seed*

The composition of the inner melt and the seed need to be such as to achieve lattice matching at seed on, this requires accurate knowledge of the effective distribution coefficient, as this may vary with growth orientation, pulling speed, and rotation rate, it cannot be predicted with accuracy.

Ternary iii-v semiconductor seeds are not readily available, but they are not necessary if the initial melt contained in the inner crucible is the appropriate iii-v binary constituent of the ternary composition required. In this case indium antimonide can be used and perfect lattice matching is assured if back diffusion does not take place in the channel between the two crucibles. As growth proceeds the composition changes according the equation [6]

$$Cs/Co = 1-(1-kCi/Co)\exp(-kM/mg)$$

in which Cs is the concentration of gallium in the crystal, Ci is the initial concentration of gallium in the inner crucible, m is the weight of melt in the inner crucible, and M is the weight of melt in the outer crucible. This equation shows that, for the case of a distribution coefficient of 2.4 and inner and outer melt weight ratios of 1/10, the crystal composition will be within 10% of the outer melt composition after one inner melt weight has been grown, it will be within 1% and remain constant after twice the inner charge has been grown. If a ternary seed is available and the distribution coefficient is known accurately enough, then an inner melt of appropriate composition can be used to match the outer melt and seed composition and a crystal of completely uniform composition can be grown.

### 2.5 *Sticking of the inner crucible*

The inner and outer crucibles are liable to stick together, especially if both are made of quartz and are allowed to contact each other. The design used in this study employed guides which operated in the graphite outer crucible support, the low friction offered by this arrangement eliminated sticking and gave good location of the crucible.

### 2.6 *Crucible damage*

If it is intended to grow crystals of constant composition then the whole of the melt cannot be grown and so some melt residue will always be left in the crucible at the end of growth. If freely floating in the melt, and not withdrawn before freezing of the melt at the end of growth, the freezing melt causes irreparable damage to the inner crucible with consequent increase in costs of the process. This was avoided in the present design by arranging for a collar on the seed shaft to lift the inner crucible clear of the melt surface at the end of growth.

## 3 **Crystal growth**

### 3.1 *Single crucible growth*

Initial attempts to grow $In_{(0.892)}Ga_{(0.108)}Sb$ utilised a single crucible and the "Bootstrapping" technique whereby an InSb seed with an initial melt of composition $In_{(0.996)}Ga_{(0..004)}Sb$ was used, a small crystal grown and small additions of GaSb made and a further small crystal grown. This was successful up to the composition $In_{(0.97)}Ga_{(0..03)}Sb$ after which the crystal became polycrystalline, despite several attempts it was not possible to proceed beyond this composition and obtain single crystal growth by this technique.

### 3.2 *Double crucible growth*

Approximately fifty ternary single crystals of compositions up to $In_{(0.88)}Ga_{(0.12)}Sb$ have been grown using the new double crucible arrangement [5] mounted in an r.f. heated, temperature controlled (not diameter controlled) "Malvern" crystal puller and employing 230 gram outer and 35 gram inner charges made from zone refined indium antimonide and gallium antimonide supplied by MCP Wafer Technology Ltd. [111] InSb seeds were used with an initial InSb inner melt, to secure lattice matching at seed on. The crystals were grown at rates from 6 to 18 mm/hour and rotation rates from 6 to 18 rpm, the higher growth rates being used with the more dilute melts. Argon, hydrogen, and 90% hydrogen-10% nitrogen atmospheres have been used, the most effective being Hydrogen. The melts were not encapsulated and some difficulty was experienced in keeping the melt surface clean in anything other than flowing hydrogen. Crystals grown in static hydrogen appeared to frequently contain precipitates or inclusions which are believed to be due to carbon incorporation. Crystals up to 150 grams have been grown.

In order to avoid cracking it was necessary to grow the crystals into an afterheater at about $400^0$C and then to cool them down over several hours. A typical 24mm diameter crystal is shown in figure 1.

## 4 **Assessment**

The composition variation along the crystal was within 10% of the target as determined by X-ray lattice parameter measurements and EDAX measurements. The composition of the crystal shown in figure 1 is depicted in figure 2 which also shows the normal and double crucible freezing curves. The data show that the double crucible design and growth procedure was successful in achieving and maintaining the desired crystal composition. The distribution of gallium corresponds to that expected from the freezing curve for double crucible growth without back diffusion. Dislocation densities of about 5.5 $10^5/cm^2$ were obtained in undoped crystals, doping with tin at $10^{19}/cm^3$ in the melt reduced the dislocation density to about 2 $10^4/cm^2$. Some low angle grain boundaries occurred after about 30% growth. Dislocation etching and X ray topographs showed evidence of terracing of the growth interface. Only a small number of crystals twinned.

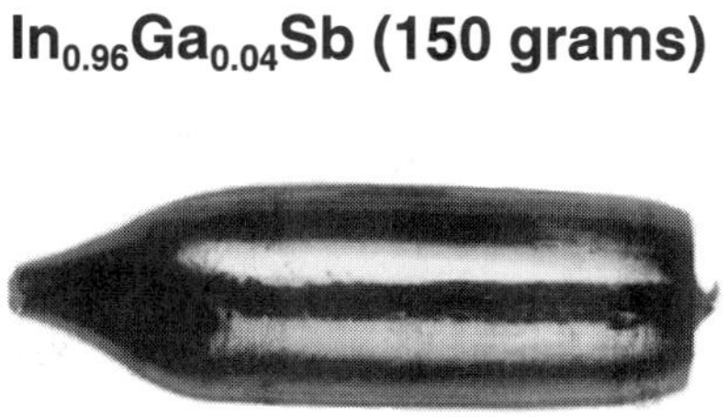


FIGURE 1

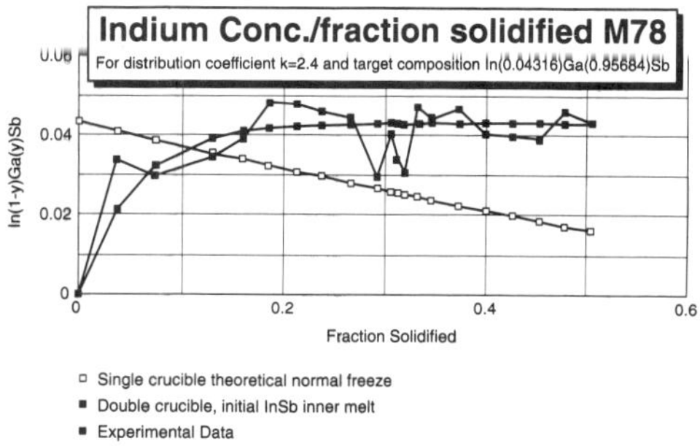


FIGURE 2

Further assessment of the crystals, MBE epitaxy of lattice matched layers, and scaling up of the equipment with a view to reducing the occurrence of low angle grain boundaries is underway.

## 5 Conclusions

A new design of double crucible and growth procedure has been developed which has enabled ternary substrates of Indium Gallium Antimonide of controlled composition to be grown, increased scale and crystal perfection should enable a new generation of narrow gap semiconductor based devices to be fabricated.

## Acknowledgements

The authors wish to thank A. M. Keir, and C. Reeve for x-ray analysis. This paper is published with the permission of the controller of Her Britannic Majesty's Stationery Office.


### References

[1] Ashley T. and Elliott C. T. 1991 *Semicond. Sci Technol.* **6** 99
[2] Pfann W. G. 1958 *Zone melting* J. Wiley and Sons p10
[3] Mullin J. B. 1958 *J. Electron. Control* **4** 358
[4] Leverton W. F. 1958 *Journal of Applied Physics* **29** [8] 1241 (1958)
[5] Beswick J. A. *Patent Application no. 9412629.9*
[6] Okada H. 1987 *Journal of Crystal Growth* **82** [4] 643 (1987)
[7] Matare H. F. *Solid State Electronics* **6** [2] 163 (1963)
[8] Blackwell G. R. *Solid State Electronics* **7** 105 (1964)

*Inst. Phys. Conf. Ser. No 144: Section 4*
*Paper presented at 7th Int. Conf. on Narrow Gap Semicond. Santa Fe 9–12 January*



# Slab-Doping of InSb:Er

Joseph P. Heremans, Dale L. Partin and Christopher M. Thrush

General Motors Research & Development Center, 30500 Mound Rd., Warren, MI 48090-9055

**ABSTRACT.** Some rare earth elements, such as Er, are donors for InSb. Carrier density and mobility data on bulk Er-doped InSb films, grown by molecular beam epitaxy on semi-insulating InP, were reported in the past. Here, we study a sequence of similar films, δ-doped with Er, and vary the spacing between the layers (or the number of layers $N_L$ in a 1.6 μm thick region of the films), and the areal Er atom density in each layer, $N_{Er}$, so that the volume density of Er averaged over the whole film, $n_{Er}$, includes the doping range of previous work. In bulk-doped films, the low-temperature electron density n is a linearly increasing function of Er concentration $n_{Er}$ up to $n \approx 1x10^{17}$ $cm^{-3}$, above which it saturates, but this is not the case in the δ-doped samples. For $N_L = 48$, n increases with $N_{Er}$ up to $n \approx 1x10^{17}$ $cm^{-3}$, just like in bulk-doped films, but instead of saturating, n decreases when $N_{Er}$ is further increased. The mobility of the δ-doped samples with $N_{Er} \leq 2.5x10^{13}$ $cm^{-2}$ is enhanced over that of bulk-doped films with similar Er and electron concentrations. This effect is more pronounced at lower temperature and at lower Er concentrations.

## 1. Introduction

Narrow-gap semiconductors, such as InSb and InAs, have large electron mobilities around room temperature, which makes them attractive materials for use in magnetic field sensors such as magnetoresistors and Hall sensors[1,2]. The influence of doping or alloying InSb with rare-earth elements such as Eu[3], Sm[4], and Er[5] has been extensively studied[6]. Er is also used as a dopant in InP and GaAs[7], mostly for optical applications. In particular, it was found[5] that Er is a donor for MBE-grown InSb films. The doping efficiency was shown to be a function of growth temperature and In/Sb flux ratio. At a growth temperature of 300 C, and a flux ratio Sb/In ≈1.06, the density $n_{extr}$ of extrinsic electrons is roughly 1/25 of that of Er atoms ($n_{Er}$), provided that $n_{Er} < 5x10^{18}$ $cm^{-3}$. At higher $n_{Er}$, $n_{extr}$ remains constant at about $1.8x10^{17} cm^{-3}$. The low-temperature mobility of homogeneously-doped films increases with Er concentration, up to a maximum at $n_{Er} \approx 1x10^{18}$ $cm^{-3}$, above which it decreases. At room temperature, the mobility decreases monotonically with $n_{Er}$. Low-temperature magnetoresistance data on those films give evidence for spin-disorder scattering.

In the present study, magnetoresistance and Hall data taken on δ-doped InSb:Er films grown under the same conditions as those in Ref. 5, are reported. Both the influence of the number of doped layers $N_L$ and of the areal Er density $N_{Er}$ in these layers are studied. δ-doping does result in improved mobilities, under the right conditions. In δ-doped samples, the extrinsic carrier density increases with the average volume Er density ($n_{Er}$), following the same relation as in homogeneously doped samples ($n_{extr} = n_{Er}/25$, for $n_{Er} < 1x10^{18}$ $cm^{-3}$) in the low Er density range, but, unlike the case of homogeneously doped films, $n_{extr}$ actually decreases if the areal Er density in the slabs is pushed higher. In the lowest-doped samples, the low-temperature mobility of the δ-doped samples can be as much as double that of the homogeneously-doped ones, but the difference decreases as $n_{Er}$ increases, and vanishes for $n_{Er} > 5x10^{18}$ $cm^{-3}$.

## 2. EXPERIMENT

InSb films were grown by molecular beam epitaxy on insulating InP substrates in a Perkin-Elmer Phi 400 MBE, with a base pressure of $10^{-10}$ torr. Two series of films were grown for this work: bulk-doped films, described in previous work[5] (2 μm thick, and grown at a substrate temperature of 300 C, with a Sb/In flux ratio of 1.06; a nominally undoped sample was grown as part of that series), and the δ-doped samples. In these, the Er flux density, which was calibrated using a quartz deposition monitor, is deposited in nominally atomically thin layers. Two parameters were explored: the areal density $N_{Er}$ of Er in each layer, and the number of layers, $N_L$, in each 2 μm thick film. The first 0.4 μm of each film grown was undoped. The Er doped layers were uniformly spaced through the remaining 1.6 μm of InSb.

Rectangular Hall bars with a length-to-width ratio sufficiently large to ensure the absence of geometrical magnetoresistance effects, were photolithographically processed from each film. The magnetoresistivity and Hall resistivity of each sample was then measured at temperatures ranging from 40 K to 400 K, in a closed-cycle helium cryostat, at fields between 0 and 2 tesla. A d.c. measurement technique was used, and the polarities of both sample current and magnetic field were reversed. From the field dependence of the Hall resistivity and the magnetoresistivity, it was apparent that two types of carriers were present in most samples. A two-carrier analysis[5] revealed high-mobility electrons , whose densities and mobilities are further reported here, and low-mobility electrons, with densities (about $2x10^{17}$ $cm^{-3}$) and mobilities (about 2000 $cm^{-2}$/V.sec) that were almost sample and temperature independent. Since these low-mobility carriers are seen in almost every lightly-doped or non-intentionally doped MBE-grown sample, it is speculated that they reside at the sample-substrate interface, which contains a large density of dislocations due to the mismatch in lattice constants. For that reason, the results reported in this paper are solely based on the properties of the high-mobility electrons.

The δ-doped samples have electron densities and mobilities shown as function of temperature in figures 1 and 2. The data on homogeneously doped films are in Ref. 5. In figure 1, we show data taken on 6 films doped each with 48 Er-containing layers, with areal densities varying from $1.7x10^{12}$ $cm^{-2}$ to $4.7x10^{14}$ $cm^{-2}$. The electron density increases with increasing Er density, up to an areal Er density of about $1x10^{13}$ $cm^{-2}$. At higher Er densities, the extrinsic electron density, as measured at low temperature, decreases again with increasing Er density. In the highest doped sample, the density at the lowest temperature starts by decreasing. This may be an artifact, due to the fact that the Hall factor r was assumed to be unity throughout the temperature range. Since there is clearly more than one scattering mechanism at work[5] it is possible that the Hall factor r changes with temperature, especially in the higher-doped samples. This would explain the apparent decrease in electron density in the 2 highest doped samples of figure 1. Between 100 and 220 K, the electron density in the samples doped at $2.5x10^{13}$ $cm^{-2}$ and above increases, but there is a break in the density at about 220K, which is not seen in the homogeneously doped samples or in the δ-doped ones with a lower Er density. In the samples with most Er, the room temperature density is even lower than that of nominally undoped InSb, suggesting that some electron trapping mechanism is at work in such samples. The temperature dependence of the mobilities is similar to that of the homogeneously doped samples, but, in general the mobility is higher.

Results measured on δ-doped samples, in which the areal density of Er is $1 \times 10^{13}$ $cm^{-2}$ per Er-doped layer, and the number of layers is varied from 12 to 96 over the 2 μm thickness, are shown in figure 2. The electron density is not appreciably different from that in homogeneously doped films, and the mobility shows only one broad maximum in its temperature dependence, though a hint of a second hump around 300K can be seen in the mobility of the sample with 12 Er-containing layers. These samples are comparable to homogeneously-doped films with low Er concentration, but with higher low-temperature mobilities.

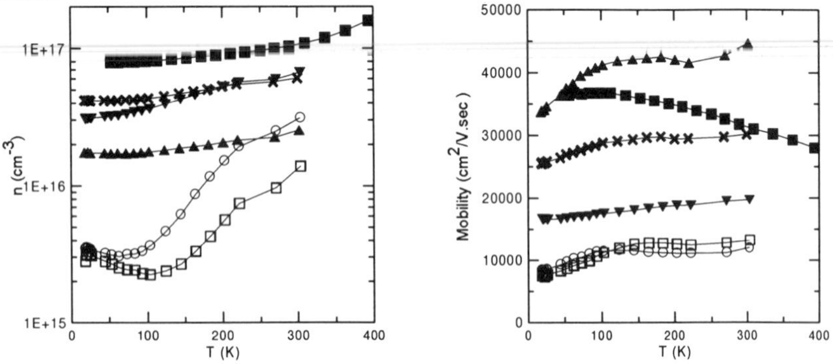


**Figure 1.** Density and mobility of δ-doped InSb:Er films, 2μm thick, with 48 layers of Er, grown at 300 C. The areal Er concentration $N_{Er}$ for the different samples is: (▲): $1.7 \times 10^{12}$ $cm^{-2}$; (■): $1.0 \times 10^{13}$ $cm^{-2}$; (x): $2.5 \times 10^{13}$ $cm^{-2}$; (▼): $7.0 \times 10^{13}$ $cm^{-2}$; (○): $2.0 \times 10^{14}$ $cm^{-2}$; (□): $4.7 \times 10^{14}$ $cm^{-2}$.

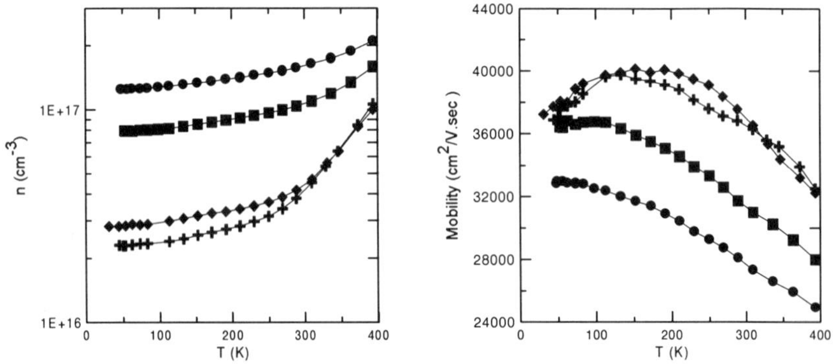


**Figure 2.** Density and mobility of δ-doped InSb:Er films, 2μm thick, with an areal density of Er of $1.0 \times 10^{13}$ $cm^{-2}$, grown at 300 C. The number of layers $N_L$ for the different samples is: (+): 12; (◆): 24; (■): 48; (●): 96.

A better understanding of the influence of δ-doping can be gathered from a study of the density and mobility of the carriers as a function of Er content, shown in Figures 3 (at 83K) and 4 (at 290 K). The non-intentionally doped sample has an electron density of $1.1 \times 10^{15}$ $cm^{-3}$ at 83K, presumably due to residual impurities and other defects, as well as to some intrinsic carriers, since there is some temperature dependence to the density[5]. In order to analyze the data further, we subtract that number from the electron density of each sample at 83K, and attribute the difference to the density of extrinsic electrons induced by Er, $n_{extr}$. A common abscissa axis for both homogeneously-doped and δ-doped samples in Figures 3 and 4 is defined to be the average volume density of Er in the samples, equal to the volume density $n_{Er}$ in the bulk doped samples, and $N_{Er} \times N_L$ / 2 µm in the δ-doped ones. The plot of $n_{extr}$ at 83 K (Fig 3) shows that, at low density (<$5 \times 10^{18}$ Er/cm$^3$), the doping efficiency in the δ-doped and the homogeneously-doped samples is the same. Above that concentration, the homogeneously-doped samples show saturation, while in the δ-doped samples, $n_{extr}$ decreases with increasing Er content. A similar behavior is observed when the total electron density at room temperature is plotted as a function of average Er density in Figure 4. The low-temperature mobility is plotted as a function of average Er density in Figure 3, and shows that δ-doping results in a remarkable improvement of the mobility over that of homogeneously-doped samples, when the Er density is below $1 \times 10^{19}$ cm$^{-3}$. The room-temperature mobility is also improved by about 25% (see Figure 4) in the same concentration range.

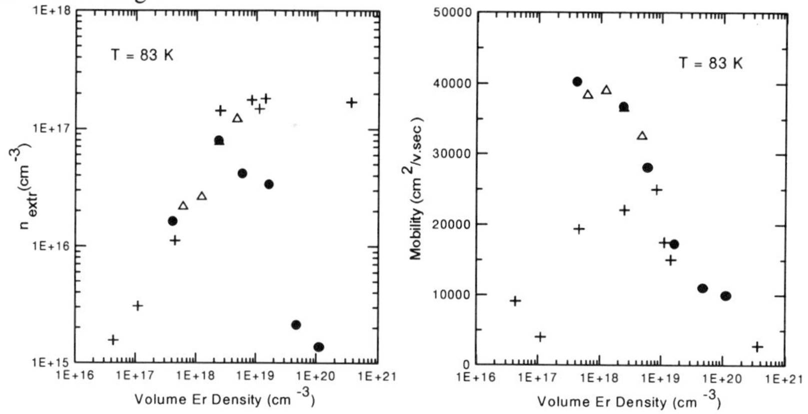


**Figure 3.** Density and mobility of electrons at 83K in the homogeneously doped film (+), in the δ-doped films with 48 Er layers (●) of different Er concentrations, and in the δ-doped films with 12 to 96 layers of Er with $N_{Er} = 1 \times 10^{13}$ cm$^{-2}$ (Δ), as a function of average Er density.

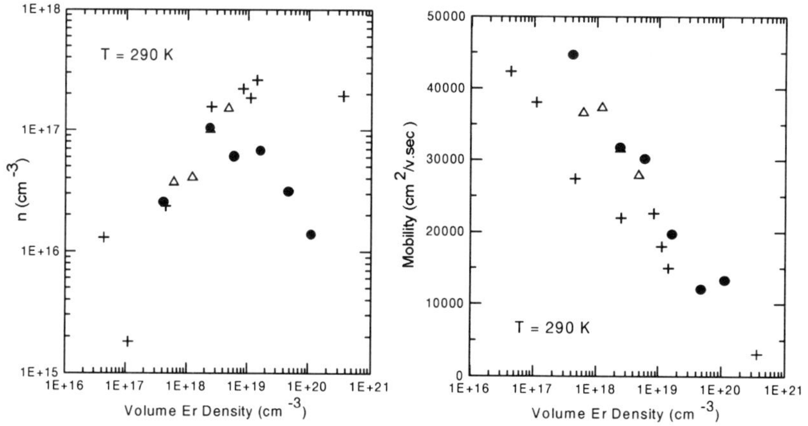


**Figure 4.** Room temperature density and mobility of electrons in the homogeneously doped film (+), in the δ-doped films with 48 Er layers (●) of different Er concentrations, and in the δ-doped films with 12 to 96 layers of Er with $N_{Er} = 1 \times 10^{13}$ cm$^{-2}$ (Δ)

## 3. SUMMARY AND ACKNOWLEDGMENTS

The efficiency of Er as a donor in InSb films grown at 300 C, in a Sb/In flux ratio of about 1.06, is similar in δ-doped samples as in homogeneously doped samples, when the average volume Er density is kept below $5\times10^{18}$ cm$^{-3}$. Above that density, the density of extrinsic electrons saturates in homogeneously-doped samples, and decreases in δ-doped ones. The mobility in δ-doped samples is improved over that of homogeneously doped films; this improvement is most noticeable at lower Er densities and lower temperatures.

The purpose of this paper is to provide experimental data about the dependence of the carrier density and mobility on the parameters that govern δ-doping of InSb with Er, but not to model these results. We can only speculate that the origin of the non-monotonic behavior of the electron density with $N_{Er}$ lies in a deepening of the potential well around the δ layer. Our experience in growing InSb homogeneously doped with rare earth elements (Er, Sm) suggests that most of the metal is electrically inactive, and thus presumably occupies the In sublattice as a trivalent ion. However, depending upon growth conditions (stoichiometry of incident fluxes and growth temperature), a small fraction of the rare earth acts as a donor, possibly because it occupies a different lattice site. Similar effects may be present in these δ-doped films, with various lattice sites being occupied as the areal Er density increases. The increased mobility in the optimally δ-doped samples may be due to the layered distribution of the Er atoms in the sample, which reduces the randomness of the location of the Er atoms. The scattering mechanisms in rare-earth doped samples include spin-disorder scattering besides the more usual impurity, dislocation and phonon scattering that are present in all MBE-grown InSb films grown on a highly mismatched substrate.

The authors acknowledge Thaddeus Schroeder, who provided the motivation for this work.

## 4. REFERENCES

[1] J. Heremans, D.L. Partin, C.M. Thrush and L. Green: Semicond. Sci. Technol. **8**, S424 (1993)

[2] J. Heremans: J. Phys. D: Appl. Phys, **26**, 1149 (1993)

[3] D.T. Morelli, D.L. Partin, J. Heremans and C.M. Thrush: J. Vac. Sci. Technol. **B10**, 110 (1992)

[4] D.L. Partin, J. Heremans, C.M. Thrush and D.T. Morelli, : J. Vac. Sci. Technol. **B10**, 873 (1992)

[5] J. Heremans, D.L. Partin, D.T. Morelli and C.M. Thrush: J. Vac. Sci. Technol. **B10**, 659 (1992)

[6] D.L. Partin, J. Heremans, D.T. Morelli, and C.M. Thrush: Mat. Res. Soc. Symp. Proc. **216**, 527 (1991)

[7] K. Uwai, H. Nakagome, and K. Takahei, J. Cryst. Growth 93, **583** (1988); P.B. Klein *et al.*, Appl. Phys. Lett. **62**, 2309 (1989).

*Inst. Phys. Conf. Ser. No 144: Section 4*
*Paper presented at 7th Int. Conf. on Narrow Gap Semicond. Santa Fe 9–12 January*



# Optical and Structural Characterisation of InAs/GaSb Superlattices

F Fuchs, J Schmitz, N Herres, J Wagner, J D Ralston, and P Koidl

Fraunhofer-Institut für Angewandte Festkörperphysik,
Tullastrasse 72, D-79108 Freiburg, Germany

**Abstract:** InAs/GaSb superlattices with InAs layer thicknesses ranging from 4 to 14 monolayers (MLs) and 10 ML thick GaSb layers have been studied. They were grown with either InSb-like or GaAs-like interfaces (IF's) applying monolayer molecular beam epitaxy for the growth of the IF's. Detailed sample characterization was carried out using Fourier transform infrared photoluminescence spectroscopy, photocurrent spectroscopy, Raman scattering, and high resolution X-ray diffraction space mapping.

## 1. Introduction

InAs/GaSb superlattices (SL's) are of current interest for infrared detector applications. Theoretical predictions [1] of optical absorption coefficients comparable to those of $Hg_{1-x}Cd_xTe$ (MCT) have recently been confirmed [2,3]. This material system benefits from the well established III-V processing techniques. Photoconductive and photovoltaic detectors have been demonstrated [2,3]. From theoretical investigations reduced thermal and tunneling noise currents relative to MCT can be expected [4].

In 1987 Smith and Mailhot [1] proposed the InAs/(GaIn)Sb materials system for infrared detector applications to be exploited. InAs/GaSb binary multilayers constitute a type II SL, with the conduction band edge in the InAs layers lying below the valence band maximum of the GaSb layers. In suitably thin InAs layers the confined electron states are pushed upwards, opening up a SL bandgap, which can be tuned by varying the thickness of the individual layers. The holes are confined in the GaSb layers, leading to spatially indirect optical transitions. Strong absorbance and efficient luminescence from such QW's thus relies on the ability to grow very thin layers with well-defined and reproducible interface (IF) properties. Because of the change of both cation and anion two types of IF's can be realized. Depending on the MBE shutter sequence, InSb-like or GaAs-like bonding can be driven [5]. To achieve a bandgap in the 8 to 12 μm spectral region, the InAs layers consist typically of 10 to 14 Monolayers (ML's) [6]. Tailoring the bandgap into the mid infrared (3 to 5 μm) requires an InAs layer thickness of 4 to 6 ML's. The effective bandgap is shown to depend on the type of IF [7], which is assumed to be due to different values of the valence band offset (VBO). However, the experimental data presented by various groups are not consistent [7,8]. In a theoretical work Hemstreet et al. [9] suggested for a SL grown with GaAs-like IF's a VBO 150 meV smaller than for structures containing InSb-like IF's. The influence of the interfaces on the electronic properties of the SL's appear therefore to be of crucial importance.

We have studied InAs/GaSb SL's with InAs layer thicknesses ranging from 4 to 14 monolayers and 10 ML thick GaSb barriers. Two series of samples were grown with either InSb-like or GaAs-like IF's applying monolayer epitaxy. The samples were grown on 2" undoped {100} GaAs substrates by solid source MBE using a valved cracker As- and a standard cracker effusion Sb-cell supplying molecular beams of $As_2$ and $Sb_2$, respectively. Details of the growth technique are presented elsewhere [10]. For the structural characterization high resolution X-ray diffraction (HRXRD) space mapping [11,12] and Raman scattering have been performed. Access to the electronic properties of the heterostructures has been achieved via Fourier-transform photoluminescence (PL) and photocurrent (PC) spectroscopy, measuring transport parallel to the layer planes.

## 2. Structural characterization

Conventional rocking curves and two dimensional HRXRD space maps have been recorded near 00L and HHL reflections [11,12]. In general, samples grown with InSb-like IF's show sharp diffraction peaks. In contrast, the SL peaks recorded from samples with GaAs-like IF's are broadened due to warping and tilting within the SL film. The analysis of the 00L rocking curves allows the determination of the actual SL periods and lattice constants in growth direction. Evaluation of the HHL rocking curves and two dimensional diffraction space maps provide information about the lattice constants in the layer plane.

In Fig. 1 both lattice constants are plotted versus the number of In-containing monolayers per SL period. The in-plane lattice constants are shown in Fig. 1 (a). With the exception of one sample they are nearly perfectly strained to the lattice constant of the underlying GaSb buffer layer. However, we note that all samples show in-plane lattice constants slightly below the value of GaSb. Furthermore the sample containing 4 ML's InAs grown with GaAs-like interfaces shows a considerably smaller in-plane lattice constant, indicating a partially relaxed film. According to continuum elasticity theory (CET), the biaxial tension (compression) in the layer plane causes an uniaxial compression (tension) in growth direction, which becomes apparent in Fig. 1 (b). A deviation of the lattice constants parallel to the growth direction from the value of GaSb is observed. Furthermore, the SL's grown with InSb-like IF's show systematically larger lattice constants parallel to the growth direction compared to GaSb. In

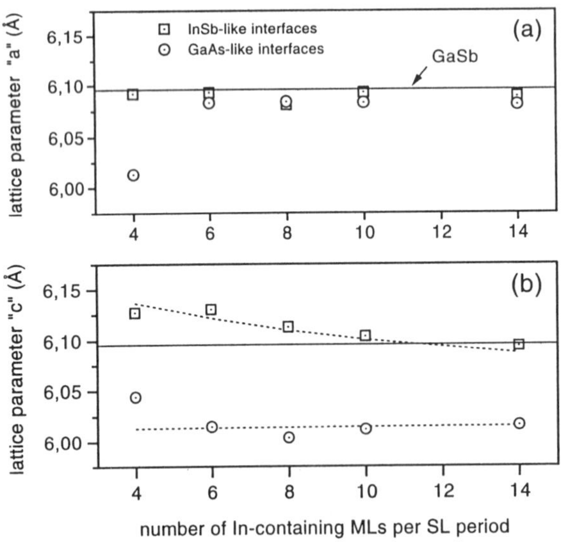


Fig. 1 Lattice constants in-plane (a) and in growth direction (b) of the InAs/GaSb superlattices versus the nominal number N of InAs monolayers.

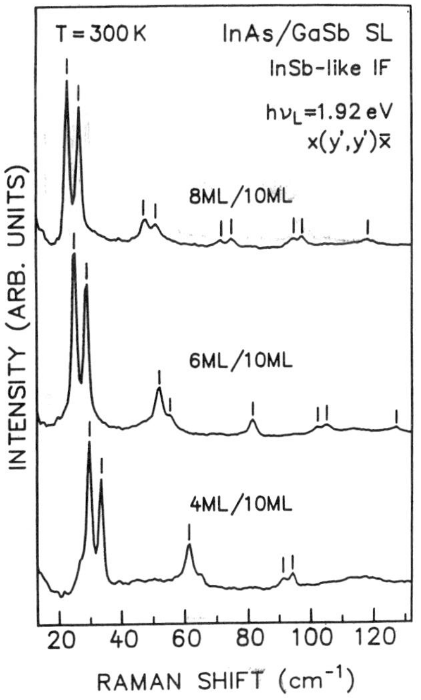


Fig. 2 Raman spectra of InAs/GaSb superlattices grown with InSb-like interfaces. Backfolded longitudinal acoustic phonon doublets are marked by vertical lines. Polarization of the incident scattered light was parallel to the (110) crystallographic direction [x(y',y')x].

contrast, samples grown with GaAs-like IF bonds show smaller lattice constants in growth direction than GaSb, and the deviation from the cubic metric is stronger than for heterostructures grown with InSb-like IF's. We find a decrease of the volume of the unit cell by ≈2 % in case of growth with GaAs-like IF bonds.

In Fig. 1 (b) the data of both series of SL's are compared with CET. The dotted lines have been obtained assuming that the films are perfectly coherently strained to the lattice constant of the GaSb buffer, given by the solid line in Fig. 1 (a). A significant deviation from CET is only observed for the samples with 4 ML InAs layer thickness.

The high structural quality of the samples is also confirmed by Raman spectra of the samples. They clearly show the InSb or GaAs interface modes depending on the type of interface bonds [13].The SL's grown with InSb-like IF's show scattering by folded longitudinal acoustic (LA) phonons up to the 5$^{th}$ order (Fig. 2). For SL's with GaAs-like IF's the folded LA phonon modes occur at higher frequencies as compared to SL's grown with InSb-like IF's, reflecting the different lengths of the Ga-As and In-Sb bonds [14].

### 3. Electronic properties

Photoluminescence provides access to information about the energetic position of the effective SL bandgap. The PL intensity is directly related to the spatially indirect overlap of the electron and hole wavefunctions across the interfaces. A decrease of the IF sharpness and the formation of non-radiative defects at the interface will result in a decrease of the PL intensity. In full agreement with the X-ray analysis, which shows better defined SL diffraction peaks for heterostructures grown with InSb-like IF's, the PL intensity of these SL's is much stronger than from structures grown with GaAs-like IF bonds.

Low-temperature PL and PC spectra of the heterostructures grown with InSb-like IF's are shown in Fig. 3. In all spectra the high energy onset of the PL matches the low-energy cut-off of the PC response. The energetic position of the SL bandgap obtained from the two series of samples is given in Fig. 4. The bandgap energy is plotted versus the nominal InAs layer thickness. The data from the series grown with GaAs-like IF's (*) has been obtained from PC measurements. These samples do not show efficient PL. Open squares correspond

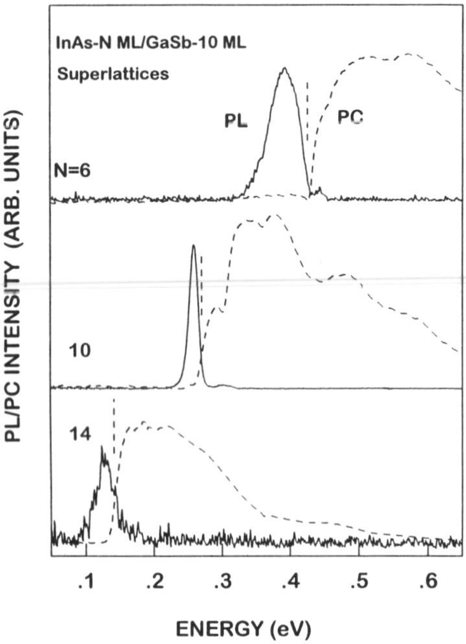


Fig. 3 Low-temperature photoluminescence (PL, full lines) and photocurrent (PC, broken lines) spectra of n-ML InAs/10-ML GaSb superlattices grown with InSb-like interfaces. The position of the effective bandgap is indicated by broken bars. The PC has been measured parallel to the SL planes.

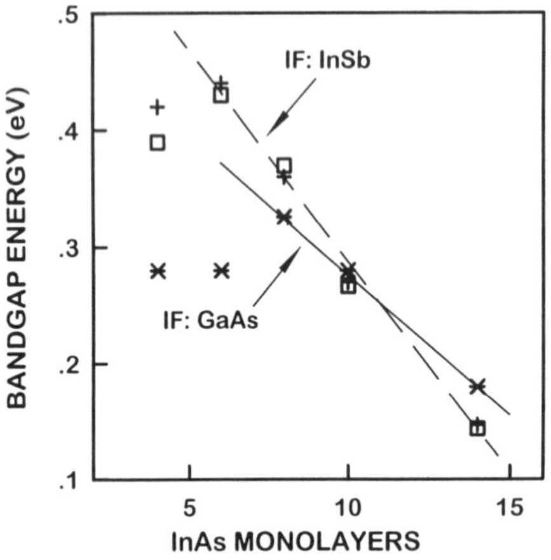


Fig. 4 Energetic position of the effective bandgap of three series of n-ML InAs/10-ML GaSb superlattices grown either with GaAs-like interfaces (PC: *) or InSb-like interfaces (PC: +,PL:□).

to data obtained by PL from SL's with InSb-like IF's, crosses are obtained from these samples by measurement of the PC.

A nearly linear dependence of the effective bandgap is observed for SL's grown with InSb-like IF's (broken line) for InAs layer thicknesses exceeding 4 ML's. The same behavior is observed for the SL's grown with GaAs-like (full line) IF's if the layer thickness exceeds 6 ML's. With X-ray diffraction techniques (preceding section) partial relaxation has been observed at the samples with the thinnest InAs layers. We conclude, that partial relaxation of the SL films induce a lowering of the effective bandgap compared to coherently strained SL's, which is stronger in the case of heterostructures grown with GaAs-like IF's. The partially relaxed 4 ML-InAs/10 ML GaSb SL with GaAs-like IF's exhibits a lowering of the effective bandgap by 150 meV compared to the value expected for a coherently strained film

Comparing the bandgap dependence on InAs layer thickness in Fig. 4 a smaller slope is observed for the SL's grown with GaAs-like IF's (full line), while the slope of the heterostructures with InSb-like IF's is larger (broken line). The intersection of the two lines is found at an InAs layer thickness of roughly 10 ML's. Extrapolating the straight lines to an InAs layer thickness of 3 ML's leads to a bandgap difference for this special structure of 100 meV. This observation is in good accordance with a higher value of the VBO in the case of InSb-like IF's as suggested by Hemstreet et al. [9]. The data shown in Fig. 4 indicate a lower bandgap of the SL's with InSb-like IF's if the InAs layer thickness exceeds 10 ML's. This behavior may be explained with a volume component induced by the increased SL unit cell (≈3.3 %) in case of growth with InSb-like IF's (Fig. 1). An increase of the volume of the unit cell in general leads to a lowering of the fundamental band gap.

In summary, two-dimensional HRXRD space mapping enables the measurement of the lattice constant, both in-plane and in growth direction of the layers. Almost all samples are coherently strained to the lattice constant of the underlying GaSb buffer. Deviation from the expected behavior has been found for InAs layer thicknesses of 4 ML's. The X-ray SL diffraction peaks of the structures with InSb-like IF's are found to be narrower than those with GaAs-like IF's, where the broadening is due to tilt and warping within the SL film. The PL and PC measurements allow us to extract the energetic position of the SL bandgap, which is found to depend linearly on the InAs layer thickness if no partial relaxation takes place. The slope of the linear dependence is found to depend on the type of interface bonding. Coherently strained heterostructures with thin InAs layers exhibit the higher effective bandgap for InSb-like IF's. The difference vanishes if the InAs layer thickness increases to 10 ML's, and even inverts for thicker InAs layers. This dependence can tentatively be explained by the different volumes of the units cells which are shown to depend on the type of IF's.

The technical assistance of K. Schwarz and G. Bihlmann is gratefully acknowledged

**References**


[1] Smith D L and Mailhiot C 1987 *J. Appl. Phys.* **62** 2545

[2] Chow D H, Miles R H, , Schulman J N, Collins D A, and McGill T C, 1991 *Semicond. Sci. Technol.* **6** C47

[3] Campbell I H, Sela I, Laurich B K, Smith D L, Bolognesi C R, Samoska A, Gossard A C, and Kroemer H 1991 *Appl. Phys. Lett.* **59** 846

[4] Youngdale E R, Meyer J R, Hoffman C A, Bartoli F J, Grein C H, Young P M, Ehrenreich H, Miles R H, Chow D H 1994 *Appl. Phys Lett* **64** 3160

[5] Chow D H, Miles R H, and Hunter A T 1992 *J. Vac. Sci. Technol.* **B10** 888

[6] Roslund J H and Andersson T G 1994 Proc. ICSMM-7 Banff 1994 to appear in *Superlattices and Microstructures*

[7] Waterman J R ,Shanabrook B V, Wagner R J, Yang M J, Davis J L, and Omaggio J P 1993 *Semicond. Sci. Technol.* **8** S106

[8] Inoue M, Yano M, Furuse N, Iwai Y 1993 *Semicond. Sci. Technol.* **8** S121

[9] Hemstreet L A, Fong C Y, Nelson J S 1993 *J. Vac. Sci.* **B11** 1693

[10] Schmitz J, Wagner J, Fuchs F, Herres N, Koidl P, and Ralston J D 1995 *Journal of Crystal Growth* , in press

[11] Van der Sluis P 1994 *J. Appl. Cryst.* **27** 1015

[12] Schuster M and Herres N 1995 *Crystallography Reviews*, Pergamon, to be published

[13] Behr D, Wagner J, Schmitz J, Herres N, Ralston J D, Koidl P 1994 *Appl. Phys. Lett.* **65** 2972

[14] Wagner J, Schmitz J, Herres N, Ralston J D, Koidl P 1995, *Appl. Phys. Lett*, to be published

*Inst. Phys. Conf. Ser. No 144: Section 4*
*Paper presented at 7th Int. Conf. on Narrow Gap Semicond. Santa Fe 9–12 January*



# Ordering and bandgap reduction in $InAs_{1-x}Sb_x$ alloys†

**D M Follstaedt, R M Biefeld, S R Kurtz, L R Dawson and K C Baucom**

Sandia National Laboratories, Albuquerque, NM 87185-1056 USA

**Abstract.** $InAs_{1-x}Sb_x$ alloys grown by MBE and MOCVD are found to have reduced emission energies due to CuPt-type order, even for Sb concentrations as low as x = 0.07 (ΔE = 25-65 meV). Cross-section TEM examination of such alloys shows the two $\{111\}_B$ variants are separated into regions 1-2 μm across with platelet domains 10-40 nm thick on habit planes tilted ~30° from the (001) growth surface. Nomarski optical images show a cross-hatched surface pattern expected for lattice-mismatched layers. The local tilt of the surface correlates with the dominant variant in each region. $InAs_{1-x}Sb_x$ /$In_{1-y}Ga_yAs$ strained-layer superlattices with low Sb content and flat surfaces also show CuPt ordering.

## 1. Introduction

Alloys of $InAs_{1-x}Sb_x$ have the lowest bandgaps of ternary III-V alloys and are of interest for potential use as infrared detectors and emitters. We have used molecular beam epitaxy (MBE) and metallorganic chemical vapor deposition (MOCVD) to grow layers of this alloy as well as strained-layer superlattices (SLS) with it as a constituent. The optical emission energies of our materials [1] are systematically below the bandgaps of bulk alloys [2]. We have examined the alloys with cross-section transmission electron microscopy (XTEM) and found that all contain CuPt-type order [3-5], which is expected to reduce bandgaps in this alloy system [6]. Ordering is found even for Sb concentrations as low as x = 0.07. Here we discuss order in mid-concentration alloys (x ≈ 0.4) of potential use as long-wavelength infrared detectors and then examine order in low-Sb alloys (x = 0.07-0.14) for mid-wavelength emitters and detectors. The latter are being used in SLSs of $InAs_{1-x}Sb_x$ /$In_{1-y}Ga_yAs$ for lasers [7], and these $InAs_{1-x}Sb_x$ layers also contain CuPt order.

Another non-random alloy structure, phase separation into two compositions, could also give reduced emission energies. Phase separation occurs on a coarse scale (30-50 nm) in $InAs_{1-x}Sb_x$ alloys grown below 400°C and even occurs with CuPt ordering [8-10]. Our alloys are grown above this temperature and the coarse separation is avoided. The x = 0.07 alloy does not show evidence of either coarse- or fine-scale (~10 nm) phase separation [5], and we therefore attribute its reduced emission energy to CuPt order. Preparation of thin XTEM specimens of these fragile alloys is discussed elsewhere [5].

We previously investigated an alloy grown by MBE at 425°C with x = 0.4 and found CuPt-type ordering reflections at ½{111} and related positions [3]. Ordering was also found by Seong et al. [8] for MBE alloys with x = 0.2 - 0.8, but was strongest near 0.5, as expected from the ideal CuPt stoichiometry. Their specimens were grown at 370°C, for which phase separation occurred. Recent evaluations by Seong et al. [10] of their alloys with x = 0.5 indicate that growth temperatures of 370-400°C give maximum ordering, and also that it occurs in the As-rich phase, $InAs_{0.74}Sb_{0.26}$; however, they did not detect order at 450°C and above. Our alloys are thus expected to have a reduced degree of order due to higher growth temperatures (425-525°C), but ordering in As-rich alloys is not unexpected.

The optical emission energy of our $InAs_{0.60}Sb_{0.40}$ alloy was 132 meV ($\lambda$ = 9.4 μm), while that expected for bulk alloys is 177 meV [2]. A SLS with a similar alloy and a type-II offset from adjacent layers gave an emission energy of 117 meV, and the photoconductive response was found to extend to 10.6 μm, or ~80 meV lower than for bulk bandgaps. The photoresponse of a phase-separated "natural superlattice" grown at 340°C extended to 12.5 μm [11]. These results indicate that ordered $InAs_{1-x}Sb_x$ alloys can have bandgaps low enough to span the 8-12 μm atmospheric window for detection.

## 2. Low-Sb Concentration Alloys

We detected CuPt order in [110] electron diffraction patterns from MOCVD alloys with x = 0.07 and 0.14 grown at 475°C on InAs substrates. In addition, the relative intensities of the two variants changed with location; regions 1 - 2 μm across could be found in which one variant was dominant or existed alone. Here we examine the domain structure of these regions in the x = 0.14 alloy.

The top of the alloy layer (1.7μm total thickness) is shown in the bright-field XTEM image in Fig. 1a. Near the center and right-hand side of the image, dark linear features are seen, which tilt downward to the left. The diffraction pattern in Fig. 1b was taken from this region, and indicates that one variant dominates. On the left side of Fig. 1a, similar linear features are seen which tilt in the opposite direction; diffraction from this region (not shown) indicates that the other variant dominates. The approximate boundary between the two regions in indicated with a dashed line extending to the surface. The surface is not flat; many areas show peaks and valleys. The peaks are 20-60 nm above the valleys, with surface tilts of 2 - 4°. Note that the direction of surface tilt in Fig. 1a above the center and right is toward the $<111>_B$ direction of the dominant variant in Fig. 1b, while the opposite tilt occurs on the left-hand side with the other variant. In several alloys, such tilting by a few degrees has been shown to favor the corresponding variant [11-13].

The surface roughness is quite evident in Nomarski optical micrographs, as seen in Fig. 3. This figure shows a cross-hatched pattern on the surface like that expected for lattice-mismatched layer/substrate combinations. Two sets of orthogonal lines are seen along the [100] and [010] directions. The "sense" of the illumination in this image is that of light coming from the upper left corner, and lines at 45° to [100] are also seen.

X-ray diffraction was used to determine the lattice constants of the layer and thereby determine its composition and degree of lattice relaxation. By using two reflections, one with a reciprocal lattice vector normal to the surface and one having an in-plane component, the lattice constants of the layer in the orthogonal and parallel directions can be obtained. This determines the composition accurately even when lattice relaxation is incomplete. For this alloy, $a_\perp$ = 0.61324 nm and $a_\parallel$ = 0.61058 nm, giving a relaxed lattice constant of $a_0$ = 0.61186 nm, which corresponds to x = 0.14 and 1.0% lattice mismatch with the InAs substrate (0.60585 nm). The $a_\parallel$ value is positioned 79% of way from the substrate $a_0$ to the unstrained $a_0$, indicating that significant relaxation has occurred, which is consistent with the Nomarski image. The layer/substrate interface shows a high density of misfit dislocations [5], and defects pass through the layer to the surface as seen in Fig. 1a.

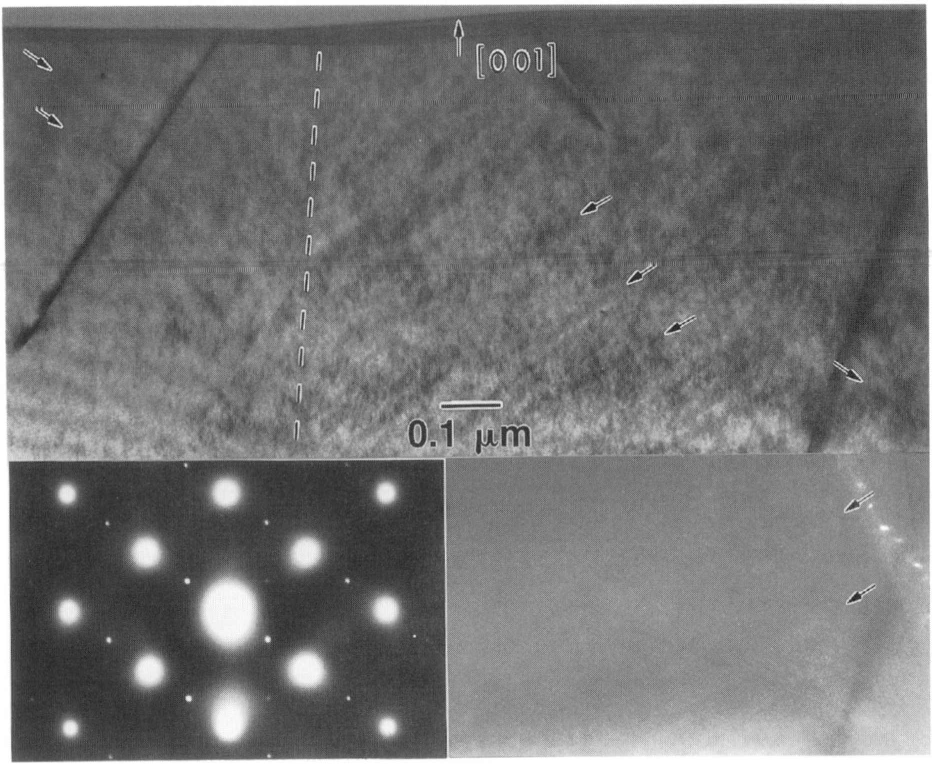


Figure 1. a) Bright-field XTEM image of $InAs_{0.86}Sb_{0.14}$ alloy grown by MOCVD on InAs at 475°C. Dashed line divides two regions with different variants (arrows). b) Diffraction pattern from region to the right of center in a). c) Dark-field image of ordered domains in the lower right side of a).

Considering strain relaxation, surface tilting, and correlation of variant with tilt, the formation of an alloy layer with regions alternating between the two $\{111\}_B$ variants is seen to be closely related to lattice mismatch of the alloy and substrate. Surface tilting results from misfit dislocations and selects one variant locally during growth. The 2 - 4° tilts are mainly due to atomic surface steps, since the angular width of the (004) x-ray reflection shows misorientations of the atomic planes by only ~0.5°. A similar separation of variants due to surface tilts has been seen in (In,Ga)P, but the tilting has a different origin [13].

The direct identification of the linear contrast in Fig. 1a with order requires dark-field imaging of the domains with the ½{111}-type reflections. These reflections are weaker than those of alloys with x = 0.4 [3] and those of $In_{0.5}Ga_{0.5}P$ [14], but we are barely able to obtain images with the domains illuminated. Figure 1c is such an image (contrast enhanced) from the lower right-hand side of Fig. 1a, with illuminated linear features similar to the dark ones in Fig. 1a. We thus conclude that the dark linear features in bright-field are the ordered domains. They appear to be platelets viewed edge-on and are 10 - 40 nm thick. They lie on habit planes tilted ~30° from (001) in the same direction as the surface tilt.

Order is incomplete in the $InAs_{0.86}Sb_{0.14}$ alloy, as expected from the ideal (1:1) CuPt stoichiometry. The separated platelets in Figs. 1a and 1c indicate that ordered domains are in a disordered matrix. Also, high-resolution images (not shown) have ordering contrast that starts and stops along $\{111\}_B$ planes, implying that order is not continuous within domains.

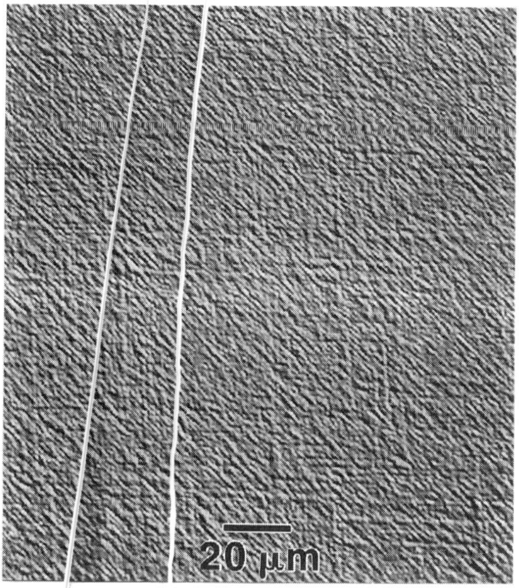


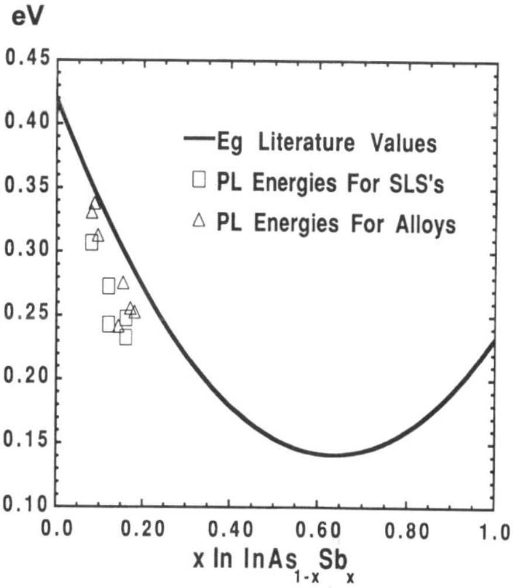


Figure 2. Nomarski image of $InAs_{0.86}Sb_{0.14}$ alloy surface grown by MOCVD at 475°C on InAs. Linear features are valleys and crests. (CVD1088).

Figure 3. $InAs_{1-x}Sb_x$ bandgap vs. x. Solid line- bulk and OMVPE from [2]. Symbols - our MOCVD results.

In Fig. 3, emission energies of our low-Sb alloys and SLSs are plotted versus Sb concentration along with a curve from Fang et al. [2] representing their OMVPE data and bulk alloy data. Our data systematically fall below the curve, evidently due to CuPt ordering. The difference from their vapor-grown alloys is not understood. It could be related to measurement of composition or to growth parameters. However the only notable difference in growth is rate (our 1.7 μm/hr vs. 0.5 μm/hr) and it does not appear substantial.

## 3. Low-Sb Concentration Strained-Layer Superlattices

We are investigating SLSs of $InAs_{1-x}Sb_x/In_{1-y}Ga_yAs$ to form improved mid-wavelength infrared emitters. By correctly choosing concentrations, such as x = 0.09 and y = 0.13 to be discussed here, the average in-plane lattice constant can be matched to InAs. The SLS also provides an in-plane strain to the $InAs_{1-x}Sb_x$ layer that reduces non-radiative Auger transitions of excited carriers [7]. The increased optical efficiency may allow lasing to be achieved at higher temperatures. However, the emission energy is still below the bandgap of bulk $InAs_{1-x}Sb_x$ alloys, as seen in Fig. 3. Figure 4a shows an electron diffraction pattern from the SLS. Very weak CuPt ordering reflections of both variants are seen, along with wavy streaks along the [001] direction. The microstructure of this order is seen in Fig. 4b to be small regions of one variant as detected by the doubled periodicity (light/dark variations) in the high-resolution lattice image along the arrowed $\{111\}_B$ planes. These two regions are in 10 nm-thick $InAs_{1-x}Sb_x$ layers of the SLS [15]. The SLS layers are flat with no tilt preference for either variant, but small single-variant regions 10-20 nm wide nonetheless form during growth at 475°C. Other alloys and SLSs grown at up to 525°C also emit at energies below bulk bandgaps, implying that order is not eliminated at this temperature.

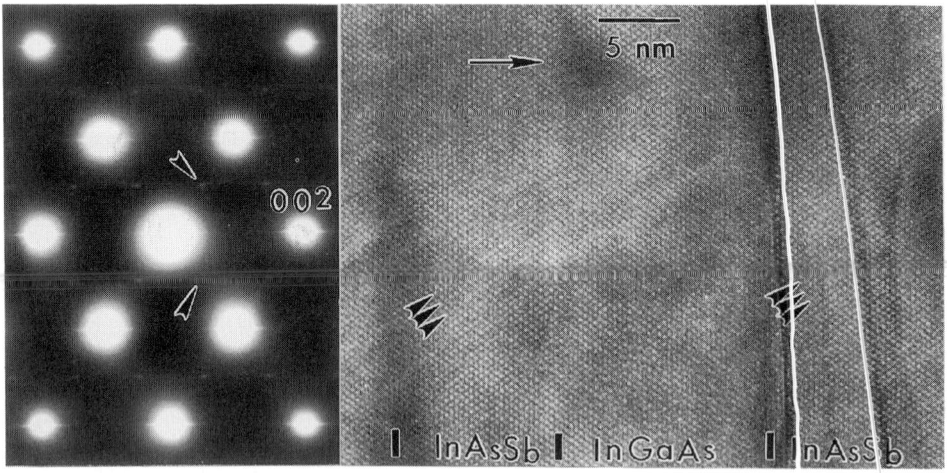


Figure 4. a) [110] diffraction pattern from $InAs_{0.81}Sb_{0.09}/In_{0.87}Ga_{0.13}As$ SLS with CuPt reflections marked. b) High-resolution lattice image with order (alternating dark/light) along $\{111\}_B$ atomic planes (arrowed). Growth direction is shown with large arrow. (Specimen CVD1172).

## References

† This work was supported by the United States Department of Energy under Contract DE-AC04-94AL85000.


[1] Biefeld R M, Baucom K C and Kurtz S R 1994 J. Crystal Growth **137** 231-234.

[2] Fang Z M, Ma K Y, Jaw D H, Cohen R M and Stringfellow G B 1990 J. Appl. Phys. **67** 7034-7039.

[3] Kurtz S R, Dawson L R, Biefeld R M, Follstaedt D M and Doyle B L 1992 Phys. Rev. **B 46** 1909-1912.

[4] Biefeld R M, Baucom K C, Kurtz S R and Follstaedt D M 1994 Mat. Res. Soc. Symp. Proc. **325** 493-498.

[5] Follstaedt D M, Biefeld R M, Kurtz S R and Baucom K C 1995 , submitted to Journal of Electronic Materials.

[6] Wei S-H and Zunger A 1991 Appl. Phys. Lett. **58** 2684-2686.

[7] Kurtz S R, Biefeld R M, Dawson L R, Baucom K C and Howard A J 1994 Appl. Phys. Lett. **64** 812-814.

[8] Seong T-Y, Norman A G, Booker G R, Droopad R, Williams R L, Parker S D, Wang P D and Stradling R A 1990 Mat. Res. Soc. Symp. Proc. **163** 907-912.

[9] Seong T-Y, Norman A G, Ferguson I T and Booker G R 1993 J. Appl. Phys. **73** 8227-8236.

[10] Seong T-Y, Booker G R, Norman A G and Ferguson I T 1994 Appl. Phys. Lett. **64** 3593-3595.

[11] Chen G S and Stringfellow G B 1991 Appl. Phys. Lett. **59** 3258-3260.

[12] Su L C, Ho I H and Stringfellow G B 1994 J. Appl. Phys. **75** 5135-5141.

[13] Friedman D J, Zhu J G, Kibbler A E and Olson J M 1993 J. Appl. Phys. **63** 1774-1776.

[14] Follstaedt D M, Schneider R P, Jr and Jones E D 1995, submitted to Journal of Applied Physics

[15] Biefeld R M, Follstaedt D M, Kurtz S R and Baucom K C 1995 , elsewhere these proceedings. This paper discusses the SLS structure in more detail.

*Paper presented at 7th Int. Conf. on Narrow Gap Semicond. Santa Fe 9–12 January*

# Remote doping of InAs/GaSb quantum wells by means of a second InAs well doped with silicon

T A Malik, S J Chung, R A Stradling, W T Yuen

*Physics Department, Imperial College, London SW7 2BZ, UK.*

J J Harris and A G Norman

*IRC in Semiconductor Materials, Imperial College, London SW7 2BZ, UK.*

**Abstract:** Remote doping of InAs/GaSb quantum wells presents a problem with molecular beam epitaxy (MBE). Silicon, which is the preferred MBE dopant for other III-Vs, acts as an acceptor with GaSb. Use of a group VI dopant risks long term memory effects. An alternative technology has been developed employing a double well structure where the second InAs well is thin ( ~ 2nm) and highly doped with silicon. Because of the high confinement energy this well acts as a source of electrons for the first well. Conductivities as high as $10^{-1}$S at 4K can be achieved with a mobility of 185,000 $cm^2$/Vs. Higher electron concentrations are possible but the mobility falls, probably due to intersubband scattering when the second subband becomes occupied. A self-consistent modelling programme has been developed to describe the electron transfer. The model includes i) the nonparabolicity of the InAs conduction band, ii) the 2D hole gas in the GaSb and iii) the surface pinning of the Fermi energy due to the presence of surface donors in the GaSb.

## 1. Introduction

N-type doping of GaSb has always presented a problem with molecular beam epitaxy (MBE). Silicon is the preferred MBE dopant for other III-Vs but acts as an acceptor with GaSb [1]. There is a need to obtain high conductances with InAs/GaSb/AlSb Quantum Wells for device applications such as High Electron Mobility Transistors [2,3] and for hybrid superconductor-semiconductor structures [4]. There is a risk of machine contamination with possible long term memory effects if a group VI element is used as a dopant [5]. In order to perform remote doping of InAs/GaSb quantum wells we have therefore developed an alternative technology exploiting the fact that InAs itself can be very heavily doped with silicon donors (up to $6 \times 10^{19}$ $cm^{-3}$). A double well structure is used where the second InAs well is thin ( ~ 2nm) and highly doped. Because of the high confinement energy and high silicon concentration, this well acts as a source of electrons for the first well. Kopev et al [6] have reported the use of a secondary thin well to improve the electrical properties of a thicker well. In this case however the second well was undoped and the main effect was to improve the mobility. This mobility increase was attributed to a smoothing effect.

## 2. Experimental Measurements

Although bulk InAs can be doped with very high concentrations of silicon donors, experiments with InAs delta doped with silicon show that the silicon can spread up to 10nm [7]. It was therefore possible that the silicon might diffuse across the InAs/GaSb interfaces and be preferentially incorporated in the GaSb, thereby producing compensating acceptors and a p-n-p dopant profile in the region of the thin well. A preliminary objective was therefore to establish that the silicon is retained within a thin quantum well. Two test structures were therefore grown, omitting the wide undoped well and spacer but otherwise identical to the double well structures, whose characteristics are discussed later. These test structures show full activation of the silicon as a donor. The detection of an n-type Hall coefficient demonstrates that the silicon had not spread significantly into the GaSb. Rather low mobilities are obtained: e.g. with a 2.5nm well the mobility obtained was 600 $cm^2$/Vs for a nominal doping concentration of $2x10^{19}$ $cm^{-3}$ ($5x10^{12}$ $cm^{-2}$) and 2000 $cm^2$/Vs for a 5nm well nominally doped at $1x10^{19}$ $cm^{-3}$ ($5x10^{12}$ $cm^{-2}$). The Hall measurements give electron concentrations larger than the nominal values, probably because of corrections amounting to 20 - 30% from the p-type background conductance of the GaSb at 77K. The measured mobilities are about 30-50% smaller than the corresponding bulk mobility at the equivalent silicon concentrations [8]. A low mobility in the thin well is an advantage for the double well structures envisaged because this reduces the parallel conduction from any electrons that do not transfer into the wider undoped well. In the double well structure the mobility in the doped well is expected to fall even lower due to the electron transfer as the ionised impurity scattering rate will increase.

A variety of double well structures were grown. All the structures were grown on GaAs substrates with a buffer of 800nm of GaAs and 800nm of GaSb. The substrate temperature was typically 450°C. An InAs doping well of thickness between 1.8 and 2.5 nm was then grown followed by a GaSb spacer of thickness between 5 and 10 nm and then by the thicker undoped InAs well with thickness between 14 and 20 nm. The interface switching sequence was chosen to produce InSb interfaces. Finally a GaSb cap of thickness 17-20 nm was grown. With this thickness of cap the GaSb "surface donors" [9] present at the air interface are expected to produce about $3x10^{11}cm^{-2}$ additional electrons in the undoped well. It should be noted however that the charge states of these defect-induced levels have not yet been determined. The details of the samples discussed in this paper are given in Table 1. As the contribution from the surface defects to the total electron concentration is significant, the results for single well samples of different cap thicknesses but without silicon doping are also included in table one and in our subsequent modelling.

Hall and mobility measurements were taken between room temperature and 77K and at 4K at low magnetic fields (~0.05T). Undoped samples consisting of a single well of thickness 20nm with InSb interfaces and a cap thickness of 20nm grown under the same conditions consistently have an electron concentration of $1.1 \pm 0.1$ $x10^{12}cm^{-2}$ at 77K and mobilities between 50,000 and 110,000 $cm^{-2}$/Vs. Samples with GaAs interfaces and otherwise identical structures have about 10% fewer electrons at 77K and mobilities within the same range. All the double well structures show much higher electron concentrations indicating that substantial electron transfer from the thin well has occurred. The enhancement of the

| Sample | Cap (nm) | Well (nm) | Spacer (nm) | Doped Well (nm) | Doping ($cm^{-3}$) |
|---|---|---|---|---|---|
| *Single Wells* | | | | | |
| IC411 | 10 | 20 | - | - | - |
| IC404 | 20 | 20 | - | - | - |
| IC410 | 50 | 20 | - | - | - |
| *Single Doped Wells* | | | | | |
| IC443 | 100 | - | - | 2.5 | 2e19 |
| IC445 | 100 | - | - | 5 | 1e19 |
| *Double Wells* | | | | | |
| IC452 | 17.5 | 20 | 10 | 2.5 | 2e19 |
| IC455 | 17.5 | 14.1 | 7 | 1.8 | 2.8e19 |
| IC463 | 17.5 | 20 | 5 | 2.5 | 2e19 |
| IC475 | 17.5 | 15 | 7 | 1.8 | 4e19 |
| IC497 | 19.0 | 14.1 | 8 | 1.8 | 2.8e19 |

**Table 1** Sample dimensions and silicon doping levels

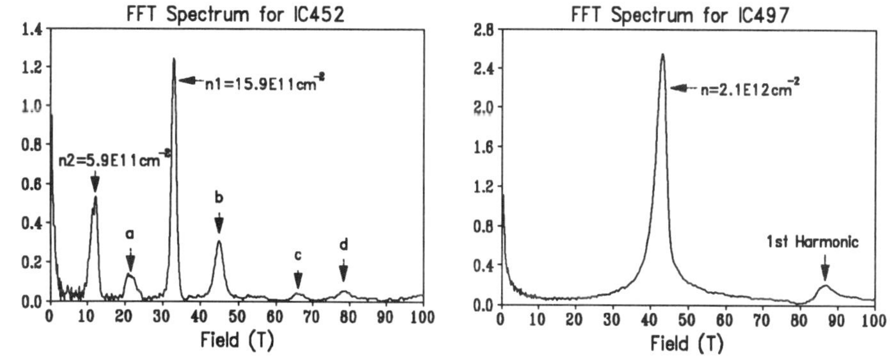


**Figure 1** FFTs for samples with two and single subband occupancies. The labelled peaks are: (a) difference peak, (b) sum peak, (c) 1st harmonic and (d) n1+sum peak.

carrier concentration is achieved without significant degradation of the mobility. However the structures for these early samples are not optimised and only about 50% of the electrons in thin well transfer. The Hall measurements show that the carrier concentration changes little until room temperature is reached. This is in contrast to the much bigger change found with single quantum well samples where the Fermi energies lie in the GaSb valence band and two dimensional hole gases are generated adjacent to the interfaces.

Shubnikov-de Haas measurements were taken at 4K to provide accurate subband occupancies for the high mobility electrons in the main well. Samples 455 and 497 have identical well thickness and doping levels and differ only slightly in spacer and cap thicknesses. Fast Fourier transforms (FFT) of the Shubnikov-de Haas measurements for these two samples show only single subband occupancy and very similar carrier concentrations in the wide well. Both of these samples have high 4K mobilities (170,000 to 185,000 $cm^2/Vs$). The Fourier analysis of the Shubnikov-de Haas recordings for the other three double well samples show that two subbands are occupied (Figure 1 shows clearly the difference between single and two subband occupancy).From the results given in Table 2 for a variety of structures it appears that the mobility falls in the undoped well when the second subband becomes occupied. The highest conductivity achieved so far ($10^{-1}$S) is with a 14nm undoped well at a distance of 7nm from a 1.8nm well doped with silicon to $5x10^{12}$ $cm^{-2}$. The low temperature mobility of this structure is 185,000 $cm^2/Vs$ with an electron concentration in the undoped well of $2.3x10^{12}$ $cm^{-2}$. As shown by Shubnikov-de Haas measurements these

| Sample | Electron Density *Theory* subband 1 | 2 | ($x1e11cm^{-2}$) Total | *Experiment* subband 1 | 2 | Total | *Difference* % | Mobility ($cm^{-2}/V.s$) |
|---|---|---|---|---|---|---|---|---|
| *Single Wells* | | | | | | | | |
| IC411 | 12.3 | 2.5 | 14.8 | 13.6 | 3.7 | 17.3 | 14 | 90,000 |
| IC404 | 9.3 | 0.7 | 10.0 | 9.5 | - | 9.5 | 5 | 200,000 |
| IC410 | 7.4 | - | 7.4 | 8.2 | - | 8.2 | 10 | 100,000 |
| *Double Wells* | | | | | | | | |
| IC452 | 16.1 | 7.4 | 23.5 | 15.9 | 5.9 | 21.8 | 8 | 155,000 |
| IC455 | 24.4 | 5.7 | 30.1 | 22.8 | (4.0) | (26.8) | 12 | 185,000 |
| IC463 | 19.2 | 9.6 | 28.8 | 21.2 | 8.9 | 30.1 | 4 | 50,000 |
| IC475 | 25.3 | 8.1 | 33.4 | 28.4 | 7.2 | 35.6 | 6 | 70,000 |
| IC497 | 23.4 | 4.8 | 28.2 | 20.8 | - | 20.8 | 36 | 180,000 |

**Table 2** Experimental and theoretical subband occupancies

electrons were all in the lowest subband. Higher electron concentrations can be achieved; e.g. $3.6 \times 10^{12}cm^{-2}$ for sample 475 but the mobility is considerably lower (70,000 $cm^2/Vs$).

The monotonic magnetoresistance is quite large for all the samples because of parallel conductance. The most likely explanation of this parallel conductance is that it arises predominantly from the thin well and this is borne out by two-carrier fits to the magnetoresistance: e.g. for sample 497 the two-carrier fit parameters are $n_1 = 2.1 \times 10^{12}cm^{-2}$ and $\mu_1 = 171{,}000\ cm^2/Vs$ for the thick well and $n_2 = 5 \times 10^{11}cm^{-2}$ and $\mu_2 = 700\ cm^2/Vs$. $n_1$ and $\mu_1$ are close to the values found from the Shubnikov-de Haas period and the low field Hall data. $\mu_2$ is identical with the value measured for the doped single well sample of thickness 2.5nm. However it should be noted that $n_2$ is considerably less than the number of electrons which should remain in the thin well. This discrepancy could be associated with localisation effects in the thin well or may be indicative of compensation.

## 3. Self-Consistent Modelling

A self-consistent modelling programme has been developed in an attempt to understand the electron transfer. The model includes the nonparabolicity of the InAs conduction band and also takes account of any holes in the GaSb near to the interfaces. The model assumes a surface pinning of the Fermi energy corresponding to the presence of the surface defects in the GaSb. Table two compares the results obtained from the Shubnikov-de Haas measurements with the predictions of the model. The model best fits the experimental data with a pinning energy of 0.2 eV above the GaSb valence band (see figure two for the band alignments). The differences between the experimental and theoretical values for the total electron concentration differ by only 10-15% except in the case of sample 497. This pinning energy is considerably closer to the valence band than that derived in reference [9] but in reasonable agreement with values deduced in references [10, 11] and with the amphoteric native defect model of Walukiewicz [12].

The model has been used to design optimised structures. Our results to date, particularly the activation of the silicon donors, encourage us to believe that thinner doping wells can be employed. The design criteria such as i) obtaining high electron concentrations, ii) maintaining high mobility at low temperatures and/or iii) eliminating parallel conductance will depend on the application and may conflict. The results of modelling a structure consisting of 2nm cap/12nm undoped well/7nm spacer/1.2nm doped well taking the doping concentration as the variable are shown in figure three. A concentration of $3 \times 10^{12}cm^{-2}$ should be achievable in the wide well for a doping concentration of $2 \times 10^{19}cm^{-3}$ without the Fermi

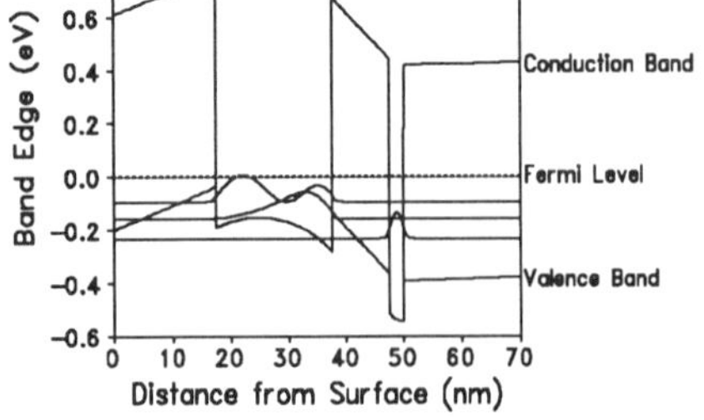


**Figure 2** Band alignments for IC452.

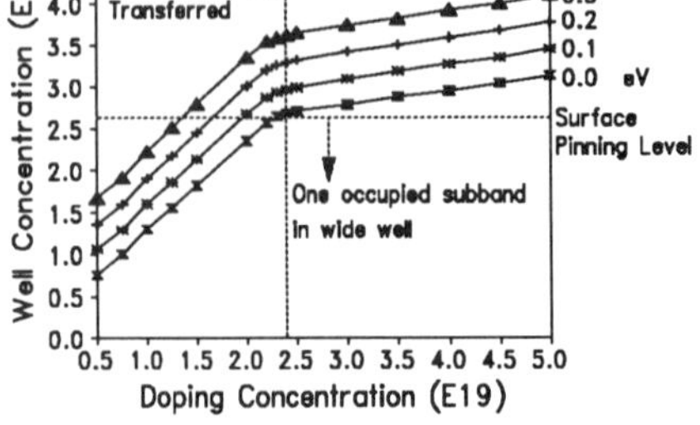


**Figure 3** Electron concentration in wide well against Si concentration for four values of surface pinning.

energy crossing into the second subband in the wide well or into the lowest subband of the doped well. Complete transfer from the doping well is still achieved up to $3.3x10^{12}cm^{-2}$ but the second subband becomes occupied. A maximum of $3.8x10^{12}cm^{-2}$ should be attainable with these design parameters before the solubility limit of Si in InAs ($5x10^{19}cm^{-3}$) becomes a problem. Even higher carrier concentrations are possible if the width of the undoped well is increased but the second subband occupancy also increases.

## References

[1] Lee M E, Poole I, Truscott W S, Cleverley I R, Singer K E & Rohlfing D M 1990 J App Phys 68 131.

[2] Bolognesi C R, Werking J D, Caine E J, Kroemer H and Hu E L 1993 IEEE Electron Device Letters 14 13

[3] Boos J B, Shanabrook B, Park D, Davis J L, Dietrich H B & Kruppa W 1993 Electronics Letters 29 1888

[4] Nguyen C, Kroemer H, & Hu E L 1992 Phys Rev Lett 69 2847 also 1993 J Cryst. Growth 127 845

[5] Williams G M, Whitehouse C R, Ward J J, Brumhead D & Ashley T 1990 J Vac. Sci Techn. B8 79

[6] Kopev P S, Ivanov S V, Ledentsov N N, Meltser B Y, Nadtochii M Y & Ustinov V M 1990 Sov. Phys Semicond 24 450.

[7] Williams R L, Skuras E, Stradling R A, Droopad R, Holmes S N & Parker S D 1990 Semicond Sci Techn 5 S338 and to be published.

[8] Wang P D, Holmes S N, Tan Le, Stradling R A, Ferguson I T, & de Oliveira A G 1992 Semicond. Sci. & Techn. 7 767

[9] Nguyen C, Brar B, Kroemer H & English J H 1992 App Phys Lett 60 1854 also 1992 J Vac Sci Techn B 10 898

[10] Altarelli M, Maan J C, Chang L L & Esaki L 1987 Phys Rev. B35 9867

[11] Furukawa A 1993 App Phys Lett 62 3150

[12] Walukiewicz W 1988 Phys. Rev. B37 4760

*Inst. Phys. Conf. Ser. No 144: Section 4*
*Paper presented at 7th Int. Conf. on Narrow Gap Semicond. Santa Fe 9–12 January*



# Metalorganic Vapor Phase Epitaxy of InSb in a Multiwafer Reactor

R Rolph and T Zielinski
Hughes Aircraft Company, Microwave Products Division
24120 Garnier Street, Torrance, CA 90509 2940, U.S.A.

E Woelk and H Jürgensen
AIXTRON GmbH, Kackertstr. 15-17, D-52072 Aachen, Germany

Abstract

Large area deposition of InSb on GaAs using Metalorganic Vapor Phase Epitaxy (MOVPE) has been developed by adapting an existing process to a multiwafer reactor. The batch size has been increased to eight 3" wafers which can be processed in the cycle time of 4 hours for the growth of a 2$\mu$m thick layer. The process was optimized to produce layers with a standard deviation of $\pm$1.4% for thickness uniformity and sheet resistivity uniformity. These data are crucial for the production of the magnetic field sensors which use these layers.

## 1. Introduction

The material system GaInAlSb has interesting applications in electronics and optoelectronics. On an R&D level, layers of this material have been prepared using Metalorganic Vapor Phase Epitaxy (MOVPE) but it was not until recently that large reactors became available to increase the scale of production. A great difficulty to overcome was the depletion of the gas phase and the resulting non-uniformity of the film. With the introduction of the Planetary Reactor$^{TM}$ [1] using a double axis rotation of the wafers, the uniformity of film properties were improved to be better than $\pm$1.5%. In a second step, the Planetary Reactor$^{TM}$ was enlarged to a batch size of five 10" wafers for the production of layer structures for advanced solar cells [2].

In the meantime, MOVPE is the method of choice for the production of large batches of GaAs and InP based film systems for optoelectronic devices. The batch size for MOVPE production has increased by about a factor of 10; from an average batch size of 1 x 2" in 1989 to about 8 x 3" in 1994. The corresponding processes have been transported from the R&D reactors to the larger production reactors. Other materials are currently following this trend.

InSb is one of the candidates for high volume production. The material shows an extremely high electron mobility in excess of 40,000 $cm^2/Vs$ at room temperature which translates to large changes in the resistivity when a magnetic field is applied. The change in the resistivity of the material is proportional to the square of the magnetic field and is thus preferred over the linearly changing Hall voltage. InSb is therefore used in devices using magnetic field sensors from InSb and permanent magnets to sense position.

## 2. The MOVPE Reactor

The Planetary Reactor™ was developed to meet the demand of the III-V semiconductor industry for the production of layered structures in the material system GaInAsP. A cutaway view of the reaction chamber is shown in Figure 1. It is a horizontal flow reaction chamber where the gases enter in the center and are exhausted at the perimeter. A collector ring made from molybdenum provides for an even radial draft and prevents the formation of jets toward the exhaust nipples. To compensate for a minute radial asymmetry that may be caused by the tolerances in the injection nozzle or the collector ring, the positions of the wafers are rotated around the center of the reactor.

Since the reaction chamber uses horizontal gas flow, the composition of the gas phase and the thickness of the diffusion barrier above the growing surface change and cause the growth rate to decreases along the radius. To integrate the different growth rates, the wafers are rotated around their center. This results in excellent uniformity of thickness and other layer properties. As found from the simulations and verified in the actual growth runs [1], the growth rate along the radius r in the reaction chamber of the Planetary Reactor™ follows a function of the form

$$g(r) = a + b\,r + \frac{c}{r}. \tag{1}$$

The constants a, b and c are functions of the total flow which is proportional to the gas velocity in the reaction chamber. The radial field expressed by equation (1) is shown in Figure 2 for three different total flows which are related to an optimal total flow ($flow_{opt}$) for which the layer properties show the maximum uniformity. Also shown in this figure is the integrated

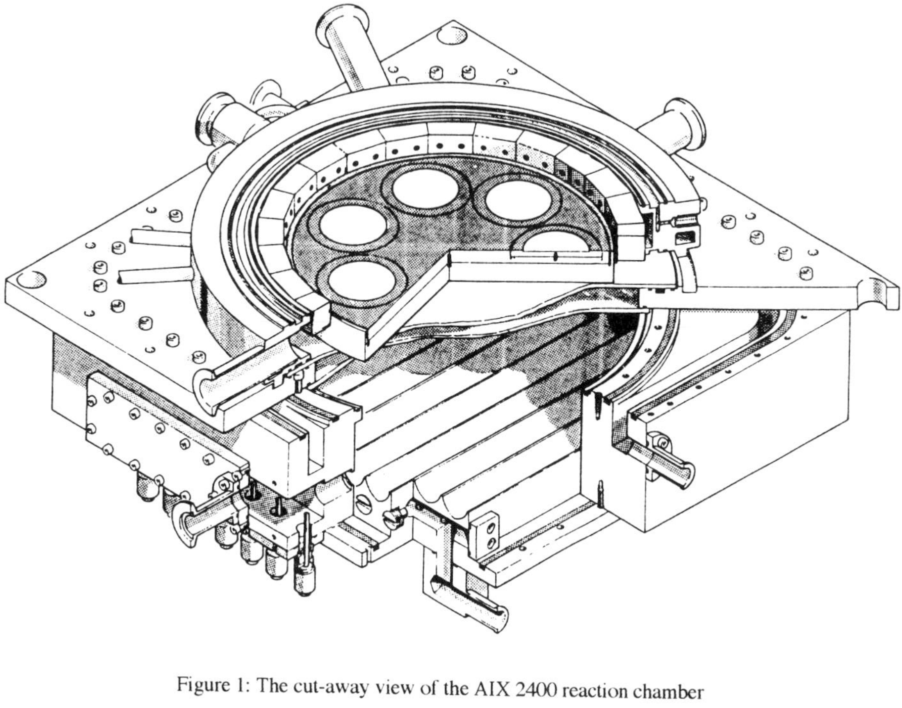

Figure 1: The cut-away view of the AIX 2400 reaction chamber

growth rate that is obtained by rotating the wafers around their center which can theoretically be as uniform as ±0.5%. If the bow of function (1) is too strong the thickness distribution resembles a bowl (Figure 2c); if function (1) is linear the thickness distribution resembles a hump (Figure 2a). Although the model was developed to describe the layer thickness uniformity for a GaAs layer, it has been proven useful for the prediction of most other layer properties and different materials grown in the Planetary Reactor™ [2].

Extremely important for the growth is also the temperature uniformity across the substrates. In a kinetically limited process, a small lateral variation in substrate temperature leads to a large variation in growth rate or in the doping of the film. The heater of the Planetary Reactor™ allows for adjustment of the temperature profile by adjusting the heating power in different areas of the susceptor. In conjunction with the main rotation, this leads to an temperature uniformity of less than ±1°C across the wafers.

## 3. The growth of InSb

InSb films have been grown in the AIX 2400 reaction chamber (capacity: eight 3" wafers) on 3" <100> oriented semi-insulating GaAs substrates using trimethylantimony (TMSb) and trimehtylindium (TMIn) as precursors. Despite the large lattice mismatch of 14.6% between the substrate ($a_{GaAs} = 5.6533$Å) and epitaxial layer ($a_{InSb} = 6.4789$Å) specular surfaces were obtained when a GaAs buffer layer was deposited at high temperature first. Subsequent growth of the InSb took place at temperatures between 450 and 500°C. As reported previously [3,4], nominally undoped InSb was found to be n-type with a carrier concentration of $N_A$-$N_D \approx$ 2-3 $10^{16}$ $cm^{-3}$. N-type material with doping levels of 3-4 $10^{16}$ $cm^{-3}$ and an electron mobility in excess of 40,000 $cm^2$/Vs was made using tellurium from a diethyltelluride (DETe) source as the dopant.

The layers were characterized by a four probe measurement for the sheet resistivity using an automatic stepper. Figure 3 shows the map of the sheet resistivity for two runs with different total flow. Figure 3a shows a sample that was grown under the conditions described in Figure 2c producing films with lower resistivity at the edge than in the center. Figure 3b

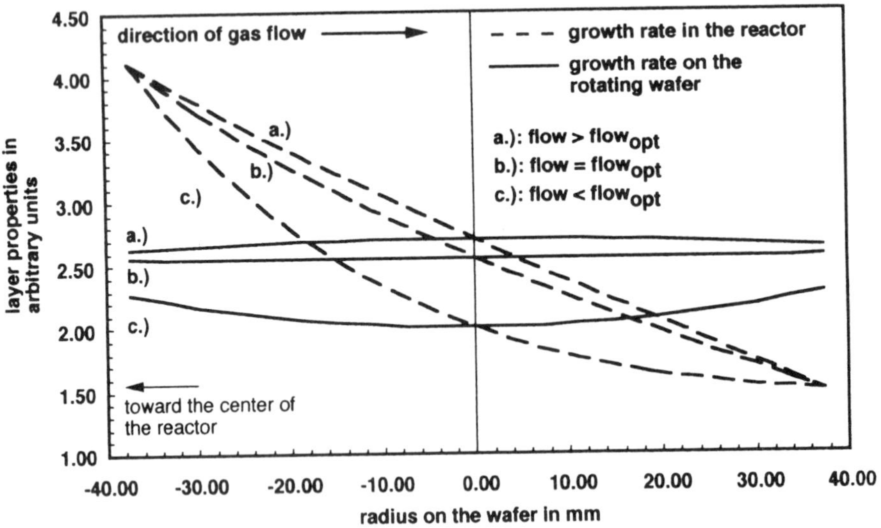


Figure 2: Distribution of growth rate and other layer properties in the AIX 2400 reaction chamber

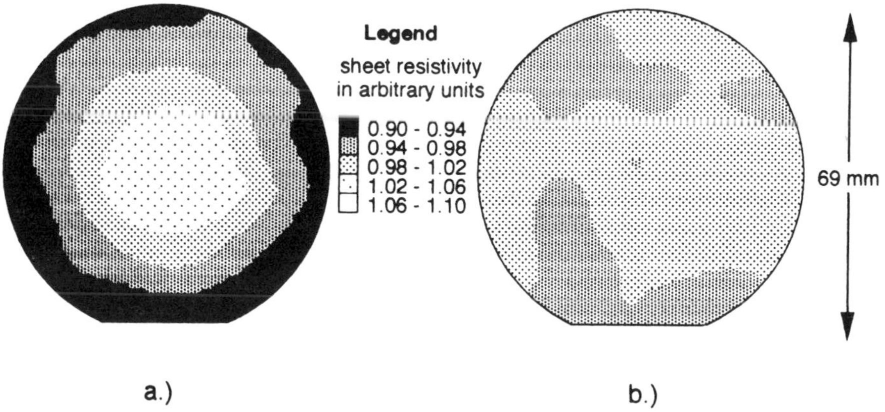


Figure 3: Distribution of sheet resistivity a.) before optimization b.) after optimization

shows the mapping of a sample that was grown under near optimal conditions (See Figure 2b.). The standard deviation is less than 1.5%. The radial non-uniformity of the layers is completely eliminated.

For production the run-to-run reproducibility of the product is of crucial importance. Figure 4 shows the standard deviation of the sheet resistivity for 32 consecutive runs. The mean and the standard deviation $\sigma$ were taken from one wafer out of the 8 wafers in the batch. The wafer-to-wafer uniformity was found to be good enough to represent the mean and standard deviation of the entire batch by one randomly selected wafer. As shown in Figure 4 the run-to-run reproducibility is better than ±1.5% which is the resolution limit of the automatic stepper. The batch to batch uniformity is sufficient for production of magnetic field sensors [5].

Dry removal of the deposits in the reaction chamber was done by using HCl gas at elevated temperature and a subsequent bakeout of the reactor. The process allows fast and efficient cleaning of the reactor without disassembling and handling parts. The building materials that are exposed to the corrosive atmosphere are stainless steel, fused quartz, high density

Figure 4: Run-to-run reproducibility of the sheet resistivity of InSb layers grown in consecutive runs

graphite (apparent density 1.84 g/cm$^3$) and molybdenum. All these materials are resilient against an attack of gaseous HCl and the deposits could be completely removed from the surfaces.

## 4. Conclusion

The growth process for InSb was successfully transferred from an R&D reactor to a multi-wafer Planetary Reactor$^{TM}$. Using well established growth data for the reaction chamber and other materials, the uniformity of the layers could be improved by process optimization to show a standard deviation of less than ±1.5% and a run-to-run reproducibility of also ±1.5%. Tellurium was used to obtain n-type layers with electron mobilities in excess of 40,000 cm$^2$/Vs at doping levels of 3-4 10$^{16}$ cm$^{-3}$. The films are being used for the fabrication of magnetoresistive sensors. Dry HCl at elevated temperatures has been used to remove deposits of material from the surfaces in the reaction chamber. The uniformity data for InSb growth fit perfectly into the data obtained so far from the Planetary Reactor$^{TM}$ with the growth of other materials.

## References

[1] Frijlink P M, Nicolas J L, Ambrosius H P M M, Linders R W M, Waucquez C and Marchal J M 1991 *J. Cryst. Growth* **115** 203

[2] Woelk E G, Strauch G, Schmitz D, Jürgensen H, Ermer J, Vijayakumar P S, Cavicchi T 1994 *to be published in Proc. State-of-the-Art Program on Compound Semiconductors, 185th Electrochemical Society Meeting, San Francisco*

[3] Behet M, Stoll B and Heime K 1993 *Proc. EW-MOVPE V, Malmö/Sweden*

[4] Behet M, Stoll B, Brysch W and Heime K 1992 *J. Crystal Growth* **124** 377

[5] Hermans J, Partin D L, Morelli D T, Fuller B K, and Thrush C M 1990 *Appl. Phys. Lett* . **57** 292

*Inst. Phys. Conf. Ser. No 144: Section 5*
*Paper presented at 7th Int. Conf. on Narrow Gap Semicond. Santa Fe 9–12 January*



# Properties of $Hg_{1-x}Cd_xTe$ grown on CdZnTe and Si substrate

**S. Sivananthan[1], Y.P.Chen[1], P.S.Wijewarnasuriya[1], J.P.Faurie[1], F.T.Smith[2] and P.W.Norton[2]**
[1] Microphysics Laboratory, Department of Physics, University of Illinois at Chicago
845, W. Taylor street (SES 2236),Chicago, Illinois 60607-7059
[2] Loral Infrared & Imaging Systems, 2 Forbes Road, Lexington, MA 02173

We present the properties of HgCdTe grown on CdZnTe and Si substrates. HgCdTe layers grown on CdZnTe substrates by MBE exhibit excellent Hall characteristics down to doping levels of $2x10^{15}$ $cm^{-3}$. Electron mobilities ranging from $(2\text{-}3)x10^5$ $cm^2/vs$ at 23K were obtained. Measured lifetime data fits very well with the intrinsic band-to-band recombination. However, below a $2x10^{15}$ $cm^{-3}$ doping level, minority carrier lifetime is limited by Schockley Reed recombination.
Single domain, twin free CdTe(111)B epitaxial layers on Si substrates are currently routinely grown by MBE with an x -ray DCRC FWHM as low as 96 arc sec. The electrical characteristics and dislocation density of HgCdTe grown by MBE on (111)B CdTe/(100)Si are comparable to those grown on bulk (111)B CdTe. HgCdTe layers grown by LPE on CdTe/Si have EPD as low as $5.3x10^5$ $cm^{-2}$.

## 1. Introduction:

$Hg_{1-x}Cd_xTe$ alloys have been extensively studied for last twenty five years for infrared device applications. Initial studies were on the growth of $Hg_{1-x}Cd_xTe$ on near lattice matched CdZnTe substrates. As the development of $Hg_{1-x}Cd_xTe$ infrared technology has progressed, so also has the search for substrate materials for epitaxial $Hg_{1-x}Cd_xTe$ that provide rugged and inexpensive alternatives to CdTe or $Cd_xZn_{1-x}Te$ substrates. The growth of alternative substrates such as CdTe/Si and $Cd_{1-x}Zn_xTe$/Si for HgCdTe IR based technology is extremely promising for the current hybrid approach as well as for a truly monolithic approach. The use of a Si substrate is especially attractive in IR photodiode technology not only because it is less expensive and available in large area wafers but also because in a FPA structure the coupling of the Si substrate with Si readout circuitry devices will allow the fabrication of very large arrays exhibiting long-term thermal cycle reliability. In this approach, since both substrate and Si read-out will have the same thermal coefficient, the fabrication of very large arrays providing high resolution will be possible. Truly monolithic device architecture will require novel approaches in heteroepitaxy, such as localized epitaxy, involving the direct growth of CdTe on prepatterned Si. In addition composite CdTe/Si can also be used in x-ray imaging and spectroscopy[1], solar conversion, and in electro-optical modulation.

Here, we are presenting the properties of HCT grown on (211)BCdZnTe and (111)BCdTe/(100)Si.

## 2. $Hg_{1-x}Cd_xTe$ on (211)BCdZnTe substrates by MBE:

$Hg_{1-x}Cd_xTe$(HCT) grown by MBE on (211)B CdZnTe substrate has been extensively studied. Incorporation of indium has been investigated[2] and it is found to be an excellent donor with an electrical activity of 100% for doping levels up to $10^{17}$ $cm^{-3}$. Indium doped HCT can be grown with precisely controlled carrier concentration and can be reproduced routinely from run to run. We do not observe any memory effect.

A series of Indium doped HCT epilayers has been grown under optimized growth conditions on (211)CdZnTe substrates[3-4]. The substrates were carefully selected to avoid impurity diffusion from the substrates. The x-ray DCRC FWHM of the HCT is comparable to that of the substrate. Etch pit density(EPD) measurements reveal that the dislocation density is low $10^5$ $cm^{-2}$ for HCT grown under optimum growth conditions.

(211)B $Hg_{1-x}Cd_xTe$ layers were grown on CdZnTe(211)B substrates in a Riber 2300 MBE machine. Layers were doped in-situ during MBE growth with indium. As-grown layers were annealed under Hg-atmosphere at 250°C in order to reduce the Hg vacancy concentration which was created during the MBE growth. The Cd composition was determined at room temperature by infrared transmission measurements. The absorption coefficient at the absorption edge was taken to be $500cm^{-1}$. The Hall coefficient $R_h(T)$ and the resistivity $\rho(T)$ of the layers were measured by the van der pauw technique for temperatures ranging from 300K to 20K and magnetic fields up to 1.0 Tesla. The lifetime measurements were carried out using the photoconductive decay technique. In both cases, samples were etched in Br-methanol solution prior to measurements. For lifetime measurements, a pulsed GaAlAs laser beam was used to generate the excess carriers on the samples and the decay signals were acquired with a low-noise cryogenic preamplifier. A detailed description of the measurement technique can be found in ref(4). Indium was used to make ohmic contacts for n-type HgCdTe layers. The biasing was kept low to avoid sweep-out effects and the light intensity also kept low in order to satisfy the low injection condition.

Fig. (1) shows the carrier concentration and mobility versus reciprocal temperature of selected layers with Cd composition between 0.22 to 0.23. These samples were annealed under Hg atmosphere at $250^0$C in order to reduce the Hg vacancy concentration resulting from the MBE growth. The expected behavior of an increase in mobility with decreasing doping levels can be seen and electron mobilities ranging from $(2\text{-}3)\times10^5$ $cm^2/vs$ at 23K were obtained with doping levels down to $\approx 1\times10^{15}cm^{-3}$. These values are comparable with the best mobilities reported for HgCdTe for this Cd composition and for this doping level.

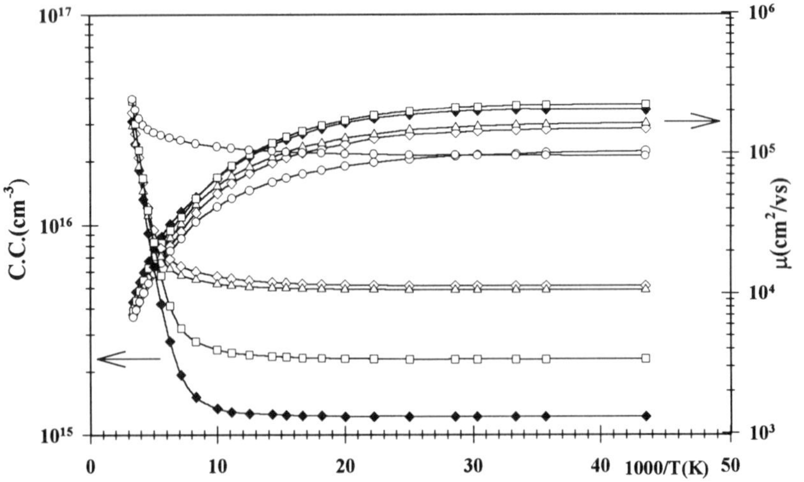


**Fig. (1) C.C. and mobility vs 1000/T(K) of indium-doped MBE-HgCdTe.**

Figure (2) shows the relation between lifetime and the electron concentration of Indium doped layers at 80K. Lifetime decreased from 1μsec to 20 nsec when doping increased from $\approx 2\times10^{15}cm^{-3}$ to $1\times10^{16}cm^{-3}$. The solid line shows $n^{-2}_0$ behavior and this confirms that the measured lifetime is limited by the Auger recombination at 80K. Figure (3) shows the lifetime versus reciprocal temperature of the MBE MCT layer which has a doping concentration of $1.4x10^{15}cm^{-3}$. The peak life time reaches approximately 1.4 μs at 130 K. The solid line represents the theoretical curve, which includes only intrinsic band to band (Auger and radiative) processes.

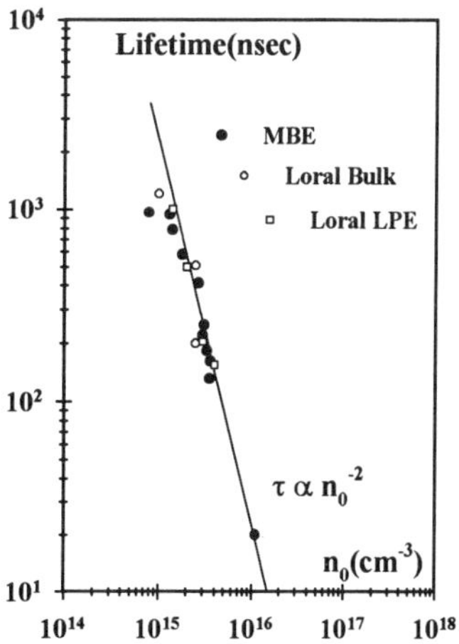


Fig. (2) Dependence of lifetime vs electron density at 80K.

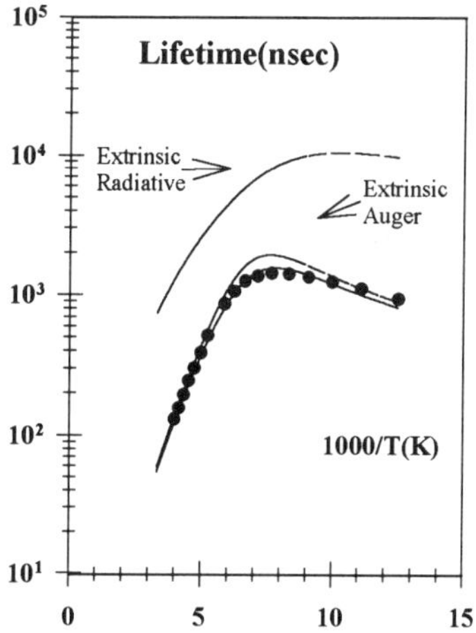


Fig. (3) Measured lifetime on a seletecd layer. Lines shows various contributions.

On the other hand, when the doping is low, data on some of the MBE layers cannot be explained by intrinsic material properties. It appears that this effect dominates when the doping level falls below $\approx 2\times10^{15}cm^{-3}$, and in order to obtain a good fit throughout the entire temperature range, contributions from SR centers must be assumed in addition to the combined Auger and radiative processes. The obtained SR energy level $E_t$ varies between 33 and 45 meV below the conduction band which is $\approx(3/4)E_g$ with respect to the valence band.

In order to find out the origin of this defect level, we carried out a lifetime measurement on an as-grown, indium-doped HgCdTe layer. The measured lifetime was again lower than the band-to-band recombination lifetime at temperatures below 200K. At 80K, lifetime was ≈150 nsec. Lifetime improved from 150 nsec to 1.4 μsec at 80K after annealing at $250^0$C. During MBE growth at $185^0$C, it is very difficult to avoid compensation due to Hg vacancies, especially when the indium doping levels are in the range of $10^{15}$ $cm^{-3}$ (we assume that Nd arises mainly from indium atoms). At $185^0$C, according to the HgCdTe P-T phase diagram, the accepter level due to Hg vacancies should be in the $10^{15}$ $cm^{-3}$ range.[5] Hence, as-grown layers are highly compensated. We obtained $N_d=1.6x10^{15}$ $cm^{-3}$ and $N_a=1.5x10^{15}$ $cm^{-3}$ from fitting of $R_h(T)$ versus temperature data. After n-type annealing under Hg pressure, we obtained $N_d=1.4x10^{15}$ $cm^{-3}$ and $N_a=1.4x10^{14}$ $cm^{-3}$. From the lifetime data, $\tau_{p0}$ increased from 162 nsec to 6 usec when Na decreased from $1.5x10^{15}$ $cm^{-3}$ to $1.4x10^{14}$ $cm^{-3}$. According to Figure (2), this SR defect level was not seen for doping levels up to $1.0 \times 10^{16}$ $cm^{-3}$. Since these samples were Indium-doped during the growth, it is safe to assume that the origin of this SR defect level is not related to extrinsic Indium[3]. Hence, we concluded that this SR level is somehow related to the Hg vacancies.

## 3. Growth of HgCdTe on CdTe(111)B / Si(001) :

CdTe can be grown onto a Si(001) substrate in two orientations. CdTe(111)B is obtained when it is grown directly onto Si(001)[5], Whereas CdTe(100) is obtained when it is grown onto Si(001) with a ZnTe buffer[5,6], or Ge buffer[7], or $BaF_2$-$CaF_2$ graded buffer layers[8]. As we pointed out in the previous publication[9], CdTe(111)B grown on nominal Si(001) always exhibits double-domain and twin structure. The formation of double domains and microtwins is strongly related to the crystal symmetry of the Si(001) substrate[10]. The mechanism of interface formation has been discussed in detail in our previous publication[10]. Since the formation of double domains and microtwins is so strongly related to the symmetry of Si(001) surface, the problem can be avoided by using a misoriented substrate whose surface symmetry is changed by introducing steps across the surface. In order to suppress formation of both double-domain and twins, the Si(001) substrate should be tilted in such a way that the steps on Si(001) surface are running along both Si[110] and Si[110], so that both $90^0$- and $180^0$-rotational symmetry of the Si(001) surface are lifted. When grown on such a Si(001) substrate, CdTe(111)B layers always exhibit a single-domain and twin-free structure. Such buffer layers are suitable for the further growth of HgCdTe by both molecular beam epitaxy(MBE) and liquid phase epitaxy(LPE).

The substrates used in this research are Si(001) wafers tilted about $1^0$ toward [170]. The CdTe layers were grown by MBE in OPUS 45, which is a multiwafer MBE system manufactured by ISA-Riber. This system can handle three 2-inch diameter substrates or one 3" to 5" substrate per growth run. The Si wafers used for epitaxial growth are prepared by a conventional chemical cleaning process, followed by thermal cleaning at around $850^0$C inside the vacuum chamber. The CdTe(111)B layers are grown on Si(001) substrates by a two-step method. CdTe(111)B is grown at about $220^0$C for several minutes followed by growth at about $260^0$C. The as-grown layer is then annealed at about $360^0$C under Te flux for 10 minutes. The remainder of the CdTe layer is grown at about $310^0$C. At the end of the growth, the layer is annealed in-situ at $360^0$C under Te flux for about 30 minutes. Details of the experimental procedure have been described in the early publication[11].

The as-grown layers are measured by double crystal x-ray diffractometer (DCD). The layer quality is evaluated by the full width at half maximum (FWHM) of its double crystal rocking curve (DCRC). DCD is also used as a major tool to detect the domain and twin contents inside the CdTe(111)B layer. This is done by examining the symmetric distribution of {224} diffraction around the CdTe[111] axis[2]. DCD measurement shows that using the optimized Si(001) substrate we can routinely obtain single-domain and twin-free CdTe(111)B layers with FWHM below 140 arc sec with the a lowest value of 96 arc sec. Some of the buffer layers were used as substrates for the further growth of CdTe and HgCdTe by LPE and HgCdTe by MBE.

A series of HgCdTe layers has been grown on CdTe(111)B/Si(100). The x-ray DCRC FWHM of the HgCdTe is consistently lower than that of the CdTe(111)B/Si(100) substrate. We have obtained a x-ray DCRC FWHM of 112 arc sec. Etch pits were obtained using Rockwell's chemical etch. The etch pit density of the as-grown layers were mostly about $3 \times 10^7$ $cm^{-2}$. Our results show a significant reduction in EPD on annealed layers. The best EPD value of $4.3 \times 10^6$ $cm^{-2}$ on an annealed layer is comparable to the EPD values reported for HgCdTe on (111)B CdTe/GaAs(100). The high dislocation density is related to the lattice mismatch between the CdTe and HgCdTe. P- and n-type HCT have been grown reproducibly by MBE on

CdTe(111)B/Si. The n-type layers were obtained using In. Electron mobility of $8x10^4$ $cm^2/vs$ at 23K has been measured before isothermal anneal. Temperature dependence of the mobility and the carrier concentration curve exhibits a classical Hall behavior (Fig. 4). The p-type $Hg_{.78}Cd_{.22}Te$ exhibited a hole mobility of 432 $cm^2/vs$. The minority carrier lifetime of this p-type as-grown HgCdTe with Cd composition of 22% is about 70 ns at 80K. These electron and hole mobility values are close to average values reported for MBE HgCdTe either grown on (111)B CdZnTe or (111)B CdTe/(100)GaAs.

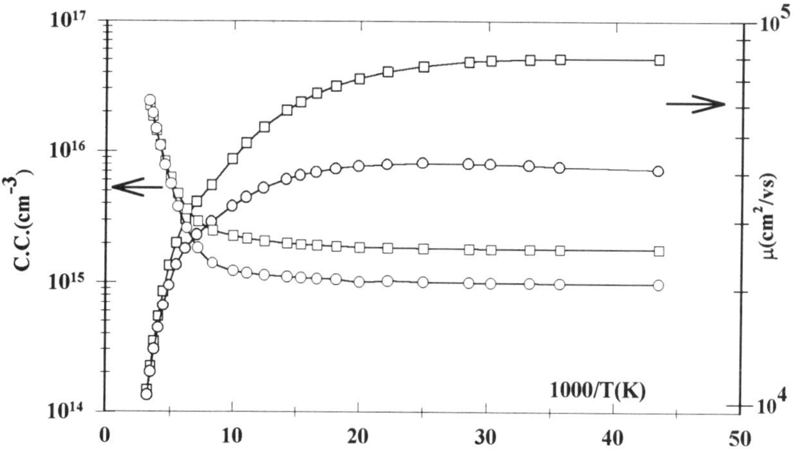


**Fig. (4) Carrier concentration and mobility vs reciprocal temperature of MBE-HgCdTe/CdTe/Si on layer.**

We have previously reported that LPE HgCdTe layers grown under Hg-melt exhibited specular surface morphology and were free of Te precipitates. No twinning was observed in LPE HgCdTe grown on CdTe(111)/Si(100) substrates. The Hg-melt did not attack the substrate around cleaved edges. More recently, we have grown CdTe on encapsulated Si substrates and evaluated the LPE growth of HgCdTe in a Te melt.

The prevention of contamination by substrate materials is difficult because Si is highly soluble in the Te melt. In order to avoid this, the wafers were oxidized in a dry oxygen flow at $950^0$C to give approximately a 150nm thick oxide coating on all surfaces. A rectangular window was opened in the front side of the wafer by masking the back side plus a 2mm wide strip around the perimeter of the front face of the wafers with resist, then removing the oxide using HF solution. These processed substrates were provided by Loral Infrared and Imaging Systems Inc. These substrates were etched using conventional chemical cleaning process and (111)B CdTe deposited by MBE. HgCdTe was grown by LPE under Te rich conditions in a slider system. The surface morphology of HgCdTe layers grown at two different temperatures showed terraced structure. This structure is similar to that observed on LPE HCT grown on tilted (111)B CdTe substrate. The SIMS measurements showed no trace of Si in the HCT layers. After isothermal annealing these In doped HCT showed a carrier concentration of about $1.0 \times 10^{15}$ $cm^{-3}$. This is consistent with the doping level of LPE HCT grown on CdZnTe under the same growth condition from the same In-doped melt. EPD chemical etching revealed that the dislocation density is about

5 x $10^5$ $cm^{-2}$ which is comparable to the best EPD reported for LPE HCT grown CdTe/GaAs/Si and did not show any evidence for the presence of twins. We have previously reported similar EPD for HCT grown in the Hg-melt.

## 5. Conclusion:

A series of **$Hg_{1-x}Cd_xTe$(x=.22)** layers have been grown by MBE on (211) CdZnTe substrates. The analysis of the influence of the growth parameters on the quality of the epilayer has been performed and optimization has been realized. Under the optimum condition x-ray DCRC FWHM duplicates that of CdZnTe substrates. Doping using indium has been investigated and it is confirmed that indium is an excellent n-type dopant in MBE HgCdTe. The doping level is controlled down to 2 x $10^{15}$ $cm^{-3}$ with carrier mobilities as high as 4 x $10^5$ $cm^2$ /vs. The EPD of these layers are in the 1-2 x $10^5$ $cm^{-2}$. Measured lifetime data fits very well with the intrinsic band-to-band recombination mechanisms above 2x$10^{15}$ $cm^{-3}$ of doping levels.

The quality of CdTe(111)B grown by MBE on Si substrate has been significantly improved as illustrated by x-ray DCRC FWHM down to 96 arc sec. This improvement has been obtained by a systematic investigation of substrate cleaning, substrate misorientation, and growth parameters. The HCT grown by MBE on (111)B CdTe/Si shows material properties which are comparable to those grown on bulk CdTe(111)B substrate. The EPD of LPE grown HCT on CdTe(111)B/Si is in the mid $10^5cm^{-2}$. The structural properties are comparable to those grown on CdTe/GaAs/Si.

## 6. Acknowledgement:
This work was supported by ARPA and monitored by airforce under contract F49620-94-C-0080. The Majority of the HgCdTe layers on (211) CdZnTe layers were grown at EPIR Ltd. Work at Loral was supported by IR&D funds.

## 4. REFERENCES

1. S.Yoo, B. Rodricks, S. Sivananthan, J.P. Faurie, and P. A. Montano SPIE Vol.2228, 201(1994)
2. M. Boukerche, P.S.Wijewarnasuriya, S.Sivananthan, I.K.Sou, Y.J.Kim, K.K.Mahavadi, and J.P.Faurie J. Vac. Sci. Tech. 6, 2830(1988)
3. P.S.Wijewarnasuriya, M.D.Lange, S.Sivananthan and J.P.Faurie, J. Appl. Phy. 75,1005(1994)
4. P.S.Wijewarnasuriya, M.D.Lange, S.Sivananthan and J.P.Faurie, in press(J Elect. mat.)
5. R. Sporken, S. Sivananthan, K.K. Mahavadi, G. Monfroy, M. Boukerche, and J.P. Faurie, Appl. Phys. Lett. Vol. 55, No. 18, pp. 1879-1881, 1989.
6. T.J. de Lyon, J.A. Roth, O.K. Wu, S.M. Johnson and C.A. Cockrum, Appl. Phys. Lett. Vol. 63, No. 8, pp. 818-820, 1993.
7. W.S. Wang and I. Bhat, Presented at 1993 U.S. workshop on the physics and chemistry of mercury cadmium telluride and other IR materials.
8. A.N. Tiwari, W. Floeder, S. Blunier, H. Zogg and M.J. Proctor, J. Cryst. Growth, Vol. 111, pp. 730-735, 1991.
9. Y.P. Chen, S. Sivananthan and J.P. Faurie, J. Electron. Mater. Vol. 22, No. 8, pp. 951-957, 1993.
10. Y.P. Chen, S. Sivananthan and J.P. Faurie , SPIE proceeding on Producibility of II-VI materials and devices V2228, p54-65, 1994.
11. Y.P. Chen, S. Sivananthan and J.P. Faurie, Presented at 1993 U.S. workshop on the physics and chemistry of mercury cadmium telluride and other IR materials.

*Inst. Phys. Conf. Ser. No 144: Section 5*
*Paper presented at 7th Int. Conf. on Narrow Gap Semicond. Santa Fe 9–12 January*




# Molecular Beam Epitaxy Growth and Characterization of $Hg_{1-x}Mg_xTe$

**S Oehling, U Lunz, H Heinke, G Plahl, C R Becker and G Landwehr**

Physikalisches Institut der Universität Würzburg, Am Hubland, 97074 Würzburg, Germany

**Abstract.** We report on the epitaxial growth and the characterization of $Hg_{1-x}Mg_xTe$ on (001), (111)B and (211)B $Cd_{0.96}Zn_{0.04}Te$ substrates by molecular beam epitaxy. The bandgap of this ternary compound is variable from -0.15 up to 3.49 eV at room temperature. Thin layers have good crystal quality and exhibit electron mobilities as high as $2 \times 10^5$ $cm^2V^{-1}sec^{-1}$. The layers are stable for Mg concentrations up to 50 % and 70 % depending on substrate orientation. Also the measured band gap differs for (001) and (211)B substrates, which may be due to atomic ordering or phase separation.

## 1. Introduction

The molecular beam epitaxial (MBE) growth of ternary II–VI semiconductors using elements of the IIb group is well established. Due to recent interest in ZnSe for blue–green laser applications binary and ternary compounds containing elements of the IIa group, such as Ca and Mg have been investigated. We report here on the growth of the $Hg_{1-x}Mg_xTe$ (MMT) material system, where band gap is in the range of -0.15 to 3.49 eV. Also the lattice constant of the two binary materials HgTe ($a_0$=6.460 Å) and MgTe ($a_0$=6.421 Å) differs only by 0.6% , which enables nearly lattice matched growth on $Cd_{0.96}Zn_{0.04}Te$ ($a_0$=6.467 Å) substrates in the whole range of composition. Consequently MMT could become an excellent material for devices involving, for example, extremely deep quantum wells and consequently large carrier confinements.

## 2. Experimental Details and Results

The samples have been grown in a Riber 2300 MBE system, using a self–made Hg effusion cell and commercial cells for solid materials. The beam equivalent pressures (BEP) have been measured with an ion gauge at the growth position. Hereafter BEP is loosely refered to as flux. For Mg this is more complicated than for the other elements because of the lower sensitivity of the gauge for this element and the high Hg background pressure, which is often larger than the Mg flux.

All substrates have been decreased and etched before growth as described elsewhere [1]. (211)B substrates were chemo–mechanically polished before etching. Prior to growth of the MMT layer, a CdTe buffer with a thickness of 100 nm at a substrate temperature of 310°C has been deposited. All the samples presented in this paper have been grown at a substrate temperature of 180°C. This value is a compromise between the high temperature needed to grow MgTe, about 300°C, and the lower value, necessary to reduce Hg desorption. Further investigations to deduce the optimum temperature for higher Mg concentrations are needed.

Before growth was initiated the mercury cell was opened for one 1 minute. Then a thin layer of pure HgTe was grown (approximately 6nm) in order to smooth the surface, as was observed by the reflection high energy electron diffraction (RHEED) pattern. As expected no oscillations could be observed during growth of HgTe because of the prevailing growth mode for this material [2] at this temperature. Upon the introduction of Mg the RHEED patterns became dimmer and oscillations could be observed as shown in Fig.1. The samples on (001) substrates all exhibited a (2×1) surface reconstruction during growth.

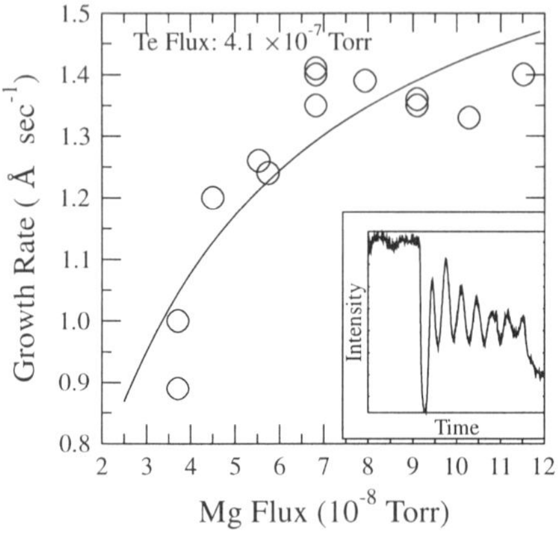


**Figure 1.** The growth rate dependence on Mg flux can not be described in the same way as that in Fig.2, possibly due to competition between Hg and Mg atoms for a lattice site.

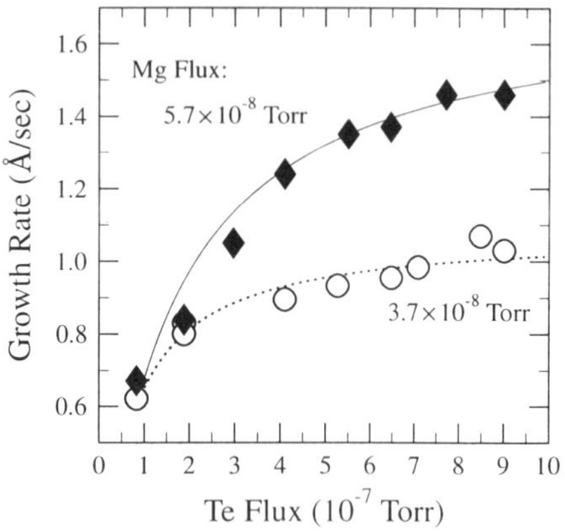


**Figure 2.** Growth rate of MMT at two fixed Mg fluxes with a Hg flux of $1.8\times10^{-4}$ Torr as measured by RHEED intensity oscillations. This behavior can be well described by a simple growth model as indicated by the curves.

We have investigated the dependence of the growth rate on the material fluxes. With increasing Hg flux the growth rate decreases slightly. This effect has also been observed for $Hg_{1-x}Cd_xTe$ [3] and can be interpretated as due to scattering in the dense Hg beam.

The dependence on Te flux for two Mg fluxes is shown in Fig.2. The curves plotted here are fits to the function

$$r_{MMT} = \frac{(s_{Hg}\,F_{Hg} + s_{Mg}\,F_{Mg}) \cdot s_{Te}\,F_{Te}}{(s_{Hg}\,s_{Hg} \cdot s_{Mg}\,F_{Mg}) \cdot s_{Te}\,F_{Te}} \qquad (1)$$

where $s_i$ is the sticking coefficient of material $i$, and $F_i$ the corresponding flux. In this case the Te flux is the rate determining factor independent of the alloy composition. The same fit fails for the Mg series which may be due to competition between Mg and Hg atoms for lattice sites which has been neglected in this simple model.

Although the samples exhibited RHEED oscillations and in most cases two dimensional RHEED patterns many of them do not have smooth surfaces. Pyramid shaped defects occur on samples grown on (001) substrates similar to hillocks prevelent on (001) $Hg_{1-x}Cd_xTe$ [4]. Layers grown on (211)B substrates exhibit star shaped defects. There are also small spherical defects, which are also visible on the best samples displaying no other defects. Here additional work to optimize the growth parameters has to be conducted.

The higher the Mg concentration the fainter the RHEED patterns during growth become and simultaneously the broader the linewidth of the reflections in x–ray diffraction (XRD) become broader, as shown in Fig.3. This may be due to the low growth temperature and at the same time to the larger lattice misfit. Thin layers can be grown with excellent crystal quality for Mg concentrations as high as 30 %. Layers with a thickness of approximately 1$\mu$m have (004) reflections whose full width at half maximum are as narrow as 34 arcsec (see Fig.4).

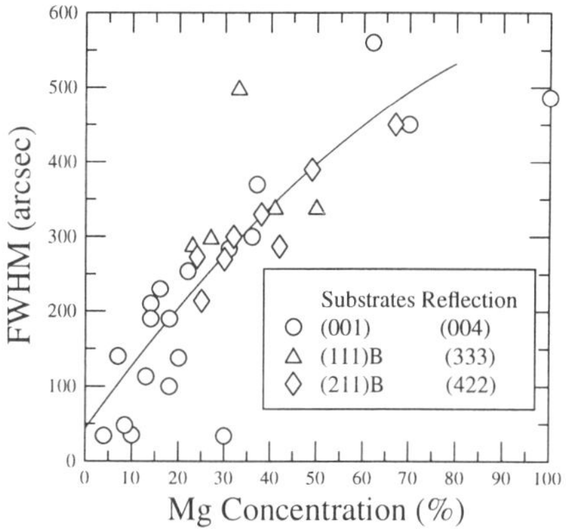


**Figure 3.** FWHM of x–ray Bragg reflections as a function of Mg concentration. The higher the Mg incorporation the broader these reflections become. This may be due to an increasing lattice mismatch, the low growth temperature or the tendency to grow in the wurzite structure.

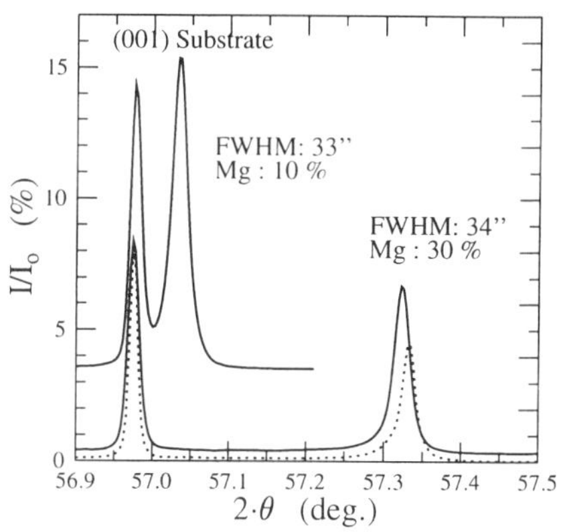


**Figure 4.** Thin layers of MMT can be grown with very good crystal quality up to 30%. The solid curve is for a 30 % sample measured directly after growth and the dotted curve measured half a year later, demonstrating its chemical stability.

Hall effect measurements in the van der Pauw configuration had been carried out on an alloy with a band gap of 120 meV at room temperature. At 85 K the electron mobility is $2 \times 10^5\ cm^2V^{-1}sec^{-1}$ and the electron concentration as low as $6.7 \times 10^{15}\ cm^{-3}$ (Fig.5). At lower temperatures no measurements were possible because of the electrical contacts.

Previously bulk $Hg_{1-x}Mg_xTe$ has been grown, but zinc blende material above 20% has not been reported in literature [5]. MgTe is known to react with air and water, and to form $H_2Te$ and MgO [6]. Our MBE grown material deposited on (001) substrates is stable up to about 50%, depending to a large extent on the crystal quality. No change in the linewidth could be observed in XRD for a good quality sample with 30% between the measurement directly after the growth and half a year later (see Fig.4). (211)B and (111)B samples were stable for Mg concentrations up to 70%.

The band gap was determined from transmission or the reflection measurements depending on whether the bandgap of the layer was below or above that of the substrate (1.62eV). The Mg concentration was estimated from the lattice constant assuming Vegards rule to be valid[7], the results are shown in Fig.6. The band gap of the samples can be described by two curves. The first which is valid for (001) samples with x$\leq$20% and all samples on (111)B or (211)B substrates, deviates somewhat from the straight line between HgTe and MgTe. The other, which is valid for (001) samples is nearly independent of Mg concentration between 15 and 50 %. Samples above 70% have not yet been investigated.

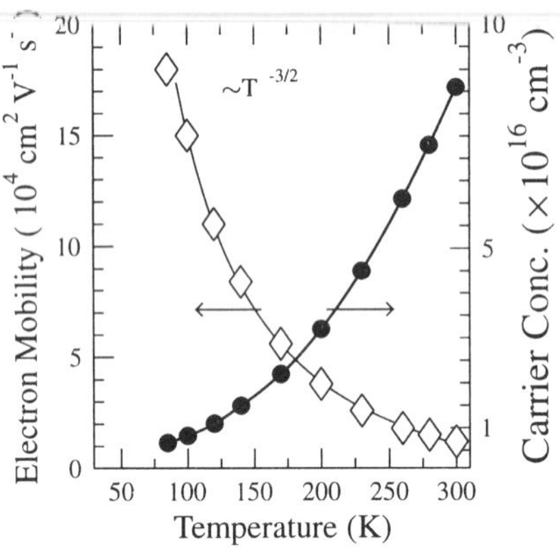


**Figure 5.** Electron mobility and concentration for a MMT sample with a bandgap of 120 meV at room temperature. The mobility is proportional to $T^{-3/2}$ as expected for scattering due to accustical phonons.

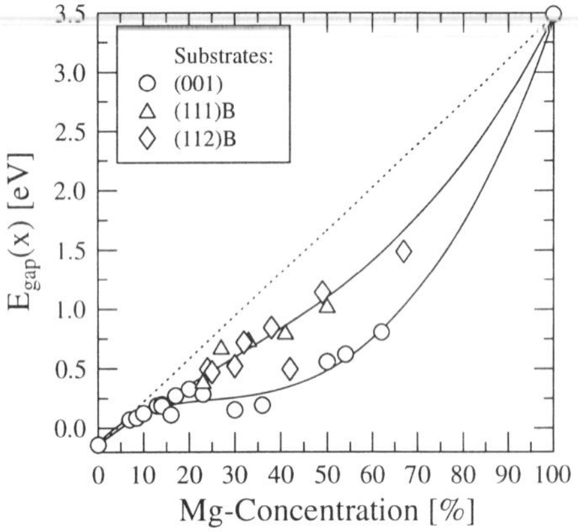


**Figure 6.** The band gap dependence on Mg concentration is described by two curves. One following roughly the straight line between the binary compounds and the other shows a large deviation starting at $x$=15% . The strong curvature of the latter was only observed for (001) substrates.

Similiar behavior is known for other materials such as GaInP, InGaAs [8] and HgZnSe [9]. InGaAs can grow in the CuPt structure, which consists of a natural superlattice along the (111) direction. In contrast separation of Hg rich and Zn rich phases occurs in HgZnSe.

At the moment we do not know which of these mechanisms is valid for MMT. In Laue x-ray diffraction no additional reflections due to atomic ordering were detected. Thus it is only possible to discuss the implications of the methods used in this investigations. First of all the reflections in XRD are quite narrow in comparison with HgZnSe which shows a phase separation and hence linewidths in the range of 800 arcsec. Another piece of evidence is the dependence on orientation. If there is a thermodynamic tendency for phase separation in MMT, this should not be correlated to the substrate orientation. In contrast CuPt atomic ordering requires a $(2\times 1)$ reconstruction, as is only the case for the (001) orientation, according to Philips *et al.* [10]. In this the radii must be sufficiently different. In our case the ionic radii are 0.82 Å and 1.27 Å for Mg and Hg respectively. On the other hand Stringfellow *et al.*[11] suggests that if both binary materials have quite similar lattice constant, there should be no ordering. In order to clarify this question transmission electron microscope measurements have to be done.

## 3. Conclusion

We have grown $Hg_{1-x}Mg_xTe$ by MBE in the composition range between 0 and 70%. Samples with low x values have excellent crystal structure and high electron mobilities. At higher Mg concentrations the crystal quality became worse. Above 50 and 70 % for (001) and (211)B substrates, respectively the layers are not stable in air. The band gap dependence on x behaves differently according to substrate orientation. The possibility for atomic ordering or phase separation has been discussed.

## 4. Acknowledgment

This work has been supported by the Deutsche Forschungsgemeinschaft, SFB no. 410. The authors are grateful to P.Wolf–Müller and T.Schumann for helping in substrate preparation and to AEG Inc. for the (111)B substrates.

## References

[1] Ehinger M, Wenzel M, Waag A and Landwehr G, *J.Cryst. Growth* **123**, 42 (1992)

[2] Oehling S, Ehinger M, Spahn W, Waag A, Becker C.R and Landwehr G. to be published

[3] Ulmer L, Magnea N, Marietta H and Gentile P, *J.Cryst. Growth* **111** , 711 (1991)

[4] He L, Becker C R, Bicknell–Tassius R N, Scholl S and Landwehr G, *Semicond.Sci.Technol.* **8**, 216, (1993)

[5] Gluzman N G, Sabirzyanova L D, Tsidil'kovsil I M, Gavalleshko N P. and Frasunyak V M. *Soc.Phys.Semicond.* **15** (8), 1981

[6] Waag A, Heinke H, Scholl S, Becker C R and Landwehr G. *J.Cryst. Growth* **131**, 607 (1993)

[7] Vegard L,*Zeitschrift f. Physik* **5**, 17, 1921

[8] Su L.C, Ho I.H and Stringfellow G.B, *Appl.Phys.Lett.* **65**, 749,(1994)

[9] Einfeldt S, Lunz U, Heinke H, Becker C R and Landwehr G, ICVGE–8, *J.Cryst. Growth* to be puplished.

[10] Philips B.A, Norman A.G, Seong T.Y, Mahajan S,Booker G.R, Skowronski M, Harbison J.P and Keramidas V.G, *J.Cryst. Growth* **140**, 250 (1994)

[11] Stringfellow G B and Chen G S *J.Vac.Sci.Technol* **B 9** (4) (1991)

[12] Handbook of Chemistry and Physics, 67th Editon, edited by Weast R C, Florida, CRC Press.

*Inst. Phys. Conf. Ser. No 144: Section 5*
*Paper presented at 7th Int. Conf. on Narrow Gap Semicond. Santa Fe 9–12 January*



# New HgCdTe VPE structures on Sapphire by using hybrid substrates

**S. Bernardi, F. Campanella**

Consorzio C.R.E.O., Via Pile N. 60, 67100, L'Aquila, Italy

**Abstract.** Mercury Cadmium Telluride ($Hg_{1-x}Cd_xTe$) layers have been grown directly on (0001) Sapphire wafers by an isothermal VPE process, starting from MOCVD CdTe/Sapphire hybrid substrates. The complete transformation of the starting CdTe films into HgCdTe alloys is controlled by two physical processes: -the HgTe Vapour Phase Epitaxy and -the HgTe/CdTe solid state interdiffusion. Working under interdiffusion rate conditions higher than the HgTe growth rate, compositionally uniform $Hg_{1-x}Cd_xTe$ layers can be obtained directly on Sapphire substrates. The final $Hg_{1-x}Cd_xTe$ films composition results completely controlled by: -the thickness of the starting MOCVD CdTe layer (grown on Sapphire), and -the HgTe thickness grown during the iso-VPE process. Morphological, optical, structural and electrical characteristics of "as-grown" and "Mercury vacancy annealed" layers will be reported, witnessing the powerful of these new HgCdTe structures.

## 1. Introduction

Mercury Cadmium Telluride ($Hg_{1-x}Cd_xTe$) is the most powerful intrinsic semiconductor for infrared photon detectors operating in the atmospheric windows for thermal imaging purposes. In spite of more than 20 years of research efforts on this particularly critical semiconductor material, its whole technological history has not been completely written. Since the end of 70's HgCdTe epitaxial structures started to be investigated, in order to overcome the wafers size limitations affecting the bulk materials generation. Since then, all the epitaxial approaches have been investigated. In particular, the big efforts devoted to Liquid Phase Epitaxy from Tellurium-rich solutions [1][2][3][4], brought this growth process to the present manufacturing stage. HgCdTe layers, grown by classic VPE on bulk CdTe, have been investigated too [5][6][7][8], but unluckily their technological exploitation was strongly limited because of the cross-section compositional gradients. According to a solid state interdiffusion process of a finite HgTe film into an infinite CdTe source (so a CdTe-bulk substrate can be considered), heavy compositional gradients into the epilayer are unavoidable.

The innovative idea, on which the present work is based on, is to perform the iso-VPE HgCdTe growth process on finite CdTe systems as CdTe/Sapphire structures are. In this way we demonstrate the possibility to prepare compositionally uniform $Hg_{1-x}Cd_xTe$ films, when the HgTe growth regime is comparable (or lower than) with the HgTe/CdTe interdiffusion rate. The final ternary alloy composition is well controlled by the starting CdTe film thickness knowing the HgTe growth rate.

## 2. Experimental

A horizontal open tube 2-zone reactor, as schematically represented in fig 1, has been used for our growth experiments. A Mercury source, made by a pure Hg-bath, kept in the first reactor

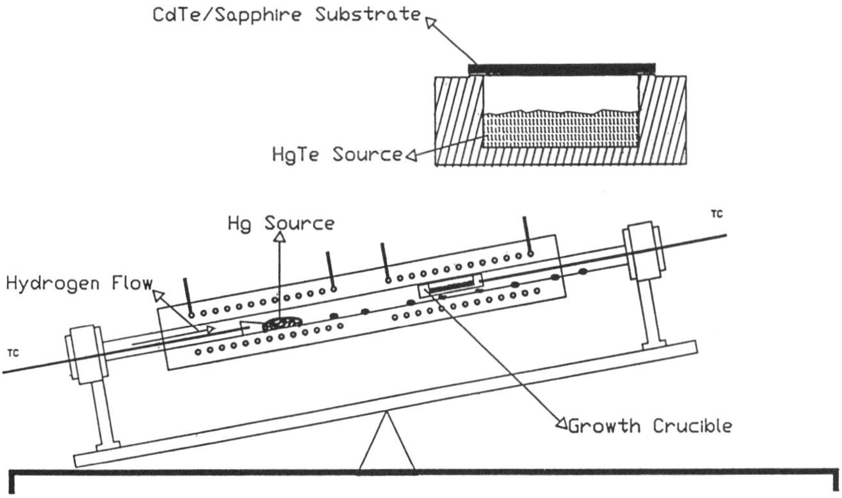


Fig. 1) Schematic representation of the 2-zone reactor used to perform our open tube iso-VPE growth processes.

zone at a temperature (~340 °C) just below the Mercury boiling point (357 °C), produces the environmental reactor Mercury overpressure. Under this Mercury-rich atmosphere the growth crucible was working, for the experiments here reported, at 530 °C. An atmospheric pressure Pd-diffused hydrogen flow plays as Hg carrier gas, beyond as anti-oxidant agent. In order to perform long time processes (50 hours or more) without Mercury source consumption, the reactor has been 10 degrees tilted toward the Hg-source to allow the Mercury recycling.

The HgTe vapours source typically used was a granular $Hg_{0.45}Te_{0.55}$ alloy. Its compositional constancy during the growth cycles, is guarantee by the environmental Mercury overpressure, also for long growth times, because of the Mercury recycling.

A graphite block having a well for containing the $Hg_{0.45}Te_{0.55}$ source, is the simple growth crucible used to perform our iso-VPE growth processes. It works with the source well exactly covered by the CdTe/Sapphire substrate downlooking positioned. Varying the quantity of granular $Hg_{0.45}Te_{0.55}$ into the well allows to obtain a variable source-substrate spacing for different growth runs.

## 3. The Physics of the process

The iso-VPE HgCdTe growth process is dominated by two main physical phenomena: -the HgTe growth on the substrate surface, and -the HgTe/CdTe solid state interdiffusion. The CdTe source is limited by the finite CdTe film thickness and the amount of HgTe grown (and interdiffused) is controlled by the growth conditions.

In principle, the same final ternary alloy could be obtained in different ways too: i.e. growing the HgTe layer on CdTe/Sapphire by LPE, MOCVD, MBE (or a VPE process with high HgTe growth rate) and interdiffusing the CdTe/HgTe structures in separate thermal treatments.

Nevertheless, the aim of our work is to find out the experimental conditions allowing the HgTe growth and interdiffusion in the same run. So, we can directly prepare technologically interesting structures by using a quite simple process. The condition to realise this one-step process is to use HgTe growth rates slower than HgTe/CdTe interdiffusion processes. In any case, it must be clear that, from the thermodynamic point of view, the single-step process here reported, is different respect to the two-step alloy preparation procedure. For the first one the complete thermodynamic equilibrium is never reached, while for the second one the equilibrium is dominating the system at the end of the interdiffusion step. So, growth performed with the present process are, in principle, affected by some compositional gradients in the direction of the films thickness. Nevertheless, this problem can be so strongly reduced by using low HgTe growth rates that it becomes practically negligible, as witnessed by the layers characteristics shown in this paper. To reliably perform our one-step iso-VPE processes on finite CdTe systems, the role of each experimental parameter has been investigated:

**a) $Hg_{1-z}Te_z$ source composition.** $Hg_{1-z}Te_z$ sources with z in the 0.50~0.85 range have been experimented. For source compositions between 0.50 to 0.65 no relevant differences were observed about the growth efficiency. What resulted really important was to use source compositions compatible with their solid state at the growth temperature. In fact, high z-composition sources are characterized by melting points lower than the typical iso-VPE growth temperatures. Their use produce the CdTe film melt-back, rather than the HgTe growth. In fact, under this situation the vapour phase is $Te_2$-rich and Tellurium overlayers deposited on CdTe/Sapphire structures, gradually dissolve the CdTe, trying to produce on the hybrid substrate a Te-solution with a Te molar fraction similar to the source, i. e. a liquid solution.

**b) Source-substrtate spacing.** It represents the most important parameter determining the HgTe growth rate. Working at 530 °C, with source compositions in the 0.65~0.50 range, growth rates from 0.4 to 3 μm/h were obtained using spacings between 3 to 34 mm wide.
The HgTe growth rate is non linearly changing with the source-substrate distance.

**c) Growth temperature.** The system is quite difficult to be analytically represented during the growth process evolution. It is equivalent to a multilayers structure with infinite layers having small compositional differences ($\delta x$), continuously changing during the process. Each interdiffusion rate of the equivalent multilayers structure, depends on : -the process temperature and -the layers composition. Working at 530 °C (800K), the complete HgTe layers interdiffusion is obtained, as a complex result of the multilayers interdiffusions, when the HgTe growth rate is $\leq 0.5$ μm/h.

**d) HgCdTe final film thickness and composition.** The growth rate is linear respect to the growth time, when all the other experimental conditions are kept constant. The final ternary film composition is approximately the starting CdTe film thickness divided by the total (CdTe + HgTe) films thickness. In conclusion, when the CdTe starting thickness is known, as well as the HgTe growth rate, any wanted ternary composition can be easily prepared by using a projected growth run.

## 4. Results

Fig 2 shows a typical FTIR transmission spectrum characterizing a Mercury Cadmium Telluride/Sapphire structure grown by the present method. The structure of fig. 2 consists of a 23 μm thick HgCdTe film, which x-composition is approximatively 0.32, supported on 450 μm thick single crystal Sapphire substrate (0001) oriented. The complete absorption at wave numbers below 1400 $cm^{-1}$ is due to the Sapphire substrate, because of the integral transmission measurement. The well defined interference fringes present in the 1400-2800 $cm^{-1}$ range and the abrupt (respect to the HgCdTe film thickness) shape of the transmission curve (left side of the transmission peak), are interpreteted as signs of good film uniformity as well as an abrupt substrate-epilayer interface. The transmission peak looks practically simmetric although the absorptions causing the two fronts of the transmission window are completely different. 23 μm of HgCdTe are producing a transmission front (left side of the peak) similar to 450 μm of bulk Sapphire (right side of the peak). This result is primarily connected to the two materials absorption coefficients. In any case it wouldn't be possible in presence of significant compositional gradients into the HgCdTe thin layer.

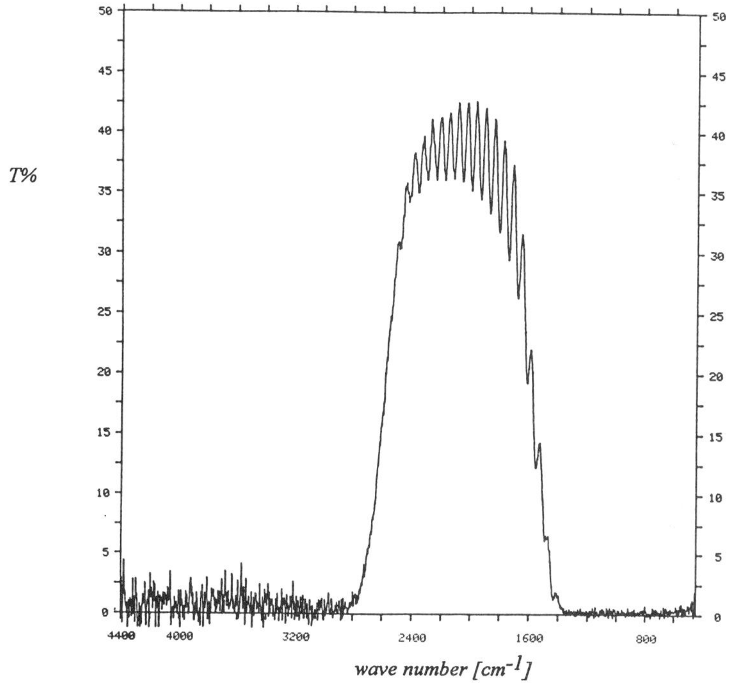


Fig. 2) FTIR transmission spectrum of an as-grown iso-VPE HgCdTe/Sapphire structure grown at 530 °C by using a single step process.

Transmittance values around 40%, on more than 20 μm thick as-grown films, demonstrate an optically high "bulk material" quality, beyond a low surface scattering connected with a good epilayer morphology.

The naked eye morphology of an iso-VPE $Hg_{0.70}Cd_{0.32}Te$/Sapphire structure is shown in fig.3 comparatively to a compositionally equivalent LPE HgCdTe/CdTe/Sapphire (with the active layer grown from Te-rich solutions) and an MOCVD CdTe/Sapphire structure. The iso-VPE layer reflectivity is better both respect to the starting substrate and to the equivalent structure which top layer was LPE grown (Te-rich solutions).

Carrier concentration and mobility measured at 77 K on as-grown x=0.32 layers, generated at 530°C, are in the 1.2-1.5×$10^{17}$ $cm^{-3}$ and 1.5-2×$10^{2}$ $cm^2$/V·sec range respectively. At 77 K the conductivity is p-type as expected for a Mercury vacancy dominated material. The behaviour of the electrical functions perfectly agrees with the equivalent data measured on $Hg_{1-x}Cd_xTe$ epitaxial structures (with similar x-composition) grown, at similar temperatures, by LPE from Te-rich solutions. Mixed conductivity with intrinsic dominant n-type transport is characterizing the electrical functions in the 300-240 K thermal range, while the lower temperature range is dominated by extrinsic p type conductivity for Hg vacancies practically completely ionizated.

In order to reduce the as-grown p-type level, for photovoltaic devices fabrication, the Mercury vacancy annealing of iso-VPE structures was investigated. The treatments were performed in closed ampoules under Hg-saturated conditions at temperatures ranging from 370 to 390 °C on the base of the experiences done for LPE (Te-rich) $Hg_{0.70}Cd_{0.32}Te$/CdZnTe structures.

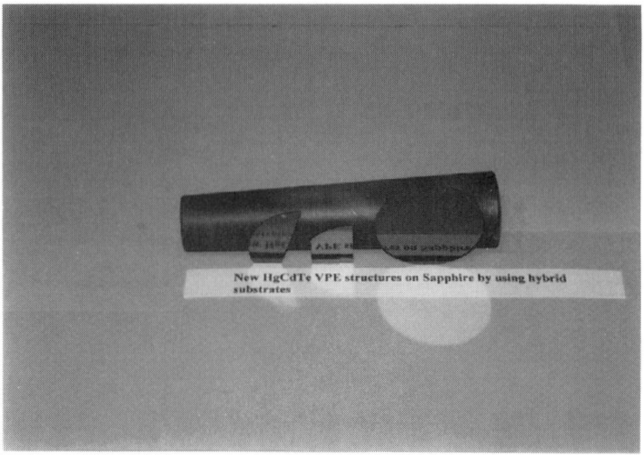


Fig. 3) Comparison between the naked eye morphology of three different Sapphire-based structures. From the right side: -2 inch. MOCVD CdTe/Sapphire, iso-VPE $Hg_{0.70}Cd_{0.32}Te$/Sapphire and LPE (Te-rich sol.) $Hg_{0.70}Te_{0.32}Te$/CdTe/Sapphire.

The iso-VPE hybrid structures p-type annealed behaviour, was verified to be quite similar to the LPE (Te-rich) classic structures. Carriers concentration in the high $10^{15}$-low $10^{16}$ $cm^{-3}$ and mobility in the 400-500 $cm^2$/V·sec range, were systematically measured at 77 K on x=0.32 films composition, under a magnetic field of 2.8 KGauss. The epilayer electrical behaviour continue to be extrinsically p-type for Mercury vacacies, but with a strong electrical characteristics improvement respect to the as-grown material. The 77 K carriers concentration reduces of more than one order of magnitude, from the first $10^{17}$ to the high $10^{15}$ $cm^{-3}$, and the holes mobility become 500 $cm^2$/V·sec against 150-200 $cm^2$/V·sec. characterizing the as grown material.

The main differences observed between the electrical characteristics of p-type annealed LPE CdZnTe-based structures and iso-VPE on Sapphire layers, concern their lateral uniformity and their run to run reproducibility. These characteristics are better for iso-VPE on Sapphire structures respect to LPE on CdZnTe layers. We attribute this behaviour to the different impurity sources at which the two HgCdTe layers are exposed during the respective growth processes and annealing treatments (internal impurity sources).

To complete our first investigation on iso-VPE HgCdTe/Sapphire structures, an X-ray characterization has been performed through double crystal rocking curves measurements. Using Cu $K\alpha_1$ radiation, (333) symmetric reflection and (111) InSb as monochromator, rocking curves as reported in fig. 4 were typically measured on ~20 μm thick

HgCdTe/Sapphire films. The symmetry peak and the full width half maximum (FWHM) values (50-70 arcsec.) are similar to the LPE (Te-rich sol.) materials on CdTe/Sapphire grown, which technological interest has been widely demonstrated [10][11].

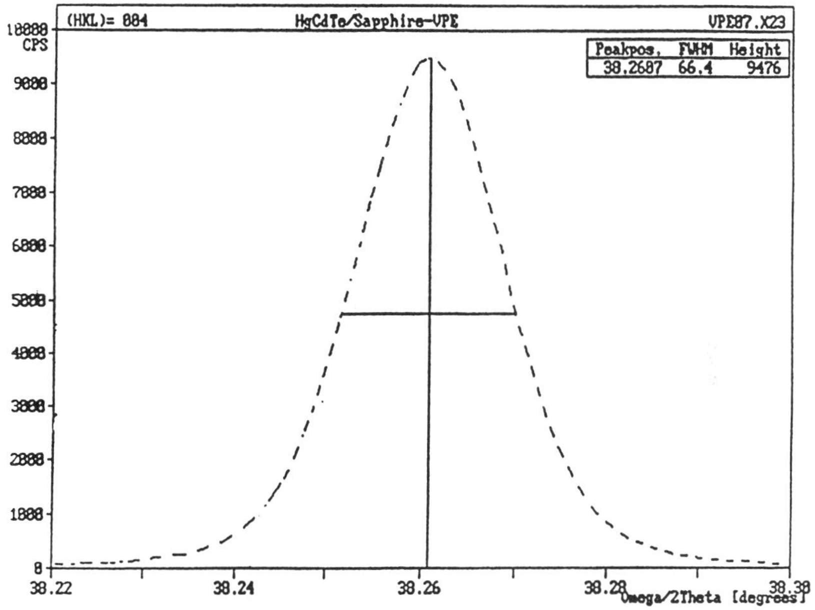


Fig. 4) Typical double crystal rocking curve measured on iso-VPE HgCdTe/Sapphire structures.

## Acknowledgements

The authors would like to thank Dr. N.V. Sotchinskii for his contribution to the results discussion, Dr. P. Franzosi for the X-ray characterizations and Dr. S. Alfuso for the structures electrical characterization.

## References

[1] Harman TC 1981 *J. Elect. Mat.* Vol. **10** No.6 1069-84
[2] Nemirovsky Y, Margalit S, Finkman E, Shacham-Diamand Y, and Kidron I 1982 *J. Elect. Mat.* Vol.**11**
[3] Pelliciari B 1986 (1988) *J. Cryst. Growth* 146-160
[4] Bernardi S. 1987 *J. Cryst. Growth* 365-371
[5] Becla P, Lagowski J, Gatos HC, and Ruda H 1981 *J. Elecrochem. Soc.* 1171-73
[6] Nemirovsky Y and Kepten A 1984 *J. Elect. Mat.* Vol. **13** No.6
[7] Fleming JG and Stevenson DA 1987 *J. Cryst. Growth* 621-27
[8] Lee SB Kim D and Stevenson DA 1991 *J. Vac. Sci. Technol.* ***B*** **9** (3) 1639-45
[9] Willardson RK, Beer CA1981, *Academic press,* Vol. **18**, 1-18
[10] Gernter ER, Tennant WE, Blackwell JD and .Rode.JP 1985 *J. Cryst. Growth* **72** 462-467
[11] Thomson J, Woodhouse NK, and Dineen C, *J. Crystal Growth* 77 (1986) 452-459

*Inst. Phys. Conf. Ser. No 144: Section 5*
*Paper presented at 7th Int. Conf. on Narrow Gap Semicond. Santa Fe 9–12 January*



# Isothermal vapor phase epitaxial growth of large-area HgCdTe film on lattice-matched CdZnTe substrate

**T S Lee*, Y T Jeoung*, H K Kim*, J M Kim*, J M Chang**, I H Park**, S B Lee**, S U Kim** and M J Park****

* Agency for Defense Development, Youseong P.O.Box 35, Taejeon, Korea
**Korea University, Anamdong 5-1, Seongbugku, Seoul, Korea

**Abstract.** $Hg_{1-x}Cd_xTe$(x=0.3) epilayers($\phi$=26 mm) have been grown by isothermal vapor phase epitaxial(ISOVPE) method on each substrate of (110) and (111)CdZnTe with area over 600 $mm^2$($\phi$=30 mm). The crystallinity of HgCdTe epilayer has been measured by double crystal x-ray rocking curve and the full width at half maximum(FWHM) was about 100 arcsec. The lateral variation of x composition over the whole HgCdTe epilayer was found to be ±0.003 by measuring through infrared transmission spectra. The cross-sectional variation over the 10 μm thickness of the film measured by electron probe micro-analysis(EPMA) is ±0.005. As-grown HgCdTe epilayers are p-type with low $10^{17}$ $cm^{-3}$ in carrier concentration and 100 $cm^2$/Vs in mobility at 77 K. After Hg vacancy annealing at 260°C for 24 hrs, the epilayer has been converted to n-type with low $10^{14}$ $cm^{-3}$ in carrier concentration and 200 $cm^2$/Vs in mobility at 77 K.

## 1. Introduction

High density array device such as IRFPA needs large-area HgCdTe structures with high lateral uniformity in properties. The area and the properties of HgCdTe epilayers, however, strongly depend on those of their substrate. Therefore, many efforts have been made to obtain the large-area and the high quality bulk(CdZnTe) and hybrid(GaAs-based, Si-based and sapphire-based) substrates. It is known that the best IR device performance have been obtained by using CdZnTe substrates whatever the epitaxial growth technique used.

The isothermal vapor phase epitaxy(ISOVPE) method used in this work has a number of attractive aspects for the "device-quality" HgCdTe epilayer growth. It is easy to implement and as-grown surface morphology obtained is excellent[1-3]. The ISOVPE growth method has the potential to produce films with high crystal quality and low impurity concentration. The elements of the source are transported to substrate via gas phase without contact between the substrate and a liquid phase as in LPE or impurity elements in the gas environment such as in the hydrocarbon environment in MOCVD. Since non-volatile impurities are not transported from the source to the substrate, the ISOVPE method may produce HgCdTe films with low carrier concentrations.

This report is on the study of the ISOVPE growth method and on their surface morphology, crystallinity, compositional uniformity and electrical properties of large-area HgCdTe epilayers on lattice-matched CdZnTe substrates.

## 2. Experiments

### *2.1. ISOVPE growth of HgCdTe on CdZnTe*

For lattice-matched substrate, CdZnTe crystals have been grown by vertical Bridgman method in our laboratory. CdTe crystals have been used as seed to obtain <110> or <111> oriented CdZnTe boules. The typical size of CdZnTe boules is 30 mm in diameter and 120 mm in length. The boules have been cut perpendicularly to their longitudinal direction for CdZnTe substrates to have (110) or (111) planes. The substrates($\phi$=30 mm) with mirror-like surface are prepared by typical polishing process. The Zn composition of substrates is obtained using the relation between Zn composition and cut-on wavelength measured by ultraviolet/visible spectrometer[4].

The $Hg_{0.368}Cd_{0.082}Te_{0.550}$ source for ISOVPE growth is prepared by evacuating and sealing the quartz ampoule containing the elements(6N purity) under $10^{-3}$ Torr in vacuum. The source compound is synthesized at 450°C for 24 hrs and heated for complete mixing of the melt at 650°C for 10 hrs followed by quenching in water to avoid a segregation. Two types of ampoules are used for the ISOVPE growth, as shown in Fig. 1. The tilted loading is easier to implement and makes it possible to obtain HgCdTe epilayer with the same area that its substrate has. The vertical loading masks the edge of a substrate and makes decrease of the epilayer area but allows better lateral composition homogeneity. In this work both loading types have been attempted.

$Hg_{1-x}Cd_xTe$ layers were grown on each substrate of (110) or (111)CdZnTe with various size ranging 50 to 500 mm$^2$. The growth temperature and time duration to obtain 30-40 μm thick epilayers are 550°C and 45-48 hrs, respectively. The heat pipe(50 cm in length and 8 cm in inner diameter) is used to obtain the constant temperature zone, where the temperature deviation over the length of 20 cm is ±1°C.

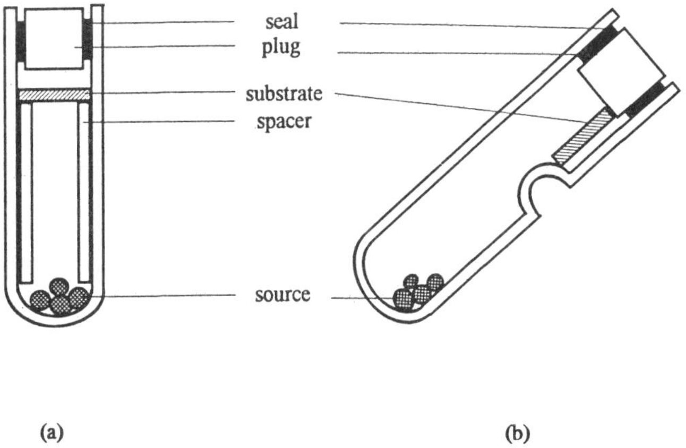


Figure 1. Experimental arrangement for ISOVPE growth ampoules: (a) vertical loading; (b) tilted loading.

### *2.2. Characterization of HgCdTe epilayers*

The surface and cross-section of as-grown HgCdTe epilayer is observed with optical microscope for the examination of morphology and epilayer thickness. To examine the crystallinity of HgCdTe epilayer, double crystal x-ray rocking curve from (333) sample reflection of Cu K$\alpha_1$ radiationis is obtained by using (111)Si crystal as monochromator. The depth profile of composition is measured by electron probe microanalysis (EPMA) in cross-section view. The lateral variation of composition over the whole epilayer is measured with infrared spectrometer using an aperture of 1x1 mm$^2$. The HgCdTe epilayers are annealed under Hg pressure at various temperatures for the type conversion and the control of carrier concentration. Electrical properties of as-grown and annealed HgCdTe epilayers are examined by using van der Pauw-Hall measurements at room temperature and 77 K.

## 3. Results and discussion

Figure 2 displays the surface morphologies of the as-grown HgCdTe epilayers on three different planes of CdZnTe substrate. The HgCdTe epilayers grown on (111)A or B substrate have mirror-like surface and the one grown on (110) substrate is less shiny. It is found that the surface becomes more mirror-like as the deviation from (111) decreases and the best morphology is obtained on (111)A±0.5° substrate.

The crystallinity of HgCdTe epilayer grown on (111)CdZnTe substrate is shown in figure 3 by measuring the full width at half maximum(FWHM) of the x-ray double crystal rocking curve. The FWHM for HgCdTe epilayer is about 100 arcsec which is larger than the value(70 arcsec) for CdZnTe substrate. It is, however, thought to be very satisfactory result in comparison with those of the epilayers grown by the other methods. It seems to result from the reduction of lattice mismatch by long graded junction between epilayer and CdZnTe substrate.

The typical compositional depth profile through the thickness of HgCdTe epilayer is flat near the surface with a deviation of ±0.005 in x value within 10 μm thickness from the surface as shown in figure 4. The lateral uniformity in composition of HgCdTe epilayer obtained with vertical loading type is ±0.003 in x value across the entire surface of 26 mm in diameter(Fig. 5). The epilayer obtained with tilted loading type has the deviation of ±0.006 in x value across the surface of 5x10 mm$^2$.

As-grown $Hg_{1-x}Cd_xTe$(x=0.3) epilayer is p-type at 77 K. At this temperature its carrier concentration and mobility are low $10^{17}$ cm$^{-3}$ and around 100 cm$^2$/Vs, respectively. The epilayers after Hg vacancy annealing at 300°C are also p-type at 77 K with low $10^{15}$ cm$^{-3}$ in concentration and around 200 cm$^2$/Vs in mobility. The low mobility seems to result from two possibility: first, the residual impurities diffused from the substrate; second, the cross-section compositional gradient. It is found that n-type conversion occurs after Hg annealing at 260°C for 24 hrs, and its carrier concentration and mobility at 77 K are low $10^{14}$cm$^{-3}$ and 26,000 cm$^2$/Vs, respectively(Table 1).

## 4. Conclusion

Large-area HgCdTe epilayers(>500 mm$^2$) with as-grown mirror-like surface are successfully

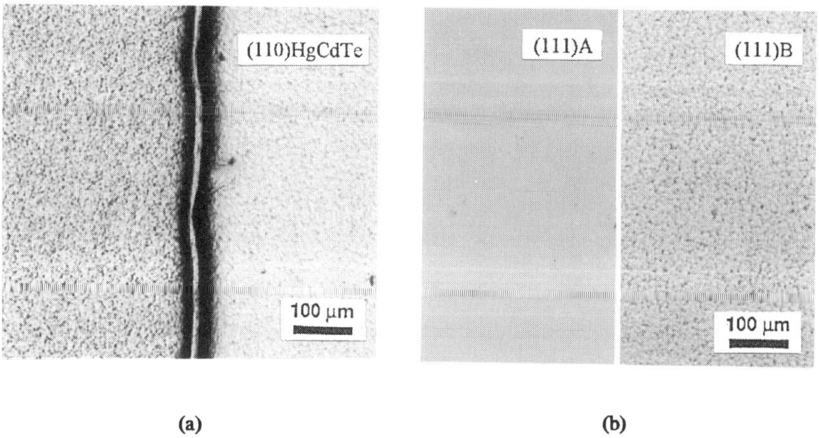


**Figure 2. The dependence of epilayer surface morphology to the orientation of the substrates: (a) surface with twin boundary on (110) CdZnTe, (b) surfaces on (111)A and B CdZnTe.**

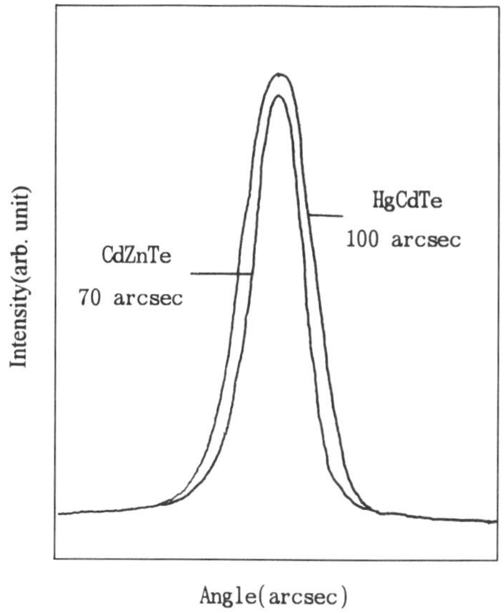


**Figure 3. Comparison of double crystal x-ray rocking surves for HgCdTe epilayer grown by ISOVPE and (111)CdZnTe substrate.**

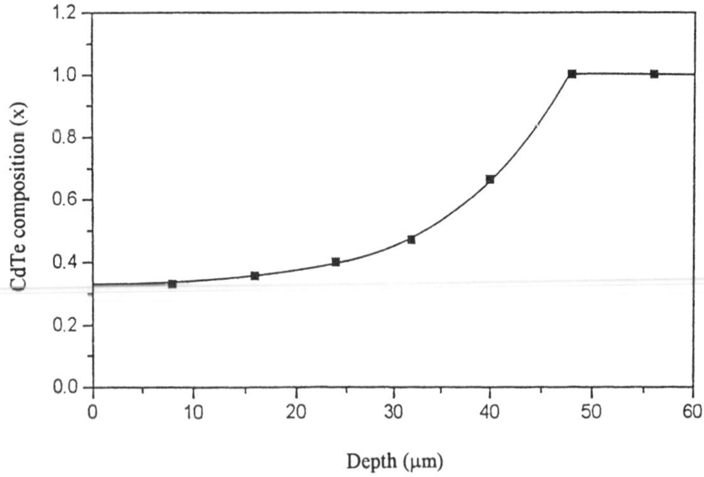


**Figure 4. A typical compositional depth profile of HgCdTe epilayer grown by ISOVPE method at 550°C for 48 hrs.**

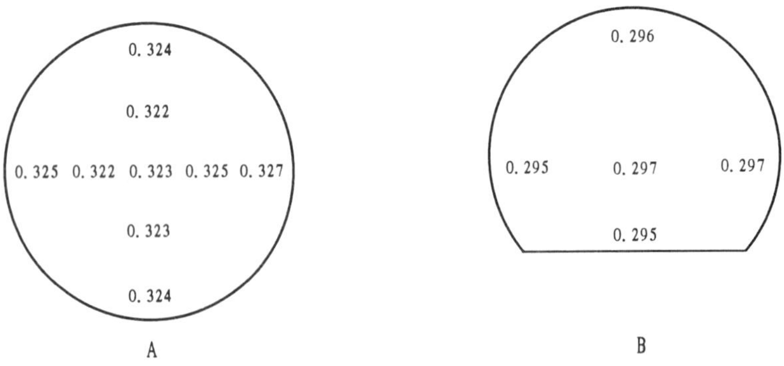


**Figure 5. The lateral compositional profile across the entire surface of HgCdTe epilayer (26 mm in diameter) obtained with vertical loading of ISOVPE growth ampoule. The infrared beam size is 1x1 mm$^2$.**

Table 1. Electrical properties of ISOVPE grown $Hg_{1-x}Cd_xTe(x=0.3)$ epilayers.

| Sample ID | Annealing Temperature(°C) / Time(hrs) | Hall Measurement Temperature(°C) | Conduction Type | Concentration n ($cm^{-3}$) | Mobility μ ($cm^2/Vs$) |
|---|---|---|---|---|---|
| MCT-1 | as-grown | 300 | p | $1x10^{18}$ | 60 |
| | | 77 | n | $1.5x10^{17}$ | 114 |
| MCT-2 | 260/24 | 300 | n | $6x10^{15}$ | 3,000 |
| | | 77 | n | $3.6x10^{14}$ | 26,000 |
| MCT-3 | 300/16 | 300 | n | $5x10^{15}$ | 1,900 |
| | | 77 | p | $3.8x10^{15}$ | 208 |

grown on CdZnTe substrate(30 mm in diameter) by ISOVPE method. The epilayers obtained with vertical loading type have excellent lateral homogeneity($\Delta x=\pm0.003$) of composition across the entire surface and also good homogeneity($\Delta x=\pm0.005$) within 10 μm thickness from the surface. FWHM of double crystal x-ray rocking curve showing a crystallinity of the HgCdTe epilayer is 140% of that for CdZnTe substrate, meaning that the crystal quality of HgCdTe epilayer could be greatly improved according to the substrate quality. The electrical properties of the epilayers could be also improved after Hg annealing although the as-grown epilayer is p-type with high carrier concentration.

It is, therefore, thought that ISOVPE method, which requires very simple and inexpensive facility, makes it possible to obtain the large-area HgCdTe epilayers with good properties and their excellent uniformity in comparison with other epitaxial growing method.

## Acknowledgments

The authors would like to thank Mr. C K Kim at Korea University for experimental assistance. This work is partly supported by the KOSEF and partly the Basic Research-Science Institute Program of the Ministry of Education of Korea 1994.

## References

[1] Cohen-Solal G, Marfaing Y, Bailly F, and Robot M 1965 *C. R. Acad. Sci. Paris.* **261** 931
[2] Fleming J G and Stevenson D A 1987 *J. Cryst. Growth* **82** 621
[3] Tung T, Golonka L, and Brebrick R F 1981 *J. Electrochem. Soc.* **128** 451
[4] Johnson S M, Sen S, Konkel W H, and Kalisher M H 1991 *J. Vac. Sci. Technol.* **B9**(3) 1897

*Inst. Phys. Conf. Ser. No 144: Section 6*
*Paper presented at 7th Int. Conf. on Narrow Gap Semicond. Santa Fe 9–12 January*



# Auger Rates in Mid-IR InAsSb Laser Structures

**H. P. Hjalmarson and S. R. Kurtz**
**Sandia National Laboratories Albuquerque NM 87185-0603**

**Abstract**

Auger rates are calculated for three InAsSb mid-infrared laser structures as a function of temperature. Compressive strain in the quantum wells reduces the mass of the holes; it is shown that this leads to a reduction in the Auger rate compared with an unstrained quantum well. The Auger rates for these structures are similar primarily due to their similar bandgap energies.

## 1. Introduction

Auger recombination is often the dominant non-radiative mechanism in narrow gap semiconductors used for infrared (IR) applications. In this work we compute the Auger rate for InAsSb optically active layers being fabricated as light-emitting diodes (LED's) and lasers.[1][2] The guiding motivation is the use of bandgap engineering to reduce the Auger rates.[3] The resultant bandstructure is highly non-parabolic which led to our development of a new theoretical technique to explore this issue. The calculation involves computation of *k.p* energies and wavefunctions for the strained quantum well structure. This information is used in the calculation of Auger rates and lifetimes as a function of doping and injected carrier density. In this paper we use this calculation to define two design rules for these structures. One, we show that a reduction of the hole transverse mass reduces the Auger rate dramatically. Two, we define the barrier thickness which is sufficient to minimize the rate due to coupling between wells.

## 2. Formalism for Auger rates

The 2D Auger rate calculation is similar to earlier work on bulk 3D structures.[4] The earliest work on 2D systems focused on quantum well laser structures.[5, 6] More recent work has focused on quantum well superlattices.[7] Fermi's Golden Rule defines the Auger rate

$$1/\tau = \frac{2\pi}{\hbar} \sum_{ii'ff'} |V_{ii'ff'}|^2 f_i f_{i'} (1 - f_f)(1 - f_{f'}) \delta(E_i + E_{i'} - E_f - E_{f'}) \tag{1}$$

in terms of Fermi functions $f_i \equiv f((E_i - E_{F_i})/k_B T)$ and matrix elements

$$V_{ii'ff'} = \frac{e^2}{4\pi\epsilon} \int d^3\mathbf{r} \int d^3\mathbf{r}' \frac{\Psi_i(\mathbf{r})\Psi_{i'}(\mathbf{r}')\Psi_f(\mathbf{r})\Psi_{f'}(\mathbf{r}')}{|\mathbf{r} - \mathbf{r}'|}. \tag{2}$$

The wavefunctions

$$\Psi_i(\mathbf{r}) \equiv \Psi_{n_i k_i}(\mathbf{r}) = \sum_{m_i} \Phi_{n_i k_i m_i}(z) u_{m_i}(\mathbf{r}) \exp(i\mathbf{k}_i \mathbf{r}) \tag{3}$$

are written in terms of an envelope function $\Phi(\mathbf{r})$ and the Bloch functions $u(\mathbf{r})$. These wavefunctions and the energies $E_{nk}$ are obtained from a strained $k.p$ calculation.[8]

By analogy to the Beattie-Landsberg expression for 3D systems, the 2D Auger lifetime $\tau$ for an n-doped structure can be written as

$$1/\tau = \frac{(n_0 + p_0 + \nu)(n_0 + \nu)}{n_i^2} \frac{1}{\tau_i} \tag{4}$$

in terms of the intrinsic lifetime $\tau_i$ defined by

$$1/\tau_i = 1.65 \times 10^{17} \frac{F(G_T)^2}{\kappa^2 (E_G/k_B T)(1 + 2m_c/m_v)(m_c/m_0)} \exp(-E_T/k_B T) \tag{5}$$

in which $E_T$ is the threshold energy, the energy which must be exceeded for an Auger transition (it is determined by energy and momentum conservation). The transition assumed involves two electrons in the conduction (C) sub-band which scatter to a conduction sub-band and a valence (V) sub-band (a CCCV transition). In these expressions, $\nu$ is the injected carrier density; $n_0$ and $p_0$ are the electron and hole equilibrium concentrations, respectively. The other quantities are the dielectric constant $\kappa$, the conduction band effective mass $m_c$, the valence band effective mass $m_v$, the bandgap $E_g$, and the temperature $T$. Only one valence and one conduction sub-band are assumed for this simplified, heuristic expression. A similar expression can be written for a VVVC transition. The calculations to be described involve a generalization of Eq. (4) in which the form factor is computed by a probabilistic average over k-vectors corresponding to the threshold energy.

The form factor

$$F(G) = \sum_{m,m'} \int dz \int dz' \varphi_{im}(z) \varphi_{i'm'}(z') \varphi_{fm}(z) \varphi_{f'm'}(z') \exp(-G|z - z'|) \tag{6}$$

represents the overlap of the initial ($i$ and $i'$) state and final ($f$ and $f'$) state wavefunctions. For a superlattice of quantum wells the form factor can be rewritten in terms of an integration over the periodic function times an exponential phase factor:

$$\varphi(z) = \chi(z) \exp(iqz). \tag{7}$$

The form factor also includes the Coulomb interaction, responsible for Auger transitions, through the momentum transfer $G_T$. This quantity controls the range

$$R \approx \frac{1}{G_T} \tag{8}$$

of the Coulomb potential. If G is small, the range is large leading to large contributions to the rate. The transverse and longitudinal contributions enter separately. A $1/G_T^2$ term describes the contributions from within each well (this term does not appear directly in the final expression). The form factor term $\exp(-G_T L)$ describes the contribution from adjacent wells in terms of the superlattice periodicity $L$.

By inspection, the Auger rate can be reduced by increasing the threshold energy. In the effective mass approximation, the activation energy

$$\Delta E \equiv E_T - E_G = \frac{m_c}{m_v + m_c} E_G \tag{9}$$

in which $E_G$ is the bandgap energy. This observation immediately leads to the use of strain to reduce the transverse hole mass thereby reducing the Auger rate.

Another design rule follows from the strength of the interaction in terms of the threshold momentum transfer $G_T$. In the effective mass approximation, the momentum transfer depends on bandgap and effective mass in a simple way:

$$G_T^2 = \frac{m_v + m_c}{m_v + 2m_c} \frac{2m_c}{\hbar^2} E_g \tag{10}$$

Equation (6) leads directly to the second design rule: Choose the barrier thickness such that $G_T L < 1$. If $L$ exceeds the range $R = 1/G_T$, then carriers in adjacent wells cannot contribute to the Auger rate. In other words, the quantum wells are decoupled with no coherent non-radiative transitions due to the long-range Coulomb potential.

The Auger calculations involve separate bandstructure and rate calculations. The $k \cdot p$ bandstructure is computed for each strained structure. The energies, wavefunctions, and dipole matrix elements are tabulated and used as input for the rate calculations. Due to the importance of strain in these structures, both energies and wavefunctions are assumed to be non-parabolic. The wavefunctions enter the calculation through the various form factor approximations. For example, the form factor can be chosen to be a constant to compare with earlier work on 3D systems. [4] Since the wavefunctions depend on spin, both Coulomb and exchange matrix elements can be included directly with no approximations. Screening can also be included.

The calculations can be done for both low and high carrier density limits. These are distinct calculations. The low density limit is useful for materials and doping studies. The high density limit applies to laser structures for which the Auger lifetime and the Auger contribution to the laser threshold current can be evaluated. This calculation requires an estimate of the external mirror losses. Given such an estimate, the total radiated optical power and its spectral dependence can be computed as a function of external electrical power.

Table I: Material characteristics of laser structures

| | Wells | | | Barriers | | |
|---|---|---|---|---|---|---|
| | Material | Thickness (nm) | Strain | Material | Thickness (nm) | Strain |
| A | $InAs_{0.86}Sb_{0.14}$ | 9.0 | -0.007 | $In_{0.87}Ga_{0.13}As$ | 13.0 | 0.007 |
| B | $InAs_{0.86}Sb_{0.14}$ | 9.0 | -0.007 | $In_{0.9}Al_{0.1}As$ | 10.0 | 0.007 |
| C | $InAs_{0.91}Sb_{0.09}$ | 10.0 | -0.0059 | InAs | 50.0 | 0.0 |

Table II: Electronic characteristics of laser structures.

| | Name | Gap (eV) | Wells (eV) | | | Barriers (eV) | | |
|---|---|---|---|---|---|---|---|---|
| | | | $E_c$ | $E_{hh}$ | $E_{lh}$ | $E_c$ | $E_{hh}$ | $E_{lh}$ |
| A | InAsSb/InGaAs | 0.329 | 0.311 | 0.057 | 0.003 | 0.499 | -0.055 | -0.001 |
| B | InAsSb/InGaAs | 0.331 | 0.311 | 0.057 | 0.003 | 0.569 | -0.110 | -0.056 |
| C | InAsSb/InAs | 0.322 | 0.350 | 0.056 | 0.010 | 0.42 | 0.0 | 0.0 |

To illustrate the design rules we examine the Auger lifetimes as a function of temperature for three proposed laser structures. Materials and electronic information about these structures is tabulated in tables I and II.[9] Because of ambiguity about the k.p parameters for these structures they are approximated by parameters for InAs. Each material is distinguished by its band offset relative to InAs and its band gap. The Luttinger parameters $g_1$ =3.34, $g_2$ = -0.23, and $g_3$ = 0.57; the optical matrix element $V_{opt}$ = 20.2 eV.[9]

We illustrate the first rule which states that a bandgap-engineered reduction in the hole effective mass leads to a reduced Auger rate. Figure 1 shows the Auger lifetime for the InSbAs/InGaAs structure with and without strain for an n-doped sample with $n_0 = 10^{10}$ cm$^{-2}$. By inspection, the strain increases the Auger lifetime dramatically by a factor of 100. The slope of the lifetime versus 1/T is increased showing that the effect has been to increase the threshold energy. By inspection, the asymptotes at high temperatures are similar. For these calculations and those to be described, the CCCH transitions from the $n = 1$ conduction sub-band were computed. The final states considered are the $n = 1, 2$ valence bands as well as the $n = 1, 2$ conduction bands.

The Auger lifetimes as a function of temperature for the three structures are compared in Fig. 2. These lifetimes are remarkably similar primarily due to their similar bandgap energies and similar thicknesses. Clearly the lifetime is sensitive to variations in bandgap energy. However, it is also sensitive to quantum well thickness. A thinner well would have a larger Auger rate than a thicker well because of the increased quantum confinement which would lead to an increased form factor. This, of course, assumes other quantities such as the threshold energy are unchanged.

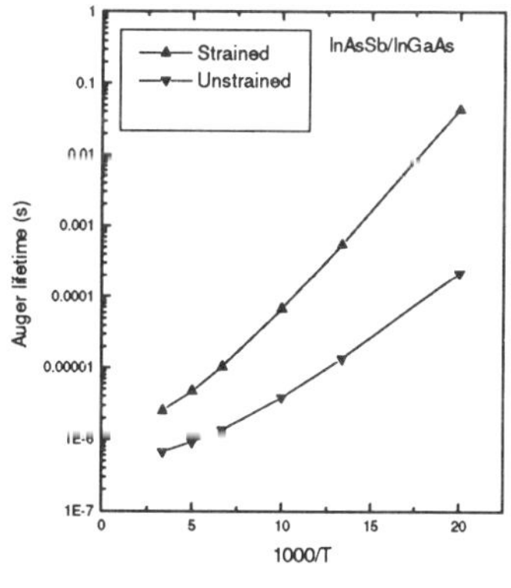


Fig. 1: Strain dependence of Auger lifetimes.

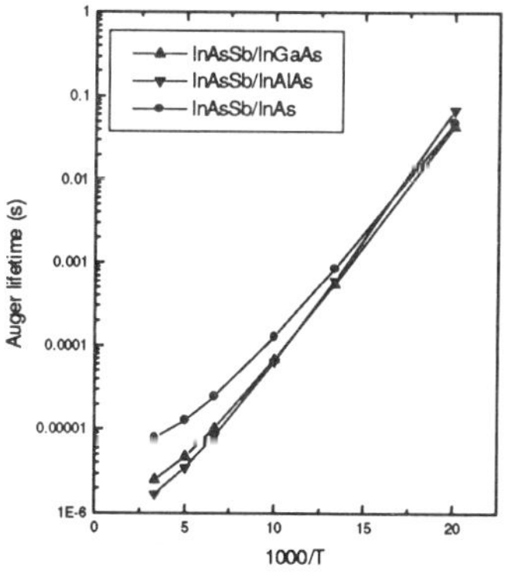


Fig. 2: Lifetimes for three structures.

Finally, these structures illustrate the second design rule. For each of these structures,

$$G_T \approx 0.02\frac{2\pi}{a} \tag{11}$$

in terms of the unit cell length $a$. This leads to a range $R$ which the barrier thickness must exceed. For these structures $R \approx 5$ nm. The barrier of each structure exceeds this range. Because of the resultant insensitivity to barrier thickness, the lifetimes for these structures are similar in spite of the variation in barrier thickness (see Table II). The second design rule would play an important role in structures with InAs/InGaSb superlattices with thin barriers (for example, 3.7 nm).[7] For these structures it has been shown that superlattice mini-band transitions can dominate the Auger rate. [7]

In general, we found the form factor to be very important. Thus it was important to compute the $k.p$ wavefunctions. At the threshold, we find

$$F(G_T)^2 \approx 10^{-4} - 10^{-3} \tag{12}$$

which is approximately 10× smaller than the form factor estimates using simple envelope functions. These $k.p$ form factors differ from more crude approximations in that the parity selection rules obtained from the most simple effective-mass wavefunctions are not valid. These selection rules are broken by inclusion of the dipole matrix element which couples the valence and conduction sub-bands. [10]. Thus the $n = 2$ conduction and valence sub-bands can contribute to the rate. We find their contributions are comparable to the $n = 1$ sub-bands. The $n = 2$ valence bands contribute due to band filling at high temperature. The $n = 2$ conduction bands contribute as final states because they reduce the threshold energy by the $n = 2$, $n = 1$ band separation. Comparisons of the calculated spectra and radiative rates to experimental data will independently test the form factors.[10] For example, the form factors and radiative rates are dependent on the optical matrix element $V_{opt}$.

Finally, we compare the Auger lifetime with a relative measurement of the lifetime for the InAsSb/InGaAs structure using photoluminescence.[9] We assume the photoluminescence intensity

$$I \approx \frac{1/\tau_r}{1/\tau_r + 1/\tau_{nr}} \tag{13}$$

in which $\tau_r$ is the radiative lifetime and $\tau_{nr}$ is the non-radiative lifetime. At high temperatures at which the Auger recombination dominates the non-radiative rate, we can assume

$$I \sim \frac{\tau_{nr}}{\tau_r} \tag{14}$$

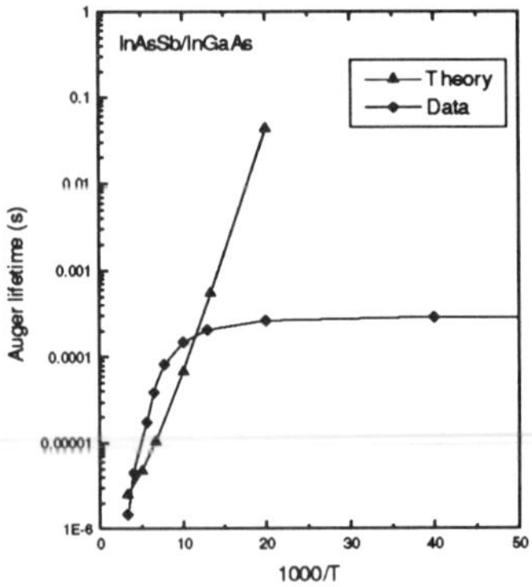


Fig. 3: Auger lifetime and photoluminescence intensity as a function of temperature.

Under this assumption the intensity is proportional to the Auger lifetime. As shown in Fig. 3, the intensity and the computed Auger lifetime show very good agreement in slope at high temperature. The measured activation energy $\Delta E \approx 0.08$ eV, in good agreement with Eq. 9.[9] Furthermore, we observe that $\tau_r, \tau_{Auger} \sim T$; this common temperature dependence cancels in Eq. 14, producing the nearly ideal Arrhenius behavior shown in Fig. 3. However, this temperature term causes all the lifetime curves to have pronounced deviations from Arrhenius behavior.

In conclusion, we have shown that bandgap engineering can reduce the Auger rates for mid-IR laser structures. The two design rules state that the hole effective mass must be reduced and the barrier thickness must exceed a critical thickness. Future work will extend these calculation to Auger and optical recombination within laser structures at elevated injected carrier densities which exceed the lasing threshold.

## 3. Acknowledgments

This work was supported by the US Department of Energy under contract DE-AC04-94AL85000. We thank T. B. Bahder, M. H. Crawford, T. J. Drummond, and Y. Zhang for useful discussions.

## References

[1] Kurtz SR, Biefeld R M, Dawson L R, Baucom K C, and Howard A J 1994 Appl. Phys. Lett. **64** 812

[2] Choi H K, Turner G W, and Le H Q (to be published)

[3] Adams A R 1986 Elect. Lett. **22** 249

[4] Beattie A R and Landsberg P T 1959 Proc. Phys Soc. A **249** 16-29

[5] Sugimura A 1983 IEEE J. Quantum Electron. **QE-19** 932

[6] Dutta N K 1983 J. Appl. Phys. **54** 1236

[7] Grein C H, Young P M, and Ehrenreich H 1992 Appl. Phys. Lett. **61** 2905

[8] Bahder T B 1990 Phys. Rev. B **41** 11992

[9] Kurtz S R, Biefeld R M and Dawson L R (accepted for publication, Phys. Rev. B)

[10] Hjalmarson H P, unpublished.

*Inst. Phys. Conf. Ser. No 144: Section 6*
*Paper presented at 7th Int. Conf. on Narrow Gap Semicond. Santa Fe 9–12 January*



# Picosecond free electron laser studies of Auger recombination in InSb and InAsSb systems

B.N.Murdin[(a)], C.R.Pidgeon[(b)], D.A.Jaroszynski[(c)], C.C.Phillips[(d)], R.A.Stradling[(d)], C.M.Ciesla[(b)], R.Praseres[(c)], C.J.G.M.Langerak[(a,b)].

(a) *FOM Institute for Plasma Physics, 'Rijnhuizen', P.O.Box 1207, 3430 BE Nieuwegein, The Netherlands.*
(b) *Department of Physics, Heriot-Watt University, Edinburgh EH14 4AS, U.K.*
(c) *LURE, Bâtiment 209d, Université de Paris-Sud, 91405 Orsay Cedex, France.*
(d) *Solid State Group, Physics Department, Imperial College, London SW7 2AZ, U.K.*

**Abstract.** Pump-probe experiments have been performed at room temperature in undoped bulk InSb, near and above the fundamental absorption edge at 180 meV using a far-infrared free electron laser. Significant bleaching was seen near the excitation frequency, with recovery times in the range 0.2-5 ns which were found to be strongly dependent on the pump photon energy. The scattering is dominated by Auger processes, which have rates following quadratic or linear carrier density dependence in low excitation and highly degenerate regimes respectively. The coefficients for Auger recombination in InSb at room temperature were found to be $1.1\pm0.5\times10^{-26}$ $cm^6s^{-1}$ and $4.0\pm0.5\times10^{-9}$ $cm^3s^{-1}$ in these two regimes. A preliminary bleaching experiment on an undoped strained layer superlattice sample of $InAs/InAs_{0.61}Sb_{0.39}$ is also reported, which demonstrates strong inhibition of the recombination process.

## 1. Introduction

It has been recognised for many years that the r.f. linac-pumped free electron laser (FEL) represents a major potential source of tuneable, picosecond, far-infrared radiation for time-resolved spectroscopy [1]. We have demonstrated the use of such far infrared systems (the Free Electron Laser for Infrared eXperiments, FELIX, in Nieuwegein and the Collaboration pour un Laser Infrarouge à Orsay, CLIO, in Orsay) for the determination of lifetimes of subband states in GaAs/AlGaAs quantum wells by saturation [2] and pump-probe [3] techniques.

Previously FEL measurements were reported of two-photon absorption and photo-conductivity in the narrow gap semiconductors InSb and InAs [4] measured over a broad spectral range - from the one-photon to two-photon energy threshold (i.e. from $E_G$ to $E_G/2$). We now report an extension of these experiments and validation of some assumptions therein: namely an identification and measurement of the Auger scattering process in InSb at room temperature over a wide range of excited carrier concentrations. As is well known, a knowledge of the Auger scattering rate in narrow gap systems is of considerable interest for many device applications since in most cases this is the limiting effect on the device performance [5,6,7]. Low threshold non-linear devices controlled by cw lasers and high sensitivity mid-infrared detectors and emitters need large excess carrier lifetimes, whereas applications such as laser mode locking, ultra fast radiation detectors and optical switches require recovery times as short as possible.

The method used here is similar to that reported previously for near-infrared, picosecond, pump-probe studies of Auger recombination in undoped InAs [7]. Band filling as the pump radiation is absorbed causes a strong dynamic blueshift in the IR absorption edge (the dynamic Moss-Burstein shift or DMS) and leads to pronounced absorption bleaching near the excitation frequency. The bandgap (180 meV or 1450 $cm^{-1}$) and effective mass of InSb have approximately half the values of those for InAs and the DMS is correspondingly significantly more pronounced, as evidenced by a wide variety of low power non-linear absorptive and refractive effects [8].

## 2. Experimental method

In the experiments described here the FEL's operated with pulse trains ('macropulses') typically of length and frequency 4 μs and 5 Hz respectively. Each macropulse consisted of a train of picosecond 'micropulses' with separation 1ns at FELIX and 32 ns at CLIO. The initial bleaching experiments were conducted at FELIX using the same experimental technique as for the earlier GaAs/AlGaAs subband bleaching experiments [2]. In order to take advantage of the larger pulse separation and shorter wavelengths the pump-probe experiment was performed at CLIO (following the experimental arrangements in Ref. [3]).

It was shown previously [7] that over the range of excited electron concentrations used throughout this study, the carrier temperatures approach the lattice temperature to within 2% in less than 3 ps. The total pump energy absorbed by the sample during each pump pulse would result in a peak temperature increase of the InSb by only ~0.02 K per pulse, and hence we neglect lattice heating effects.

Undoped 3.25 μm-thick InSb epilayers were grown by low-temperature MBE on semi-insulating GaAs substrates [9]. The as-grown wafers had measured 77 K Hall carrier concentrations of $\sim 10^{14}$ $cm^{-3}$, and were intrinsic ($\sim 2\times 10^{16}$ $cm^{-3}$) at room temperature.

The sample was mounted on a 2 mm pinhole at 300 K. The pump beam was normally incident, was unfocussed and had a full-width-half-maximum spot size of 6 mm. A small part of the beam was diverted with a $CaF_2$ beam splitter into a movable retro-reflector which acted as a delay line. The probe was then directed (also unfocussed) onto the sample at 15° to the pump. The effective pump and probe energy fluences per micropulse were estimated to be 50 and 5 μJ/$cm^2$ respectively, including losses due to beamsplitters and windows. The signal was collected with a cooled (77K) MCT detector and the transmittance of the probe, T, was measured as a function of optical delay, $t_{delay}$, between pump and probe pulses. In addition the probe beam was monitored with a second beamsplitter and detector, so that drift in the laser power (up to 5% over 10 minutes) could be divided out.

## 3. Experimental Results and Analysis.

The pump-probe data - i.e. probe transmission versus probe delay after the pump pulse - are shown for three different photon energies in Fig. 1. The absorption is completely bleached near zero delay in each case.

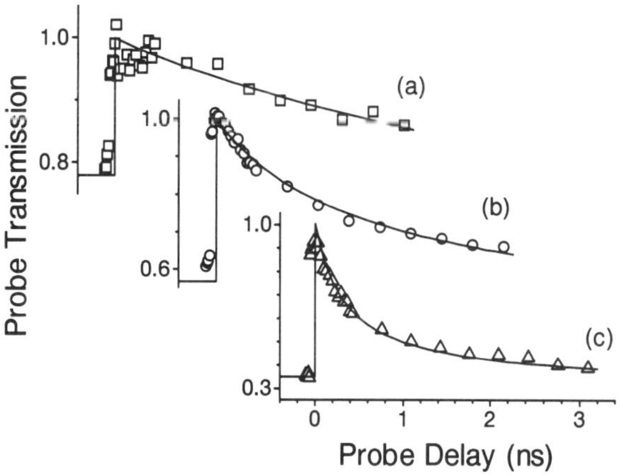


**Figure 1.** Pump-probe experimental results (points) for three pump energies a) $\hbar\omega$=198 meV, b) 226 meV and c) 263 meV. The reflection loss has been divided out of the transmission. The theoretical curves are obtained from a fit of the rate equation model described in the text, which includes the carrier density dependent Auger lifetime.

In order to interpret the pump-probe experiment (probe transmission versus delay) we need to convert the measured transmission into values of excess carrier concentration.

The recovery time of the bleaching varies from around 2 ns at a photon energy of 198 meV, to 0.2 ns at 263 meV. This is due to the carrier concentration dependent relaxation process. These recovery times are much longer than the pulse duration (2 ps) and we can write the rate equation for the recombination process (i.e. for times after the pump pulse):

$$\frac{dN_e}{dt} = -\frac{N_e}{\tau} \qquad [1]$$

where the 'average' recombination rate, $\gamma=\tau^{-1}$ is some function of the total electron concentration, $N_e$. We may also neglect the pulse duration (i.e. the probe pulse shape function I(t) is a delta function) when calculating the sample transmission:

$$T = \frac{\int I(t-t_{delay}).(1-R)^2.\exp[-\alpha(t)d].dt}{\int I(t-t_{delay}).dt} \approx (1-R)^2.\exp[-\alpha(t_{delay})d] \qquad [2]$$

We have assumed that the absorption coefficient, $\alpha$, while time dependent, is constant through the sample (the initial concentration gradient is zero for complete bleaching). The reflection loss term $(1-R)^2 \approx 0.41$. $\alpha$ is related to the carrier density via the carrier distributions:

$$\alpha = \sigma \begin{Bmatrix} \left[1-f_e(E_e^{lh})-f_{lh}(E_{lh})\right]\left[g_e^{-1}(E_e^{lh})-g_{lh}^{-1}(E_{lh})\right]^{-1} \\ + \text{ similar electron - heavy hole term} \end{Bmatrix} \qquad [3]$$

Here $f_{e,lh,hh}$ are the occupation probability functions for each band, and $g_{e,lh,hh}$ are the density of states functions. $E_e^{lh}$ and $E_{lh}$ are the electron and light hole energies for e-lh pair creation and similarly for e-hh pairs (at the same excitation energy). The occupancies can be integrated to give the total excited carrier concentration.

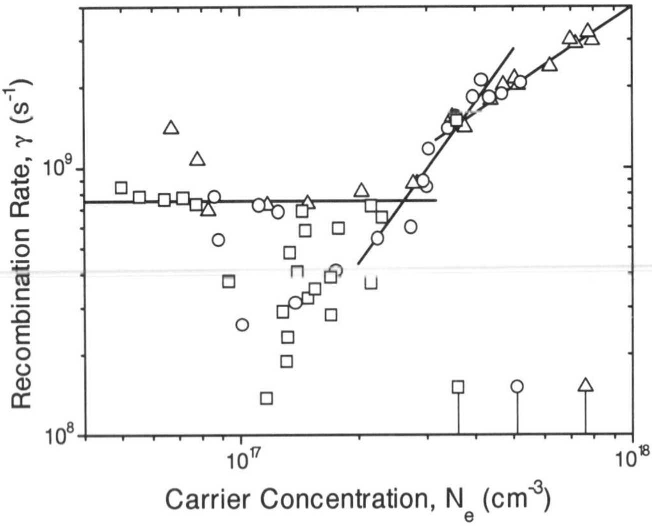


**Figure 2.** Inverse carrier lifetime versus non-equilibrium electron density. The experimental points are taken by differentiating pump-probe data, $N_e(t_{delay})$ (from Fig. 1 and Eqs. 2 and 3), to obtain the recombination rate. The pump photon energies were (□) $\hbar\omega$=198 meV; (○) 226 meV, (Δ) 263 meV. The starting carrier concentrations corresponding to complete bleaching are indicated at bottom right for each of the three photon energies. The three guides to the eye (solid lines) indicate from left to right: $\gamma$=constant, $\gamma \propto N_e^2$, and $\gamma \propto N_e$.

In order to calculate the bandstructure and the density of states, we use a Kane model (i.e. with non-parabolicity included) with a momentum matrix element $E_p$=23.4 eV. We have also taken the heavy hole mass as $m^*_{hh}$=0.4. The cross-section, $\sigma$, is determined by fitting the theoretical transmission obtained with equilibrium Fermi distributions to the small-signal transmission spectrum, measured with a Fourier Transform spectrometer.

We assume that carriers thermalise rapidly compared with the recombination rate in which case the concentration becomes determined: every value of transmission corresponds to a unique quasi-Fermi energy for electrons and thus a unique excess carrier concentration (the number of holes is also fixed by conservation of charge). We can now find the carrier concentration, $N_e(t_{delay})$, from the transmission, $T(t_{delay})$, by numerically inverting Eqs. 2 and 3. By smoothing, differentiating and using Eq. 1 we find the recombination rate $\gamma(N_e)$, shown in Fig. 2.

Figure 2 suggests three regimes. Below about $N_e$=2×10$^{17}$ cm$^{-3}$, the recombination rate varies widely and is generally governed by non-radiative, defect and interface related recombinations [7]. Secondly, between $N_e$=2 to 3×10$^{17}$ cm$^{-3}$, an $N_e^2$ dependence with coefficient $\gamma_2$=1.1±0.5×10$^{-26}$ cm$^6$s$^{-1}$ may be seen which is typical of an Auger process in the non-degenerate case. Finally, at higher carrier densities (above about 3×10$^{17}$ cm$^{-3}$), the rate goes as $N_e$ with coefficient $\gamma_1$=4.0±0.5×10$^{-9}$ cm$^3$s$^{-1}$. This is characteristic of an Auger recombination in a degenerate semiconductor [10], and in broad agreement with previous plasma reflection experiments [11].

The pump-intensity dependence of the transmission of the InSb sample is shown in Fig. 3. Also shown is a fit from our rate equation model. This technique has been applied to an undoped strained layer superlattice (SLS) sample of $InAs/InAs_{0.61}Sb_{0.39}$ (period 21.5 nm, equal well and barrier thickness, total thickness 2 μm), grown by MBE at Imperial College. This type-II SLS system offers bandgaps much less than the parent

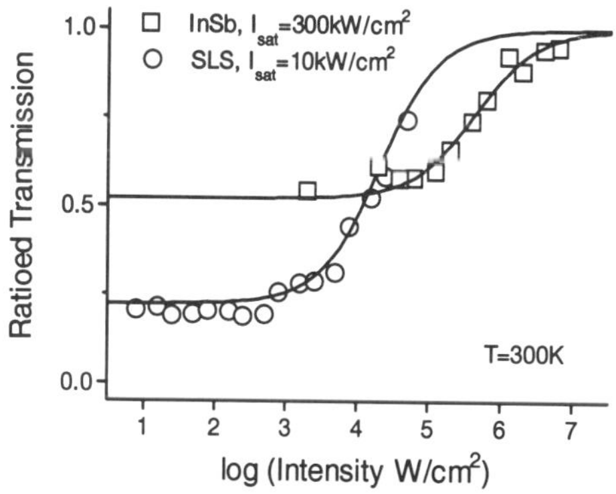


**Figure 3.** Sample transmission versus average pump intensity for the InSb sample at $\hbar\omega$=247meV (□) and the strained layer superlattice sample at 161meV (○). The photon energy is about 35% above the energy gap in both cases. The intensity scale is 'raw' and uncorrected for reflection/window losses etc.

alloys and our sample has $E_G \approx 120$ meV [12], and is expected to show inhibited Auger rates due to the increased difficulty of satisfying both energy and momentum conservation. A preliminary result is shown in Fig. 3, and it can be seen that the saturation intensity is between 1 and 2 orders of magnitude less than for the InSb sample, indicating a much smaller recombination rate. (The bleaching recovery times are greater than the micropulse repetition rate (1ns) in these cases and a quasi-steady state is obtained. The relevant intensity is therefore the average macropulse intensity). For a complete analysis of this experiment it may be necessary to take into account the fact that surface morphology causes considerable scattered light loss in the SLS samples, hampering absolute transmission measurements. Pump-probe measurements on this and other SLS samples are planned for the near future to explore more precisely this inhibition of Auger scattering.

**Acknowledgements.** The authors would like to thank P. Fauchet for helpful discussions. This work was performed as part of the research programme of the 'Stichting voor Fundamenteel Onderzoek der Materie' (FOM) with financial support from the 'Nederlandse Organisatie voor Wetenschappelijk Onderzoek' (NWO). One of the authors (CJGML) is grateful to the EPSRC for financial support, and CMC is grateful to the DRA, Malvern, for a CASE studentship.

## References


[1] SR Geenfield *et al*, *Chem. Phys. Lett.*, **170**, 133 (1990); D Zimdars *et al*, *Phys. Rev. Lett.*, **70**, 2718 (1993).
[2] M Helm *et al*, *Appl. Phys. Lett.*, **63**, 3315 (1993).
[3] BN Murdin *et al*, *Semic. Sci. Tech.*, **9**, 1554 (1994).
[4] BN Murdin *et al*, *Opt. and Quant. Electr.*, **24**, 22 (1993); BN Murdin *et al*, *Optical Materials*, **2**, 89 (1993); BN Murdin *et al*, *Nucl. Instr. and Meth. in Phys. Res.*, **A341**, 165 (1994).
[5] AM White, *Infrared Phys.*, **25**, 729 (1985).
[6] T Ashley *et al*, *Semic. Sci. Tech.*, **8**, 386 (1983).
[7] KL Vodopyanov *et al*, *Phys. Rev.*, **B46**, 13194 (1992).
[8] BS Wherrett *et al*, Chapter 10 in "Optical Nonlinearities and Insabilities in Semiconductors", ed. H Haug, Academic Press, pp239-272 (1988)
[9] SD Parker et al, *Semic. Sci. Tech.*, **4**, 663 (1989).
[10] A Haug, *Solid State Electron.*, **21**, 1281 (1978).
[11] PM Fauchet, *Phys. Stat. Sol. (b)*, **110**, K11 (1982).
[12] MJ Pullin *et al*, this conference (1995).

*Inst. Phys. Conf. Ser. No 144: Section 6*
*Paper presented at 7th Int. Conf. on Narrow Gap Semicond. Santa Fe 9–12 January*




# Spin-conserving and Spin-flip Transitions between hybrid magneto-electric Subbands in InAs/AlSb Quantum Wells

C. Gauer†, A. Wixforth†, J.P. Kotthaus†, M. Kubisa‡, W. Zawadzki§, B. Brar*, and H. Kroemer*

†Sektion Physik der LMU München, D-80539 München, Germany
‡Institute of Physics, Polish Academy of Sciences, 02-668 Warsaw, Poland
§Institute for Experimental Physics, Univ. of Innsbruck, A-6020 Innsbruck, Austria
permanent adress: Institute of Physics, Polish Academy of Sciences, 02-668 Warsaw, Poland
*Department of Electrical and Computer Engineering, UC Santa Barbara, CA 93106, USA

**Abstract.** Resonances between hybrid magneto-electric subbands in InAs/AlSb quantum wells have been studied in the Voigt configuration. The simultaneous observation of both spin-conserving and spin-flip transitions allows us to determine experimentally the influence of the depolarization effect on the resonance energy and the lineshape. We demonstrate that the spin-conserving transitions are driven by the in-plane magnetic field and that the relevant optical matrix elements can be deduced from the magnetic field dependence of the oscillator strength. The spin-flip resonances, on the other hand, are made possible by the bulk inversion-asymmetry of InAs.

## 1. Introduction

The collective and single-particle excitations of a two-dimensional electron gas in semiconductor heterostructures have been the subject of sustained investigations both experimentally and theoretically during the last two decades. Narrow-gap semiconductors offer interesting possibilities because the strong interband coupling results in pronounced band nonparabolicity and strong spin-orbit interaction leading to large spin g-values. Also, the effects of inversion asymmetry, as present in zinc-blende materials [1], are particularly pronounced in narrow-gap systems. It was found that this inversion asymmetry can lead to a zero-field spin-splitting [2] and to an enhanced probability for the excitation of spin-flip resonances [3].

Here, we report on the investigations of optical transitions between hybrid magneto-electric subbands (HMES) on InAs/AlSb multi-quantum wells. The HMES are formed when the magnetic field has a component transverse to the direction of the electric confinement. It is well known that intersubband transitions (IST) are usually allowed only for the light polarization parallel to the direction of electric confinement. We demonstrate that an in-plane magnetic field breaks this polarization selection rule and makes possible the observation of intersubband transitions for the polarization perpendicular to the magnetic field. Moreover, the optical matrix element can be determined by measuring the magnetic field dependence of the oscillator strength. The bulk inversion asymmetry of InAs allows the excitation of combined intersubband spin-flip transitions. The energetic difference between both resonances directly gives the depolarization shift. We also demonstrate that the depolarization strongly reduces the line-broadening caused by the non-parallel dispersion of the subbands in narrow-gap materials.

## 2. Experiment

The multi-quantum well described here was grown on a GaAs substrate followed by a 1μm AlSb buffer and a 10 period superlattice [4] to accomodate the strain between the substrate and the active layers. The 20 period multi-quantum well consisting of 15 nm InAs layers and Te-δ -doped AlSb barriers of 10 nm width had a topmost barrier of 55nm width in order to minimize band-bending effects induced by the surface charge [5]. The carrier density per well is $2.5 \cdot 10^{12} cm^{-2}$ at T = 4.2K as determined by the classical Hall effect.

Mid infrared (MIR) spectra were taken using a rapid scan Fourier transform spectrometer with the sample mounted in Voigt-geometry. Experimentally, we determined the relative change in MIR transmission $T(B \neq 0)/T(B=0)$ of the unpolarized radiation for different in-plane magnetic fields B. We also measured transmission spectra under oblique incidence ratioing against a reference substrate. All experiments were performed at low temperatures T = 4.2K with a Cu-doped germanium photoconductor as a detector.

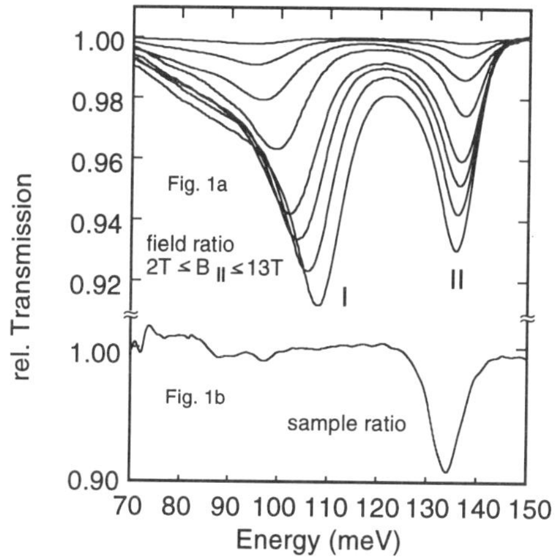


**Fig. 1:** Fig. 1a shows transmission spectra of our InAs/AlSb multi-quantum well in the magnetic field ratio at in-plane magnetic fields B=2T, 4T, 6T, 8T, 10T, 11T, 12T, and 13T. Fig. 1b is a spectrum of the same sample ratioed against a reference substrate with the radiation incident under an oblique angle.

In Fig. 1a we show the experimental data for magnetic fields between B=2T and B=13T measured in the Voigt geometry (inset of Fig. 2). The reference spectrum is the one taken at zero magnetic field. We observe two resonances: one at around E=100meV (I) and a second at E=135meV (II). As confirmed by ratioing against a reference substrate there is no resonant absorption in the quantum well at B=0T. We therefore conclude that both resonances are induced by the in-plane magnetic field and strongly gain oscillator strength at high fields. The position of line I shifts to higher energies with increasing magnetic field, whereas the position of line II remains almost constant. Line II is also much narrower and more symmetric than line I.

Fig. 1b shows a spectrum of the same sample taken at an oblique angle of 57° ratioed against a reference substrate. The resonance observed at E=135meV is identified as an IST from both its polarization dependence and theoretical calculations. Therefore we interpret resonance II in Fig. 1a also as an IST as the energetic position and lineshape are almost identical in both cases. However, no resonance corresponding to line I of Fig. 1a is observed in Fig. 1b at zero magnetic field.

## 3. Interpretation

Spin-conserving IST are usually shifted to energies above the single particle energy differences by depolarization effects whereas spin-flip excitations are not affected by the depolarization field as revealed by inelastic light scattering experiments [7]. Calculating the depolarization shift for our system we find that it roughly corresponds to the energetic difference between lines I and II for B->0T. As the oscillator strength of both resonances I and II depends on the magnetic field in a very similar manner we conclude that line I is a combined spin-flip intersubband resonance. This interpretation is further corroborated by the linewidth of the resonances. As shown in [8] the non-parabolicity induced line-broadening is greatly reduced by the depolarization field. This is exactly what we find experimentally: Line I (interpreted as the spin-flip transition) is much broader than the spin-conserving resonance II at all fields.

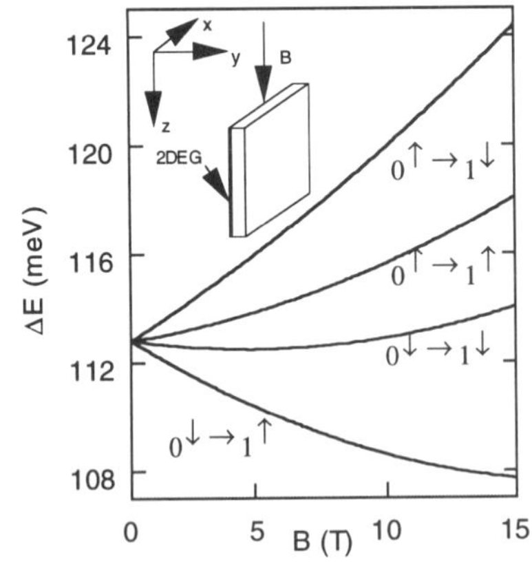


**Fig. 2:** Single particle transition energies between the HMES. The inset shows schematically our experimental geometry.

First, we explain the possibility of exciting an IST in the Voigt geometry by in-plane polarized light. Choosing the electric confinement to be in the y-direction and the magnetic field in the z-direction the electronic wavefunction for the n-th subband can be written as $\Psi_n = e^{i(k_x x + k_z z)}\varphi_n(y)$ [6]. From this general form of the wavefunction it can be seen that the interaction Hamiltonian must be a function of y in order for the intersubband matrix element to be non-zero. With the gauge $\vec{A} = (-By, 0, 0)$ the interaction Hamiltonian for light polarized in the x-direction $\alpha_x$ is found to be $H_{\text{int}} = \frac{e}{m^*}\alpha_x(p_x - eBy)$. Here, the last term is y-dependent for B≠0, resulting in a matrix element proportional to the magnetic field

$$\langle 1|H_{\text{int}}|0\rangle_x = \frac{e}{m^*}\alpha_x eB\langle 1|y|0\rangle \quad (1)$$

For the polarization parallel to the magnetic field $\alpha_z$ the IST matrix element still remains zero at finite fields. It follows from the above reasoning that the intensity I of the IST should vary as $I \propto B^2$ which is what we observe experimentally for line II. This magnetic field dependence provides an alternative way to measure the optical matrix elements of the IST as shown in [9] for GaAs quantum wells. In the case of InAs, however, a quantitative analysis requires that the narrow band gap situation and in particular the energy dependent electronic mass be taken into account.

To this end we calculate the electron energies for a potential well U(y) in the presence of an external magnetic field $\vec{B} \| z$. The band-strucure of InAs is accounted for within the three-level $\vec{k} \cdot \vec{p}$ model as described in [10]. From this calculation we find that all subbands are spin-split at finite magnetic fields. Therefore four transitions become possible between the two lowest electrical subbands: Two are spin-conserving and two involve a spin-flip. In Fig. 2 we plot the magnetic field dispersion for these transitions for $k_\|=0$. In this calculation the screening effects have been taken into account by fist order perturbation theory. It is seen that the energy of the two spin-conserving excitations has only a very weak magnetic field dependence while the spin-flip transitions are more widely separated at high fields. At high values of $k_\|$ ($k_F=4\times10^6 cm^{-1}$ in our sample) all resonance energies are strongly reduced (about 35meV at $k_\|=k_F$) as a result of the non-parallel subband dispersions. Furthermore, the magnetic field dependence also decreases drastically at high $k_\|$. This explains why we are not able to resolve all four transitions in our experiment: The nonparabolicity-induced broadening is much larger than the energetic separation of the spin-flip resonances which results in one broad line. The spin-conserving transitions, however, show only very weak magnetic field dependence and are subject to the depolarization shift, hence resulting also in a single symmetric and narrow resonance.

We now determine the oscillator strength of the spin-conserving and spin-flip transitions by calculating the matrix elements for the electric dipole excitations with the interaction Hamiltonian in the matrix form $H_{int} = eA_0 \sum_i \alpha_i \frac{\partial H_{Kane}}{\partial P_i}$. Here, $A_0$ is the amplitude of the vector potential, $P_i$ the canonical momentum with $i \in (x,y,z)$ and $H_{Kane}$ corresponds to the 8-band matrix Hamiltonian. The multi-band calculation yields for the matrix element $\langle 1|y|0\rangle$=31.0Å. Measuring the absorption under an oblique angle as shown in Fig. 1 we obtain a somewhat lower value of 24.5Å from the experiment. Evaluating (1) for the magnetic field induced IST $\langle 1|y|0\rangle$ is found to be 25.0Å thus giving almost the same value as in the conventional absorption measurement. This very good agreement proves that in the Voigt configuration an accurate determination of the matrix elements is indeed possible.

The question remains as to why the excitation of spin-flip transitions is so remarkably effective in our experiment. It is well known that strong interband $\vec{k} \cdot \vec{p}$-coupling and spin-orbit interaction result in a possibility of spin-flip transitions in narrow-gap materials. However, this mechanism is too weak to account for the strong spin-flip resonances as observed experimentally. On the other hand, it was shown theoretically by Rashba and Sheka [3] and later confirmed experimentally by Chen et al. [11] that the inversion asymmetry of the crystal also gives rise to spin-flip excitations. This mechanism should predominate at large $\vec{k}$-values as present in our sample.

In our 8x8 $\vec{k} \cdot \vec{p}$ matrix the dominant inversion asymmetry terms have the schematic form $G(P_iP_j+P_jP_i)$, where G is a material constant. These terms give only small corrections to the electron energies, but they strongly enhance the possibility of spin-flip transitions at large $\vec{k}$-values. Since the strength of the inversion asymmetry terms is not yet known for InAs we take a value similar to that of InSb $G = 3\hbar^2 / 2m_0$ [12]. We quantitatively describe the depolarization effects in our system by generalizing the approach of Ando [13] for mixed spins. In the computation we assume a level broadening $\hbar / \tau = 5meV$. While the detailed results of the theory are given in [14] we would like to present here the general features: The inversion asymmetry gives rise to spin-flip IST excited by the same light polarization $\alpha_x$ that also causes the spin-conserving transitions. Both types of excitations - spin-flip and spin-conserving - have roughly the same strength. Although we deal with four transitions, our calculations predict only two peaks in agreement with the experiment. Both the oscillator strength and the position of the resonances agree quantitatively with the experimental data.

We may add that in a recent experiment carried out in a strip-line geometry we also found spectroscopic evidence for a zero-field spin-splitting in the sample described here. The magnitude of the splitting is roughly 10meV which is in reasonable agreement with the predictions made in [2].

## 4. Conclusion

To summarize, we have observed both spin-flip and spin-conserving intersubband optical transitions in InAs/AlSb quantum wells subjected to an in-plane magnetic field. The external field makes possible the observation of spin-conserving resonances although the incident light is polarized in the plane of the two-dimensional electron gas. This mechanism also allows us to measure the matrix elements relevant for IST. The spin-flip intersubband excitations are induced by the bulk inversion asymmetry of InAs. Our results demonstrate that the depolarization field drastically reduces the linewidth of the spin-conserving transitions while the depolarization shift can be deduced from the energetic difference between the spin-flip and spin-conserving resonances.

The work in Munich was sponsored by the Volkswagen Stiftung. The Santa Barbara group gratefully acknowledges support from the Office of Naval Research and from QUEST, the NSF Science and Technology Center for Quantized Electronic Structures (Grant DMR 91-20007).

## References

[1] Kane E O 1957 *J. Phys. Chem. Solids* **1** 249
[2] Lommer G, Malcher F and Rössler U *Phys. Rev. Lett.* **60**, 728 (1992)
[3] Rashba E I and Sheka V I 1961 *Fiz. Tverd. Tela* **3** 1735; ibid. **3** 1863 (1961 *Sov. Phys. Solid State* **3** 1257)
[4] Tuttle G, Kroemer H and English J H 1989 *J. Appl. Phys:* **65** 5239
[5] Nguyen C, Brar B, Kroemer H and English J H 1992 *Appl. Phys. Lett.* **60** 1854
[6] Zawadzki W 1987 *Semicond. Sci. Technol.* **2** 550
[7] Pinczuk A and Abstreiter G 1989 in *Light Scattering in Solids V,* edited by M. Cardona and G. Güntherodt (Berlin: Springer Verlag) p 153
[8] Zaluzny M 1991 *Phys. Rev.* **B 43** 4511
[9] Gauer C, Wixforth A, Kotthaus J P, Abstreiter G, Weimann G and Schlapp W, submitted for publication
[10] Zawadzki W, Klahn S and Merkt U 1986 *Phys. Rev.* **B 33** 6916
[11] Chen Y F, Dobrowolska M, Furdyna J K and Rodriguez S 1985 *Phys. Rev.* **B 32** 890
[12] Kane E O 1980, in *Narrow Gap Semiconductors. Physics and Applications*, edited by W. Zawadzki (Berlin: Springer Verlag) p 13
[13] Ando T 1978, *J. Phys. Soc. Jpn.* **44** 475
[14] Gauer C, Wixforth A, Kotthaus J P, Kubisa M, Zawadzki W, Brar B and Kromer H, submitted for publication

*Inst. Phys. Conf. Ser. No 144: Section 6*
*Paper presented at 7th Int. Conf. on Narrow Gap Semicond. Santa Fe 9–12 January*



# Bistable Switching between Dissipative and Quantum Hall Conduction in InSb, InGaAs and GaAs

**G Nachtwei*, A Linke*, M Thiede*, C Breitlow+, H Scherer+, F-J Ahlers+, L Bliek+, H Künzel# and J Böttcher#**

*Humboldt-Universität zu Berlin, AG Magnetotransport, Invalidenstr.110, D-10115 Berlin, Germany

+Physikalisch-Technische Bundesanstalt, Bundesallee 100, D-38116 Braunschweig, Germany
#Heinrich-Hertz-Institut für Nachrichtentechnik GmbH, Einsteinufer 37, D-10587 Berlin, Germany

**Abstract.** We have investigated the non-linear current transport near the Quantum Hall Effect (QHE) breakdown at a variety of AIII-BV semiconductor interfaces (GaAs/GaAlAs and InGaAs/InAlAs heterojunction, InSb grain boundaries).

Although the material parameters as band gap and effective electron mass are rather different in the systems investigated, the switching from non-dissipative QH conduction to dissipative current transport has very striking similarities. In particular, a hysteresis of the longitudinal resistance $R_{xx}$ dependence on the filling factor near the flanks of the QH plateaux was observed in a distinct current range. The sudden changes in $R_{xx}$ were accompanied by sudden drops of the Hall resistance below quantized values in all systems.

Time resolved measurements reveal the bistable character of the dissipative and non-dissipative states. We attribute this effect to hot electrons generated in current- carrying filaments. These filaments develop below a critical electron density of about $3\times10^{15}m^{-2}$ due to impurity potential fluctuations and/or by geometrical patterning of narrow current channels.

## 1. Introduction

Since the discovery of the Quantum Hall Effect (QHE) [1] a lot of work has been done to investigate the physical limits of the effect. The precise quantization of the Hall resistance $\rho_{xy} = h/ie^2$ (h - Planck's constant, e - electron charge, i=1,2 .. integer) under quantum Hall conditions is accompanied by a vanishing resistivity component $\rho_{xx}$. However, above certain sample currents and lattice temperatures, the precise quantization of $\rho_{xx}$ breaks down and $\rho_{xx}$ takes finite values [2]. In a variety of cases, this breakdown was observed to occur suddenly at certain values of the filling factor $\nu = n_s h/eB$ ($n_s$ - sheet carrier concentration, B - magnetic field) characterised by jumps of $\rho_{xx}$ for high enough currents [3-5]. Some authors found these jumps to be quantized in units of $h/e^2$ [4,5].

Several models have been developed to explain these jumps, mainly the Quasi Inter-Landau-Level scattering (QUILLS) [6], the hot electron relaxation model [7] and the assumption of inhomogeneous carrier distribution [8]. As most of the critical current densities measured at ordinary Hall samples are considerably lower than the values predicted by the theories [6-8], a non-uniform current distribution can be assumed leading to a local enhancement of the electric field. The mechanisms leading to this development of rather narrow current paths are not fully understood, yet. However, the possible mechanisms of carrier redistributions in inhomogeneous samples [9] or below a critical carrier density [10] appear much more likely to explain the local jump-like breakdown than the edge current picture [11] (for an overview see [12]).

In this work, we observe breakdowns of the QHE accompanied by jumps in $\rho_{xx}$ and $\rho_{xy}$ at GaAs, $Ga_{0.47}In_{0.53}As$ and InSb interfaces with some similarities, especially concerning the bistable character of switching between QH- and dissipative conduction (or different QH states). We will show that these phenomena are related to geometrical constructions and/or an effective confinement of the current path below a critical carrier density.

## 2. Experimental results

First, we want to present measurements of bistable switching between QH- and dissipative conduction at narrow constrictions in GaAs/GaAlAs Hall bars ($n_s$=3.8×10$^{11}$ cm$^{-2}$, $\mu_H$=2.7× 10$^5$ cm$^2$/Vs). Near the constriction, the Hall electric field is enhanced by the ratio of the bar to constriction width. This leads to a heating effect of the conduction electrons. As shown by Komiyama et al. [7], the power balance between the power gain at the electron system by the electric field (which is mainly the Hall field $E_y \gg E_x$ ) and the power loss due to the relaxation of the overheated electron system (hot electron temperature $T_e$) to the lower lattice temperature $T_1$ :

$$\sigma_{xx}E_y^2 \cong \rho_{xx}j_x^2 = \frac{\varepsilon(T_e)-\varepsilon(T_1)}{\tau(T_e)} \qquad (1)$$

($\sigma_{xx}$, $\rho_{xx}$ are the longitudinal conductivity and resistivity, $E_y$ is the Hall electric field, $j_x$ is the current density, $\varepsilon$ (T) is the energy of the electron system per unit square at temperature T and $\tau(T_e) \cong c/T_e^2$ is the relaxation time of the heated electron system. [7])

leads to an S-shaped $U_x$-$I_x$-characteristic (or $\rho_{xx}$ vs. $j_x$ dependence). Under real conditions, this results in two different states of the system distinguished by the temperatures $T_e$ and $T_1$. If $\sigma_{xx}$ is finite at $T_e$ (overheating until breakdown), the electron system is decomposed into two states (dissipative at $T_e$ and non-dissipative at $T_1$). (Fig. 1) shows the dependence of $R_x$ vs. $j_x$ for a constriction of 200 μm and filling factors slightly below 2. A clear hysteresis between

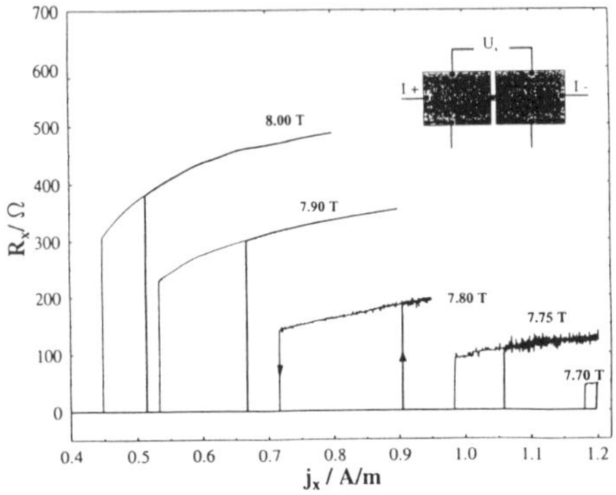


(Fig. 1) Hysteresis of $R_x(j_x)$ at different filling factors near 2 (T=50 mK, w=200 μm).Arrows mark the hysteresis direction.

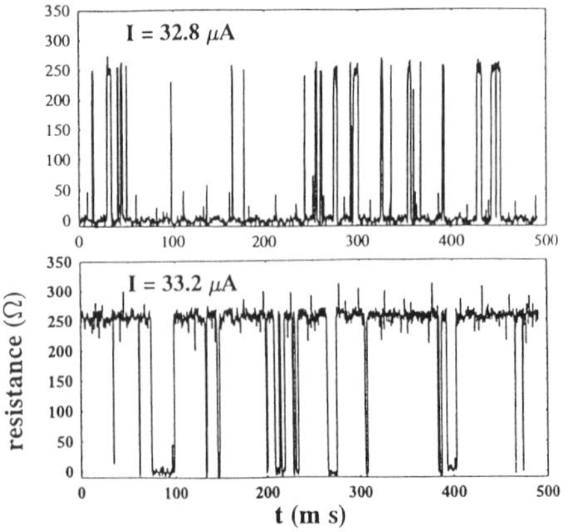


(Fig. 2) Time resolved fluctuations of $R_x$ (w=50 μm) at T=100 mK and filling factor 1.93 near the critical current.

the sudden jumps is visible and due to the coexistence of two different electron states. The $R_x$ value of the dissipative state was found to increase nearly linearly with the deviation from the integer filling factor (2-ν). No relation of the jumps in $U_x$ to $\hbar\omega_c/e$ or multiples could be proven in contradiction to [5]. Time resolved measurements revealed the bistable switching between QH and dissipative conduction (fig. 2). Near the critical current densities, the probability of the system being found in the non-dissipative states turns from about 1 to about 0. The typical times between the switches increase from a few ms (w=50 μm) to seconds (w=300 μm) for increasing width. The latter case can be explained by a "greater choice" of the system to find a non-dissipative current filament in wider samples [9]. We will return to this point later in the description of the results at $Ga_{0.47}In_{0.53}As/Al_{0.48}In_{0.52}As$ heterojunctions.

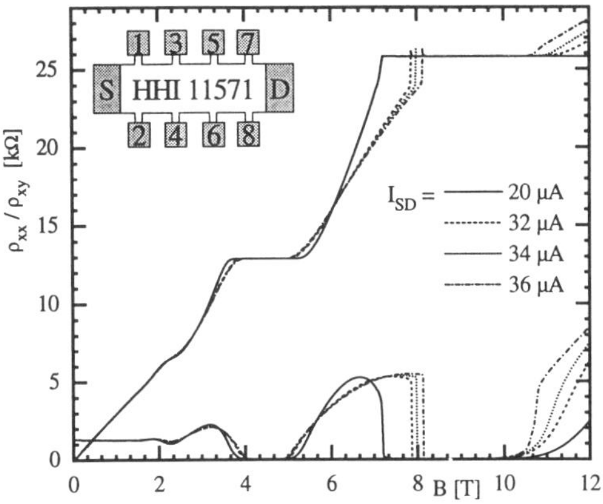


(Fig. 3) SdH (contact pair 4-6) and QHE (contact pair 3-4) for HHI 11571 (Hall-bar, $n_s$=2.2·10$^{11}$cm$^{-2}$) at T=1.3 K.

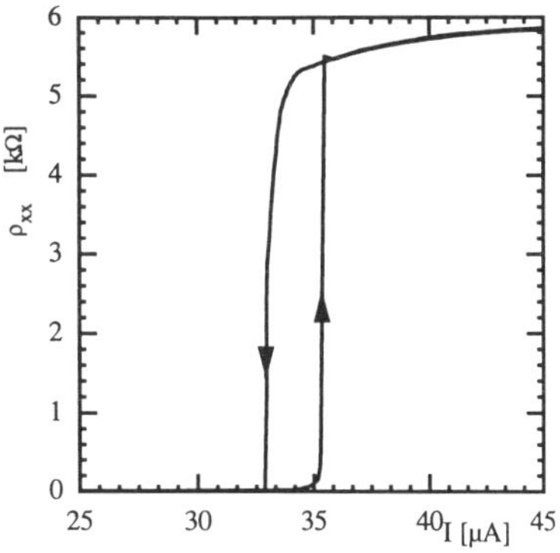


(Fig. 4) $\rho_{xx}$ in dependence on current at B=7.8 T (T=1.3 K). The hysteresis proves the bistable character of switching between QH and dissipative conduction.

We made Hall bars of 300 μm width and 400 μm spacing between the potential probes of these heterojunctions (see inset of fig. 3). Although no geometrical constriction confines the current path, sudden jumps of the resistances $\rho_{xx}$ and $\rho_{xy}$ occur at currents above 16 μA (fig. 3). These jumps mark a bistable switching between QH and dissipative conduction linked to slight changes of the filling factor ν > 1 at the low field flank of the i=1 QH plateau. The bistability is visible by a hysteresis of $\rho_{xx}$ in dependence on the magnetic field being swept up or down and a hysteresis in the $\rho_{xx}$(I) dependence similar to the GaAs structures (fig. 4). The jumping behaviour depends on the contact pair chosen. If the contacts are close to the sample corners near the current contacts where the potential drop of the source-drain resistance ($R_{SD}$) in the QH-regime occurs, the low resistance state is non zero due to local overheating of the electron system. Otherwise, the resistance $\rho_{xx}$ switches down to a value close to zero.

The carrier concentration of all GaInAs samples we investigated at which switching was observable had electron densities below $n_s$=3.03×10$^{11}$ cm$^{-2}$. The model proposed by Dorda [10] introduced a critical carrier concentration of

$$n_c = \frac{\alpha^2}{2\pi a_0^2} = 3.03\times10^{11}\mathrm{cm}^{-2} \qquad (2)$$

(with α=137, $a_0$=5.29×10$^{-11}$ m (Bohr radius)) below which the 2D system decomposes into current-carrying filaments and localized electron states. This generates an effective confinement of the current path leading to a local enhancement of the Hall field and heating effects of the electron system. The sample shown in (fig. 3) has a carrier concentration of $n_s=2.2\times10^{11}$ cm$^{-2}$ and jumps in $\rho_{xx}$ and $\rho_{xy}$ occur above I=16 µA. Another sample with $n_s=1.9\times10^{11}$ cm$^{-2}$ showed switching above I=13 µA. According to Dorda's theory, this could explained by a decreasing number of current filaments with decreasing electron density. Interestingly, we observed another type of switching (between the quantized states i=1 and i=2) in strongly inhomogeneous 2D systems at InSb grain boundaries if the carrier concentration is reduced below $3\times10^{11}$ cm$^{-2}$ by the application of hydrostatic pressure [13]. The inhomogeneity of these samples was evident from transport measurements at different pairs of contacts. As visible from (fig. 5a), a rather 'unusual' kind of breakdown occurs at a current above 18 µA. This switching is also bistable (see the hysteresis at I=20 µA) and reduces the Hall resistance from 25.813 kΩ (i=1) to a value close to the i=2 plateau. Simultaneously, a S-shaped U-I-characteristic was observed (fig. 5b), probably due to a field induced breakdown between domains of different carrier concentrations separated by depleted zones [13]. This means, the sample becomes partially insulating for an averaged $n_s=2.4\times10^{11}$ cm$^{-2}$ and strong inhomogeneities. The idea of a smooth Metal-Insulator-Transition below $n_c=3.03\times10^{11}$ cm$^{-2}$ suggested by Dorda [10] is also supported by these findings.

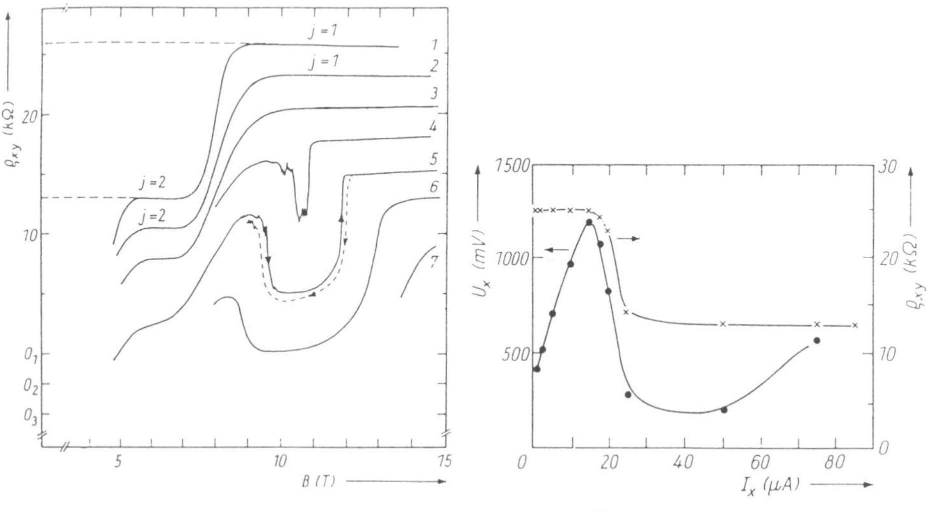


(Fig. 5) a) Hall resistance at an InSb grain boundary ($n_S=2.4\times10^{11}$ cm$^{-2}$) under pressure (580 MPa) at different currents (I=0.5, 5, 15, 18, 20, 25 and 50 µA with numbers 1-7), T=1.46 K.
b) $U_x$-$I_x$ characteristic and $\rho_{xy}(I_x)$ at B=10 T, sample as in a).

To summarize, we have observed bistable switching between QH- and dissipative conduction (GaAs/GaAlAs and GaInAs/AlInAs heterojunctions) and different QH states (InSb grain boundary). The effect leads to sudden jumps in the U-I-characteristics or the $\rho_{xx}$(B)- and $\rho_{xy}$(B)-dependences, respectively. The observed hysteresis in the corresponding dependences proves the bistability of the states involved. The effect can be explained by local overheating of the electron system either by geometrical constrictions or/and the development of narrow current-carrying filaments in inhomogeneous samples below a certain carrier density of $3\cdot10^{11}$ cm$^{-2}$.

**References :**


[1] Klitzing K v, Dorda G and Pepper M 1980 *Phys.Rev.Lett.* **45** 494

[2] Ebert G, Klitzing K v, Ploog K and Weimann G 1983 *J.Phys. C : Solid State Phys.* **16** 5441

[3] Bliek L, Braun E, Hein G, Kose V, Niemeyer J, Weimann G and Schlapp W 1986 *Semicond. Sci. Technol.* **1**, 110

[4] Bliek L, Hein G, Kose V, Niemeyer J, Weimann G and Schlapp W 1986 *Springer Series in Solid-State Sciences* **71,** 113

[5] Cage M E, Marullo-Reedtz G, Yu D Y and Van Degrift C T 1990 *Semicond. Sci. Technol.* **5**, 351

[6] Eaves L and Sheard F J 1986 *Semicond. Sci. Technol.* **1**, 346

[7] Komiyama S, Takamasu T, Hiyamizu S and Sasa S 1985 *Solid State Comm.* **54**, 479

[8] Sachrajda A S, Landheer D, Boulet S and Moore T 1989 *Phys. Rev. B* **39**, 10460

[9] Woltjer R, Eppenga R and Schnurmans M F H 1986 *Solid State Sci.* **71**, 104

[10] Dorda G 1990 *Supperlattices and Microstructures* **7**, 103

[11] Nachtwei G, Heide S, Breitlow C, Svoboda P and Cukr M 1994 *Phys.Rev.B* **50**, 8488

[12] Haug R J 1993 *Semicond. Sci. Technol.* **8**, 131 (1993)

[13] Kraak W, Nachtwei G, Herrmann R and Glinski M 1988 *phys. stat. sol. (b)* **148**, 567

*Inst. Phys. Conf. Ser. No 144: Section 6*
*Paper presented at 7th Int. Conf. on Narrow Gap Semicond. Santa Fe 9–12 January*



# Far Infrared Free Electron and "Impurity-Shifted" Cyclotron Resonance in Te-doped GaSb*

R.J. Wagner, B.R. Bennett, B.V. Shanabrook, and J.R. Waterman,

Naval Research Laboratory, Washington, DC 20375-5347, USA.

M.D. Goldenberg,

SFA, Inc., Landover, MD 20785, USA.

Abstract: Far infrared magneto-absorption studies have been performed on MBE-grown Te-doped GaSb films. At sample temperatures above 50K, free electron cyclotron resonance is observed with an effective mass of $(0.040\pm0.001)m_e$. At lower temperatures, an electron-related bound state absorption line is observed. The energy variation of this transition agrees with theoretical predictions for the 1s to 2p(m=+1) transition.

## 1. Introduction

Experimental studies of the conduction band of GaSb have been limited due to incorporation of large ($10^{16}$ $cm^{-3}$ or greater) native defect densities during both bulk and epitaxial growth processes. These defects appear as double acceptors and pin the Fermi level near the valence band edge, act as scattering centers for free carriers and seriously degrade carrier mobility. Earlier measurements of free electron cyclotron resonance were performed on p-type material either using microwave pumped cyclotron resonance [1] or magnetic field dependent radiative recombination [2]. In the work reported here, electron cyclotron resonance was observed directly in lightly Te-doped GaSb films which exhibited electron transport at temperatures below 77K. At lower temperatures, evidence of an electron-related bound state absorption could be seen.

## 2. Experimental techniques

### *2.1 Thin film growth.*

Two different molecular beam epitaxy (MBE) systems were used to prepare the films for the present study. In both cases effusion cells loaded with 8N Ga, 6N Sb and 6N GaTe with added Ga were used to provide the molecular beams. All growths were performed on semi-insulating GaAs with buffer layers of undoped GaAs and AlSb. The doped GaSb films were grown at temperatures of 450-520°C to a thickness of 4-7 μm. Typically, the Ga flux was sufficient to produce a monolayer/s of GaSb while the Sb flux was adjusted to produce an Sb-stable surface [3]. The GaTe cell temperature was between 275°C and

520°C depending on the MBE system and desired Te doping concentration [4]. In the most lightly doped material, this yielded samples with n ~ $2 \times 10^{15}$ $cm^{-3}$ and μ ~ 3000 $cm^2/V \cdot s$ at 300K. Electrons appeared to be completely frozen out at 77K. For samples with electron densities in the mid-$10^{16}$ $cm^{-3}$ range, electron mobilities of up to 6000 $cm^2/V \cdot s$ were found at 77K.

*2.2 Far infrared magneto-optical apparatus*

The far infrared apparatus included both a fast-step interferometer and a optically-pumped laser system. Most of the data presented in this paper was obtained using the interferometer in the 80-260 $cm^{-1}$ spectral region. However, the laser system was useful for circular polarization experiments and for experiments at radiation frequencies below 80 $cm^{-1}$. Magnetic fields were developed by a 0-13 T superconducting solenoid. Light pipe optics were used and a sample heating stage made it possible to vary the temperature between 4.2K and 80K. In order to emphasize the magneto-optical absorption lines, the spectra obtained using the interferometer at fields between 3 and 12 T were divided by a zero field spectrum taken with the sample in the optical beam and at the same temperature.

## 3. Experimental results

The data shown in Fig. 1 was obtained for a 3.5 μm GaSb sample with an electron concentration of $1.5 \times 10^{15}$ $cm^{-3}$ and mobility of 3000 $cm^2/V \cdot s$ at 300K. Each of the spectra

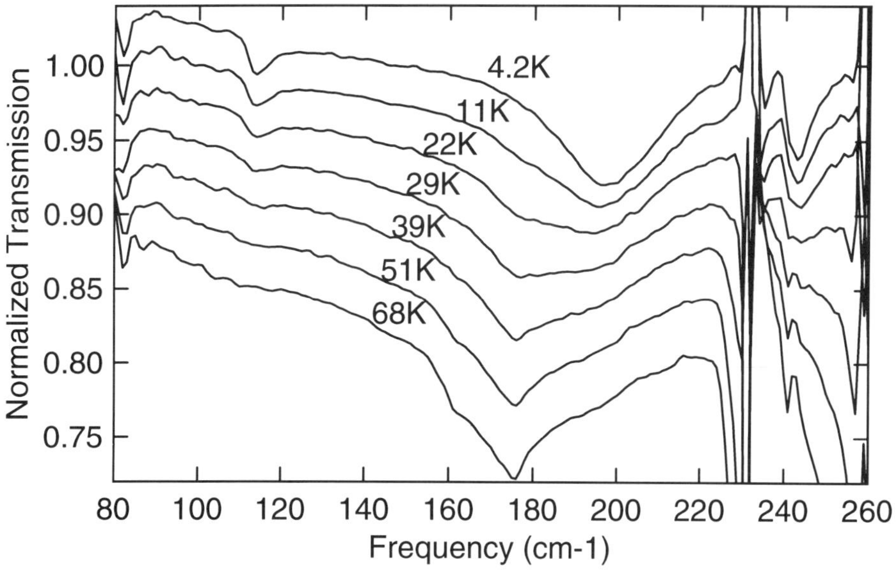


Figure 1: Absorption spectra of n-doped GaSb with a fixed magnetic field of 8T applied to the sample. Approximate sample temperatures are indicated. Each of these spectra was obtained by dividing the "raw" spectrum by a zero field spectrum. The spectra are displaced vertically for ease of viewing.

[5] for the Landau level energies: $E_L = -E_g/2 + [(E_g/2)^2 + E_g \hbar eB(L+1/2)/m_e^*]^{1/2}$. Here $E_g$ is the energy gap of GaSb, 0.81 eV, and a band edge effective mass, $m_e^*$, of $(0.040\pm0.001)m_e$, gave good agreement with the data and with earlier results [1,2]. As is apparent, the model does not fit the data perfectly. The low field interferometric data may be distorted by the normalization procedure. At higher fields, electron phonon effects may be depressing the resonance frequency relative to the calculated resonance frequency. The contributions of these various effects can only be resolved with higher quality GaSb films.

The characteristics of the absorption line observed at 4.2K resemble the behaviour of shallow donors in other small gap semiconductors such as InSb [6]. In these cases, the 1s to 2p(m=1) transition of the hydrogenic donor moves almost parallel to free electron cyclotron resonance in a high field limit. The distinction between low and high field limits is based on a comparison of the shallow donor binding energy or effective Rydberg, Ry*, to the zero point magnetic field-dependent cyclotron energy, $\hbar\omega_c/2$. The effective Rydberg is equal to $Ry_{hydrogen} m^*_e/m_e\varepsilon^2$ and the cyclotron energy is $\hbar\omega_c = \hbar eB/m^*_e$. With $Ry_{hydrogen}$= 13.6 eV, $m^*_e = 0.040m_e$, and the dielectric constant of GaSb, $\varepsilon = 15.7$, Ry* is 2.2 meV ($17.8cm^{-1}$). An effective magnetic field can be defined: $\gamma = \hbar\omega_c/2Ry^*$. At a magnetic field of 8 T, the cyclotron energy is about 20 meV ($160cm^{-1}$) and $\gamma = 4.5$. In this effective field range, two variational calculations have been performed in order to obtain the hydrogen-like energy levels. Praddaude [7] calculated the energies for 14 states at $\gamma = 1, 2$, and 3. Larsen [8], focussing on shallow donors in InSb, calculated the energies for the lowest s- and p-states down to $\gamma = 5$. The particular transition that is observed in the present experiment is from the 1s-like ground state (below the lowest Landau level) to the 2p(m=1)-like excited state (below the second Landau level). The transition energies using the results Refs. 7 and 8 are shown in Fig. 2 along with the 0 field energy. While the data compares favorably with theory at $\gamma = 5$, this is not the case for $\gamma = 1$, 2 and 3.

In addition to the energy dependence presented in Fig. 2, we have used a circular polarizer with the fixed-frequency far infrared laser system to determine that this bound state transition does indeed have the same selection rule as applies to free electron cyclotron resonance. In addition, we have searched, using relatively heavily doped samples, for other hydrogenic transitions (1s to 2p(m=-1) and 1s to 2p (m=0)) obeying different selection rules. At present, no other transition has been observed.

An alternate approach to analyzing the bound state data is to assume that the electrons are localized in potential wells due to both donor and native acceptor impurities. In undoped GaSb, hole concentrations of 1 to 2 x $10^{16}/cm^3$ are typically observed [9]. These holes are thought to arise from doubly ionizable native acceptors [10]. In order to achieve n-type behavior, comparable densities of Te donors are required. Excess donor densities in the $10^{15}$ $cm^{-3}$ to $10^{16}$ $cm^{-3}$ range contribute electrons to the two absorption lines that we report here. However, it is possible that the strong fields produced by the charged impurities can confine electrons, leading to freeze-out at relatively high temperatures (compared to shallow donor freeze-out) and to the impurity-shifted transition which we observe at 4.2K. The confining potential can be simply modeled as a 3D harmonic oscillator. With the application of a magnetic field, the harmonic oscillator level splittings proportional to $\omega_c$ develop. For the case of the lowest levels, resonant absorption should occur at $\omega(\pm) = [\omega_{ho}^2 + (\omega_c/2)^2]^{1/2} \pm \omega_c/2$ where $\omega_{ho}$ represents the spacing between adjacent harmonic oscillator energy levels. This behavior has been observed in a variety of reduced dimensional structures such as dots [11] and disks [12]. To the best of our knowledge, this phenomena has not been observed in bulk semiconductors. Nonetheless, was taken with the magnetic field at 8 T in order to avoid, at low fields, line broadening due to the fact that the angular cyclotron frequency becomes comparable to the scattering rate and, at high fields, absorption line distortion which occurs near the GaSb optical phonon band at 230 $cm^{-1}$. The data shows that the strong absorption line at 200 $cm^{-1}$ looses strength to a new absorption line at 175 $cm^{-1}$ as the sample is heated toward 68K. While the two lines are not well-resolved at intermediate temperatures, it is apparent that two distinct lines exist rather than a single line shifting from 200 $cm^{-1}$ to 175 $cm^{-1}$ as temperature is increased. The high temperature line, which will be identified as electron cyclotron resonance, does not appear to have a simple Lorentzian line shape. This may reflect sample homogeneities which occur during the growth process. An additional line at 115 $cm^{-1}$ can be seen in Fig. 1. This absorption line is near the position of electron cyclotron resonance for bulk, lightly n-doped GaAs. We have found that this line is occasionally observed in samples that have epilayers and/or heterostructures of GaSb, AlSb and InAs. The reason for its occurrence is not understood.

The magnetic field dependence of the two absorption lines which are attributed to electrons in GaSb have been studied at 4.2K and at elevated temperatures. This data is summarized in Fig. 2. The position of the absorption line that occurs for sample temperatures above 25K (□ and ◆) extrapolates to nearly zero frequency at zero magnetic field. Thus we identify this line as free electron cyclotron resonance. The dashed line near the data was calculated assuming that the resonance is due to transitions from the lowest, L = 0, to the second Landau level, L = 1, and using a two-band non-parabolic band model

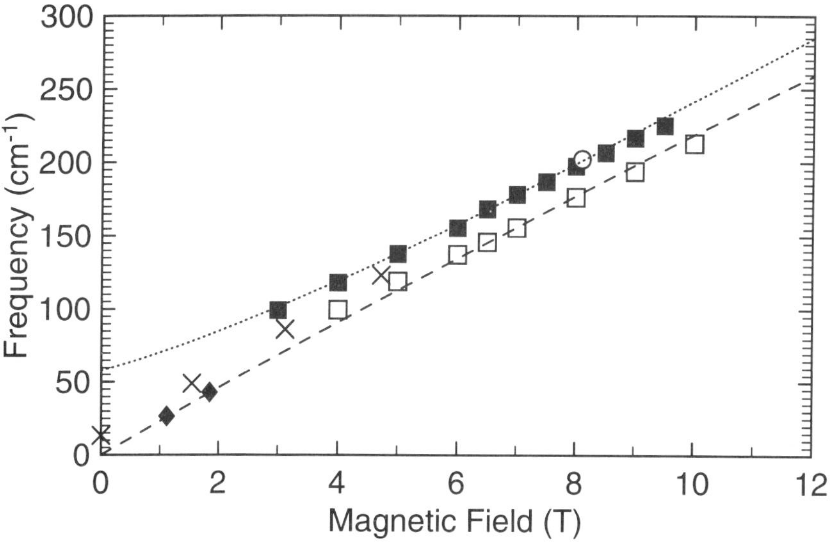


Figure 2: The field dependence of the absorption lines which occur at 4.2K (■) and above 25K (□). Included in this figure are two data points obtained with the far infrared laser system (◆). The dashed line represents the anticipated variation of cyclotron resonance for a non-parabolic electron with band edge effective mass of 0.040$m_e$. The calculated values for the 1s to 2p(m=1) hydrogenic transition by Praddaude [7] and Larsen [8] are shown by ×'s and an **O**, respectively. An x also represents the 0 field transition energy.The dotted line was calculated by assuming electron localization in a parabolic potential (See text).

we calculated the $\hbar\omega(+)$ line position (dotted line in Fig. 2) with $\hbar\omega_{ho}$ = 58 $cm^{-1}$ (7.2 meV) and $\hbar\omega_c$ determined with a constant effective mass of $0.041m_e$. It is apparent that the parabolic well localization model provides a much better overall fit to the data than the shallow donor model. Furthermore, the strong binding of the parabolic well model helps to explain the strong electron freeze-out that occurs at 77K for samples doped into the low $10^{16}$ $cm^{-3}$ range. It is surprising however, that an irregular, possibly random confining potential could lead to a well-defined resonance absorption line.

## 4. Conclusion

This experiment has shown that, in spite of high concentrations of native acceptor defects, Te-back doping can be used to explore conduction band-related properties of electrons in GaSb. While this first report of the direct observation of free electron cyclotron resonance has not yet yielded detailed knowledge of conduction band properties, it does represent a starting point for further efforts to improve GaSb material quality for magneto-optical studies. The observation of the bound electron spectroscopic feature is interesting in its own right.

## 5. Acknowledgements

This work was supported by the Office of Naval Research. The authors wish to thank T.L. Reinecke, NRL, and H.D. Drew, University of Maryland for useful discussions concerning electron confinement in parabolic wells and W.J. Moore for his assistance with the fast-step interferometer.

## References

[1] Bimberg D and Rühle W 1974 *Proc. 12th Intl. Conf. on the Phys. of Semicon.* (Stuttgart: B.G. Teubner) 561-5
[2] Hill D A and Schwerdtfeger C F 1974 *J. Phys. Chem. Solids* **35** 1533-43
[3] Waterman J R, Shanabrook B V and Wagner R J 1992 *J. Vac. Sci. Technol.* **B10** 895-7
[4] Turner G W, Eglash S J and Strauss A J 1993 *J. Vac. Sci. Technol.* **B11** 864-7
[5] For example, see McCombe B D and Wagner R J 1975 *Adv. in Electronics and Electron Physics, Vol. 37* (New York: Academic Press) 1-78
[6] Kaplan R 1969 *Phys. Rev.* **181** 1154-1162
[7] Praddaude H C 1972 *Phys. Rev.* **A6** 1321-4
[8] Larsen D M 1968 *J.Phys. Chem. Solids* **29** 271-80
[9] Johnson G R, Cavenett B C, Kerr T M, Kirby P B, and Wood C E C 1988 *Semicond. Sci. Technol.* **3** 1157-65
[10] Noack R A, Ruhle W, and Morgan T N 1978 *Phys. Rev.* **B18** 6944-56
[11] Sikorski Ch and Merkt U 1989 *Phys. Rev. Lett.* **62** 2164-7
[12] Allen S J Jr., Störmer H L and Hwang J C M 1983 *Phys. Rev.* **B28** 4875-7

*Inst. Phys. Conf. Ser. No 144: Section 6*
*Paper presented at 7th Int. Conf. on Narrow Gap Semicond. Santa Fe 9–12 January*



# Long wavelength photoresponse of short period InAs/GaSb superlattices

M. Lakrimi, T.A. Vaughan, R.J. Nicholas, N.J. Mason, and P.J. Walker

Department of Physics, Clarendon Laboratory, Oxford-University, Parks Road, Oxford OX1 3PU, UK

**Abstract.** We have performed interband photoconductivity experiments at 2 K on a series of superlattices with different layer thicknesses. We show that the band gap is indeed tunable and measure values ranging from 5 to 20 $\mu$m for the first time on samples grown by metal-organic vapour phase epitaxy. An 8 band *k.p* calculation is used to estimate the band gap and also to fit the interband Landau level transitions throughout the semiconducting layer thicknesses range.

## 1. Introduction

Antimonides have attracted much interest recently due to their applications as environmental sensors and long wavelength detectors and emitters [1]. Using different material combinations, band gaps ranging from 1.7 $\mu$m to beyond 20 $\mu$m can easily be covered. In this report, we focus on the InAs/GaSb system which has a staggered type II band line-up. The conduction band of InAs lies below the valence band of GaSb and, for short period superlattices (SLs), the energy quantisation is such that the ground electron confinement level in the InAs layer is pushed above that of the holes in the valence band of GaSb. A positive band gap is thus established and the structure behaves like a narrow gap semiconductor, thus containing no intrinsic carriers at low temperatures. In this work, a photoconductivity study is used to demonstrate that the band gap is easily tuned to long wavelengths (small energies) by simply varying the SL period. In the second part, we discuss interband magneto-optical experiments and their quantitative interpretation using an 8 band *k.p* model and a modified two band model.

## 2. Bandgap variation with layer thickness

The samples consist of 20, 80, or 100 period superlattices grown by metal-organic vapour phase epitaxy on GaAs substrates [2]. The samples were mounted in the Faraday configuration in a superconducting magnet and photoconductivity experiments were performed at 2 K at constant magnetic field, using a Bruker infrared spectrometer [3]. At low temperatures, the SL and GaSb buffer and cap layers all freeze out, leading to high resistances and so large photoconductivity signals. Spectra from a series of samples with different period thicknesses are shown in figure 1. The traces have been ratioed against the background and the zero baselines are shown. It is clear that the absorption edge changes with layer thickness and ranges from ~5 $\mu$m (221 meV, sample 1625) to 20 $\mu$m (61 meV, sample 1698) although the spatially indirect nature of the band gap leads to only a slowly rising edge without a clear band edge feature. The layer thicknesses are given in table 1.

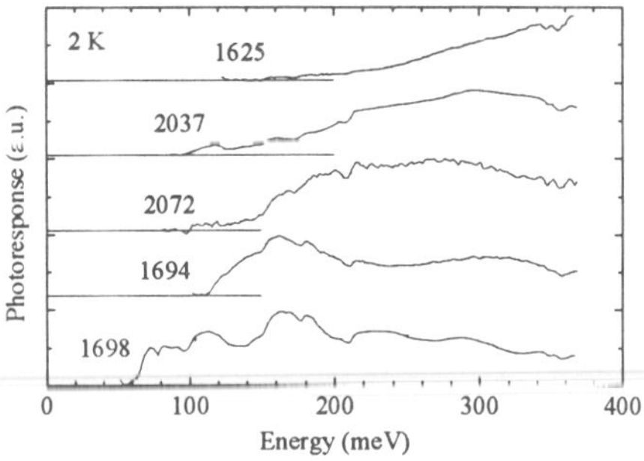


**Figure 1.** 2 K photoconductivity spectra taken at zero field for a series of samples.

**Table 1.** Superlattice period as determined by the various structural characterisation techniques along with the measured band gap at 2 K. The individual layer thicknesses for InAs and GaSb are fitted using the 8-band *k.p* model as described in the text. The interface type is also indicated.

| Sample No | No of periods | Interface type | SL period (Å) | Observed $E_g$ (meV) | *k.p* fit InAs/GaSb (Å) |
|---|---|---|---|---|---|
| 1625 | 20 | GaAs | 91 | 221 | 41/50 |
| 1694 | 20 | Alternating | 87 | 138 | 60/27 |
| 1697 | 100 | Alternating | 148 | 125 | 55/93 |
| 1698 | 100 | Alternating | 146 | 61 | 75/71 |
| 2037 | 71 | InSb | 113 | ~90-140 | 63/50 |
| 2072 | 80 | GaAs | 75 | ~150-180 | 57/18 |

## 3. 8 band *k.p* theoretical model

Early theoretical models of the InAs/GaSb system consisted of solving a two band Kane model [4, 5]. Whilst this approach provided a good general understanding of the band gaps, the treatment does not give a good description of the valence band. A three band envelope function model, incorporating the effects of strain, has been used by Roslund and co-workers to calculate the cut-off wavelength and absorption coefficient of thin period InAs/(Ga,In)Sb superlattices [6].

In the light of the present experiments, it has become important, particularly when a magnetic field is applied, to include a fuller treatment of the valence band. The details of the calculation will be published elsewhere. In short, we followed the work of Ram-Mohan *et al* [7] and used the Weiler III-V Hamiltonian [8] with the band parameters taken from reference [9]. The SL period thickness was determined by X-ray analysis, transmission electron microscopy (TEM), and from the folded longitudinal acoustic phonons [10], all methods agreeing to within 4%. Keeping the total SL period constant, adjustments are made to the individual layer thicknesses to match the measured zero-field energy gap, hence the only fitting parameter is the ratio of the layer widths. We find that layer thicknesses derived from *k.p* fitting match TEM thicknesses to ±2Å. When a magnetic field is applied, the Landau levels of the valence band are complex, as illustrated in figure 2 where several Landau levels anti-cross and repel each other.

A common problem in this materials system is the presence during growth of step bunching in which individual layer thicknesses vary laterally, while maintaining an approximately constant periodicity [2]. We have used the *k.p* model to investigate the sensitivity of the band gap to the InAs thickness for a constant SL period, the result of which is displayed in figure 3 assuming a band overlap of 140 meV. It is particularly striking that the plot exhibits a broad minimum between about 60 and 80% where the band gap varies very slowly with the InAs thickness. This result is technologically important since layer width fluctuations will have a minimal effect on the band gap in this region, and hence making growth reproducibility of these thin structures a much less critical problem.

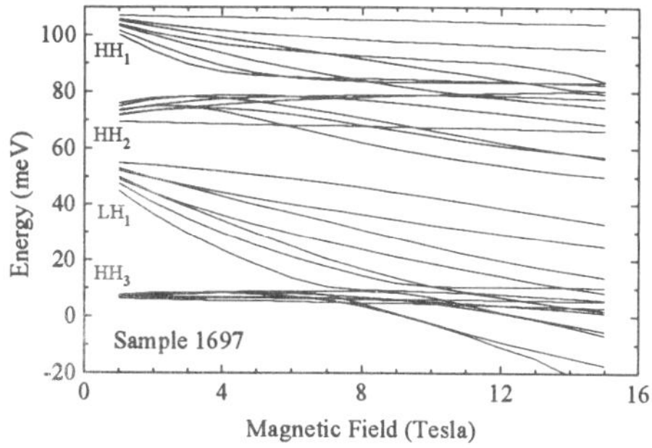


**Figure 2.** 8-band *k.p* calculation for sample 1697 showing the complexity of the valence band.

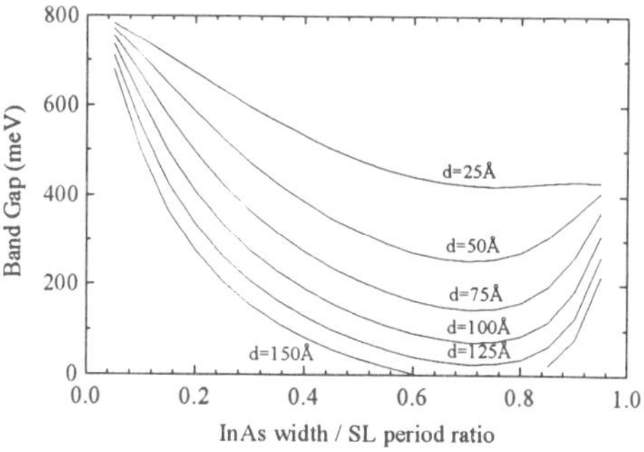


**Figure 3.** *k.p* Calculation of the bandgap as a function of InAs to period ($d$) thickness ratio.

A further possibility for optimising interband response comes from the use of the different interface configurations. Both experimental [11, 12] and theoretical [13-15] evidence suggests that the band overlap can be influenced by suitably biasing the switching sequence of the gases at the interface to engineer, ideally, a monolayer of 'InSb' or 'GaAs'. The band overlap is 40 meV larger for an 'InSb' biased interface than for 'GaAs' [11]. It follows that for the same layer thicknesses the 'GaAs' type samples will have a larger band gap compared to the 'InSb'.

In this work, in addition to having superlattices with a fixed switching sequence throughout the SL, a series of samples have also been grown with the interface type alternated so that the structural pattern is ...InAs:GaAs:GaSb:InSb:InAs..., where the growth order is from left to right. Since the band overlap at the 'InSb' interface is larger than that at the 'GaAs' interface, this leads to an asymmetric potential profile, equivalent to an internal electric field even for [001] growth. This is similar to the strain induced piezoelectric field [16] which occurs in [111] growth. However, in the resulting SL the electrons and holes are pushed towards the same interface as opposed to piezoelectric [111] structures where the carriers are confined at opposite interfaces. This should lead to significant increases in the overlap between the electron and hole wavefunctions, leading to enhanced optical matrix elements and hence increased absorption coefficients. The three structures grown with alternating interfaces showed significantly sharper band absorption edges.

## 4. Interband magneto-optics

A more accurate determination of the band gap can be achieved by studying the interband magneto-optical transitions. Typical photoconductivity spectra consisting of many clear higher order Landau level excitations, in addition to the strong band edge $n = 0$ transition, are shown in figure 4. The Landau levels are highlighted by plotting the ratio of the spectra taken with and without magnetic field. The existence of sharp features is indicative of the high structural quality of the superlattice. Samples of poor quality, for example with some degree of step bunching do not show any Landau level features, although a broad band edge can still be identified from the raw zero magnetic field spectrum.

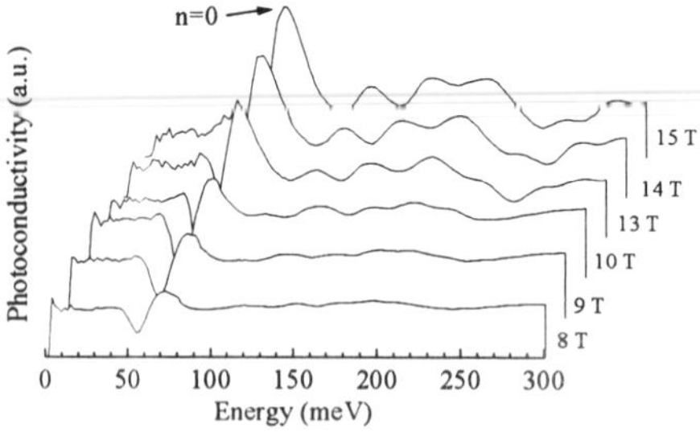


**Figure 4.** Typical spectra taken at fixed field (indicated on the right hand side) ratioed against the zero field response. The traces are offset for clarity.

Initially, we fitted the resonance energies to a simple two band model [3, 17] in which most of the non-parabolicity is due to the conduction band. The fitting parameters are thus the electron and hole effective masses and the zero-field bandgap. For longer period superlattices the experimental data could be fitted well, as shown in figure 5. The analysis for this sample yields a band gap of 61 meV corresponding to a wavelength of 20 $\mu$m. This sample has been studied by magnetotransport in the high temperature range where carriers are thermally activated and the data was found to extrapolate very well to this low temperature band gap value [18]. However, for some samples valence band interactions give rise to Landau level transitions that the two band model cannot reproduce.

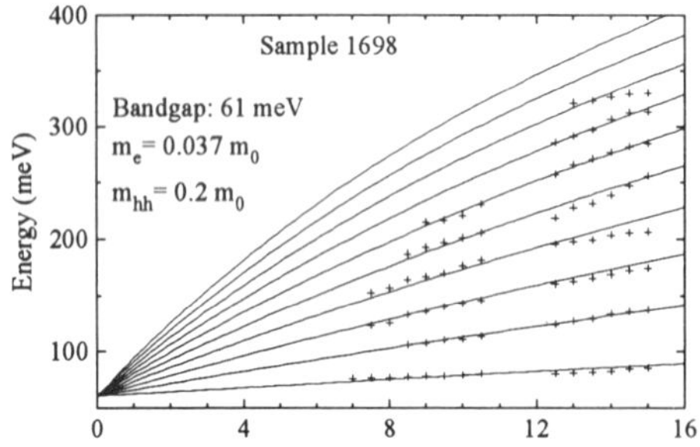


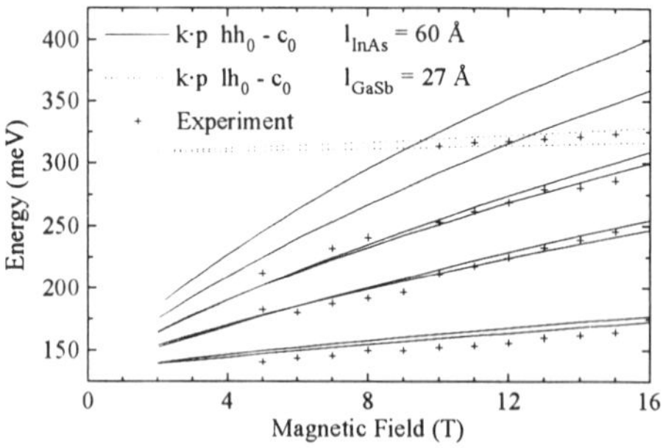


**Figure 5.** Fitted Landau level resonances using the modified bulk exciton model.

**Figure 6.** $k.p$ fit to the data for SL 1694 which also gives the electron to light hole transition.

The 8 band calculation described above which included the considerable interaction between the light and heavy holes in the GaSb valence band was extended to include the effect of magnetic field. This was found to fit the data from both the shorter and longer period superlattices. In addition, the 8 band model allowed us to identify the first electron to light hole transition, as shown in figure 6, providing further confirmation of the accuracy of the model. The double-line denotes spin-splitting of the transitions, although this was not resolved in the experimental data.

## 5. Conclusion

In conclusion, we have shown that the bandgap in InAs/GaSb superlattices is tunable and can be measured accurately using interband magneto-optics. An 8 band *k.p* calculation shows that the effect of layer width fluctuations and reproducibility in the growth can be improved by growing the InAs layer with a thickness of between 60 and 80% of the superlattice period. Mixed interfaces increase the optical matrix elements and thus may improve the efficiency of antimonide based detectors and emitters.

## Acknowledgements

The authors wish to thank the EPSRC (UK) for continued financial support, and K. Belcher and S. Moulder for technical assistance. One of us (TAV) wishes to acknowledge financial support from the National Science Foundation (USA). The authors are also indebted to B.I. Lake, R.A. Cowley for X-ray diffraction, and I. Murgatroyd and G.R. Booker for transmission electron microscopy characterisations.

# References

[1] For a review see Mikhailova M P and Titkov A N 1994 Semicond. Sci. Technol. **9** 1279. Aardvark A, Allogho G G, Bougnot G, David J P R, Giani A, Haywood S K, Hill G, Klipstein P C, Mansoor F, Mason N J, Nicholas R J, Pascal-Delannoy F, Pate M, Ponnampalam L, and Walker P J 1993 Semicond. Sci. Technol. **8** S380.

[2] Booker G R, Klipstein P C, Lakrimi M, Lyapin S, Mason N J, Nicholas R J, Seong T Y, Symons D M, Vaughan T A and Walker P J 1995 J. Cryst. Growth **145** 778.

[3] Vaughan T A, Lakrimi M, Nicholas, R J, Mason N J, and Walker P J 1995 *The physics of semiconductors*, editor Lockwood D J (Singapore: World Scientific), in press.

[4] Sai-Halasz G A, Esaki L, and Harrison W L 1977 Appl. Phys. Lett. **30** 651.

[5] Miles R H, Chow D H, Schulman J N, Collins D A, and McGill 1990 Appl. Lett. **57** 801.

[6] Roslund J H and Andersson T G 1994 Superlatt. and Microst. **15**, in press

[7] Ram-Mohan L R, Yoo K H and Aggarwal R L 1988 Phys. Rev. **B38** 6151.

[8] Weiler M H 1981 *Semiconductors and semimetals, vol 16*, edited by Willardson R K and Beer A C (New York: Academic Press) p119.

[9] Landolt-Börnstein 1982 *Numerical data and functional relationship in science and technology, New series group III, vol 17a*, edited by Madelung O (Berlin: Springer-Verlag), p258-278 and p297-310.

[10] Lyapin S G, Klipstein P C, Mason N J, and Walker P J 1994 Superlatt. and Microst. **15**, in press.

[11] Daley M S, Lubczyński W, Warbuton R J W, Symons D M, Lakrimi M, Nicholas R J, Mason N J, and Walker P J 1995 J. Phys. Chem. Sol., in press.

[12] Meyer J R, Hoffman C A, Shanabrook B V, Bennett B R, Wagner R J, Waterman J R, and Youndale E R 1995 *The physics of semiconductors*, editor Lockwood D J (Singapore: World Scientific), in press.

[13] Dandrea R G, Duke C B, and Zunger A 1992 J. Vac. Sci. Technol. **B10** 1744.

[14] Foulon Y and Priester C 1992 Phys. Rev. **B45** 6259.

[15] Hemstreet L A, Fong C Y, and Nelson J S 1993 J. Vac. Sci. Technol. **B11** 1693.

[16] For a review see Smith D L and Mailhiot C 1990 Rev. Mod. Phys. **62** 173.

[17] Makado P C and McGill N C 1986 J. Phys. **C19** 873.

[18] Symons D M, Lakrimi M, Nicholas, R J, Mason N J, Vaughan T A and Walker P J 1995 Phys. Rev. **B51**, in press.

*Inst. Phys. Conf. Ser. No 144: Section 6*
*Paper presented at 7th Int. Conf. on Narrow Gap Semicond. Santa Fe 9–12 January*



# Far-IR Spectroscopy of Semiconducting and Semimetallic InAs/Al$_x$Ga$_{1-x}$Sb Quantum Wells in High Magnetic Fields

J Kono and B D McCombe

Department of Physics, State University of New York at Buffalo, Buffalo, New York 14260, USA

J P Cheng

Francis Bitter National Magnet Laboratory, Massachusetts Institute of Technology, Cambridge, Massachusetts 02139, USA

I Lo*, W C Mitchel and C E Stutz

WL/MLPO and WL/ELRA, Wright Laboratory, Wright-Patterson Air Force Base, Dayton, Ohio 45433-6533, USA

**Abstract.** We present results of a detailed far-infrared magneto-optical study on a series of high-mobility InAs/Al$_x$Ga$_{1-x}$Sb ($0.1 \le x \le 1.0$) type-II single quantum wells. Semiconducting samples ($x \ge 0.4$), in which only 2D electrons exist in the InAs wells, exhibited cyclotron resonance (CR) splittings due to large nonparabolicity. Semimetallic samples ($x = 0.2$ and 0.1), in which both 2D electrons and 2D holes are present, showed a variety of absorption lines, including "X-lines" that we have ascribed to intra-exciton transitions and electron-CR with strongly oscillatory linewidth, amplitude, and mass, part of which are interpreted in light of peculiar "anti-nonparabolic" band structure resulting from band overlap and coupling between conduction-band states in InAs and valence-band states in AlGaSb.

## 1. Introduction

Recently there has been renewed interest in the 6.1-Å family of III-V compounds—InAs, GaSb, and AlSb—due to their interesting band-edge alignment as well as considerable improvements in the growth of these materials by molecular beam epitaxy. Interest in heterostructure systems composed of these substances and their alloys has also been stimulated by the large nonparabolicity, small mass, and large *g*-factor of the 2D carriers contained, as well as possible applications as IR detectors and sources. Recent progress in the fabrication of InAs/Al$_x$Ga$_{1-x}$Sb quantum-well (QW) structures with high-mobility (> $10^5$ cm$^2$/V-s) 2D carriers (e.g. Tuttle *et al* 1989) has led to various interesting magneto-transport and magneto-optical observations: e.g. magnetic-field-induced semimetal-semiconductor transitions (SST) (Smith *et al* 1988; Lo *et al* 1993, 1994; Barnes *et al* 1994), cyclotron resonance (CR) splittings due to nonparabolicity (Scriba *et al* 1993; Yang *et al* 1993), strong CR oscillations (Kono *et al* 1994), and new FIR lines attributed to intra-excitonic resonance (Cheng *et al* 1995).

In this report we present results of FIR magneto-optical studies on a series of high-mobility InAs/Al$_x$Ga$_{1-x}$Sb single QW's having $x$ between 0.1 and 1. Our recent studies on these samples have revealed several new phenomena (Kono *et al* 1994; Cheng *et al* 1995). For semiconducting samples ($x \ge 0.4$), we have observed spin-resolved CR, similar to results of other groups (Scriba *et al* 1993; Yang *et al* 1994). For semimetallic samples, we have observed additional FIR lines (X-lines) and strong filling-factor-dependent oscillations in the CR linewidth, amplitude and mass only when there is substantial band overlap and low enough chemical potential at low enough temperatures and high enough magnetic fields. The X-lines have been attributed to internal transitions ($1s \rightarrow 2p_+$-like) of excitons composed of spatially separated electrons and holes. We have extended these studies to higher magnetic fields (≤15T), to tilted fields (≤45°), and to include laser magneto-spectroscopy. Both CR and an electron-like-X-line shifted down in energy with a small tilt angle but the X-line was found to lose intensity rapidly with increasing angle of tilted field, while CR intensity is maintained and the CR position follows the cosine dependence on angle typical of 2D confinement. Detailed studies with an optically-pumped FIR laser have revealed several new lines, but we focus our attention on the anomalous behavior of CR in this paper. Our studies have shown that the new CR structures in laser spectroscopy and some aspects of CR previously observed in Fourier-transform spectroscopy, at least in part, result from band structure effects. Namely, the hole states in the AlGaSb barriers are coupled to the electron states in the InAs wells. This leads to a unique band structure in a magnetic field and FIR lines whose transition energy goes to zero at finite magnetic field but rapidly become cyclotron resonance-like transitions with increasing magnetic field, qualitatively consistent with the theory of Fasolino and Altarelli (1984).

## 2. Samples and experimental methods

Six samples investigated in the present work were grown by molecular beam epitaxy under the same growth conditions. Only the Al-composition $x$ of the $Al_xGa_{1-x}Sb$ barrier layers was varied: $x$ = 1.0, 0.8, 0.5, 0.4, 0.2, and 0.1. Nominal sample structure consists of a 15-nm InAs-QW sandwiched between two $Al_xGa_{1-x}Sb$ barrier layers grown on a semi-insulating GaAs substrate, with the thicknesses of the upper and lower barrier 15 nm and 3 μm, respectively. A 10-nm GaSb cap-layer was grown on top of the entire structure. Samples were not intentionally doped, but a large density of free electrons was invariably found in the InAs wells. Samples with $x \geq 0.4$ are "semiconducting," containing only 2D electrons, whereas samples with $x = 0.1$ and 0.2 are "semimetallic," containing both 2D electrons and holes. All the samples showed the negative persistent photoeffect (e.g. Tuttle *et al* 1989), which was used to control carrier density in the samples *in situ*. Fourier-transform spectrometers and an optically-pumped molecular gas laser were used in conjunction with superconducting magnets and a Bitter-type magnet to carry out the FIR magnetospectroscopy on these samples.

## 3. Experimental results and discussion

### *3-1. Fourier-transform FIR magnetospectroscopy*

Figure 1(a) displays typical transmittance spectra at several magnetic fields with $\boldsymbol{B}$ normal to the sample surface for a semimetallic sample with $x = 0.1$. At 4T CR is dominant, but it rapidly loses intensity and broadens with increasing field and becomes a shoulder of the X-line at 6T. However, CR regains intensity with further increasing field and quickly becomes dominant again at 7T. This general behavior can be also seen when the field is tilted 15° with respect to the surface normal [Fig. 1(b)] but with slightly different line shape and reduced frequencies both for CR and X, indicating the quasi-2D nature of both lines. At 45°, as shown in Fig. 1(c), the X-line dramatically shifts down in energy and loses intensity, while CR shifts down in energy as the cosine of the angle between the field and surface normal and increases in intensity. At 7 and 8T, the X-line is not detectable. Although we cannot clearly pinpoint the detailed mechanism at this point, these results strongly suggest that a large in-plane magnetic field tends to destroy the X-line in favor of CR, in contrast to a large perpendicular magnetic field, which stabilizes it (Cheng *et al* 1995).

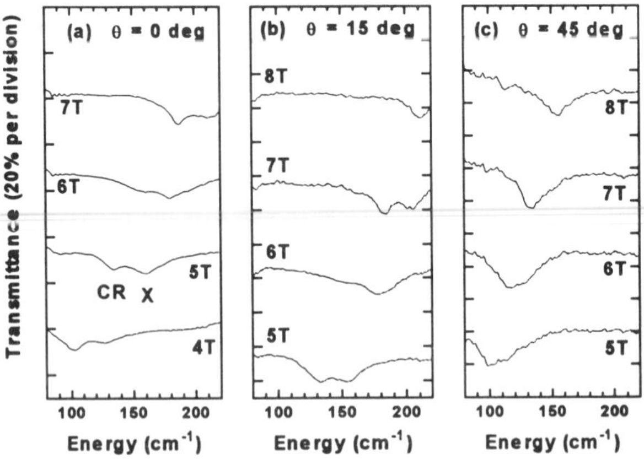


**Fig. 1** FIR transmittance spectra for a semimetallic sample ($x = 0.1$) at several magnetic fields and at different angles of the field with respect to the normal to the sample surface: (a) 0°, (b) 15°, and (c) 45°.

Figures 2(a)-2(c) show results of high-field studies on two samples ($x = 0.1$ and $0.2$). In the $x = 0.2$ sample, the overlap between the conduction-band of InAs and the valence-band of $Al_{0.2}Ga_{0.8}Sb$ is small (~50 meV), so that a magnetic-field-induced semimetal-semiconductor transition (SST) occurs at very low fields. Magnetotransport measurements (Lo *et al* 1993, 1994) have shown that the SST occurs at 2-3T, above which only 2D electrons exist in the InAs well. The plot of transition energy vs field [Fig. 2(b)] exhibits nearly linear dependence in the low-field range 1-7T, corresponding to an effective mass of $0.035m_0$, much higher than the band-edge mass ($0.023m_0$) due to nonparabolicity. In the range between 250 and 300 cm$^{-1}$, transmission experiments were impossible because of the reststrahlen band of the GaAs substrates. Figure 2(a) shows transmittance spectra for this sample at fields between 13 and 15T, where the system is semiconducting. CR shows splittings due to strong conduction-band nonparabolicity, similar to other semiconducting samples of ours and also to earlier reports

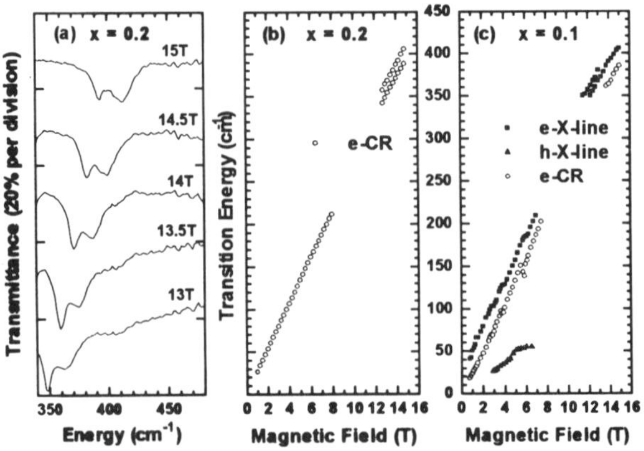


**Fig. 2** Results of high-field studies on two samples ($x = 0.2$ and $0.1$). (a) Typical transmittance spectra in the field range between 13 and 15T. Nonparabolicity-induced CR-splittings are observed. Transition energy vs field for (b) $x = 0.2$ and (c) $x = 0.1$.

(Scriba *et al* 1993; Yang *et al* 1993) on InAs/AlSb single QW's. In Fig. 2(c) transition energy is plotted against field for the $x = 0.1$ sample. X-lines, both electron- and hole-like, are observed, and the e-X-line has been studied over the entire field range. This is consistent with the fact that this sample is more strongly semimetallic than the $x = 0.2$ sample and the SST cannot occur in this field range.

### 3-2. *FIR laser magnetospectroscopy*

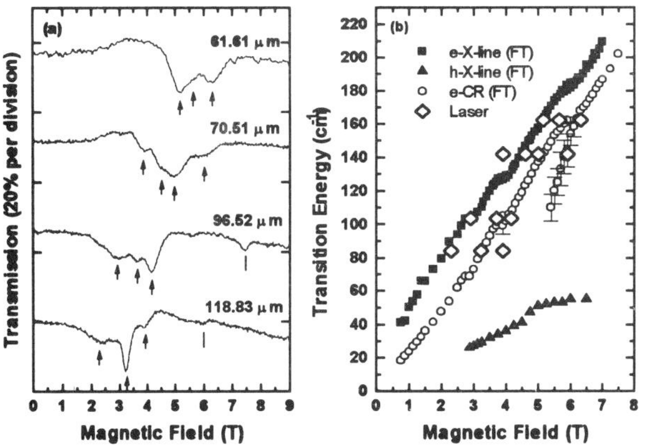


**Fig. 3** (a) FIR laser spectra for a semimetallic sample ($x = 0.1$) at four different wavelengths. High-field lines observed in lower two traces are bulk CR in GaAs. (b) Summary of peak positions vs magnetic field for both Fourier-transform (FT) and laser spectroscopy of the $x = 0.1$ sample.

FIR laser magnetospectroscopy of the $x = 0.1$ sample exhibits rich spectra as shown in Fig. 3(a). Here four traces are shown corresponding to four different FIR laser wavelengths. At least three absorption lines are clearly observed for each wavelength; we have attached arrows to only those features whose resonance fields can be easily determined. In order to determine the origin of the lines unambiguously, positions of the transmission minima are plotted in Fig. 3(b), together with results obtained by Fourier-transform magneto-spectroscopy of the same sample at the same temperature. It is clearly seen that some of the features are CR and others correspond to the e-X-line. The most important point to note is that the lines that are observed as broad low-energy tails in Fourier-transform spectra are observed clearly as minima in laser spectra. This is because these lines (one at 3.5-4T and another at 5.5-6T) have extremely large slope vs field.

Existence of lines whose transition energies extrapolate to zero at finite magnetic field but rapidly approaches CR with increasing magnetic field has been predicted by Fasolino and Altarelli (1984). However, such lines have not been unambiguously observed experimentally to our knowledge. The basic idea (Altarelli 1983) is that conduction-band states in InAs and valence-band states in (Al)GaSb are coupled through boundary conditions on the envelope functions at the InAs-(Al)GaSb interface. This coupling opens up an energy gap in the energy vs in-plane wavevector relations, making nominally "semimetallic" systems semiconducting. The Landau level spectrum based on this picture is very complex, reflecting the "anti-nonparabolic" nature as a consequence of the band-overlap, and observable features sensitively depend on the position of $E_F$ of the system, which are mostly determined by extrinsic effects such as surface donors and deep barrier states. If, as is usually the case, the system is *n*-type and $E_F$ is far above the gap, only normal CR can be observed and no unusual behavior can be expected in transition energy vs magnetic field. We believe that the small overlap in our system coupled with the position of $E_F$ has led to a situation in which the predicted lines are observed.

The strong filling-factor-dependent oscillations in CR linewidth (Kono *et al* 1994; Heitmann *et al* 1986) and the appearance of the X-lines only in "semimetallic" systems cannot be understood in terms of single-particle band-structure effects; further investigations are necessary to clarify the origin of these effects.

## 4. Conclusions

In conclusion, we have carried out FIR Fourier-transform and laser magnetospectroscopy on a set of $InAs/Al_xGa_{1-x}Sb$ single QW samples in magnetic fields up to 15T in the energy range between 10 and 450 $cm^{-1}$. Semiconducting samples, containing a high-mobility 2D electron gas, exhibited sharp CR with splittings due to large conduction-band nonparabolicity. Semimetallic samples, containing both 2D electrons and 2D holes, showed a variety of interesting phenomena, including the appearance of lines that behave like CR at high fields but whose transition energy extrapolates to zero at finite magnetic field, with an extremely large slope vs field. The observation of this behavior in CR is the first conclusive evidence of the strong coupling between conduction-band states in InAs and valence-band states in (Al)GaSb in this heterostructure system.

## Acknowledgements

This work was supported in part by the Office of Naval Research under Grants No. N00014-89-J-1673 and N00014-91-J-1939. We would like to thank M. O. Manasreh and R. K. Evans for sample preparation, R. J. Wagner for providing us with unpublished results, and B. V. Shanabrook for an illuminating question. One of us (J.K.) acknowledges financial support from the Ishizaka Foundation, Japan.

## References

*Present address: Department of Physics, National Sun Yat-Sen University, Kaohsiung 80424, Taiwan, Republic of China.

[1] Altarelli M 1983 *Phys. Rev. B* **28** 842

[2] Barnes D J, Nicholas R J, Warburton R J, Mason N J, Walker P J and Miura N 1994 *Phys. Rev. B* **49** 10474

[3] Cheng J P, Kono J, McCombe B D, Lo I, Mitchel W C and Stutz C E 1995 *Phys. Rev. Lett. in press*

[4] Fasolino A and Altarelli M 1984 *Surf. Sci.* **142** 322

[5] Heitmann D, Ziesmann M and Chang L L 1986 *Phys. Rev. B* **34** 7463

[6] Kono J, McCombe B D, Cheng J P, Lo I, Mitchel W C and Stutz C E 1994 *Phys. Rev. B* **50** 12242

[7] Lo I, Mitchel W C and Cheng J P 1993 *Phys. Rev. B* **48** 9118; 1994 *ibid* **50** 5316

[8] Scriba J, Wixforth A, Kotthaus J P, Bolognesi C R, Nguyen C, Tuttle G, English J H and Kroemer H 1993 *Semicond. Sci. Technol.* **8** S133

[9] Smith III T P, Munekata H, Chang L L, Fang F F and Esaki L 1988 *Surf. Sci.* **196** 687

[10] Tuttle G, Kroemer H and English J H 1989 *J. Appl. Phys.* **65** 5239

[11] Yang M J, Wagner R J, Shanabrook B V, Waterman J R and Moore W J 1993 *Phys. Rev. B* **47** 6807

# First observation of sharp excitonic photoluminescence from high purity epitaxial InAs

Y. Lacroix, S.P. Watkins, C.A. Tran and M.L.W. Thewalt

Department of Physics, Simon Fraser University, Burnaby, B.C., Canada, V5A 1S6

Well resolved emission lines characteristic of high purity material have been identified in the photoluminescence of epitaxial InAs at 1.4K. Exciton-polariton, neutral-acceptor-bound exciton (principal and two-hole), donor-acceptor pair band, and phonon assisted transitions are seen in InAs grown by metalorganic chemical vapour deposition on InAs substrates using tertiarybutylarsine and trimethylindium . InAs layers simultaneously grown on GaAs substrates for transport studies resulted in high Hall mobilities of up to $1.2x10^5 cm^2/Vs$ at 50K. These growths, resulting in perfectly specular surfaces, were obtained by a two step method consisting of a 100nm prelayer deposited at 400°C followed by the bulk layer at a higher growth temperature.

## 1. Introduction

While InAs is of considerable technological importance as a narrow-gap quantum well material and as a constituent of ternary and quaternary III-V alloys, surprisingly little is known of its optical and impurity properties. Few photoluminescence studies have been undertaken [1][2][3], and none have shown the sharp, readily identifiable transitions typical of InP and GaAs, hence little is known regarding impurity content or binding energy. We report here the growth, photoluminescence characterisation as well as Hall transport measurements of high purity InAs grown by metalorganic chemical vapour deposition (MOCVD). The studied samples were grown on both InAs and GaAs substrates to provide photoluminescence and Hall transport characterisation respectively [4][5].

## 2. Procedure

### *2.1 Growth*

The InAs epilayers were grown simultaneously on both GaAs and InAs substrates using a low pressure (50 Torr) vertical MOCVD reactor. The total carrier gas flow is 2.3 standard litres per minute. The gases enter a mixing chamber at the top of the reactor and then flow through a quartz nozzle at normal incidence to the wafer. A hydrogen purge flow surrounding the gas column preserves a constant velocity profile inside the reactor. The substrates are mounted on a molybdenum susceptor heated radiatively by a graphite resistance heater. The substrate temperature is controlled by an optical pyrometer which monitors the inside of the susceptor through a quartz light pipe. The quoted temperatures are based on calibrations using a thermocouple touching the susceptor top surface under typical flow conditions. Tertiarybutylarsine (TBA) at 10°C and trimethylindium (TMI) at 18°C were used as the source compounds.

Figure 1. The two step sample growth profile. A 1000 Angstrom prelayer grown at 400°C is deposited before the actual epilayer growth at 540°C. Both InAs and GaAs substrates are in the reactor chamber during growth.

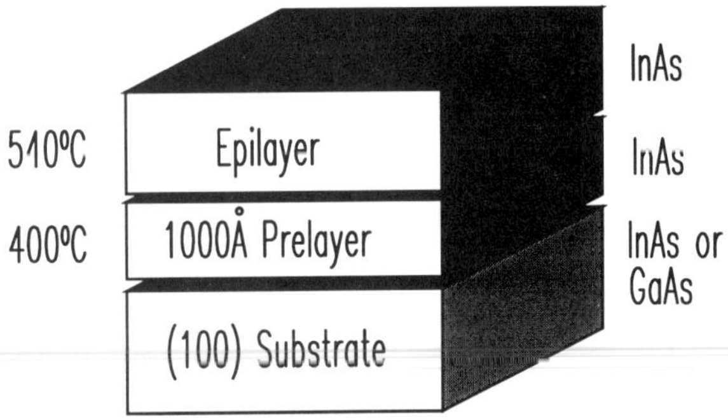


Growth conditions were found which produced optically featureless surfaces on several micrometer thick layers grown on the GaAs substrates. A two step procedure (see Figure 1) was used in which a thin 100nm nucleation prelayer was grown at 400°C at a growth rate of 0.28 nm/s followed by the thick bulk layer at a rate of 0.62 nm/s at the final growth temperature. TEM pictures show clearly that misfit dislocations are distributed in the prelayer and very few continue into the bulk layer. Using this process, several samples were grown for varying TBA partial pressures and growth temperatures [5].

*2.2 Transport measurements*

Pure indium Hall contacts were alloyed under hydrogen at 250°C to the samples grown on the semiinsulating GaAs substrates in a van der Pauw configuration. The Hall mobility and carrier sheet concentration were obtained for each growth whose parameters resulted in mirror-like surfaces. These measurements were obtained for a range of temperatures between 1.8 and 293K at a field strength of 100 Gauss.

*2.3 Optical measurements*

For photoluminescence the samples grown on InAs substrates were typically illuminated by $1mW/cm^2$ of argon ion laser light while immersed in superfluid $^4He$ at 1.4K. A Bomem DA8 Fourier transform interferometer with a $CaF_2$ beam splitter and a liquid nitrogen cooled InSb detector were used. A 2.5μm to 3.5μm band pass filter in contact with the cold detector absorbed the radiation in all but the region of interest. The sections of the light path that were not evacuated were purged with dry $N_2$ to eliminate $H_2O$ absorption. The low and high spectral resolutions used were 0.06meV and 0.025meV respectively.

Reflectivity spectra were collected using the same set-up by reflecting light from a globar source into the interferometer using the sample surface.

**3. Results**

Our results from both Hall mobility and photoluminescence measurements show that the growth procedure outlined above may be used to achieve high purity material. This section interprets these data.

*3.1 Hall mobility data*

Figure 2 shows the Hall mobility as a function of temperature for samples of different thicknesses from 0.5μm to 10μm. A 50K mobility as high as $1.2x10^5$ is seen in the thickest

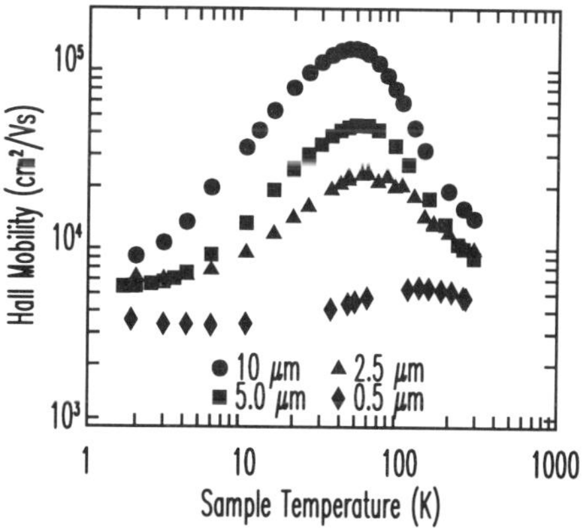


Figure 2. Hall mobility data from samples 299Ga (0.5μm), 300Ga (2.5μm), 269Ga(5.0μm) and 302Ga (10μm) between 1.8K and 293K.

sample, but the value at the low and high temperatures are more than an order of magnitude lower. At first glance, one might wrongly explain this temperature dependence as due to a change in the scattering mechanism, as is the case in some other semiconductors, and extract some impurity ionisation energy. But the fact that this temperature dependence of the mobility disappears as the layer thickness is reduced points to a different interpretation of these data [6].

The InAs surface contains a high density of surface states which causes the Fermi level to bend upon charge redistribution and form a thin conducting electron accumulation layer. Therefore the Hall measurements, neglecting the lesser effect of the substrate interface, depend on the properties of two carriers: surface electrons and bulk electrons. The measured properties will be that of either or both depending on whether the current flows at the surface, through the bulk, or a combination of the two as a function of temperature. A detailed interpretation of the transport measurements may be found in reference [5].

The actual bulk mobility is not being measured but since it is surely larger than that of the surface, we may conclude that the high mobilities obtained were underestimates of the bulk mobility since they were a combination of the accumulation layer mobility with the bulk mobility even at 50K.

### *3.2 Photoluminescence data*

Figure 3 shows a photoluminescence spectrum collected over the entire range of observed InAs band gap emissions at 1.4K and 1mW/cm$^2$ of argon illumination, for sample #269In. At lower excitation intensity, or at slightly higher temperatures, the DAP band peak was the dominant feature in the luminescence. The expected red shift and narrowing of the DAP band with decreasing excitation was observed.

A well defined step in the sample reflectivity was seen at the polariton (X) energy in a reflectance spectrum (not shown), confirming our identification of this luminescence feature. The peak of the polariton luminescence, which defines the excitonic band gap energy $E_{ex}$ occurs at 415.28±0.02meV and the acceptor-bound exciton binding energy in these samples is thus 2.11±0.03meV, the shift from $E_{ex}$ to the lowest energy principal $A^o$,X feature. The doublet signature was also observed at 383meV, although much weaker, corresponding to a longitudinal optical (LO) phonon assisted recombination of the $A^o$,X. The measured Brillouin zone centre phonon energy was 30.05±05meV which is consistent with previously reported results by Raman scattering measurement. The same doublet seen

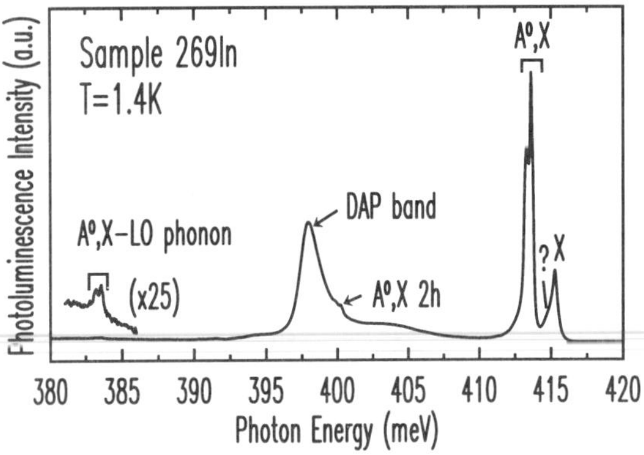


Figure 3. The photoluminescence spectrum of sample 269In at 1.4K with 1mW/cm$^2$ excitation at a spectral resolution of 0.06meV. The labelled features are the exciton-polariton (X), the neutral-acceptor-bound exciton (A°,X), its two-hole transition (A°,X 2h), its phonon-assisted transition (A°,X-LO phonon), and the donor-acceptor pair band (DAP band). The shoulder which is thought to result from unresolved neutral-donor-bound exciton transitions is marked with a question mark (?).

at 400meV on the high energy side of the DAP band must result from a two-hole transition (A°,X 2h), leaving the acceptor in an excited $2S_{3/2}$ final state.

A high resolution spectrum of the A°,X and the A°,X 2h features of sample #275In shows that the doublet can be resolved into a more complicated structure. Figure 4 shows both features superimposed on shifted energy scales to reveal that they clearly originate from the same process except for an energy shift of 13.39±0.01 meV. We attribute this shift to the $1S_{3/2}$ to $2S_{3/2}$ transition of an impurity bound hole during the exciton recombination, in good agreement with effective mass theory calculations which give 11.31 meV. The low and high energy peaks of the doublets of figure 3 are in figure 4 further resolved into A,B and C,D peaks, having energies of 413.17, 413.32 and 413.65, 413.78 meV (±0.01meV) respectively. The exact match between the A°,X and A°,X 2h fine structure, and the absence of any other observable A°,X 2h transitions, indicates that only one acceptor species is present and that all of the fine structure must originate in the initial state of both transitions, the acceptor-bound exciton.

Donor impurity content could not be investigated under the present conditions. The question mark (?) in figure 3 points to an unresolved shoulder on the low energy side of the polariton feature which we suspect is due to neutral-donor-bound exciton recombination. Lower temperatures, or possibly even lower donor concentrations, may be required in order to obtain well resolved donor-bound exciton transitions, since the carrier concentrations in these samples may still be above the density required for localisation of excitons to individual donor sites.

## 4. Conclusion

A two-step MOCVD growth process of epitaxial InAs on both InAs and GaAs resulted in optically flat surfaced high purity samples. A 100nm prelayer served to absorb almost all the misfit dislocations between the GaAs substrate and the epilayer.

Careful interpretation of the data was required due to a conductive accumulation skin layer on the surface of the InAs, but a 50K mobility of $1.2\times10^5$cm$^2$/Vs was obtained,

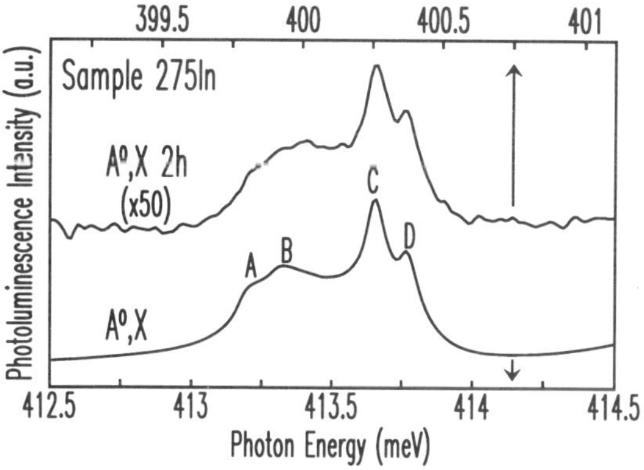


Figure 4. High resolution (0.025meV) spectra of sample 275In at 1.4K with 3mW/cm$^2$ excitation. The bottom spectrum shows the neutral-acceptor-bound exciton (A°,X) principal transition. The top spectrum show the two-hole transition of the A°,X with its energy axis shifted by 13.39meV with respect to the bottom curve. The smooth sloped background due to the high energy edge of the DAP band was subtracted from the top curve to facilitate the comparison. The labels A, B, C and D depict the main features, which are seen to be identical in the principal and two-hole transitions.

surpassing the highest reported mobilities to date by samples grown by molecular beam epitaxy [7].

The photoluminescence data identified for the first time well resolved excitonic feature that were used to extract precise information on band gap, impurity and phonon parameters. An excitonic band gap of 415.28±0.02meV was measured at 1.4K. An acceptor with a bound exciton binding energy of 2.11±0.03meV and a $1S_{3/2}$ to $2S_{3/2}$ energy difference of 13.39±0.01meV was identified in all samples studied. The $LO_{\Gamma}$ phonon energy was measured at 30.05±0.05meV.

Further photoluminescence work using magnetic fields, and possibly lower temperatures will be required to study donor-related transitions. Intentional dopings with various acceptor impurities are currently being performed so that the observed acceptor can be identified.

This work was supported by the Natural Science and Engineering Research Council of Canada.

**5. References**

[1] Mooradian A and Fan H 1964 *Proceedings of Seventh International Conference on the Physics of Semiconductors* (Paris: Academic) Vol. 4, 39
[2] Grober R D and Drew H D 1989 *J. Appl. Phys.* **65** 4079
[3] Fang Z M, Ma K Y, Cohen R M and Stringfellow G B 1991 *Appl. Phys. Lett.* **59** 1446
[4] Lacroix Y, Watkins S P, Tran C A and Thewalt M L W 1995 *Appl. Phys. Lett.* (accepted for publication)
[5] Watkins S P, Tran C A, Ares R and Soerensen G 1995 *Appl. Phys. Lett.* (accepted for publication)
[6] Weider H H 1974 *Appl. Phys. Lett.* **25** 206
[7] Wang P D, Holmes S N, Tan Le, Stradling R A, Ferguson I T and Oliveira A G 1992 *Semidond. Sci. and Technol.* **7** 767

*Inst. Phys. Conf. Ser. No 144: Section 6*
*Paper presented at 7th Int. Conf. on Narrow Gap Semicond. Santa Fe 9–12 January*



# Piezoresistance and long time relaxation effect in a two dimensional hole system at a Be-doped $GaAs/Al_{0.5}Ga_{0.5}As$ heterojunction under uniaxial stress

**O.P. Hansen[a], V. Kravchenko[b], J.S. Olsen[a], N. Minina[b], and A. Savin[b]**

[a]Oersted Laboratory, Niels Bohr Institute, University of Copenhagen, Denmark
[b]Department of Low Temperature Physics, Moscow State University, Russia

**Abstract.** The samples were grown by MBE, and uniaxial compression up to 2 kbar in the [110] direction was applied at temperatures from 1.5 K to 150 K. Long time relaxation effects in the resistivity were detected in the course of compression (loading) and unloading. By heating to 180 K and subsequent cooling, equilibrium values of resistance versus stress were established. The compression causes a decrease in resistance. At 4.2 K and 2 kbar the resistance is 35% of its value without compression. Hall effect measurements indicate a free carrier density which is only weakly (less than a few percent) dependent on the compression.

## 1. Introduction

Since the first successful MBE-growth of a two dimensional hole system at a GaAs/AlGaAs heterojunction (Stormer et al 1983) there has been an increasing interest focussed on the GaAs valence band. The biaxial stress, arising due to lattice mismatch at the heterojunction, is a useful engineering parameter in the field of semiconductor lasers (Adams 1986), and has attracted considerable theoretical interest as well.

The influence of uniaxial stress on the structure of the valence band has been investigated theoretically (Sanders and Chang 1985), and by optical measurements (Zitouni et al 1992). As far as we know, no transport measurements have been reported.

Long time relaxation effects, which resemble those we observe, are well known phenomena in the conductivity of inhomogeneous semiconductors. These effects appear after rapid cooling, after illumination, and after application of pressure (Shik and Vul' 1975, Vul' and Shik 1975, Sheinkman and Shik 1976).

## 2. Samples and experimental technique

All samples were obtained from one and the same MBE-grown wafer. The substrate was

[001] GaAs, and the following sequence of layers was grown: 1 $\mu$m GaAs, 70 Å $Al_{0.5}Ga_{0.5}As$, 500 Å Be-doped 1 x $10^{18}$ $cm^{-3}$ $Al_{0.5}Ga_{0.5}As$, and 50 Å Be-doped 2 x $10^{18}$ $cm^{-3}$ GaAs.

Samples were cracked from the wafer, after scratching with a diamond, to bars 3.0 x 0.8 x 0.625 $mm^3$.

Two types of contacts were prepared. One of these had just four parallel Mg/Au stripes evaporated directly on the sample; alloyed 1 minute at 450 $^0$C. On the other one a Hall bar mesa was first made and afterwards Zn/Au contact pads were evaporated; alloyed 3 minutes at 380 $^0$C. In both cases Au wires were bonded to the contact pads. The samples are sketched in Fig. 1a.

In Fig. 1b we show a sample placed in a stainless steel ring. Prior to the placement, the sample corners are rounded by grinding to make the sample fit tightly inside the ring. One 'ear' of the ring is fixed to the outer part of a co-axial stretching device, while a force can be applied to the other 'ear'.

The stretching device fits into a variable temperature cryostat with a superconducting magnet. The force can be varied continuously by a handle at room temperature.

The stainless steel ring - stretching device - technique was developped by Brandt et al 1985.

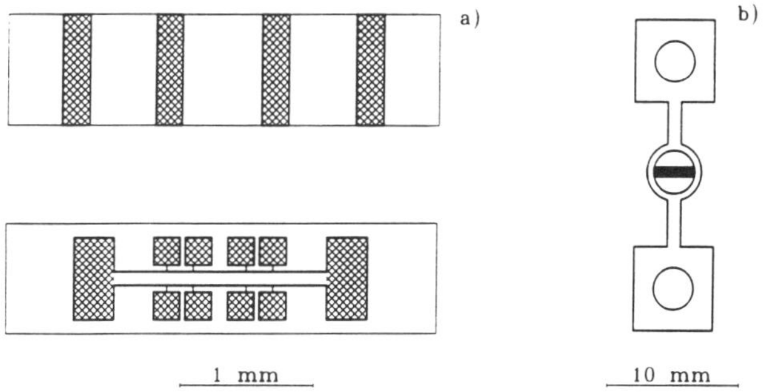


Fig. 1 a) Stripe and mesa samples. The hatched parts are the contact pads. b) Sample in stainless steel ring with 'ears' for stretching.

## 3. Experimental results

Long time relaxation effects of resistance were observed in samples with both types of contacts. They appear after compression (loading) as well as after unloading. In Fig. 2 we show the effects for a sample with Hall bar mesa. The temperature is 77 K with sample immersed in liquid nitrogen in Fig. 2a, while it is 4.2 K with sample in gaseous helium in Fig. 2b. The time sequence presents loading-unloading processes: Upon loading with 0.36 kbar at time $t_1$ the resistance displays a decrease, which overshoots and then relaxes. Unloading at time $t_2$ makes an increase in resistance, overshoot and relaxation. The sequence is repeated with a 0.73 kbar load at times $t_3$ and $t_4$. Finally, during the time interval $t_5$-$t_6$, the sample is heated to 180 K and cooled again to the measuring temperature. This thermal cycling restores closely the resistance to its value at the start of the sequence. The extend to which restoration is achieved depends on the cooling rate: The slower the cooling rate the better the restoration.

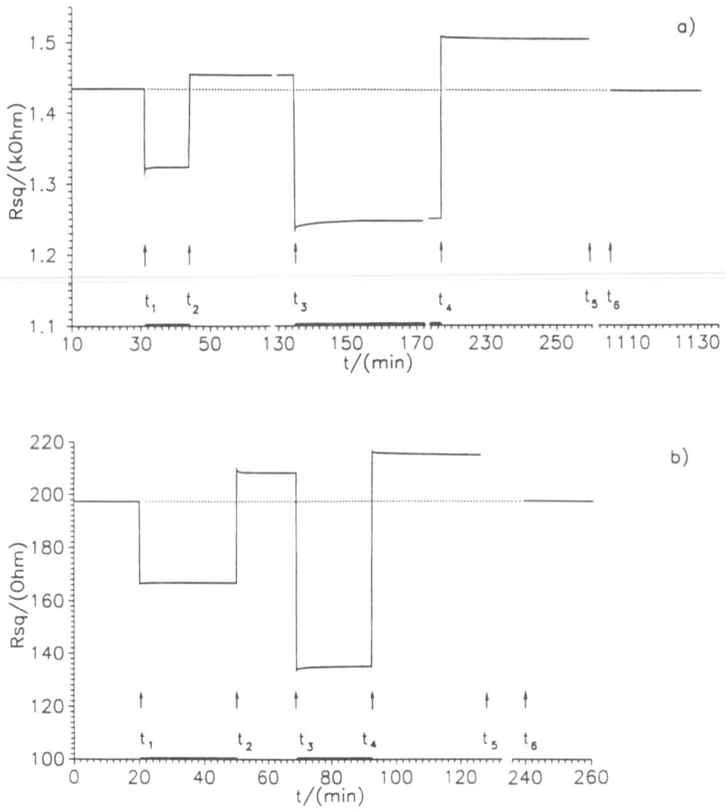


Fig. 2. Time dependence of square resistance during a 0 - 0.36 kbar - 0 - 0.73 kbar - 0 - thermal cycling sequence. a) 77 K, b) 4.2 K. Heavy lines on time axis denote periods with stress applied. See text for more details.

By the use of thermal cycling, equilibrium values of resistance versus stress were established. Results are displayed in Fig. 3 for measurements at 4.2 K on the same sample as for Fig. 2 (Hall bar mesa). In both of the figures, a) and b), the measurements start after a slow cooling from room temperature (open circle, square resistance 197 Ω). The device is then loaded stepwise. In a) each loading step was followed by unloading; thermal cycling was made only in the unloaded state. In b) unloading was used only after the first and the last step, while thermal cycling under load was made at each step in between; resistance values obtained after thermal cycling are shown by open circles. The cooling time during these thermal cyclings was approximately half an hour, i.e. much shorter than during the very first cooling from room temperature. The two curves a) and b) give closely the same values of resistance versus stress, and we consider them to represent (quasi) equilibrium values. The square resistance drops from 200 ohm at zero stress to 70 ohm at 2.2 kbar.

In order to see if the drop in resistance was caused by an increase in the density of free carriers, or by an increase in the mobility, we made Hall measurements in the liquid helium temperature range. Figure 4 shows the Hall resistance versus magnetic field at 1.5 K. Curve 1 is for zero stress, while curve 2 is for 1.6 kbar; the sample is the same as for the two previous figures. We note two features: First, the appearance of quantum Hall steps which become more pronounced on application of stress, and second, the initial slopes which are the same within a few percent. The two clear steps seen on curve 2 correspond to 8, respectively 10 filled Landau levels, while the initial slope gives a free carrier density of 9.9 x $10^{15}$ $m^{-2}$.

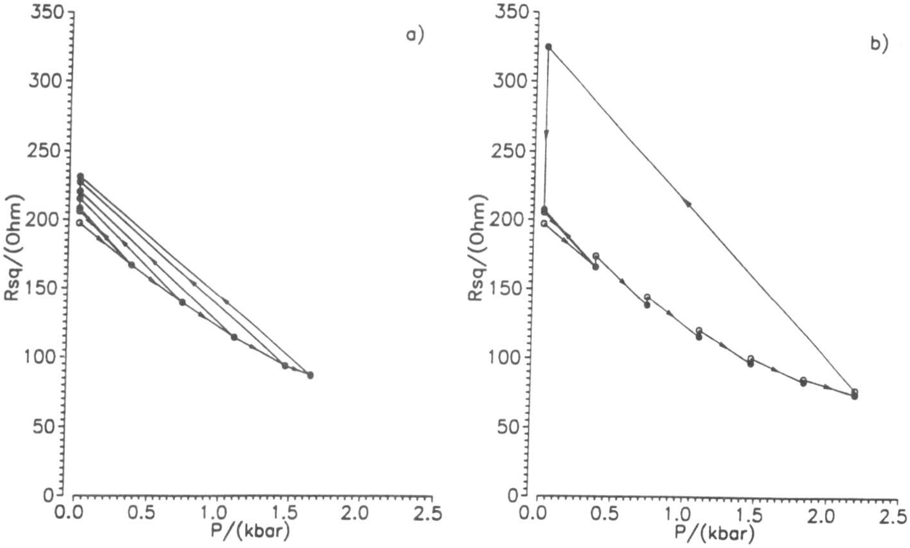


**Fig. 3. [110] stress dependence of square resistance at 4.2 K. a) Filled points at zero stress were obtained after unloading from different stresses, open points after thermal cycling. Filled points for the stressed state were obtained after loading without thermal cycling, cf. Fig. 2. b) Filled points were obtained after loading (unloading) and open points after thermal cycling without the change of load.**

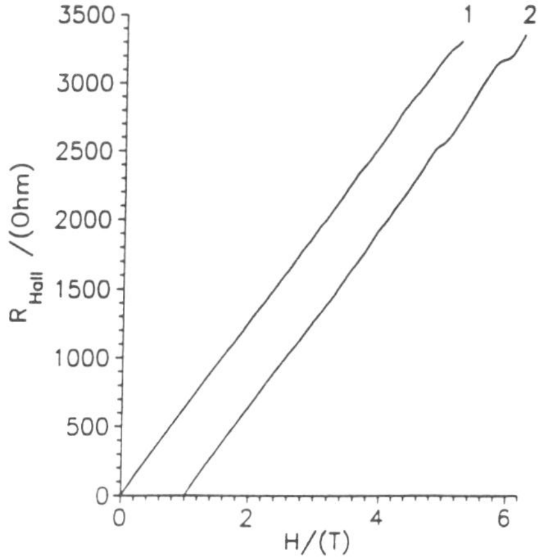


Fig. 4. Hall resistance versus magnetic field. Curve 1 is for zero stress, curve 2 for 1.6 kbar. Curve 2 is shifted 1T to the right in order to present both curves on the same graph.

## 4. Discussion

As for the long time relaxation effect we note that the ring-stretching technique has previously been used for transport investigations on a number of semimetals and narrow gap semiconductors. In non of these did long time relaxation show up (e.g.: Brandt et al 1985). Further, as it was possible by thermal cycling to establish (quasi) equilibrium values of resistance versus stress, it is our conclusion that the effect should be ascribed to the Be-doped heterostructure sample.

A possible mechanism causing long time relaxation was analysed by Shik and Vul' 1975 in terms of potential barriers at inclusion boundaries, and chemical potentials shifting differently in the inclusions and in the surrounding material when pressure was applied. However, the starting point for this type of analysis was, that long time relaxation basically represents a change in the carrier density and not in the carrier mobility. In our case, the magnitude of the change of slope of Hall resistance versus field, when the stress state is changed, is less than a few percent.

## Acknowledgements

This work has benefitted by NATO Linkage Grant 931579 and by grants 11-1080-1 and 11-1124-1 from The Danish Research Council. C.B. Sørensen is acknowledged for the sample growth and contact preparation.

## References

Adams A R 1986, *Electronic Letters* **22** 249
Brandt N B, Egorov V S, Lavrenyuk M Yu, Minina N Ya, and Savin A M 1985, *Sov. Phys. JETP* **62** 1303.
Sanders G D and Chang Y-C 1985, *Phys. Rev.* **B32** 4282
Sheinkman M K and Shik A Ya 1976, *Sov. Phys. Semicond.* **10** 128
Shik A Ya and Vul' A Ya 1975, *Sov. Phys. Semicond.* **8** 1085
Stormer H L, Schlesinger Z, Chang A, Tsui D C, Gossard A C, and Wiegmann W 1983 *Phys. Rev. Lett.* **51** 126
Vul' A Ya and Shik A Ya 1975, *Sov. Phys. Semicond.* **8** 1264
Zitouni K, Saidi N, Kadri A, El Khalifi Y, Lefebvre P, and Gil B 1992, *High Pressure Research* **9** 93

*Inst. Phys. Conf. Ser. No 144: Section 6*
*Paper presented at 7th Int. Conf. on Narrow Gap Semicond. Santa Fe 9–12 January*



# The electronic states of GaAs/AlGaAs Fibonacci quasi-periodic superlattice

**H. B. Mao, W. Lu, Z. H Ma, J. M. Zhang, S. Jiang, and S. C. Shen**
National Laboratory for Infrared Physics, Shanghai Institute of Technical Physics, Shanghai 200083, P.R.China

**Abstract.** In this paper the photoluminescence (PL), photoreflectance (PR) modulation spectroscopy and photocurrent (PC) spectroscopy were used to study the interband transitions of GaAs/AlGaAs Fibonacci quasi-periodic superlattice of order 11, and three groups of interband transitions were observed in experiments. The transfer-matrix method was used to calculate the sublevels and subbands of the Fibonacci quasi-periodic superlattice and corresponding periodic superlattice. The sublevels of Fibonacci quasi-periodic superlattice can be divided into several groups, every one of which corresponds to its relevant subband of the corresponding periodic superlattice. The theoretical calculation is in good agreement with the experimental results.

## 1. Introduction

With the development of molecular beam epitaxy (MBE) and other materials growth techniques, the periodic semiconductor superlattice (SL) has been the most intensively studied up to now. But physical systems with controlled nonperiodicity along one direction(the growth axis) can also be produced easily with such techniques, among which Fibonacci quasi-periodic superlattice is simple and more works have been done on it[1,2]. The electronic states of the nonperiodic systems are not invariant by Bloch translation, therefore the investigation on their special band structure has much significance both in theories and in experiments.

Up to now, the experimental research on Fibonacci quasi-periodic superlattice mainly concentrates on their structure[3] and lattice dynamics[1], while few research on their optical and electronic states properties has been done[4], especially the electronic states of Fibonacci superlattice of high order.

In this paper several experimental methods, such as photoluminescence(PL), photoreflectance modulation spectroscopy and photocurrent spectroscopy, were used to study the interband transition of the Fibonacci SL, and then the transfer-matrix technique was used to study its electronic states structure.

## 2. The structure and fabrication of samples

Let us recall that a Fibonacci sequence $S_l$ of order $l$ is obtained by $l$ successive applications of the transformation rule $A \to AB$ and $B \to A$[3,4]. Starting from $S_0 = A$ we obtain $S_1 = AB$, $S_2 = ABA$, $S_3 = ABAAB \cdots$ and shall also consider $S_{-1} = B$. The most important consequence of this rule is self-similarity: we obtain again sequence $S_{l-k}$ just by replacing in $S_l$ each complete $S_{k+l}$ sequence by A and each $S_k$ by B. When $k = 1$ this gives the Fibonacci recurrent relation:

$$S_l = S_{l-1}S_{l-2} \tag{1}$$

The sequence $S_l(l > 2)$ comprises $F_l$ elements A and $F_{l-1}$ elements B, $F_l$ being the $l$th Fibonacci number given by the recurrent law $F_l = F_{l-1} + F_{l-2}$ starting with $F_0 = F_1 = 1$. As $l$ increases the ratio $F_l/F_{l-1}$ converges toward $\tau = (1 + \sqrt{5})/2$ which is known as the golden mean.

The samples in our experiments were grown by Riber 32R&D MBE system on semi-insulating GaAs(001) substrate under As-rich conditions. At first a buffer layer of 5000Å thickness was grown to compensate the deformation of the substrate, and the growth was interrupted between the wells and barriers to pursue recovery of the interface flatness. The epitaxial growth rate and the aluminum composition in the barrier were determined by RHEED (reflection high energy electron diffraction) intensity oscillation. In this work each of the two basic elements

A and B of the Fibonacci SL are composed of a $GaAs$ well and a $Al_{0.31}Ga_{0.69}As$ barrier. The thicknesses of the GaAs well and the $Al_{0.31}Ga_{0.69}As$ barrier of the elements A are 40Å and 25Å, respectively. The thicknesses of the GaAs well and the $Al_{0.31}Ga_{0.69}As$ barrier the element B are 64.7Å and 40.5Å, respectively. The order of our Fibonacci SL sample is 11. Two periodic SL composed of the element A and the element B were prepared simultaneously as two reference samples, and we will name them superlattice 1 and superlattice 2, respectively, in following. To fabricate electrodes, a highly $n^+$–doped (Si, $3 \times 10^{18} cm^{-3}$) GaAs contact layer of 5000Å thickness was grown before the structure, and a highly $n^+$–doped GaAs contact layer of 500Å thickness was grown after the structure for each sample.

## 3. Experimental results

The photoluminescence (PL) spectra were measured on a SPEX 1403 double spectrometer at the temperature of 77K, and the samples were excited by the 5145Å laser line of an Ar-ion laser. Figure 1 shows the PL spectra of the Fibonacci SL sample and two reference samples. It can be seen from figure 1 that the peaks of the 11H (here 11H means the transition between the first electron sublevel or subband in the conduction band and the first heavy hole sublevel or subband in the valence band) recombination of the three samples are located at 1.568 meV, 1.619 meV and 1.570 meV, respectively. The PL spectra of the Fibonacci SL sample and superlattice 2 sample both have two peaks, in which the small peaks located above the 11H peaks can be identified as the 11L (here 11L means the transition between the first electron sublevel or subband in the conduction band and the first light hole sublevel or subband in the valence band) recombination. For the Fibonacci SL sample the width at the half-maximum of the main peak is 4.5meV, and the energy splitting between the light holes and heavy holes is 16meV.

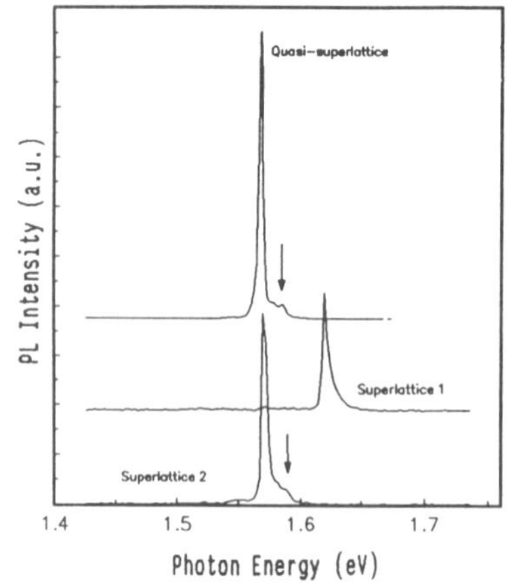


Fig.1 The photoluminescence spectra of Fibonacci quasi-superlattice and two superlattice samples at 77K.

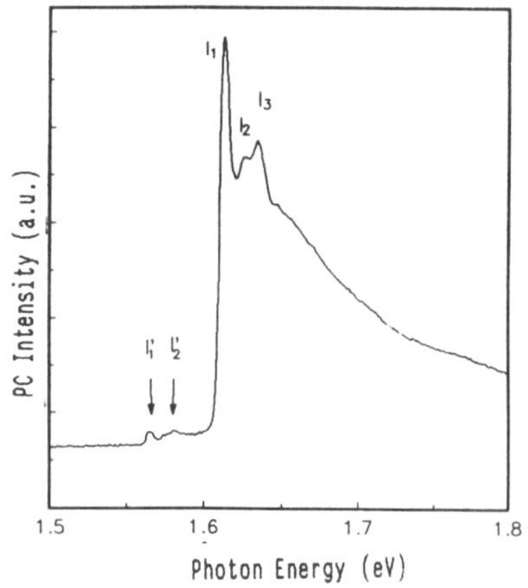


Fig.2 The photocurrent spectrum of Fibonacci quasi-superlattice at 77K.

To measure the photocurrent (PC) spectrum, which is perpendicular to the layer, the Fibonacci SL sample was processed into a mesa of $1 \times 1mm^2$ area with ohmic contact on the two highly $n^+$-doped GaAs layers. The photocurrent spectrum was measured on Nicolet Fourier Spectrometer. Figure 2 is the PC spectrum of the Fibonacci SL sample at the temperature of 77K. From figure 2 it can be seen that the PC spectrum is composed of two groups of peaks. For the first group, two small peaks ($I_1'$ and $I_2'$) located at 1.562 eV and 1.578 eV can be identified as the 11H and 11L interband transitions. For the second group three strong peaks

($I_1$, $I_2$ and $I_2$) located at 1.609 eV, 1.622 eV and 1.630 eV are observed. The behavior of this group of peaks is considerably different from the first one. Obviously, for this phenomenon there is a discrepancy between Fibonacci SL and the periodic superlattice[5].

The photoreflectance (PR) measurements were taken at room temperature from 6000Å to 9000Å, using the 5145Å line from an Ar-ion laser as a modulation source (pump beam). A tungsten lamp was used as a probe after a monochrometer, while a Si photodiode was used to detect the light reflected by the sample[6].

Figure 3 is the PR spectra of the Fibonacci SL sample and two reference samples. The signal-to-noise ratio is not very well for the PR spectra, but following results can still be obtained from the PR spectra. There is only one subband in the conduction band for superlattice 1 sample, and two structures (11H and 11L) can be identified from the PR spectra. For superlattice 2 sample there are two subbands in the conduction band, and two structures( 11H and 11L) can be identified for the first subband, while only the 22H (here 22H means the transition between the second electron subband in the conduction band and the second heavy hole subband in the valence band) structure can be identified for the second subband. The PR spectra of the Fibonacci SL sample are composed of three groups of sublevels. The signal of the first group of sublevel is not strong, from which only one structure can be identified. The signal of the second group of sublevels is strong, and three structures can be identified by line shape fit, whose energies are 1.527eV, 1.541eV and 1.550eV, respectively. The energy gap of the semiconductors will decrease with the increase of the temperature. If the effect of the temperature is taken into account, the locations of these three peaks are consistent with the three structures of the second group sublevels of the PC spectrum. And three structures can also be identified for the third group of sublevels.

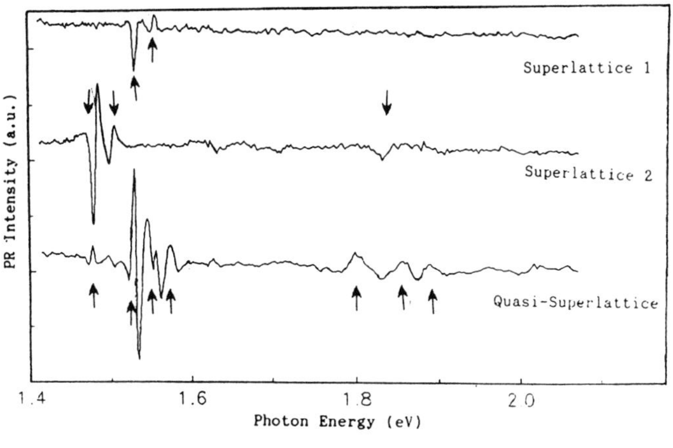


Fig.3 The PR spectra of Fibonacci quasi-superlattice and two superlattice samples at room temperature.

## 4. The theoretical calculation of the sublevels of Fibonacci SL and discussions

In the periodic superlattice system, the conduction band and the valence band are split into several subbands because of the periodic variation of the potential along the growth direction. In the Fibonacci quasi-periodic systems, though there is no periodic potential, the conduction band and the valence band will be divided into some sublevels because of the confinement of the barrier, and the property of these sublevels has much difference from the subbands of the periodic superlattice, meanwhile it relates close to the order of the Fibonacci sequence. The transfer-matrix technique is a simple but effective method for the study of the sublevels of nonperiodic systems[7]. In this paper the transfer-matrix method is used to study the sublevels of the conduction band of our Fibonacci SL sample. Laruelle *et al* have done much research on the electronic states of Fibonacci SL of low order ($l \leq 5$) both in theory and in experiment[4,8].

Therefore we will study the electronic states of Fibonacci superlattice with high order and the relation between electronic states and the parameter of the system.

Figure 4 is the electronic states of Fibonacci SL of order $l = 5, 6, 7$. From figure 4 it is seen that although the sublevels structure is complex for low order, with the increase of Fibonacci order, the distribution of the sublevels structure becomes consistent, while the width of the sublevles becomes more and more narrower. We will discuss this problem in detail by taking the following example, Fibonacci SL of order 7, which agrees with the structure parameter of our experimental sample. According to the tunneling coefficient, all sublevels can be divided into three groups. The first group is composed of several states, however, the energy splitting between them is less than 1 meV, which is hard to be distinguished in experiment. This sublevel is located at 52 meV above the bottom of the conduction band. The second group is composed of three sublevels, which are located at 82 meV, 95 meV and 109 meV, respectively, above the bottom of the conduction band. The three sublevels are also degenerate states, whose degeneracy are 8, 4 and 8, respectively. The third group is also composed of three sublevels, which are located at 202 meV, 207 meV and 212 meV, respectively, above the bottom of the conduction band.

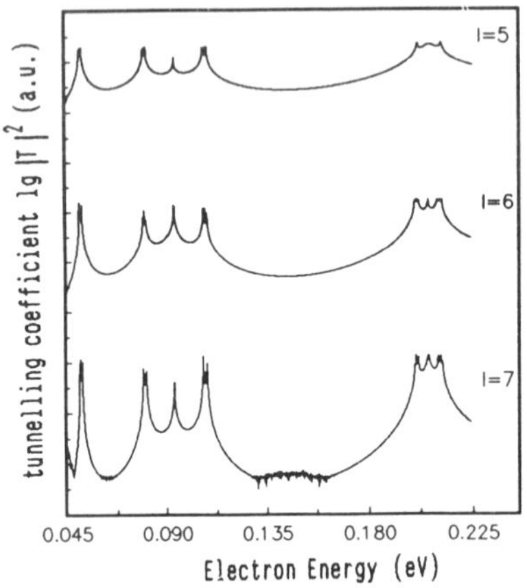


Fig.4 The conduction band structures of Fibonacci quasi-superlattice of order 5, 6, and 7.

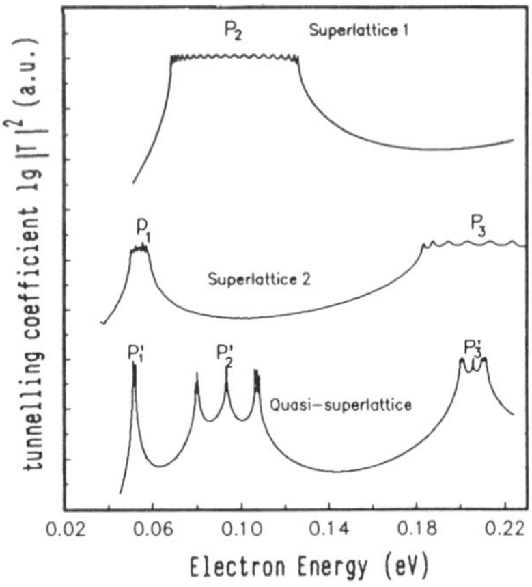


Fig.5 The conduction band structures of Fibonacci quasi-superlattice of order 6 and two superlattice samples. The sublevels groups $P_1'$, $P_2'$ and $P_3'$ correspond to subband $P_1$, $P_2$ and $P_3$, respectively.

The relation between the sublevels of the Fibonacci SL and the subbands of the periodic superlattice with corresponding parameter is an interesting question. Figure 5 is the sublevels of the Fibonacci SL of order 6 which has structure parameter of our experimental sample and two reference samples. From the results of the theoretical calculation, it is known that every group of sublevels of Fibonacci SL corresponds to its relevant subband of the periodic superlattice. For the Fibonacci SL the electronic properties are not invariant by Bloch translation along the growth direction, the subband of the periodic superlattice is split, and becomes a group of sublevels. Some of the groups of sublevels have been obviously split as the second group, while some of the groups of sublevels are not split such as the first group.

To identify the physics origin of the peaks observed in the experiments, the sublevels of the light and heavy holes are also calculated with the transfer-matrix method by making an assumption that the coupling between the light holes and heavy holes can be neglected. The calculation results show that states of the light holes and heavy holes are also composed of some groups of sublevels like the electronic states. For the heavy holes the first group of sublevels is located at 13.5 meV below the top of the valence band, while the second group of sublevels is located at 26.4 meV below the top of the valence band. Because the heavy holes have larger effective mass, the energy splitting of the second group of sublevels is still very small. For the light holes the first group of sublevels is located at 38.6 meV below the top of the valence band, while the second group of sublevels includes three split sublevels, which are located at 59meV, 72 mev and 84 meV, respectively, below the top of the valence band. The higher states of heavy holes and light holes will not be discussed in this paper.

It is the primary question to identify the three strong peaks $I_1$, $I_2$ and $I_3$ in PC spectrum of the Fibonacci SL. From above calculations we think that the three peaks are the interband transition between the second group of sublevels of the heavy holes and the three split sublevels of the second group of the electrons. Because the theoretical energy of the $I_1$ peak is 1.616 eV, which is very close to the experimental value 1.609 eV. However, the minimum energy difference between the second group of sublevels of the light holes and electrons is 1.649 eV, which has been beyond the maximum energy of the above three transition observed in experiments. Therefore the origin of the three peaks can be identified.

There are still some disagreements between the theoretical and experimental results. PL and PC spectra show that the energy splitting between the light holes and heavy holes is 16meV, however, the theoretical energy splitting between them is 23 meV. The main factor which cause this disagreement is the neglecting of the coupling between the light holes and heavy holes. In fact, this coupling is strong. In addition, the energy splitting between the 3 strong peaks in experiments does not coincide well with the theoretical results. Some important properties of the Fibonacci SL still need to be studied further. The PC spectrum of the Fibonacci SL shows that the second group of peaks (three strong peaks) are quite different from the first group of peaks: the photocurrent is abnormal strong in this energy range compared with periodic superlattice. But the physics origin is not known.

## 5. Conclusions

In this paper three spectroscopic methods (PL, PR and PC) were used to study the electronic states of the Fibonacci SL with order 11, and some new spectroscopic phenomena were observed as compared with the periodic superlattice. In the Fibonacci SL three groups of transition between the sublevels of conduction band and valence band were observed, in which the first group is not split, while the second and the third group have been split. The transfer-matrix method was used to calculate the sublevels and subbands of the Fibonacci SL and corresponding periodic superlattice. The calculated results show that with the increase of Fibonacci order, the distribution of the sublevels becomes unanimity. Its sublevels can be divided into some groups, and every group of sublevel corresponds to its relevant subbands of the corresponding periodic superlattice. For some groups of sublevels the energy splitting is very small, and in fact they degenerate into one sublevel such as the first group of the conduction band, while some group of sublevels is obviously split such as the second group of the conduction band. these results are confirmed by the experimental results very well.

## References


[1] D. C. Hurley, S. Tamura, J. P. Wolfe, K. Ploog and J. NaGle, *Phys. Rev.* B **37**, 8829(1988)

[2] Mahito Kohmoto,Bill Sutherland and Chao Tang, *Phys. Rev.* B **35**, 1020(1987)

[3] J. Todd, R. Merlin, R. Clarke,, K. M. Mohanty and J. D. Axe, *Phys. Rev.Lett.* **57**, 1157(1986)

[4] Francois Laruelle and Benard Etienne, *Phys. Rev.* B **37**, 4816(1988)

[5] R. Tsu, L. L. Chang,G. A. Sai-Halasz and L. Esaki, *Phys. Rev. Lett.* **34**, 1509(1975)

[6] S. C. Shen, *Chinese Journal of Infrared and millimeter waves*, **6**, 369(1987)

[7] M. Ya. Azbel, *Phys. Rev.* B **28**, 4106(1983)

[8] S. Ostlund and R. Pandit, *Phys. Rev.* B **29**, 1394(1984)

*Inst. Phys. Conf. Ser. No 144: Section 6*
*Paper presented at 7th Int. Conf. on Narrow Gap Semicond. Santa Fe 9–12 January*



# SUBBAND LANDAU LEVELS FROM THE MAGNETO-CAPACITANCE SPECTROSCOPIES

Kun Liu Junhao Chu Shaoling Guo Dingyuan Tang

National Laboratory for Infrared Physics
Shanghai Institute of Technical Physics, Academia Sinica
420 Zhong Shan Bei Yi Road, Shanghai 200083, China

**Abstract** The capacitance spectroscopies of an n-type InSb metal-insulator-semiconductor (MIS) structure are measured at the temperature of 1.2K with sweeping the magnetic field from 0 to 8 Tesla perpendicular to the sample surface. With a newly presented experimental model, the subband Landau levels are studied.

## 1. Introduction

Investigations of Landau levels are very important for understanding the properties of materials as well as the basic characteristics of physics[1]. They can be studied in many ways such as the magneto-transport measurement, the magneto-optical measurement, spin resonance and cyclotron resonance measurements, etc. In this paper, a newly developed experimental method, i.e. the magneto-capacitance technique, is introduced to study the Landau levels in the accumulation layer of an n-type InSb MIS structure, which proves quite simple and feasible.

## 2. Experiment

A MIS diode of n-type InSb material is fabricated by sputtering a layer of $SiO+SiO_2$ of thickness about 1800Å as an insulator with the photo-enhanced CVD method, followed by photographing Au electrodes on the insulator. Ohmic contacts are formed by using the conduction glue and Au wires. With a home-established magneto-capacitance spectrometer, the magneto-capacitance spectroscopies are measured as shown in Fig.1 at the temperature of 1.2K and in different magnetic fields from 0 to 8 Tesla. It can be seen obviously that there exist step-like capacitance oscillations in the magneto-capacitance spectroscopies. As we know, electron subbands are formed in the accumulation layer of the n-type InSb MIS structure in the electric quantisation conditions and degenerated into Landau levels when a magnetic field is applied perpendicular to the surface of the MIS diode. Thus, as each Landau level passes to the Fermi level, electrons will flush into the Landau level and contribute to the capacitance, which hence lead to the oscillations of capacitance, from which, the subband Landau levels can be calculated as described in the next section. The step-like capacitance oscillations measured in the Fig. 1 may be resulted from the poor qualities of the material and the increasing capacitance of the device in the region. The obvious capacitance oscillations have been observed in other InSb MIS devices.

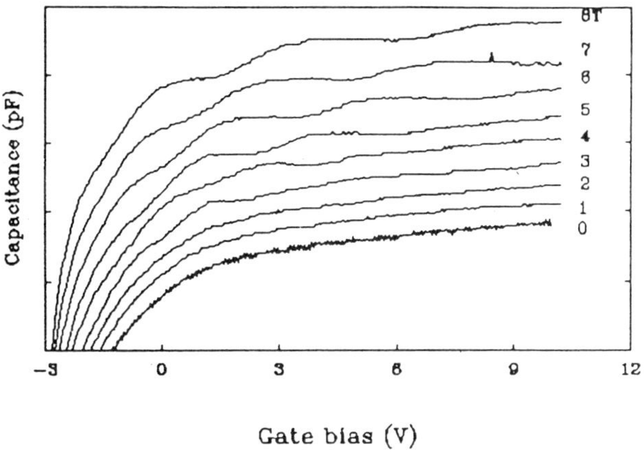


Fig.1 The measured magneto-capacitance spectra of n-type InSb MIS structure at 1.2K.

### 3. Results and discussion

Fig. 2 shows the band-bending for the n-type InSb MIS structure in the accumulation case, from which, simple equations can be figured as[2]:

$$e\Phi_s = E_{Fs} + (E_g - E_{Fb}) \qquad (1)$$

$$e\Phi_s(Vg) = \int_{V_{fb}}^{Vg} \left(1 - \frac{C(V)}{C_i}\right) dV \qquad (2)$$

where $e\phi_s(Vg)$ denotes the band-bending at the gate bias Vg and can be obtained by using formula (2). $E_{Fs}$ is the Fermi level at the semiconductor surface, while $E_{Fb}$ is the Fermi level in the bulk and can be derived from the material dopant. C(V) is the

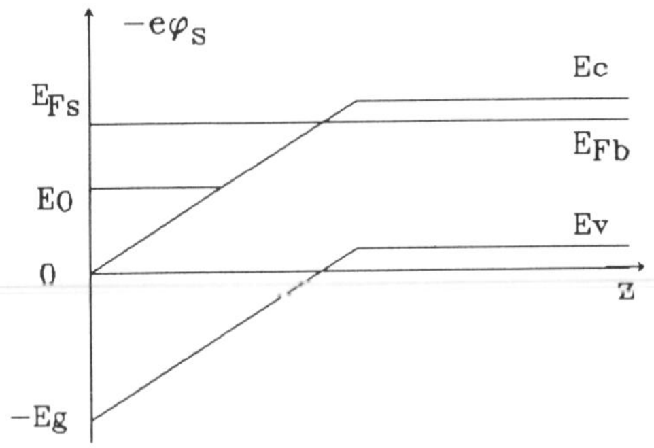


Fig.2 Band-bending structure of the n-type InSb MIS structure

capacitance of the device at the gate bias V, and $C_i$ is the capacitance of the insulator. $V_{fb}$ is the flat-band voltage of the device, and Eg is the band gap energy of the semiconductor. Thus, by using the above two formula and the measured magneto-capacitance spectra, the value of $E_{Fs}$ can be calculated, which represents different Landau levels as the capacitance oscillates. In this way, Landau levels can be obtained.

Fig. 3 shows the magnetic field dependence of gate bias when the capacitance oscillate. It is clear that the gate bias congregates to the point of (0,-3.2) as the magnetic field decreases to zero, this means that different Landau levels observed becomes to be generated into subbands. Fig.4 shows the magnetic field dependence of the electron density in the accumulation layer. Dots are the results calculated from the measured capacitance spectroscopies, lines are theoretical results with quantum numbers of 1/2, 3/2, 3, 5, etc. With this identification, the experimental results coincide well with the theoretic ones. The reason why Landau levels with other quantum numbers such as 1,2,4..., have not been observed may be that there exist resonant defect states in the conduction band and interactions among different Landau levels. Thus, Landau levels are re-composed into what we observed. Poor qualities of the material may also be one of the reason of the quantum numbers

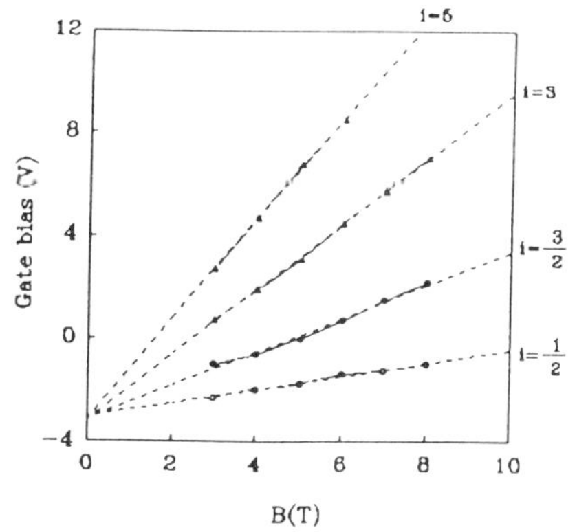


Fig 3. The magnetic dependent gate bias at the capacitance peak

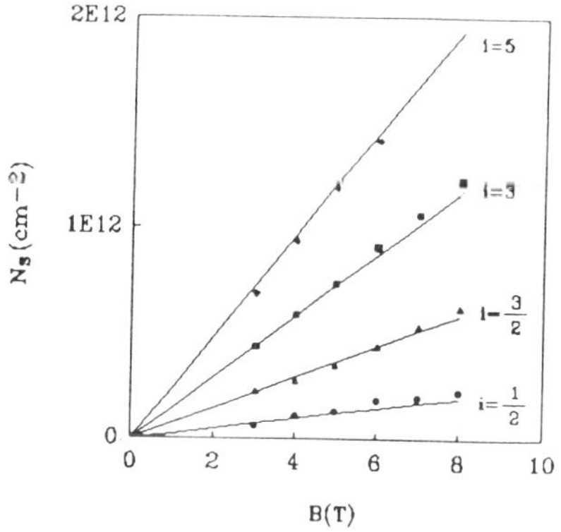


Fig. 4 The magnetic dependent electron density in accumulation layer

derived. Fig. 5 is the Landau levels calculated by using formula (1-2) and the magneto-capacitance spectroscopies shown in the Fig.1. The results show that different Landau levels congregate to the energy of 25meV as the magnetic field tends to zero, since the gate bias at the oscillation peak congregates to -3.2V with decreasing magnetic field to zero.

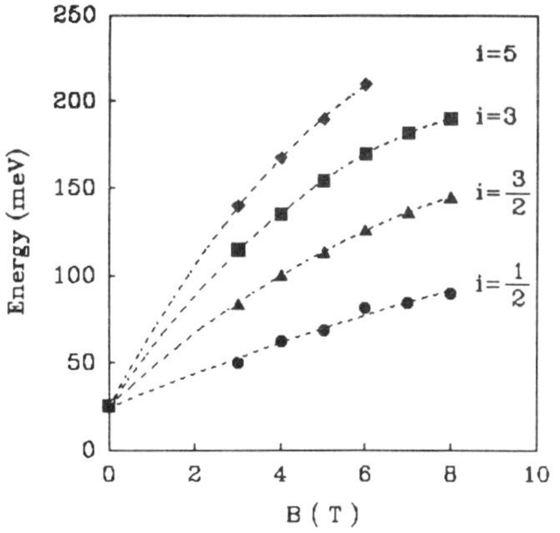


Fig. 5 The calculated Landau levels in the accumulation layer

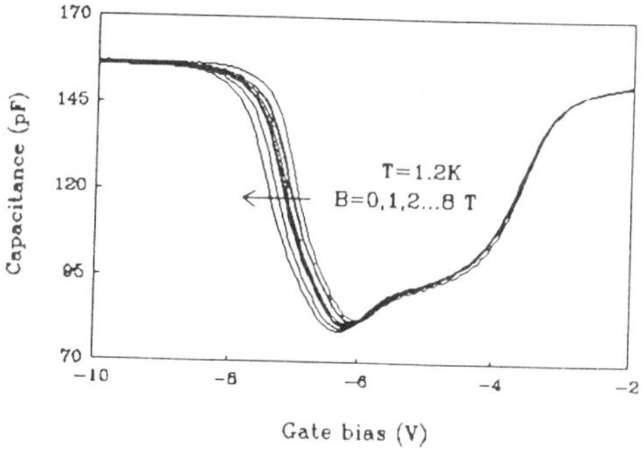


Fig. 6 The capacitance spectroscopy for the p-type inversion layer

Fig. 6 shows the magneto-capacitance spectroscopies in the p-type inversion region. As the magnetic fields increases, the capacitance edge moves to a more negative gate bias, which is due to the increment of the bottom hole Landau level with increasing the magnetic field. Thus, the onset voltage increases and the capacitance onset edge moves to a more negative gate bias. The energies of the bottom Landau level are also calculated versus the magnetic fields in reference 3.

## 4. Conclusion

The electron Landau levels in the accumulation layer of an n-type InSb MIS structure were investigated with a newly developed magneto-capacitance spectroscopy method. Landau levels with quantum numbers of 1/2, 3/2, 3 and 5 were observed, while others were not observed, which is probably due to the resonant defect states in the conduction band and interactions among Landau levels. In addition, the movement of the hole Landau levels with the magnetic field is also observed.

## References


[1] T. Ando, Review of Modern Physics, 54(2) (1982)437.

[2] Kun Liu, Junhao Chu, Haijiang Ou et al. J. Appl. Phys. 76(12) (1994)7998.

[3] Kun Liu, Junhao Chu, Haijiang Ou et al. Appl. Phys. Letts, (Unpublished).

*Inst. Phys. Conf. Ser. No 144: Section 7*
*Paper presented at 7th Int. Conf. on Narrow Gap Semicond. Santa Fe 9–12 January*



# Device physics of quantum well infrared photodetector with optical output

**V Ryzhii and M Ryzhii**

Computer Solid State Physics Laboratory, University of Aizu, Aizu-Wakamatsu, 965-80 Japan

**Abstract.** A quantum well infrared photodetector which converts long-wavelength infrared radiation into short-wavelength infrared (or visible) radiation is proposed and physics of its operation is considered. The quantum well infrared photodetector utilizes an intraband absorption of radiation in a quantum well in the emitter and interband emission in either classical or quantum well in the collector. It is shown that the efficiency of energy conversion of long-wavelength radiation into short-wavelength radiation can be about unity.

Much interest has been emerging in quantum well infrared photodetectors (QWIPs), which utilize intraband (intersubband) absorption of radiation (see, for example, review article by Levine[1] and references therein). The intraband absorption in the QWIPs is associated with electron (or hole) transitions from bound state in a quantum well into continuum states above the barriers. High performance QWIPs based on single- or multiple-QW structures have been fabricated and investigated.

For many applications, it would be desirable to integrate QWIPs with other semiconductor devices. This integration will allow to achieve compact, reliable optoelectronic components with wide functional capabilities. A QWIP integrated with a QW laser was fabricated recently[2]. However, in this integrated device the QWIP and laser operate separately.

This report deals with the QWIPs based on a heterostructure with doped emitter contact (wide gap $N^+$ region), emitter barrier (undoped wide gap $N$ region), emitter QW (narrow gap $n^+$ region), collector barrier (undoped wide gap $N$ region), collector QW (narrow gap $p^+$ region) and doped collector contact (wide gap $P^+$ region). The emitter QW is intended for absorption of long-wavelength infrared radiation connected with photoexcitation of the electrons from the bound state in the QW into continuum states. The photoemission of the electrons from the emitter QW results in lowering of the emitter barrier and injection of extra electrons from the emitter through the barriers into the collector QW. The collector QW serves as a source of short-wavelength (infrared or visible) radiation which arises due to radiative recombination of the injected electrons and holes located in this QW (see Fig.1).

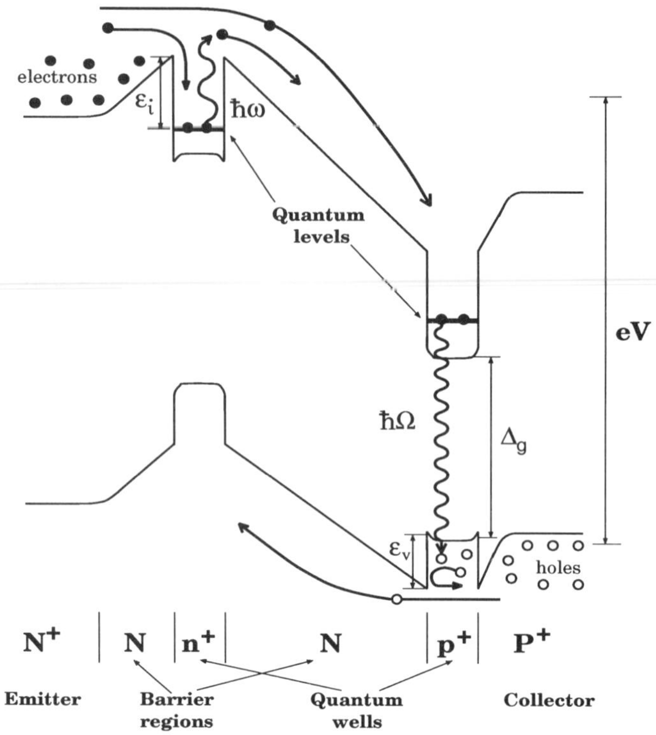


**Figure 1.** Schematic diagram illustrating the sequence of layers, band structure and nonradiative (solid trajectories) and photon-assisted (wavy trajectories) electron processes in the QWIP.

Thus input infrared signal with photon energy $\hbar\omega \sim \varepsilon_i$ (long-wavelength radiation) can be converted by the QWIP into optical signal with the photon energy $\hbar\Omega \geq \Delta_g$ (short-wavelength radiation). Here $\Omega$ and $\omega$ are frequencies of short- and long-wavelength radiation, respectively, and $\hbar$ is Planck's constant, $\varepsilon_i$ is the energy of the emitter QW ionization, $\Delta_g$ is the energy gap in the collector QW. It is of value that due to the effect of photocurrent gain input long-wavelength radiation can be efficiently converted into short-wavelength radiation. By the principle of operation the proposed QWIP as an integrated converter of the radiation is similar to the devices utilizing interband optical transitions reported by Beneking et al[3] and Sasaki et al[4].

The purpose of this work is to discuss device physics of the QWIPs and evaluate their performance.

For the QWIP under consideration one can suppose that the main mechanism of the electron injection from the emitter is the thermionic injection over the barrier between the heavily doped emitter region and the QW.

The equations for the density of the total collector current $j$ and the potential $\varphi$ of the emitter QW with respect to the emitter contact can be expressed as[5]

$$j = j_T + \beta\, j_m \cdot \exp\Big(\frac{e\varphi - \varepsilon_i - \varepsilon_0}{kT}\Big) + \frac{e}{2}\,\Sigma\,\sigma\, I_\omega, \tag{1}$$

$$(1-\beta)\, j_m \cdot exp\Big(\frac{e\varphi - \varepsilon_i - \varepsilon_0}{kT}\Big) = 2 j_T + e\,\Sigma\,\sigma\, I_\omega. \tag{2}$$

Here $j_T$ is the density of the current caused by thermal excitation of the electrons from the emitter QW, the second term in the right side of eq.(1) is the density of the current injected from the emitter through the QW into the collector region and the third one is the density of the current created by the electrons photoexcited from the QW by long-wavelength radiation. These contributions to the total current are included also in eq.(2), which describes the balance of the electrons in the emitter QW. In eqs (1) and (2) $e$ is the electron charge, $\beta$ is the efficiency of the transport of injected electrons through the QW (so that $(1-\beta)$ is the probability of the capture of an injected electron in the QW), $j_m$ is the preexponential factor in the expession for the thermionic current, $k$ is Boltzman's constant, $T$ is the temperature, $\varepsilon_i$ is the energy of the electron transition between the bound state in the QW and continuum states, $\varepsilon_0$ is the difference between Fermi energies in the emitter contact and emitter QW with respect to the ground states, $\Sigma$ is the sheet density of the electrons in the QW which is typically close to the sheet density of the donors $\Sigma_D$ in it, $\sigma$ is the cross-section of the electron photoexcitation from the QW into continuum states due to absorption of long-wavelength photon, and $I_\omega$ is the intensity of long-wavelength radiation.

It is supposed that applied voltage $V$ is large enough so that electron injection from the collector can be neglected. We neglected also the hole injection, supposing that the potential barrier in the QW in the collector confines the holes in the collector region effectively. For the sake of simplicity we suppose also that the heights of the potential barriers in the QW are equal to each other (symmetrical QW). Equations (1) and (2) immediately lead to the following expression for the current density

$$j = \Big(\frac{1+\beta}{1-\beta}\Big) j_T + \frac{1}{2}\Big(\frac{1+\beta}{1-\beta}\Big)\, e\,\Sigma\,\sigma\, I_\omega \tag{3}$$

The first term in the right side of eq.(3) represents the dark current density. The second one is connected with the photoexcitation of the electrons from the QW (photocurrent density). It is seen that the photocurrent density exceeds the current density of photoexcited electrons by the factor $g = \dfrac{1+\beta}{1-\beta}$, which is in fact the photoelectric gain. In the case of relatively small probability of the capture of the injected electrons $(1-\beta) \ll 1$, the photoelectric gain $g \gg 1$. It is naturally to suppose that all electrons injected into the region behind the emitter QW recombine in the collector QW. In this case the intensity of short-wavelength spontaneous radiation $I_\Omega$ connected with the interband radiative recombination can be expressed as

$$I_\Omega = \frac{j}{e(1+\frac{\tau_r}{\tau_n})} = \Big(\frac{1+\beta}{1-\beta}\Big)\frac{j_T + \frac{e}{2}\Sigma\sigma I_\omega}{e(1+\frac{\tau_r}{\tau_n})} \tag{4}$$

Here $\tau_r$ and $\tau_n$ are the recombination times due to radiative and nonradiative mechanisms, respectively. The variation of short-wavelength radiation $\Delta I_\Omega$ owing to absorption of long-wavelength radiation can be expressed using eq. (4) as

$$\Delta I_\Omega = \frac{1}{2}\cdot\frac{1+\beta}{1-\beta}\cdot\frac{\Sigma\sigma}{1+\frac{\tau_r}{\tau_n}}\cdot I_\omega \tag{5}$$

The efficiency of the optical energy conversion $K = \frac{\Omega}{\omega} \cdot \frac{\Delta I_\Omega}{I_\omega}$ is given by

$$K = \frac{1}{2}\left(\frac{\Omega}{\omega}\right) \cdot \frac{1+\beta}{1-\beta} \cdot \frac{\Sigma\sigma}{1+\frac{\tau_r}{\tau_n}} \tag{6}$$

In the case $\hbar\omega = 0.1eV$, $\hbar\Omega = 1eV$ (conversion of the radiation with wavelength $\lambda \simeq 10\mu m$ into radiation with wavelength $\lambda \simeq 1\mu m$), $\sigma = 2 \cdot 10^{-15} cm^2$ , $\Sigma = 10^{12} cm^{-2}$, $\tau_r = \tau_n$ and $1 - \beta = 10^{-2}$ we obtain $\frac{\Delta I_\Omega}{I_\omega} \approx 0.1$ and $K \approx 1$ (values of $\sigma$ and $1 - \beta$ correspond to usual cross-section of the electron photoexcitation from the QW and usual recombination velocity of injected electrons[6]).

Here we neglected the hole injection from collector region. This is valid if the density of the hole current injecting into $N$ region is small in comparison to the dark electron current density. Our estimates show that this assumption is correct if $\varepsilon_v + \dfrac{\pi\hbar^2\Sigma}{m} - \varepsilon_i \gg kT$, where $\varepsilon_v$ is the discontinuity of the valence band. It means that the value $\varepsilon_v$ should be large enough. This restricts a selection of materials suitable for fabrication of the QWIPs in question. This condition is met, for instance, for the QWIPs utilizing $Al_{0.25}Ga_{0.75}As-GaAs$ heterostructures with $\varepsilon_v = 0.14$ eV and $\varepsilon_i = 0.10$ eV, ($\hbar\omega \sim 0.1$eV or $\lambda \sim 10\mu m$) at least at a temperature when the electron gas in the QW is degenerate.

## References


[1] Levine B F 1993 *J. Appl. Phys.* **74** R1

[2] Tsang J S, Lee C P, Tsai K L and Chen H R 1994 *Electronics Lett.* **30** 450

[3] Beneking H, Grote N and Svilans M N 1981 *IEEE Trans. Electron Devices* **ED-29** 404

[4] Sasaki A, Matsuda K, Kimura Y and Fujita S 1982 *IEEE Trans. Electron Devices* **ED-29** 1382

[5] Ryzhii V 1994 *Japan. J. Appl. Phys.* **33** 78

[6] Vinter B, Luc F, Bois P, Thibaudeau L and Rosencher E 1994 *Solid State Electron.* **37** 773

*Inst. Phys. Conf. Ser. No 144: Section 7*
*Paper presented at 7th Int. Conf. on Narrow Gap Semicond. Santa Fe 9–12 January*



# Infrared photodetectors using capacitance variation in $InAs_{1-x}Sb_x$ MIS devices

Y. Zhang and J. E. Furneaux

Department of Physics and Astronomy
and
Laboratory for Electronic Properties of Materials
University of Oklahoma, Norman OK 73019

**Abstract.** We report a demonstration of a photo-capacitive response in *p*-type $InAs_{1-x}Sb_x$ metal-insulator-semiconductor (MIS) devices to normally incident interband infrared radiation. Strong infrared response covering a wide spectral range from 2 μm to 10 μm has been observed at 77 K. This spectral response depends on the MBE growth composition $x$ in the $InAs_{1-x}Sb_x$ compound. A peak response of $R_P(\lambda) = 8.4 \times 10^5$ V/W at 5.5 μm was obtained. The performance of such an $InAs_{1-x}Sb_x$ MIS device is relatively insensitive to material growth parameters. Thus such devices are promising for use as IR photodetectors and in arrays.

In recent years many infrared photodetector studies based on III-V compound semiconductors have focused on quantum well infrared photodetectors (QWIPs) [1-3] and strong claims have been made regarding the performance of this IR detection scheme. However, considerable effort still remains to be expended before QWIPs are viable devices. Substantial improvements are necessary in a number of areas: devices exhibit excess noise due to majority-carrier operation [4], they are insensitive to normally incident radiation, and absorption is limited due to the limited number of quantum wells which can be produced for a given device. The $InAs_{1-x}Sb_x$ alloy is potentially a very useful material for the fabrication of intrinsic bandgap infrared detectors [5]. Up to now this promise has not been realized because producing quality *p-n* junctions in $InAs_{1-x}Sb_x$ has proven difficult. In this paper we demonstrate a new approach to the problem of interband $InAs_{1-x}Sb_x$ infrared detectors, MIS devices biased in depletion. The idea of a MIS infrared detector was first proposed by M. Kinch using (HgCd)Te. This idea has been extensively investigated in the past few years [6-8]. Although MIS and photoconductive (HgCd)Te infrared detector technologies have made impressive progress recently, they still exhibit material drawbacks, *e. g.* inhomogeneity and irreproducibility. A MIS $InAs_{1-x}Sb_x$ photodetector has many advantages: it is easy to fabricate, and it can cover a wide spectral range. Moreover, its performance is relatively insensitive to the material procesng technology, a dominant advantage.

The samples studied here were grown on $p$-type GaAs (100) substrates by molecular-beam epitaxy (MBE) in an Intevac Gen II system. A 1 μm $p$-doped buffer layer of GaSb was grown on the substrate, followed by a 2 μm $p$-doped InAs-$InAs_{1-x}Sb_x$ buffer layer which is compositionally step graded from $x = 0$ to the final Sb composition. The active region, a 3.5 μm thick ~$10^{17}$ $cm^{-3}$ Be doped, $InAs_{1-x}Sb_x$ layer, then is grown. Using MBE to grow the samples resulted in high quality surfaces and interfaces as demonstrated by the sharp patterns observed via *in situ* reflection high energy electron diffraction (RHEED). From these $InAs_{1-x}Sb_x$ layers, MIS devices were fabricated. A bilayer insulator consisting of a 20 nm anodic $InO_x$ layer and a 60 nm PECVD $SiO_2$ layer were deposited. The capacitor MIS devices were then defined by evaporating 8 nm thick, 1 mm diameter, semi-transparent NiCr gates. Finally, Ohmic contacts to the metal gate and to the $p$-type substrates were made. Here we report studies on 3 devices which have different Sb compositions, $x_1 = 0.13$, $x_2 = 0.26$, and $x_3 = 0.41$.

A Fourier-transform-infrared, FTIR, spectrometer was used to measure the infrared absorption edge for each of the three different sample compositions. The GaAs substrates were lapped to significantly reduce their thickness and hence the free carrier absorption which could mask absorption features due to the $InAs_{1-x}Sb_x$ layer. A lapped sample, with the active $InAs_{1-x}Sb_x$ layer removed by selective etching, was used as the reference for absorption measurements. The infrared absorption spectra measured at 77 K are shown in Fig. 1. The absorption edges of the three samples are at 4.2 μm, 6.3 μm, and 8.1 μm respectively. Photoresponse measurements of the $InAs_{1-x}Sb_x$ MIS infrared devices at 77 K were then performed using an infrared monochrometer system with a glowbar used as the infrared source. Instead of chopping the radiation, we applied a small AC signal to the MIS $InAs_{1-x}Sb_x$ capacitor which was DC biased into deep depletion (~-2.5 V in our situation). A lock-in amplifier was then used to measure the change in the AC conductivity produced by the infrared radiation. This conductivity is due to the replacement current through the MIS capacitor. In fact, the spectral response $R(\lambda)$ of the

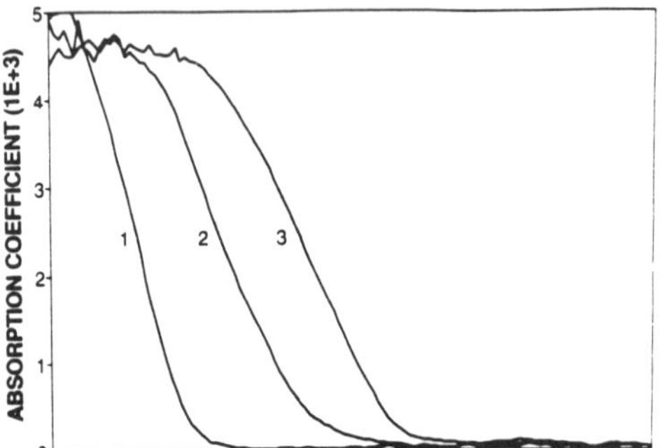


Fig. 1 The infrared absorption spectra measured at 77K. $x_1 = 0.13$, $x_2 = 0.26$, and $x_3 = 0.41$.

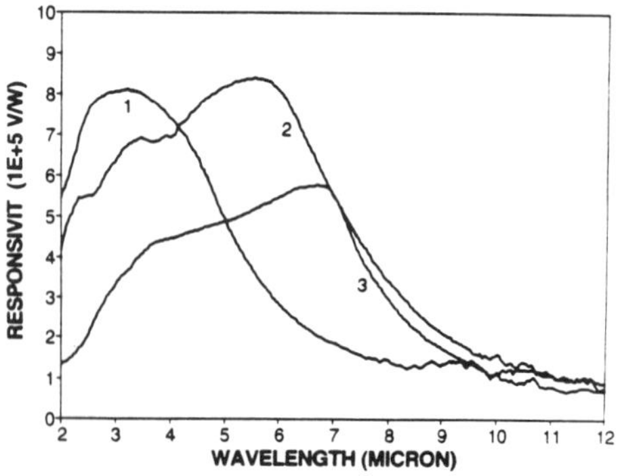


Fig. 2 The spectral response $R(\lambda)$ of the three devices with different Sb composition. The three peaks $R_p(\lambda)$ are $R_p(\lambda, x_1) = 3.2$ μm, $R_p(\lambda, x_2) = 5.5$ μm, and $R_p(\lambda, x_3) = 6.7$ μm, $R(\lambda)$ covers a spectral region from 2 μm to 10 μm. The maximum peak response at the wavelength of 5.5 μm is about $8.4 \times 10^5$ V/W.

MIS devices measured by the lock-in is the photo-capacitive signal, $C(\lambda)$. Fig. 2 shows the measured spectral response $R(\lambda)$ of the three devices. The peak responses are found to be at 3.2 μm, 5.5 μm, and 6.7 μm, while the devices demonstrate useful response from 2 μm to 10 μm tuned compositionally by $x$ in the $InAs_{1-x}Sb_x$ compound. The maximum response of $8.4 \times 10^5$ V/W was found at 5.5 μm, corresponding to a detectivity, $D^*$, of $\sim 10^{11}$ cmHz$^{1/2}$/W. We deduced this value by using a pyroelectric detector and a bolometer as calibration standards.

The $p$-type $InAs_{1-x}Sb_x$ MIS devices operate as photocapacitors when they are biased in depletion. Because there is considerable positive fixed oxide charge in these devices, a $V_g = -2.5$ V is necessary to set the operating point. By considering the equivalent circuit of the $p$-type MIS capacitor biased into depletion, we can understand the physics of the photo-capacitive response. One Ohmic contact to the MIS capacitor is to the gate, and the other is to the substrate. The depletion layer capacitance $C_d$ in the circuit is related to the valence band because $C_d$ is charged and discharged by the flow of majority holes in and out of the depletion layer edge region through the Ohmic substrate contact. $C_d$ is determined by $\varepsilon_s/w$, where $w$ is the depletion layer width and $\varepsilon_s$ is the dielectric constant of $InAs_{1-x}Sb_x$ material. For a simplified model, we suppose the capacitor works at normal AC frequencies and ignore the charging and discharging of the interface traps. The resulting equivalent circuit of the MIS detector is then shown in Fig. 3. $R_d$ is a phenomenologically equivalent resistor connecting the interface and the substrate, through which both majority holes and minority electrons pass. Fig. 4 shows a C-V curve characteristic of our $InAs_{1-x}Sb_x$ MIS devices. In the dark when the device is biased into depletion ($V_g \leq -2.5$ V), the total capacitance of the MIS is expressed as the series combination of the insulator capacitance and depletion capacitance $C = C_{ox}C_d/(C_{ox} + C_d)$. When the infrared radiation is introduced to the device, the interband transition and tunneling increase the carrier (electrons and holes) density in the

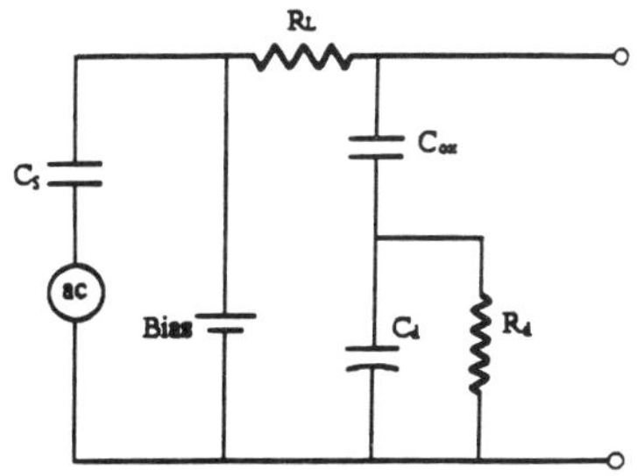


**Fig. 3** A simplified equivalent circuit for the MIS detector, $R_d$ is a phenomenologically equivalent resistor connecting the interface and the back substrate, through which both majority and minority carriers pass. $C_d$ is the depletion layer capacitance and $C_{ox}$ is the oxide layer capacitance.

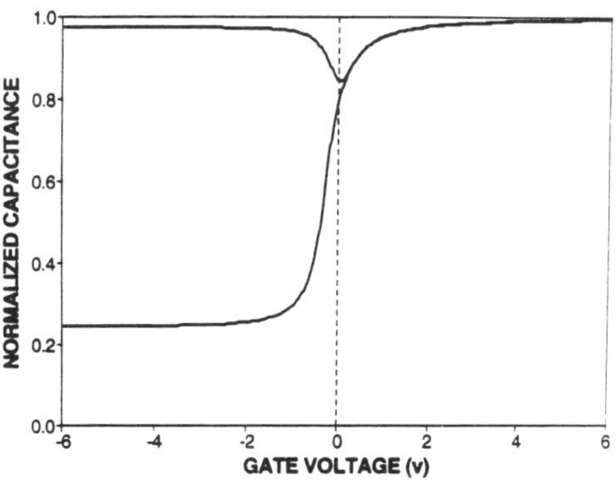


**Fig. 4** The C-V characteristic curve of an $InAs_{1-x}Sb_x$ MIS device, the C-V curve normally exhibits a "low-frequency" inversion feature even at higher frequencies. In the depletion region ($V_g = -2.5$ v) when the device is isolated from infrared radiation the total capacitance is $C = C_{ox}C_d/(C_{ox} + C_d)$, when the infrared radiation is introduced, the total capacitance is drastically increased to the that of $C_{ox}$.

vicinity of interface. In this situation, the carriers generated in that region effectively reduce the value of $R_a$ or even short the $C_d$ (depending on the intensity of radiation introduced). This results in a drastic change to the C-V curve - total capacitance as big as $C_{ox}$ as shown in Fig. 4.

In conclusion, we present experiments on $p$-type $InAs_{1-x}Sb_x$ MIS infrared photocapacitors. When these devices are biased into deep depletion, interband transitions induced by IR radiation, result in a variation of the total capacitance. The infrared spectral response $R(\lambda)$ can be compositionally tuned to cover a wide spectral region from 2 μm to 10 μm. The peak response $Rp(\lambda) = 8.4 \times 10^5$ V/W, corresponding a detectivity $D^* = 10^{11}$ cmHz$^{1/2}$/W, has been achieved at 5.5 μm. As compared to $p$-$n$ junction $InAs_{1-x}Sb_x$ photodiodes, MIS infrared devices offer simple fabrication technology and is insensitivity to defects introduced during growth. Our experiments indicate that $InAs_{1-x}Sb_x$ MIS devices for use as IR photodetectors and arrays is highly promising.

It is a pleasure to thank Ryan Doezema for helpful discussions. This work was partially supported by the National Science Foundation through grant number OSR-9108771 and by the Army Research Office through grant number DAAH04-G-94-0248.

## References

[1] F. Levine, R. J. Malik, J. Walker, K. K. Choi, C. G. Bethea, D. A. Bleinman,and J. M. Vandenbery, Appl. Phys. Lett., **50**, 273 (1987).

[2] F. Levine, K. K. Choi, C. G. Bethea, J. Walker, and R. J. Malik, Appl. Phys. Lett., **51**, 934 (1987).

[3] K. Choi, B. F. Levine, C. G. Bethea, J. Walker, and R. J. Malik, Appl. Phys. Lett., **52**, 1979 (1988).

[4] K. Kinch, and A. Yariv, Appl. Phys. Lett. **55**, 2093 (1989).

[5] O. Bubulac, A. M. Andrews, E. R. Gertner, and D. T. Cheung, Appl. Phys. Lett., **36**, 734 (1980).

[6] K. Kinch, and S. R. Borrello, Infrared Phys., **15**, 111 (1975).

[7] L. Polla, and A. K. Sood, J. Appl. Phys., **51**, 4908 (1980).

[8] L. Teszner, Laser Focus World, **Jan.**, 117 (1994).

*Inst. Phys. Conf. Ser. No 144: Section 7*
*Paper presented at 7th Int. Conf. on Narrow Gap Semicond. Santa Fe 9–12 January*



# Photocurrent Spectra of $InAs/Ga_{0.75}In_{0.25}Sb$ Superlattices With Infrared Detector Structure

**X. L. Huang, W. Z. Shen, W. Lu, S. C. Shen.**
National Laboratory for Infrared Physics, Shanghai Institute of Technical Physics, Shanghai 200083, P. R. China

**T. G. Andersson**
Department of Physics, Chalmers University of Technology, S-41296 Göteborg, Sweden.

**Abstract.** The $I-V$ characteristics, the photocurrent spectra ( PC ) and PC phases of $InAs/Ga_{0.75}In_{0.25}Sb$ superlattices ( SLs) with infrared detector structure are investigated. The band structure alignment, the peaks and phases of photocurrent spectra and the corresponding transport processes of photoexcited electrons are discussed. In addition, the optimum models for detector output with signal voltage or current are also discussed.

## 1. Introduction

In recent years, $InAs/Ga_{1-x}In_xSb$ SLs has been studied extensively due to their potential to application of infrared detector[1]. The structures of epitaxial material[1,3] and minibands[4,9], infrared absorption and PL specta[1−4], Hall effect[5], metal–semimetal transition[6] , and PC spectra for both photoconductive and photovoltaic type structure have been investigated[1,2]. In this article, the $I-V$ characteristics, the PC response and its phases for a detector structure of $InAs/Ga_{1-x}In_xSb$ SL are measured, the minibands and the processes of photoelectric response for the detector are discussed.

## 2. Experiment

The sample ( #1752 ) was prepared with Varian GEN II MBE system[7], a GaSb buffer layer with thickness of $0.5\mu m$ was grown on GaSb substrate at first, followed by 20 periods of $InAs(2.5nm)/Ga_{0.75}In_{0.25}Sb(2.5nm)$ SL. The sample was cut into $1 \times 3mm^2$ chip, and then two $Au(100nm)/AuGeNi(100nm)$ electrodes were deposited on the top of chip, connected with $Au$ wires to outside circuits[10].

The $I-V$ characteristics of the detector structue were measured with two multimeters. The PC spectra were measured with Nicolet 800 Fourier Transform Infrared spetrometer ( FTIR ), the light beam was focused on the anode area of the detector structure, the output signal was pre-amplified, and fed into the data processing system of FTIR[10]. The PC spectra were also measured with a SPM2 monochromator, the light beam from a Nernst glower reflected through a chopper ( 1kHz ) and the slit ( 1mm ) of SPM2, is focused on the detector structure. Biased with a current supplier, the photoelectric signal and its phase of the sample were measured with EG &G 5209 lock in amplifier[10].

### 3. Results and discussions

Fig. 1 shows the $I-V$ characteristics of the sample at 77K. The forward current $I^+(V_b)$ increases with bias almost linearly in the region of $V_b = 0$ to $0.40V$ , $I^+(V_b)$ has a wave-like shape in 0.40 to $0.60V$, which results in a M-like shape for the differential resistance $R_d$. $I^+(V_b)$ increases with bias superlinearly in the $V_b$ region of 0.60 to $0.80V$. Since the resistance of GaSb substrate is relatively low, and the distance between two electrodes is quite large, the detector structure investigated can be equvalent to two SL layers in series connected with GaSb substrate. The serial resonance tunneling ( SRT ) and disconnection of minibands might occur at the wave-like shape area of $I-V$ characterisics in comparison with the SRT of the usual multiple quantum wells. Since the separations of valence minibands ( $HHn$ and $LHn$ ) are much smaller than those of conduction minibands ($En$ ), the wave-like shape component of $I^+(V_b)$ mainly due to the SRT of $HHn$, the total $I^+(V_b)$ should also include the contributions of the tunneling current ( $I_T$ ) of $En$ and the currents ( $I_O$ ) of thermoexcitation and photoexcitation over the barriers. The reversal current $I^-(V_b)$ increases with bias slower than $I^+(V_b)$ in the region of $V_b = 0$ to $-0.40V$ , but faster than $I^+(V_b)$ in the region of $-0.40$ to $-0.60V$, the differences between $I^-(V_b)$ and $I^+(V_b)$ may be due to the asymmetry of the detector structure.

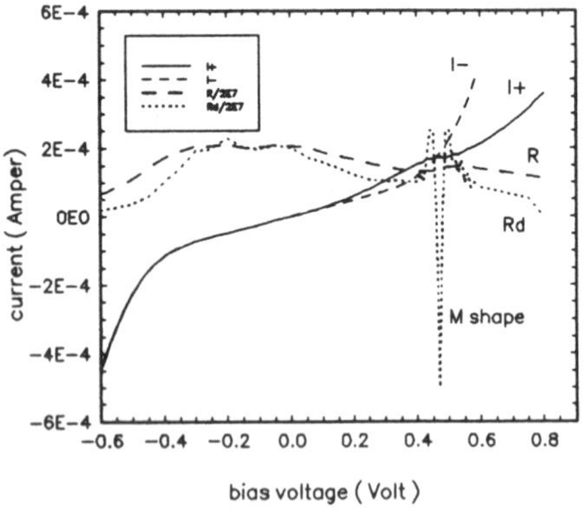


Fig. 1 The $I-V$ characteristics of $InAs(2.5nm)/Ga_{0.75}In_{0.25}Sb(2.5nm)$ superlattices at 77K; the current I, resistance R and differential resistance $R_d$ versus bias voltage $V_b$ are shown in the figure.

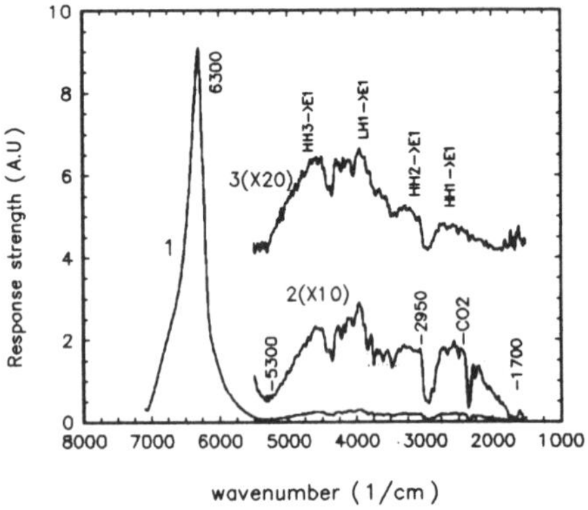


Fig.2 The photocurrent spectra of $InAs(2.5nm)/Ga_{0.75}In_{0.25}Sb(2.5nm)$ superlatices at 77K, measured with a Nicolet 800 FTIR spectrometer.

Fig. 2 shows the PC spectra at zero bias which is measured with a FTIR spectrometer. Curve 2 is 10 times enlarged of curve 1 in 5500 $\sim 1500cm^{-1}$. Curve 3 is an enlarged ( 20 times ) PC spectrum, measured with a GaSb filter for cuting of the radiation above $5000cm^{-1}$ and divided by InSb detector response of the spectrometer for normalization. The structure near $2350cm^{-1}$ of curve 2 is due to the absorption of $CO_2$ in comparison with curve 3, thus the PC spectra of the sample is composed of 3 adjacent peaks and a separate peak in the region of 5500 $\sim$ $1500cm^{-1}$.

The strong peak of curve 1 in Fig. 2 has Lorentz line shape and locate at $6300cm^{-1}$, the full width at half maximum ( FWHM ) is $300cm^{-1}$, which should result from the intrinsic interband transition of GaSb substrate and GaSb buffer layer [10]. The excited electrons diffused in the substrate will recombine with the holes , while the electrons diffused in the superlattices layers can be transported to the sample surface area. Thus the electrons may be accumulated at the surface area ( including the inner interfaces ) to form built-in polarized electric field ( $E_{bi}$ ) and produce photoelectric signal via electrical contacts.

The 4 peaks of curve 2 in Fig. 2 are located at 4500, 3950, 3200 and 2550$cm^{-1}$ respectively, the corresponding energy separations of the neighbouring peaks are 600, 750 and 650$cm^{-1}$ ( 73, 94 and 78$meV$ ) , respectively. The 4 peaks of curve 2 can attributed to the transitions of $HH_3$, $LH_1$, $HH_2$ and $HH_1$ to $E_1$ by comparision the transition energies with the theorectical calculation shown in Ref. [9]. The peak strength of $HH_2 \rightarrow E_1$ is relatively weak due to the forbidding of the transition [10]. The PC response of $HH_1 \rightarrow E_1$ cut off at $1750 \pm 50cm^{-1}$ ( 217$meV$ ), which is also close to the experimental results ( $220 \pm 20meV$ ) of Ref. [3]. The FWHM and full width ( FW ) of $HH_1 \rightarrow E_1$ are 700$cm^{-1}$ ( 87$meV$ ) and 1300$cm^{-1}$ ( 160$meV$ ), respectively, they are smaller than that obtained by PL [3] , thus the line shape of $HH_1 \rightarrow E_1$ of curve 2 should be close to the density of states ( DOS ) of $E_1$, and the transitions of $HH_3, LH_1$ and $HH_2$ to $E_1$ form the adjacent line shapes.

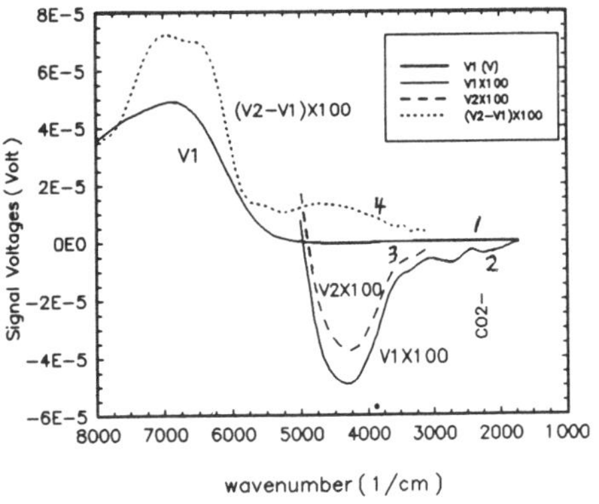


Fig. 3 The photocurrent spectra of $InAs(2.5nm)/Ga_{0.75}In_{0.25}Sb(2.5nm)$ superlattices structure at 77K, measured with SPM2 monochromator and EG&G 5209 lock in amplifier.

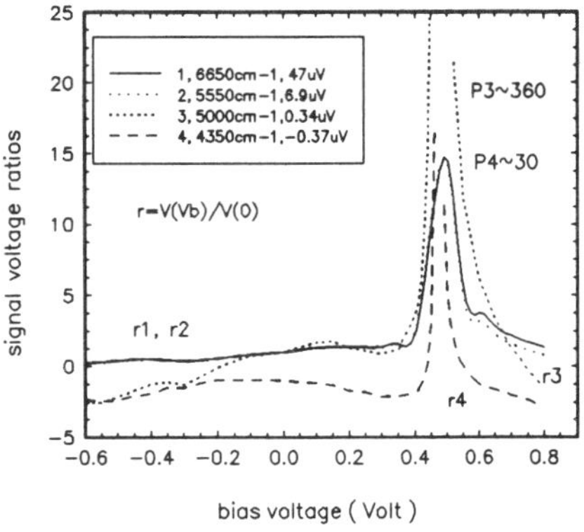


Fig. 4 The signal voltage ratios versus bias voltage at 77K, measured with SPM2 monochromator and EG&G 5209 lock in amplifier.

Fig. 3 shows the PC spectra at zero bias and measured with SPM2 monochromator. Curve 2 is 100 times enlarged of curve 2, while curve 3 is same as curve 2, but illustrated with an extra light beam ( 30W lamp, 4cm away ), again 100 times enlarged, curve 4 is the subtraction of curves 3 and 2. It is known from curves 1 and 2, that for the wavenumber higher than 4930$cm^{-1}$, PC response is positive with a phase of $49^o$ respecting to the chopper signal, and for the wavenumber smaller than 4930$cm^{-1}$, the PC response is negative with a phase of $-128^o$ respecting to the chopper signal. The change in phase $\approx 180^o$ implies that the signal current and the photoexcited carriers in the two regions might be opposite. Since the wave function of $E_1$ in the wells can penetrate into the barrier, the excited and diffused electrons of GaSb substrate may form forward PC and

$E_{bt}$ while the photoexcited electrons from $HHn$ to the superlattice subband of $E_1$ might form reversal PC by $E_{bt}$ ( tunneling and drifting ).

Fig. 3 shows also that an extra light beam illustration makes curve 2 shifting upwards ( shown as curve 3 ), and the phase reverse point ( $V_S = 0$ ) moves from $4930cm^{-1}$ to $4900cm^{-1}$. The line shape of $\Delta V = V_3 - V_2$ at wavenumber region larger than $5000cm^{-1}$ is similar to that of $V_1$, and $\Delta V \approx 1.5\% V_1$ ( at $6300cm^{-1}$ ), thus $\Delta V$ results mainly from $\Delta R_d$ ( $\Delta R_d \approx 1.0\% R_d$ ) in this region. $\Delta V$ decreases gradually at wavenumber region smaller than $5000cm^{-1}$ , while $\Delta V \approx 25\% |V_1|$ at $4350cm^{-1}$ ( the peak of $V_1$ ).

Fig. 4 is the response voltage ratio $r(V_b) = V_S(V_b)/|V_S(0)|$ for different fixed wavelength. The wavenumber of incident light beam for the curves 1, 2, 3 and 4 are 6670, 5550, 5000 and $4350cm^{-1}$ , and the corresponding $V_S(0)$ are 47, 6.9, 0.34 and $-0.37\mu$V, respectively. Curve 1 is almost same as curve 2, due to the similar photoelectric transportation process, though their $V_S(0)$, are 7 times diffenent. In minus bias region of $V_b = 0$ to $-0.6V$, $r(V_b)$ decreases gradually but is kept positive. At forward bias, $V_S$ increases first, then increases fast and forms a peak at $V_b = 0.40$ to $0.60V$, and decreases while $V_b > 0.60V$ ( due to $R_d$ ).

The $V_S(0)$ of curves 3 and 4 shown in Fig. 4 are similar ( 0.34V and $-0.37V$ ). In minus bias region of $V_b = 0$ to $-0.60V$, the $r(V_b)$ of curve 3 decreases from 1 to $-2.5$ ( closes to that of curve 4 ), because of the PC intensity shifting downward with bias ( see in Fig. 3 ). At positive bias of $V_b = 0.50V$, the peak $r(V_b)$ of curve 3 is 12 times as large as that of curve 4, since the PC intensity shifting upwards. In positive bias region of $V_b$ larger than $0.70V$, $R_d$ decreases quicker than R, the signal current transitting over barrier ( $I_{SO}$ ) ( due to $E_1 \rightarrow E_2$, $HH_n \rightarrow E_C$ and the tail of GaSb ) is going to saturate, $E_{bt}$ is proportional to $R$, threrfore the signal current of tunneling ( $I_{ST}$ ) ( due to $HHn \rightarrow E_1$) is proportional to $R/R_d$ and surpasses $I_{SO}$ finally, thus the total $V_S$ of curve 3 becomes reversal at $V_b > 0.7V$ ( see in Fig. 4 ).

It is seen from Fig. 4 that for the bias at SRT peak, $R_d$ becomes very large and $I_{ST}$ becomes very small, but $I_{SO}$ is less affected by $R_d$ and surpasses $I_{ST}$ finally, thus the PC signal of curve 4 changes $180^o$ in phase and forms forward signal voltage ( $V_S$ ) at $0.45 \sim 0.55V$. If a detector is operated at SRT peak, $R_d$ is relatively large ( even if the well is doped ), one can use the forward $I_{SO}$ transportation and the carrier might be accumulated at the surface area thus giving a relatively large forward $V_S$ output. If a detector is operated far from SRT , $R_d$ is relatively small ( favour for reducing the dark current ), one can use the reversal $I_{ST}$ to obtain relatively large signal current output.

## 4. Conclusion

In conclusion, the $I - V$ characteristics, the photocurrent responses and their phase behavior of $InAs(2.5nm)/Ga_0.75In_0.25Sb(2.5nm)$ SLs with infrared detector structure are investigated. Four peaks corresponding to intersubband transitions $HHn(LHn)toE_1$ were observed from PC spectra. The phase behaveior of the PC response and the extra light illumination are used for discusssion of the excitation and transportation processes of photo-excited carrier.

**Acknowledgement**: The authors thanks for the supports and helpful discussions that given by Dr. Y. Chang, Z. Y. Li, Prof. J. H. Zhu , Prof. S. W. Chen, Dr. Y. M. Mu, Dr. Y. Cai and Dr. H. B. Mao.

**References**

[1] D. H. Chow, R. H. Milles, J. N. Schulman, D. A. Collins and T. C . McGill, *Semicond. Sic. Technol.* 6 (1991) C47.

[2] R. H. Miles, D. H. Chow, J. N. Schulman and T. C. McGill, *Appl. Phys. Lett* , **57**, 801 (1990).

[3] D. H. Chow, R. H. Miles, R. H. Soderstrom and T. C. McGill, *Appl. Phys. Lett*, **56**, 1418 (1990).

[4] J. Katz, Y. Zhang and W. I. Wang, *Appl. Phys. Lett* **62**, 609 (1993).

[5] C. A. Hoffman, J. R. Meyer, E. R. Youngdate, F. J. Bartoli and R . H. Miles, *Appl. Phys. Lett*, **63**, 2210 (1993).

[6] C. A. Hoffman, J. R. Meyer, F. J. Bartoli and W. I. Wang, *Phys. Rev.* **B 48**, 1959 (1993)

[7] W. Q. Chen, S. M. Wang, T. G. Andersson and J. Thordson, *Phys. Rev.* **B 48**, 14264 (1993)

[8] V. Berger, E. Rosencher, N. Vodjdani and E. Costard, *Appl . Phys. Lett*, **62**, 378 (1993).

[9] D. L. Smith and C. Maihiot, *J. Appl. Phys*, **62** , 2545 (1987).

[10] X. L. Huang, W. Z. Shen, W. Lu,Y. M.Mu and S. C. Shen, *J. Infrared Millim. Waves*, **13**, 15 (1994).

*Inst. Phys. Conf. Ser. No 144: Section 7*
*Paper presented at 7th Int. Conf. on Narrow Gap Semicond. Santa Fe 9–12 January*



# Electro-Optical Filters and Modulators Based on Intersubband Processes in InAs-GaSb-AlSb-Family Double Quantum Wells

J. R. Meyer, C. A. Hoffman, and F. J. Bartoli
Code 5613, Naval Research Laboratory, Washington, DC 20375

L. R. Ram-Mohan
Worcester Polytechnic Institute, Worcester, MA 01609

**Abstract.** We discuss a new family of electro-optical intensity and phase modulators based on normal-incidence intersubband transitions in GaSb-based asymmetric double quantum wells. An applied electric field transfers the electron population from Γ-valley states to L-valley states, thereby radically altering the resonance energies, strength and polarization dependences of the optical matrix elements, and in-plane effective masses which govern the plasma contribution to the dielectric constant.

## 1 Introduction

Recent years have seen rapid advances in the state-of-the-art for near-infrared electro-optical (EO) modulation devices based on the excitonic quantum-confined Stark effect (QCSE). By contrast, the development of fast EO modulators, filters, and switches for the mid-wave and long-wave infrared spectral regions remains at a primitive stage. Important MWIR and LWIR technologies such as remote chemical sensing, pollution monitoring, laser surgery, and future ultra-low-loss fiber communications will require EO modulators of various types as key elements. Electrically-tunable filters with fast switching speeds are also urgently needed in passive IR sensor systems, as a means of enabling multi-spectral detection. In this work, we demonstrate that most of the device types now being implemented in the near-IR can be readily adapted for longer wavelengths by employing intersubband-based EO processes in place of the excitonic QCSE. While the usefulness of previous intersubband modulators[1, 2, 3, 4] has been severely limited by the requirement that the optical electric field have a component along the quantum well growth axis, the devices proposed here are designed for normal-incidence operation.

Previous intersubband-based EO modulators have employed heterostructure systems such as GaAs-$Al_xGa_{1-x}As$, in which the electrons occupy the isotropic Γ-valley. However, it has been shown that the highly anisotropic electrons populating an L-valley or X-valley can display strong intersubband interactions at normal incidence as long as the growth axis is misaligned with the principal axes of the constant-energy ellipsoids.[5] Several groups have recently observed large normal-incidence absorption coefficients in GaSb/AlSb-based structures,[6, 7, 8] and a normal-incidence EO absorption filter employing $Ga_{1-x}Al_xSb$ stepped quantum wells was studied theoretically.[9]

In the present work, we consider an asymmetric double quantum well (ADQW) configuration whose most novel feature is that the application of an electric field transfers virtually the entire electron population from a well in which the conduction-band minimum is at the Γ-point to another well in which it is at the L-point. Thus we can radically alter the intersubband resonance energies, the strength and polarization dependences of the optical matrix elements, and the in-plane effective masses (which govern the plasma contribution to the refractive index). This high degree of control and tunability provides all the elements required for a wide range of EO devices. Not only are normal-incidence modulators now possible, but the predicted EO coefficients are larger than for most previously-discussed mechanisms because rather than simply shifting the resonances with electric field, we can effectively *eliminate* them if desired.

## 2 Γ-L Asymmetric Double Quantum Wells

Figure 1 illustrates the Γ and L band edge profiles and quantized energy levels for an AlSb-GaSb-$Ga_{1-x}Al_xSb$-$Ga_{1-x}In_xSb$-AlSb asymmetric double quantum well (ADQW). Note that the GaSb well on the left has its conduction-band minimum at the L-point while the $Ga_{1-x}In_xSb$ well on the right (which can alternatively be composed of InAs) has its minimum at the Γ-point. However, the structure is designed such that the lowest quantized levels in the two wells lie close enough in energy that the net ground state for the ADQW can be modulated by reversing the direction of the applied electric field from positive [Fig. 1(a)] to negative [Fig. 1(b)]. This dynamically-tunable transfer of electrons from L states (which strongly couple to the optical field at normal incidence) to Γ states (whose transition matrix elements vanish at normal incidence) provides an extremely effective mechanism for modulating the IR absorption coefficient and refractive index. The $Ga_{0.6}Al_{0.4}Sb$ spacer layer separating the two quantum wells is low enough that interwell tunneling will occur on a sub-nanosecond time scale following reversal of the applied voltage. On the other hand, the large GaSb-AlSb conduction-band offset allows us to employ the same basic scheme for EO devices spanning the entire LWIR and MWIR spectral regions (1.5 – 20 μm).

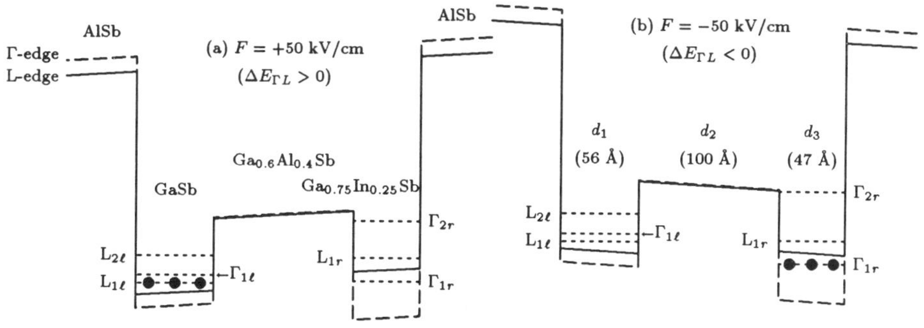


Figure 1. Γ-point (dashed) and L-point (solid) conduction band profiles and quantized energy levels for the ADQW at applied electric fields of (a) +50 kV/cm and (b) -50 kV/cm.

In order to quantitatively model the ADQW EO response, we have derived the quantized energy levels, wavefunctions, and optical matrix elements using a multi-band finite-element method (FEM) $k \cdot p$ formalism.[10] This highly-flexible algorithm allows for an arbitrary number of constituent layers with arbitrary band profiles, as well as externally-applied electric fields. Results for the Γ-valley states were obtained from the full 8-band FEM calculation, while subbands for the anisotropic L-valley electrons were obtained from a 1-band version of the algorithm using the quantization effective masses.

The relative shift of the energy minima for the Γ and L wells ($\Delta E_{\Gamma L}$) is governed primarily by the product of the field and the distance between the wells. Figure 2 plots $\Delta E_{\Gamma L}$ for a series of ADQWs with spacer-layer thicknesses $d_2$ ranging between 0 and 100 Å. The energy shift is clearly much stronger for large $d_2$, although it does not vanish even when the spacer layer is removed entirely.[11]

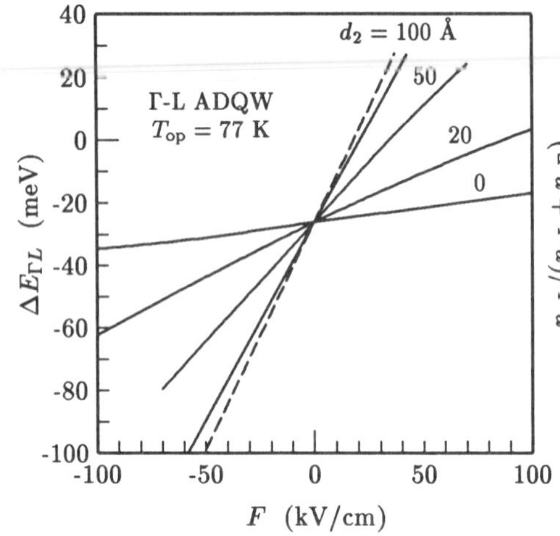


Figure 2. Relative energy difference between the Γ and L ground states vs electric field, for ADQW structures with InAs (solid curves) and $Ga_{0.75}In_{0.25}Sb$ (dashed curve) Γ-valley wells.

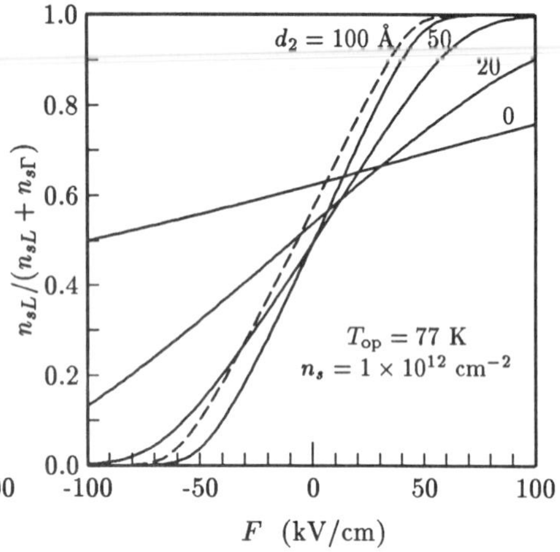


Figure 3. Fraction of the electrons populating L-valley states vs applied electric field, for the same ADQWs whose $\Delta E_{\Gamma L}$ are given in Fig. 2.

By tuning $\Delta E_{\Gamma L}$, we can efficiently modulate the relative electron populations of the two wells. This is illustrated in Fig. 3, which plots the population fraction occupying L-states in the left well as a function of applied field for the same structures as in Fig. 2. The sheet concentration per period ($n_s$) is taken to be $1 \times 10^{12}$ cm$^{-2}$, and we have accounted for the internal field induced by charge separation. The curves were obtained using dispersion relations from the FEM calculation and employing Fermi-Dirac statistics. Note that since the L-valley density of states is 10-20 times greater due to its 4-fold degeneracy and larger effective masses, most of the electrons occupy L-states whenever the two bands are evenly aligned in energy ($E_{\Gamma L} = 0$). As expected, the modulation is found to be much stronger for large $d_2$. For the InAs-well structure with $d_2 = 100$ Å, we predict that a relatively modest field of ±50 kV/cm is sufficient to switch from <4% to >96% population in the L-well, and slightly higher fields produce near-complete emptying and filling. While an operating temperature $T_{op}$ of 77 K has been assumed, similar modulation behavior is predicted for $T_{op}$ = 300 K except that somewhat higher voltages must be applied to achieve the same population-transfer efficiency.

## 3 Intensity and Phase Modulation

At incident photon energies near resonance with the L-valley intersubband separation, the normal-incidence absorption coefficient at a given field [$\alpha_z(F)$] will approximately follow the fraction of the electron population occupying L-valley states (Fig. 3). The magnitude of the modulation $\Delta\alpha_z$ can be quite high, since the L-valley absorption coefficients measured for GaSb-based quantum wells at normal incidence are typically on the order of $10^4$ $\text{cm}^{-1}$.[6–8]

The Kramers-Kronig transformation specifies that any resonant modulation of the absorption coefficient must be accompanied by a corresponding resonant modulation of the refractive index. For the ADQW structure illustrated in Fig. 1 (see also the dashed curves of Figs. 2 and 3), Fig. 4 plots the field-reversal (+40 → −60 kV/cm) shift of the refractive index as a function of normally-incident photon energy, where an experimental broadening parameter of $\Gamma \approx 10$ meV[7] has been assumed. The dashed curve represents the resonant (Kramers-Kronig) contribution by itself, which changes sign when the photon energy is equal to the intersubband transition energy and has its largest magnitude when $|\hbar\omega - E_{12}^L| \approx \Gamma$. However, a significant advantage of the Γ-L ADQW structure is that the interwell transfer also modulates the plasma contribution to the dielectric constant: $\epsilon_p \propto \sum_j N_j/m_{xj}\omega^2$, where $N_j$ is a given valley's electron population and $m_{xj}$ is its in-plane effective mass.[12] Previous single-valley EO designs resulted in $\Delta\epsilon_p \approx 0$, since the total density was fixed by the doping level and the field modulations of the wavefunctions and energy levels were not accompanied by any appreciable variation of the in-plane mass. However, the present devices will yield large $\Delta\epsilon_p$, because the relative valley populations are modulated by the field-induced transfer and the Γ-valley mass is much lighter than the L-valley mass. This is illustrated by the solid curve of Fig. 4, which represents the net index change due to the resonant and plasma contributions combined.

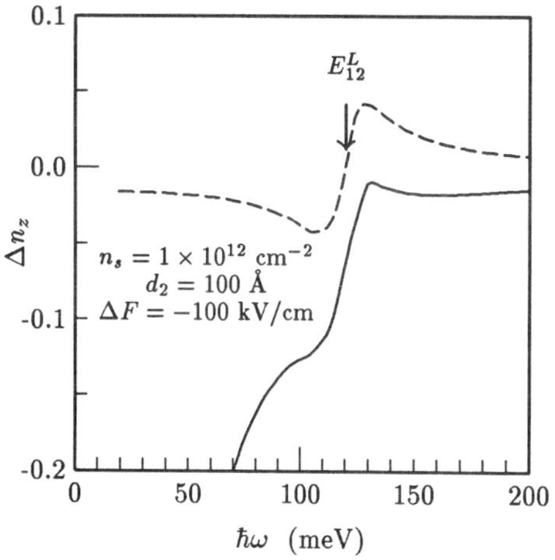


Figure 4. Field-reversal shift of the refractive index vs photon energy for the ADQW illustrated in Fig. 1. The dashed curve is the resonant term alone, while the solid curve also includes the plasma contribution.

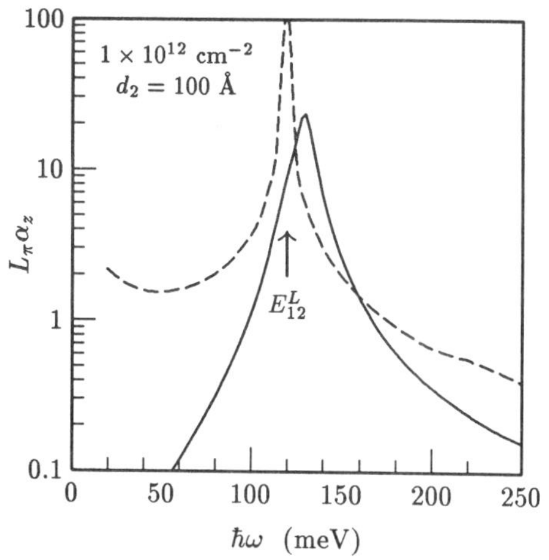


Figure 5. Propagation length required for a phase shift of $\pi$ (normalized by the absorption depth) vs photon energy. The dashed curve again represents the resonant term alone while the solid curve includes $\Delta\epsilon_p$.

The plasma mechanism is particularly attractive because it is not accompanied by any significant absorption as along as the device is operated far off resonance ($|\hbar\omega - E_{12}^L| \gg \Gamma$). This non-resonance feature will be crucial in the design of practical phase modulators with minimal insertion losses. If we define $L_\pi$ to be the propagation length required to achieve a phase shift of $\pi$, the important figure of merit $L_\pi \alpha_z$ must be kept less than unity. Calculated values for this parameter as a function of photon energy are given in Fig. 5, where the Γ-L ADQW structure is the same as that in Fig. 4 and again the dashed curve corresponds to the resonant modulation alone. Note that whereas the performance is marginal for such a device, the additional plasma index shift leads to $L_\pi \alpha_z \ll 1$ at a wide range of both small and large photon energies.

## 4 Conclusions

We have used band structure engineering to design ADQW structures in which an applied bias transfers virtually the entire electron population from Γ states to L states. The modulations of both the absorption coefficient and the refractive index are particularly strong because the resonances are effectively turned on or off by the applied electric field, rather than simply shifted to a different energy.

Unlike previous intersubband devices which had no response at normal incidence, the intensity and phase modulators discussed here are suitable for insertion into Fabry-Perot devices (e.g., using monolithic Bragg reflectors above and below the ADQWs) to significantly enhance the modulation per unit active-layer thickness.[13] Taken together, these effects provide all of the elements required for a wide range of EO devices, including absorption modulators, phase modulators, Fabry-Perot devices, EO switches, beam-steering components, and frequency-conversion elements with voltage-tunable phase matching.[14] Although we have specifically considered Γ-L transfer in GaSb-based ADQWs, the concepts discussed above are equally applicable to Γ-L and Γ-X processes in other III-V heterostructure systems such as GaAs-$Al_xGa_{1-x}As$. We also note that further possibilities emerge when additional substructures (steps, etc.) are incorporated into one or both arms of the ADQW design.

Work at NRL was supported by ONR. We thank Hui Xie for useful discussions and Quantum Semiconductor Algorithms for the use of finite-element band structure software.

## References

1. P. F. Yuh and K. L. Wang, IEEE J. Quant. Electron. 25, 1671 (1989).
2. D. A. Holm and H. F. Taylor, IEEE J. Quant. Electron. 25, 2266 (1989).
3. N. Vodjdani, B. Vinter, V. Berger, E. Bockenhoff, and E. Costard, Appl. Phys. Lett. 59, 555 (1991).
4. E. B. Dupont, D. Delacourt, and M. Papuchon, IEEE J. Quant. Electron. 29, 2313 (1993).
5. C.-L. Yang and D.-S. Pan, J. Appl. Phys. 64, 1573 (1988).
6. E. R. Brown, S. J. Eglash, and K. A. McIntosh, Phys. Rev. B 46, 7244 (1992).
7. L. A. Samoska, B. Brar, and H. Kroemer, Appl. Phys. Lett. 62, 2539 (1993).
8. Y. Zhang, N. Baruch, and W. I. Wang, Appl. Phys. Lett. 63, 1068 (1993).
9. H. Xie, W. I. Wang, J. R. Meyer, C. A. Hoffman, and F. J. Bartoli, J. Appl. Phys. 74, 1194 (1993).
10. L. R. Ram-Mohan and J. R. Meyer, J. Nonlinear Opt. Phys. Mat. 4 (in press).
11. H. Xie, W. I. Wang, and J. R. Meyer, J. Appl. Phys. 76, 92 (1994).
12. E. R. Youngdale, J. R. Meyer, F. J. Bartoli, and C. A. Hoffman, Int. J. Nonlinear Opt. Phys. 1, 493 (1992).
13. R.-H. Yan, R. J. Simes, and L. A. Coldren, IEEE J. Quant. Electron. 27, 1922 (1991).
14. J. R. Meyer, C. A. Hoffman, F. J. Bartoli, H. Xie, and L. R. Ram-Mohan, to be submitted to IEEE J. Quant. Electron.

*Inst. Phys. Conf. Ser. No 144: Section 7*
*Paper presented at 7th Int. Conf. on Narrow Gap Semicond. Santa Fe 9–12 January*




# TWO COLOR $Hg_{1-x}Cd_xTe$ PHOTOCONDUCTIVE IR DETECTORS SENSITIVE TO RADIATION FROM 3−5 μm AND 8−14 μm BANDS

Xing-Can Zhang, Cheng-Cai Huang, Ze-Xiu Wang,
Wei Li, Zheng-Lun Tian, Min Zhang

Kunming Institute of Physics, Yunnan, P.R.China

**ABSTRACT**: Two color $Hg_{1-x}Cd_xTe$ Photoconductive IR detectors which are concentrical sandwich have been fabricated by using n-type bulk crystals in our laboratory. The composition values of $Hg_{1-x}Cd_xTe$ are $x \approx 0.3$ and $x \approx 0.2$, respectively, and hence, the two color detectors are sensitive to radiation from 3−5 μm and 8−14 μm bands. As an example, operating at 77 K with a 300 K, 180° FOV background, peak detectivities $D^*(\lambda_p, 980, 1)$ as high as $5.9 \times 10^{10}$ cmHz$^{1/2}$/W and $2.3 \times 10^{10}$ cmHz$^{1/2}$/W have been achieved for top and bottom elements, respectively, the top element has peak wevelength $\lambda_p$ of 5.2 μm and cut-off wavelength $\lambda_c$ of 5.6 μm, whereas the bottom one has $\lambda_p$ of 12.5 μm and $\lambda_c$ of 13.2 μm. Excepting mentioned above, we also discuss some problems relative to the fabrication and performance of the two color $Hg_{1-x}Cd_xTe$ photoconductive IR detectors.

## 1. INTRODUCTION

The two color IR detectors are widely applied in arears such as acquisition and identification of the target as well as clutter and decoy rejection. Such detectors which are concentrical sandwich configuration allow simultaneous recording of spectral and spatial information without using beam splitting or dispersive optical systems, in consequence, they are extensively studied. The two color detectors responding to radiation from 3−5 μm and 8−12 μm bands, have been reported previously, such as $PbTe/Pb_{0.8}Sn_{0.2}Te$ [1], $InSb/Pb_{0.79}Sn_{0.21}Te$ [2], $InSb/Hg_{0.8}Cd_{0.2}Te$ and $Hg_{0.71}Cd_{0.29}Te/Hg_{0.8}Cd_{0.2}Te$ and so on [3,4].

Two color $Hg_{1-x}Cd_xTe$ photoconductive IR detectors which are concentrical sandwich have been fabricated by using n-type bulk materials with composition values of $x \approx 0.3$ and $x \approx 0.2$, respectively, in our laboratory. Choosing $Hg_{1-x}Cd_xTe$ materials to fabricate two color IR detectors is due to following reasons: the energy gap of $Hg_{1-x}Cd_xTe$ and hence cut-off wavelength, can be adjusted by varying the alloy composition $x$; also, $Hg_{1-x}Cd_xTe$ is readily prepared with n-type doping concentration $< 10^{15}$ cm$^{-3}$ which is ideal for photoconductors, it is possible to obtain high performance photoconductive devices; in addition, they are a direct gap semiconductor with absorption edge so sharp as to allow sufficient transmission beyond the cut-off wavelength. This paper describes the fabrication technique and performance of the two color $Hg_{1-x}Cd_xTe$ photoconductive IR detectors. We also disscuss some problems relative to fabrication technique and performance of two color $Hg_{1-x}Cd_xTe$ photoconductive IR detectors.

## 2. FABRICATION OF TWO COLOR IR DETECTORS

The two color IR detectors are fabricated from n-type $Hg_{1-x}Cd_xTe$ bulk materials grown by either Bridgman method or solid state recrystallization technique. The composition values of materials are $x \approx 0.3$ and $x \approx 0.2$ respectively, and hence, responding to $3-5\,\mu m$ and $8-14\,\mu m$ bands at 77 K. The carrier concentration is generally less than $10^{15}\,cm^{-3}$, whereas carrier lifetime is about $10^{-6}\,s$ at 77 K.

The n-type $Hg_{1-x}Cd_xTe$ ingots are cut into the wafers of about 1 mm thick using a wire saw. Having investigated the composition and electrical properties of these wafers, that satisfying parameter requirements are selected and again cut into smaller chips, then they are carefully polished and thinned to thickness of about $50\,\mu m$. Through suitable cleaning and etching, these chips are further processed to form the elements with a sensitive area of $0.25\,mm^2$ by using photolithographic techniques. Then, gold leads are applied on two ends of elements so as to measure their performance.

For the elements to operate at 77 K, they are first assembled on the cold finger of the dewar which has a Ge transmitting window with antireflection coating on it and then measured. Subsequently, high performance elements are selected for fabrication of two color IR detectors. The bottom element sensitive to $8-14\,\mu m$ band is first affixed to the cold finger of the dewar, then the top one responding to $3-5\,\mu m$ wavelength range is positioned exactly above. The two elements are concentric sandwich, they are either spaced at a distance of about $80\,\mu m$ or directly adhered together by means of a transmitting $8-14\,\mu m$ band adhesive, thus producing the two color IR detector.

## 3. PERFORMANCE OF TWO COLOR IR DETECTORS AND DISCUSSION RELATIVE TO PROBLEMS

The spectral response of a two color $Hg_{1-x}Cd_xTe$ photoconductive detector at 77 K is shown in Fig. 1. Values of cut-off wavelength $\lambda_c$ of for the top and bottom elements are $5.72\,\mu m$ and $11.56\,\mu m$, respectively. They are detemined by the composition of $Hg_{1-x}Cd_xTe$ and operating temperature of detector. Values of cut-on wavelength $\lambda_{con}$ of the detector are $3.1\,\mu m$ for the top element and $6.1\,\mu m$ for the bottom one, respectively. Except for characteristics of oneself element, the former is also concerned in transmission property of the Ge IR window which is shown in

Fig.2; whereas the latter is relative to cut-off wavelength of the top element. We can see from the shape of curves shown in Fig.1 that they decrease sharply than ideal ones with decreasing of wavelength, this may be due to the surface recombination of elements and the effect of transmission property of Ge window on the elements.

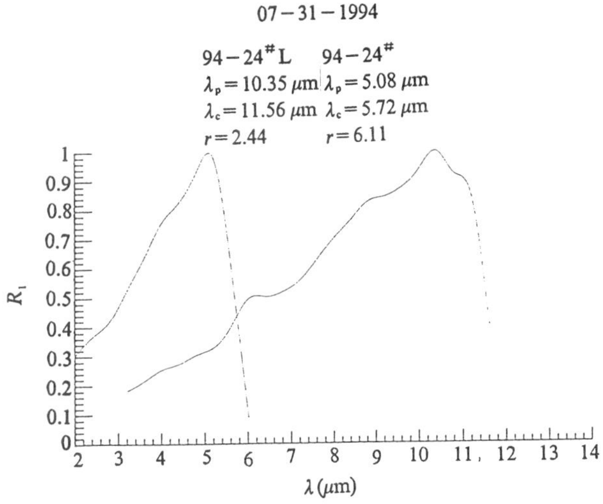


Fig.1 Relative responce spectaum

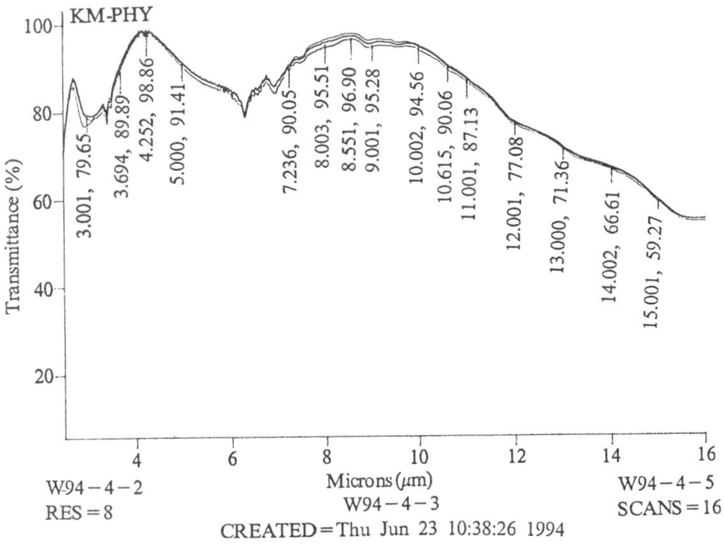


Fig.2 The transmission property of Ge window

In order to understand more the spectral response of the two color IR detectors, Fig.3 shows the spectral curves of the other device, among them, that of the bottom element exhibits a obviously oscillatory behavior, this is to be attributed to optical interference effects.

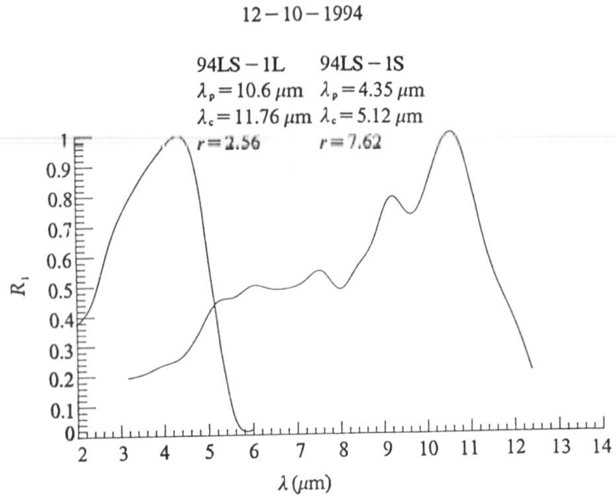


Fig.3 Relative responce spectaum

The performance parameters of a two color $Hg_{1-x}Cd_xTe$ IR detector and their test conditions are summarised in table 1.

Table 1 Performance of a two color $Hg_{1-x}Cd_xTe$ photoconductive IR detector

| Test conditions | | |
|---|---|---|
| Blackbody temperature | 500 K | |
| Chopping frequency | 980 Hz | |
| Incident flux density | $8.76 \times 10^{-6}$ Wcm$^{-2}$ | |
| Test bandwidth | 15.4 Hz | |
| Background temperature | 300 K | |
| Detetor temperature | 77 K | |
| Window material | Ge evaporated antireflection coating | |
| Field of view | 180 ° | |
| **Performance parameters of elements (uncoated)** | | |
| Element | top | bottom |
| Material | $Hg_{1-x}Cd_xTe$ ($x=0.3$) | $Hg_{1-x}Cd_xTe$ ($x=0.2$) |

| | | |
|---|---|---|
| Sinsitive area $A_D$(cm$^2$) | $2.5\times10^{-3}$ | $2.5\times10^{-3}$ |
| Resistance $R_D$($\Omega$) | 129 | 70 |
| Cut-on wavelength $\lambda_{con}$($\mu$m) | 3.3 | 9.8 |
| Cut-off wavelength $\lambda_{coff}$($\mu$m) | 5.6 | 13.2 |
| Peak wavelength $\lambda_p$($\mu$m) | 5.16 | 12.48 |
| Blackbody detectivity $D^*$(500,980,1) (cmHz$^{1/2}$/W) | $9.9\times10^9$ | $7.3\times10^9$ |
| Peak detectivity $D^*(\lambda_p$,980,1) (cmHz$^{1/2}$/W) | $5.9\times10^{10}$ | $2.3\times10^{10}$ |

The $D^*_\lambda$ value of the photoconductive detector is theoretically less than that of the photovoltaic one by a factor of $\sqrt{2}$ for same detector material at 77K. Moreover, taking into account the transmission loss of the Ge window, reflection loss of the top element and bottom one(uncoated), as well as the transmission loss through the top element for the bottom one, one can see from these data listed in table 1 that the detectivity of the top element is close to background noise limited one, whereas, that of the bottom one has attained higher level as well.

## 4. CONCLUTION

The two color $Hg_{1-x}Cd_xTe$ photoconductive IR detectors have been fabricated in our Laboratory by using bulk materials with $x$ values of about 0.3 and 0.2, The detectors are concentrical sandwich. This paper reports the spectral response characteristics and performance parameters of the detectors, and further discusses some problems relative to their performance.

Operating at 77K with a 300K, 180 °FOV background; the $D^*(\lambda_p$, 980, 1) values of $5.9\times10^{10}$cmHz$^{1/2}$/W at 5.2$\mu$m and $2.3\times10^{10}$cmHz$^{1/2}$/W at 12.5$\mu$m peak wavelength have been achieved for the top element and bottom one, respectively.

**ACKMOWLEDGEMENTS** The authors would like to thank Shao-Ze Xiao, Wan-Xing Ren for providing $Hg_{1-x}Cd_xTe$ materials, Bo-Qi Gu, Kai-Fang Cheng for the performance measurement of two color IR detectors, Guo-Zhen Li for technical assistance given us in measuring $D^*_{bb}$, and Hong-Bo Zhang, Xu Hu for typing this paper.

## REFERENCES

[1] A.H.Lockwood, J.R.Raion, P.S.Chai, F.J.Renoa. Infrared Physics Vol.16, p.509–514 (1976)

[2] W.Rolls. Infrared Physics Vol.17, p.419–421, (1977)

[3] P.C.Su, X.B.Yi, H.Ma, G.L.Zen, X.C.Zhang. Infrared Technology Vol.13, No.2 p.3–8 (1990)

[4] G.Fiorito, G.Gasparrini, F.Svelto. Infrared Physics Vol.16, p.531–534 (1976)

*Inst. Phys. Conf. Ser. No 144: Section 7*
*Paper presented at 7th Int. Conf. on Narrow Gap Semicond. Santa Fe 9–12 January*



# Random Telegraph Noise in HgCdTe Photodiodes

Iwao Sugiyama, Toshiyuki Ueda, Nobuyuki Kajihara, and Yoshihiro Miyamoto

Fujitsu Laboratories Ltd. 10-1 Morinosato-Wakamiya, Atsugi 243-01, Japan

**Abstract.** Random telegraph noise (RTN), a discrete fluctuation in the electron current, is common in submicron devices. We observed bistable fluctuations in the dark current of long-wavelength HgCdTe photodiodes. The RTN had several anomalous characteristics but we found some results that suggest an RTN mechanism in the photodiodes. We also found the structures in the reverse bias current-voltage characteristics in the form of increasing steps in the leakage current. The temperature dependencies of both the RTN and the steps on the leakage current suggests that both phenomena originated from the same tunneling. We propose that this tunneling occurred at inclusion-defects in the depletion layer. The capacitance-voltage characteristics of the diode suggested that the inclusions exist in the depletion layer. The kind of defects is still unclear. We propose that the most likely mechanism for RTN is the modulation of the barrier height of the tunneling by the change in the charge state near the inclusion-defects.

## 1. Introduction

Random telegraph noise (RTN) is a discrete fluctuation in the current in electron devices. This fluctuation is generally attributed to the capture and emission of electrons by the single trap. Hence, it is commonly observed in submicron devices[1]. The fluctuation between two discrete levels makes a Lorentzian distribution in the noise power spectrum. Since the superposition of such spectrums with a set of various time constant shows 1/f noise characteristics, study of RTN has centered on the origin of the 1/f noise[2].

Excess low-frequency noise, including 1/f noise in HgCdTe photodiodes, often limits the performance of infrared-focal-plane arrays. However little attention has been paid to discrete noise in HgCdTe photodiodes. We found bistable fluctuations in the dark current of HgCdTe photodiodes. Although the photodiodes were large devices, there is nothing unusual in this case because several papers have reported such noise in large devices. These papers proposed models that shows that RTN is attributed to single defects[3]. We present the RTN characteristics in HgCdTe photodiodes. Although RTN had several anomalous characteristics, some results suggested a mechanism exists.

## 2. Experiments

The diodes we used in this experiment were fabricated on liquid-phase-epitaxy grown $Hg_{0.78}Cd_{0.22}Te$. As illustrated in Figure 1, the photodiodes were gate-controlled $n^+$/p type. The p-type layer was Hg-vacancy doped, while the $n^+$ region was formed by the standard $B^+$ implantation technique. The area of the implanted region was $230 \times 230$ μm.

The measurements were made without illumination. Although the gate electrode can adjust the surface potential near the p-n junction, it was normally connected to the anode. Since the surface fixed charge density was as low as $10^9$ cm$^{-2}$, this condition provided a minimal leakage current. The noise current was amplified by a low-noise current amplifier and viewed on a spectrum analyzer. A DC-leakage current was measured by source monitors and both capacitance and conductance were measured by a precision LCR meter.

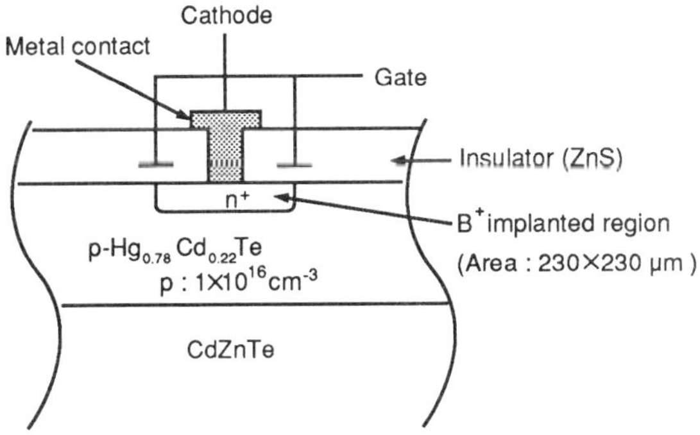


Figure 1. Structure of photodiode.

## 3. Results and Discussion

### *3.1. RTN characteristics*

RTN even appeared occasionally in the reverse-bias diode with a high dynamic resistance. No diode in particular has RTN. We confirmed that this fluctuation did not come from the electrode contacts or from any other chatter in the circuits. We often observed large variations in the RTN amplitude. Each type of RTN had a different current amplitude and fluctuation frequency, and typically RTN amplitudes were several hundred picoamperes. We confirmed that the RTN amplitude did not depend on the photocurrent, so the RTN is only appeared in the dark current.

The RTN frequencies also covered a wide-range from several hundreds kHz to below 1 Hz. The transition time showed a Lorentzian distribution and the mean transition time was reduced by increasing the temperature in a thermally-activated manner. This characteristic is consistent with RTN in other materials. Figure 2 shows an example of RTN that had a very clear waveform and a large amplitude. We focused on this RTN for the characterization.

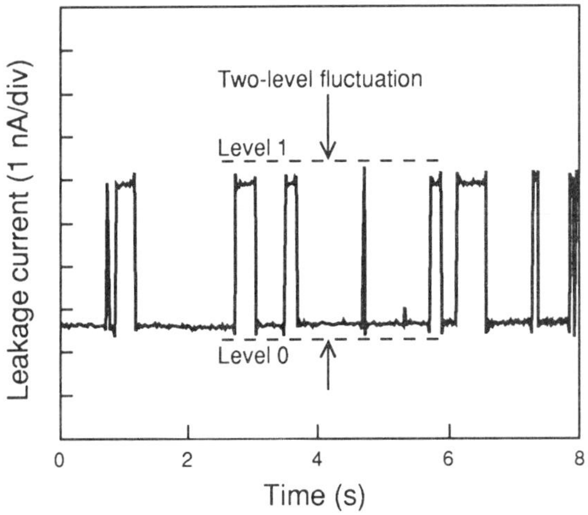


Figure 2. RTN waveform.

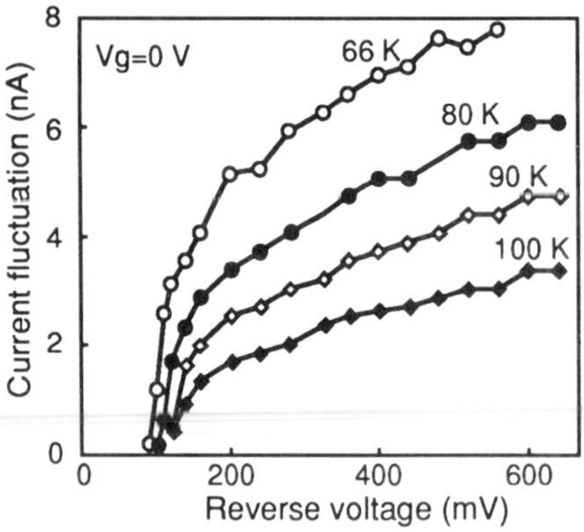


Figure 3. RTN current amplitude against reverse bias.

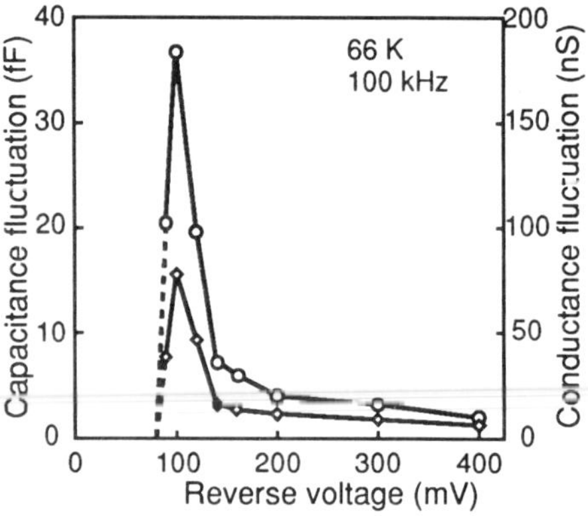


Figure 4. RTN in both capacitance and conductance. Upper line shows capacitance fluctuation and lower line shows conductance fluctuation.

The RTN-current amplitude strongly depended on the reverse voltage. Figure 3 shows the reverse characteristics of RTN. We found that the RTN current had a threshold voltage, about 100 mV in this case, and it increased abruptly just above this voltage. The temperature dependence of RTN was negative. Only tunneling currents can have a negative temperature coefficient in the leakage current component of HgCdTe photodiodes. Therefore the RTN originates from tunneling. Also, we confirmed that RTN did not depend on the gate voltage. This means that RTN did not have any surface-leakage component.

In contrast to the RTN in the current, we found that bistable fluctuations synchronized with the RTN current occurred in both the junction capacitance and conductance. Since these were very small fluctuations, a large time constant was needed to get a clear waveform. We can only detect such fluctuations at a low temperature(66 K) so that the transition is very slow.

The fluctuation in capacitance represents the change in the space-charge density in the depletion layer, or the fluctuation of the depletion layer width. Clearly from Figure 4, the reverse characteristics of these two were different from those of the RTN current. Both fluctuations had peaks near or above the threshold voltage.

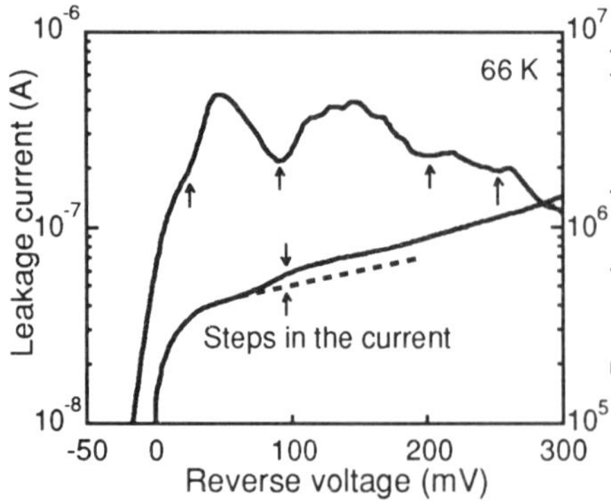


Figure 5. Structures in I-V. Upper line shows dynamic resistance and lower line shows leakage current.

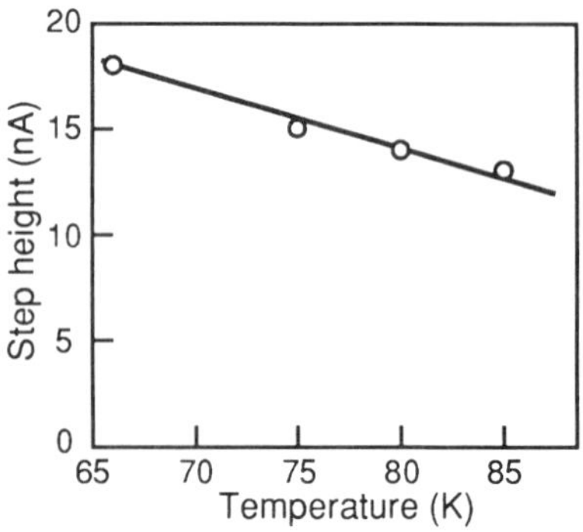


Figure 6. Step height against temperature.

### 3.2. *Current-voltage characteristics*

Since the RTN was only observed in the reverse dark current, the current-voltage (I-V) characteristics give us some information about the origin of the RTN. Figure 5 shows the reverse bias I-V characteristics at 66 K. We found that several humps, indicated by arrows, exist in the dynamic resistance at reverse bias. These humps correspond to the steps in the current. Such structures have been reported for cases of low forward bias[4]. It was concluded that the structures in the forward bias originated from impurity-to-band tunneling.

Notice that one of the steps occurred at the reverse voltage of 100 mV, corresponding to the threshold voltage of the RTN current. Also, one of the humps in the dynamic resistance near or above this bias was consistent with the characteristic of conductance fluctuation of the RTN. We measured the height of this step between 66 and 85 K. As indicated in Figure 6, the step height is decreased with increasing the temperature. This means that the temperature dependence of the step height was negative. Considering RTN current also have negative temperature coefficient and that the other dark-current components, -diffusion and g-r, have positive temperature coefficients, we can conclude that both the RTN and the structures in the I-V originate from the same tunneling phenomena. This is to say that the step in the leakage current was modulated by a defect and generated RTN.

### 3.3. *Inclusions in the depletion layer*

We calculated the mean depletion layer width from the capacitance-voltage characteristics of our diodes. It was about 0.82 μm at a bias voltage of zero, and expanded to about 1.2 μm at 300 mV. The wide depletion layer suggests that an $n^-$ layer exists between the $n^+$ and p-type layers. This $n^-$ layer was created by post-implantation annealing[5]. So far, based on this result, the calculated tunneling current was very small because of the small electric field. Tunneling can be generated if $n^+/p$ junctions exist. Therefore we propose that small inclusions exist in the depletion layer, producing a kind of $n^+/p$ dipole. As shown in Figure 7, if the depletion layer crosses the inclusion, the tunneling current would occur at this small pn junction. Although we can not identify the type of defect for such an inclusion, some reports proposed similar inclusions or inhomogeneity to the anomalous Hall effect in HgCdTe[6]. According to our inclusion model, when the depletion region crosses the inclusion-defect, a high-concentration region near the defects holds the depletion layer and causes bending in the diode's capacitance-voltage characteristic. This phenomenon was presented in Figure 8.

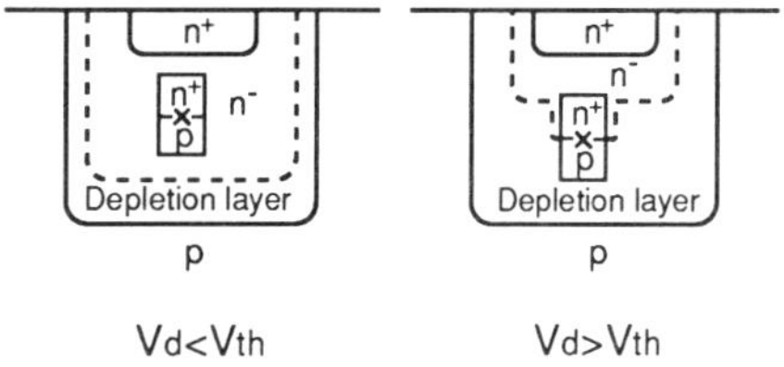


Figure 7. Inclusion model of photodiode.

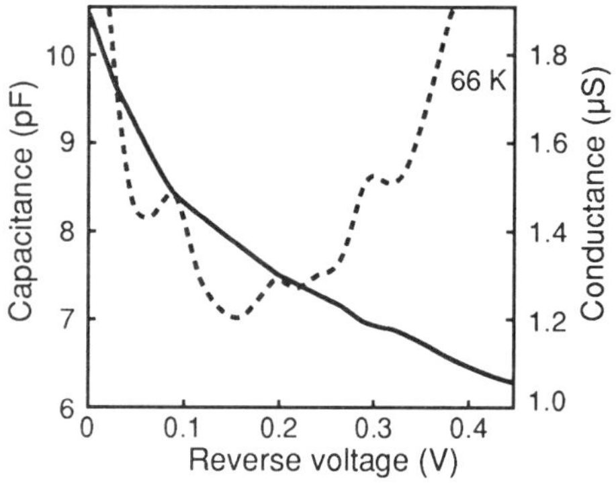


Figure 8. Capacitance-voltage characteristics. Solid line shows capacitance and dotted line shows conductance.

The peaks in the conductance shown in Figure 8 correspond to the structures in the dynamic resistance. Tunneling would occur at this bias voltage. Therefore these results suggest that the inclusion-defect produces leakage paths between the cathode and anode when the depletion crosses the defect.

If the charge state around the defect is changed, the barrier height is reduced or increased by the trapping of electrons, thus producing a fluctuation in the tunneling current. This kind of enhanced tunneling was numerically analyzed for the case of the $Al\text{-}Al_2O_3\text{-}Al$ barrier[7]. Also, the existence of activation energies in the transition rate of RTN suggests that a single trap produces the fluctuations.

## 4. Conclusions

The RTN in HgCdTe photodiodes shows several anomalous properties but some results suggest a mechanism exists. We found that structures exist in the I-V characteristics of diodes. Since both the RTN and the structure in the I-V have similar properties, we suggest that a step in the leakage current was modulated by a defect and became RTN. We also found bending phenomena in the capacitance-voltage characteristics of the diode. This indicates the existence of small inclusion-defects in the depletion layer. The excess tunneling current at the inclusion was the origin of the RTN. The kind of defects producing such inclusions is still unknown.

## Acknowledgments

We thank H. Nakamura for her assistance, and K. Ozaki and M. Tanaka for fabricating the photodiodes. We also thank K. Yamamoto and M. Dohi for their support during this work.

## References

[1] Ralls K S, Skocpol W J, Jackel L D, Howard R E, Fetter L A, Epworth R W, and Tennant D M 1984 *Phys. Rev. Lett.* **52** 228-31

[2] Uren M J, Day D J, and Kirton M J 1985 *Appl. Phys. Lett.* **47** 1195-7

[3] Campbell P M, Snow E S, Moore W J, Glembocki O J, and Kirchoefer S W 1991 *Phys. Rev. Lett.* **67** 1330-3
Judd T, Couch N R, Beton P H, Kelly M J, Kerr T M, and Pepper M 1986 *Appl. Phys. Lett.* **49** 1652-3
Cavicchi R E and Panish M B 1990 *J. Appl. Phys.* **67** 873-7

[4] DeWames R E, Pasko J G, Yao E S, Vanderwyck A H B, and Williams G M 1988 *J. Vac. Sci. Technol.* **A6** 2655-63
Anderson W W and Hoffman H J 1982 *J. Appl. Phys.* **53** 9130-45

[5] Bubulac L O 1988 *J. Crystal Growth* **86** 723-34

[6] Chen M C, Parker S G, and Weirauch D F 1985 *J. Appl. Phys.* **58** 3150-3
Ivanov-Omskii V I, Berchenko N N, and Elizarov A 1987 *Phys. Stat. Sol.* **(a)103** 11-28

[7] Schmidlin F W 1966 *J. Appl. Phys.* **37** 2823-32

*Inst. Phys. Conf. Ser. No 144: Section 7*
*Paper presented at 7th Int. Conf. on Narrow Gap Semicond. Santa Fe 9–12 January*



# Electronic and optoelectronic devices in narrow-gap semiconductors

T. Ashley

Defence Research Agency, St. Andrews Road, Malvern, Worcestershire, WR14 3PS, UK.

**Abstract.** We review recent work at the DRA to produce a range of electronic and optoelectronic devices made from the narrow-gap semiconductor indium antimonide. The aim of the research has been to remove the general requirement for cryogenic cooling and, as far as possible, enable ambient temperature operation. Heterostructure devices which include detectors, light emitting diodes, negative luminescence devices and transistors are described.

## 1. Introduction

The narrow-gap semiconductors indium antimonide (InSb) and mercury cadmium telluride ($Hg_{1-x}Cd_xTe$) have formed the basis of infrared detection systems for many years. The potential of these materials for use in many applications in addition to that of infrared detection, such as infrared emitters and ultra high speed transistors, has also been known for a long time. At room temperature, however, the narrow-gap materials have an intrinsic carrier density which is between $10^{16}cm^{-3}$ and $10^{17}cm^{-3}$, and a thermal generation rate of the order of $10^{23}cm^{-3}s^{-1}$ to $10^{24}cm^{-3}s^{-1}$, the consequence of which is a large leakage current and noise in detectors and transistors, and a potentially poor radiative efficiency in emitting devices. The solution which is employed for detectors is to cool to 80K, or below, which necessitates the use of expensive encapsulations and cooling mechanisms, and leads to additional bulk and weight. This prevents their penetration into several potential civil markets, where cost is of prime importance, and also limits their use in many military areas. The need for cryogenic cooling largely negates the advantages of the narrow-gap materials in the other application areas where alternatives, albeit inferior, do exist.

The concept of 'non-equilibrium' operation [1] and the ability to grow by epitaxial techniques the multi-layer structures needed to achieve it, now offer the possibility of uncooled devices in several of these areas. In this paper we review recent work at the DRA aimed at achieving a range of devices based on InSb which will operate with zero or minimal cooling. The basic structure which is common to all of the devices is described briefly in section 2, and its impact on the performance potential of infrared photodiodes is discussed in section 3. The electroluminescence behaviour of the diodes is dscribed in section 4. Results from field-effect transistors, which confirm their potential for extremely high frequency operation with low power dissipation, are given in section 5.

## 2. General device structure

We have made so-called 'non-equilibrium' or 'HOT' (Higher Operating Temperature) devices from InSb and $Hg_{1-x}Cd_xTe$. The structures differ in detail, however the principles of operation for both material systems are similar, so results will be given for the InSb devices only. In general we have an 'active' region which has a low doping level such that it is intrinsic or near intrinsic at room temperature, to which n- and p-type contacts are made via regions which have either a high doping level or a wider energy-gap or both, such that they give minimal transport of minority carriers into or out of the active region under the appropriate bias conditions.

The basic structure which is employed for all of the InSb devices is shown in figure 1 together with an energy band diagram. It has the form $p^+\underline{p}^+\pi n^+$ or $p^+\underline{p}^+\upsilon n^+$, where $\pi$ and $\upsilon$ refer to material which is near intrinsic p- or n-type respectively and which forms the 'active' region, and the underline represents material of a wider energy-gap. The active region can range in thickness from about 0.3$\mu$m to 5$\mu$m, depending on the type of device, whilst the $n^+$ and $p^+$ regions are typically about 1$\mu$m thick. The $n^+$ InSb is heavily degenerate and therefore has a very low hole density and results in a comparatively large step ($\sim$180meV or 7kT at room temperature) in the valence band at the $\pi n^+$ or $\upsilon n^+$ junction. In contrast, $p^+$ InSb is not highly degenerate therefore we use a region of indium aluminium antimonide ($In_{1-x}Al_xSb$ with $x \sim 0.15$) to give a large step ($\sim$250meV or 10kT) in the conduction band at this junction. The $In_{1-x}Al_xSb$ has a lattice constant which is smaller than InSb such that with $x=0.15$ there is a lattice mismatch of 0.6%, so the thickness of the wider-gap region is limited to a few hundred Angstoms to avoid strain relaxation [2]. A thickness of 200Å is used typically, which is sufficient to prevent electron tunnelling and so form a barrier to electron flow at this point. The $p^+$ InSb then forms a low resistance and relatively low recombination velocity contact to the $In_{1-x}Al_xSb$.

The layers are grown by molecular beam epitaxy in a VG Semicon V80H system onto (001) InSb substrates using procedures described previously [3]. The substrate temperature is 420°C, calibrated with reference to a surface phase transition monitored with RHEED, the growth rate is 0.5$\mu$mhr$^{-1}$, and the group V to group III absolute flux ratio is between 1.5:1 and 2:1. The $n^+$ material is doped with silicon to a chemical level of $2x10^{18}cm^{-3}$, whilst the $p^+$ material is doped with beryllium to a similar level.

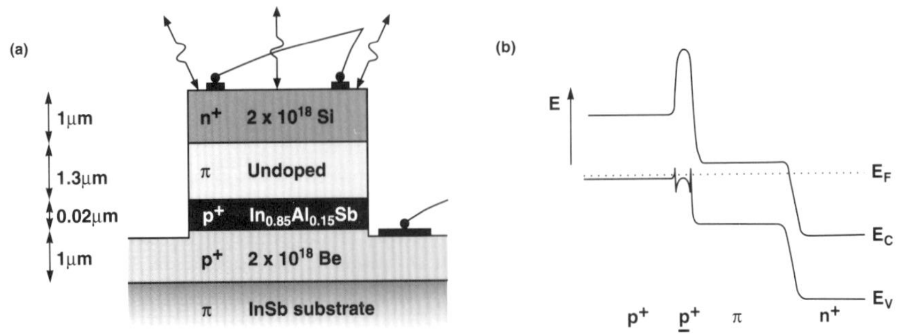


*Figure 1.* Schematic structure (a) and energy-band diagram (b) of basic InSb epitaxial device.

### 3. HOT detectors

Very high performance 2D arrays of detectors fabricated by ion implantation or diffusion of a p-type dopant into an n-type InSb substrate and cooled to 80K are now available. By careful choice of doping levels, the diffusion length of holes in the n-type material can be made similar to the absorption depth of the infrared radiation, so that the quantum efficiency and sensitivity can be optimised. As the temperature is raised, this optimisation becomes difficult to achieve, and also the leakage current from the $p^+$ material becomes very significant so the performance degrades rapidly. The diodes cease to be background limited at temperature above about 150K, and at ambient temperature they are virtually ohmic.

Our epitaxial devices provide a means of optimising the active part of the device to achieve a high quantum efficiency within the minimum volume of material over a wide temperature range. At zero bias, which is the usual state for detector operation at present, the band structure described above ensures that there is little transport of the respective minority carriers from the contact regions, and hence little additional noise. The effect which this has on detector performance can be seen in figure 2, which compares theoretical curves for detectivity, calculated from the zero bias resistance, for a conventional $p^+n$ diode and an epitaxially grown $p^+\underline{p}^+\nu n^+$ structure with a 3$\mu$m thick active region. We see, for example, an increase in operating temperature of about 40K in the vicinity of 200K. At ambient temperature (294K) the D* of the epitaxial devices is $2.5 \times 10^9$cmHz$^{0.5}$W$^{-1}$, calculated from measured values of $R_0A$, which is an order of magnitude higher than that of commercially available single element thermal detectors.

An additional factor which determines the detector sensitivity as the operating temperature of the InSb detectors is increased is that the energy-gap of the semiconductor decreases, such that the cut-off wavelength extends from 5.5$\mu$m at 80K to 6.0$\mu$m at 200K and 7.0$\mu$m at 300K. This means that the detector ceases to be well matched to the 3-5$\mu$m atmospheric transmission window, leading to additional noise without a corresponding increase in signal. One solution is to form the active region from $In_{1-x}Al_xSb$ with a composition such that the cut-off wavelength is decreased to the optimum value. To obtain a 5$\mu$m cut-off would require $x \sim 0.023$ at 200K and $x \sim 0.039$ at room temperature, and the mismatch of this material to an InSb substrate would be too large for a useful thickness to be grown.

To overcome this problem, we have chosen to attempt to grow crystals of $In_{1-y}Ga_ySb$ by a double crucible Czochralski technique to provide a source of lattice matched substrates

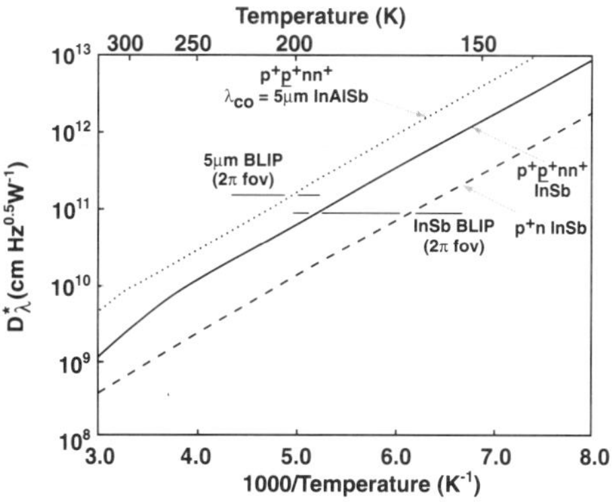


*Figure 2.* Calculated detectivity versus temperature for photodiodes comparing conventional InSb (dashed), epitaxial InSb (solid) and epitaxial InAlSb with 5$\mu$m cut-off wavelength (dotted).

for the growth of the $In_{1-x}Al_xSb$, details of which are reported in another paper presented at this conference [4]. A gallium fraction of y=0.035 is required to match to the lattice constant of $In_{1-x}Al_xSb$ with a cut-off wavelength of 5$\mu$m at room temperature, and single crystals of this composition have been grown. The detector performance which is expected to ensue from this technique is shown in figure 2, where the predicted detectivity for material with a constant 5$\mu$m cut-off wavelength is plotted. The improvement in $D^*$ is about a factor of 2.5 over epitaxial InSb devices and a factor of 12 over conventional InSb detectors. Background limited performance, in a $2\pi$ field of view, is indicated up to a temperature of 200K, which is achievable with Peltier coolers, leading to the possibility of high performance, compact, comparatively inexpensive imaging systems.

## 4. Electroluminescent devices

Sources of mid to long wavelength infrared radiation which operate efficiently at ambient temperature would find numerous applications in spectroscopy and communications. Convenient, compact sources would be of particular interest for use in infrared gas analyzers, as there are many noxious industrial gases which have absorption spectra in this wavelength range. Examples include $CH_4$ (3.3$\mu$m), HCl (3.54$\mu$m), $CO_2$ (4.2$\mu$m), CO (4.6$\mu$m), NO (5.3$\mu$m), $NH_3$ (6.1$\mu$m), $SO_x$ (7.4$\mu$m) and $O_3$ (9.5$\mu$m). Current point gas sensors generally employ filament lamps as the source, however these have an extremely low output at wavelengths in excess of 4.5$\mu$m, have a limited life and are not intrinsically safe in inflammable environments. Electroluminescence from diodes at room temperature has been reported from a variety of materials in the 3$\mu$m to 4.6$\mu$m range, including PbSSe at 4.6$\mu$m [5], $PbSe/Pb_{1-x}Sr_xSe$ at 4.4$\mu$m [6], $Hg_{1-x}Cd_xTe$ at 3.5$\mu$m [7], $InAs_{1-x-y}Sb_xP_y$ at 4.6$\mu$m [8], $InAs_{1-x}Sb_x/GaSb$ at 4$\mu$m [9] and $InAs_{1-x}Sb_x/InAs$ at 4.5$\mu$m [10]. We have recently reported preliminary results on InSb and $Hg_{1-x}Cd_xTe$ diodes in the 5-7$\mu$m range [11,12]. Luminescence has also been observed from InSb and $Hg_{1-x}Cd_xTe$ structures without a diode junction by the use of crossed electric and magnetic fields [13,14] or due to carrier accumulation/exclusion effects [15].

Our epitaxial device structure affords very good confinement of carriers in the active region to facilitate efficient electroluminescence. Electrons injected from the $n^+$ contact are contained by the barrier in the conduction band at the $\underline{p}^+\pi$ junction, and holes injected from the $p^+$ contact are contained by the large step in the valence band at the $\pi n^+$ junction. Numerical modelling indicates that more than 99% of the current through the device involves recombination in the active region, even at very high injection levels where the electron quasi fermi energy is several kT above the conduction band edge.

The experimental devices were made from material with a 1.3$\mu$m thick active region and with the $n^+$ region uppermost. They were etched down to the $p^+$ region to form mesas 300$\mu$m in diameter, then passivated with an anodic oxide. Chromium/gold contacts were applied separately to each $n^+$ region in the form of an annulus, with internal diameter of 180$\mu$m and external diameter of 240$\mu$m, and in common to the $p^+$ region. The devices were then anti-reflection coated with 0.7$\mu$m of tin oxide.

Infrared microscope images of the device under forward bias at ambient temperature show uniform electroluminescence from the whole surface. The spectral emission characteristics of the device under low forward bias are shown in figure 3, and the emission is seen to peak at 5.8$\mu$m. Even near equilibrium there is a significant population of electrons in the conduction band which leads to a small Moss-Burstein shift in the emission, compared with that predicted solely from the energy-gap. High forward injection, ie a

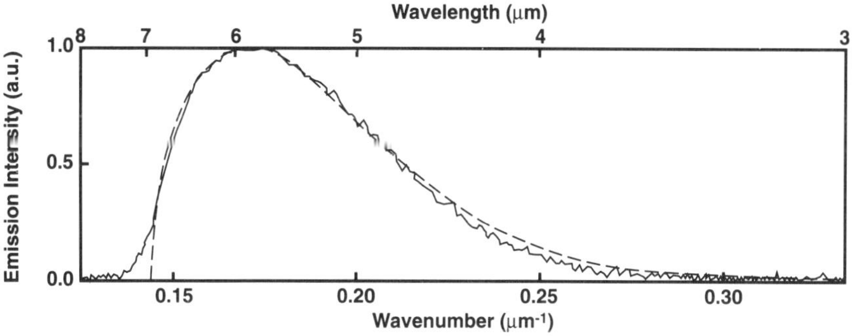


*Figure 3.* Electroluminescence spectrum of an InSb diode at 294K, showing experimental data (solid) and theoretical fit (dashed).

current approximately fifty times the saturation leakage current, lead to a further shortening in the peak wavelength by 0.3$\mu$m to 5.5$\mu$m. Figure 3 also shows a theoretical spectrum which was calculated including the effects of re-absorption, band nonparabolicity and degeneracy, with only the electron quasi fermi level as a fitting parameter. Very good agreement with the experimental spectrum is obtained.

The output power of the electroluminescence as a function of current is shown in figure 4, and it can be seen that values in excess of 10$\mu$W (14mWcm$^{-2}$) are obtained. A pulse bias with a duty cycle of 40:1 was used to minimise heating effects. The relationship is approximately of the form power $\propto$ current$^z$ with z=0.81, in reasonable agreement with that expected from the ratio of radiative recombination rate (light-hole plus heavy-hole) to total recombination rate (radiative plus Auger-1 and Auger-7) where z$\sim$0.67. At low forward bias, ie near equilibrium, the measured external efficiency (which is defined as $\eta_{ext}$ = 2 x photon flux / current) is 0.06%, which corresponds to an internal efficiency of 5% in excellent agreement with the predicted value.

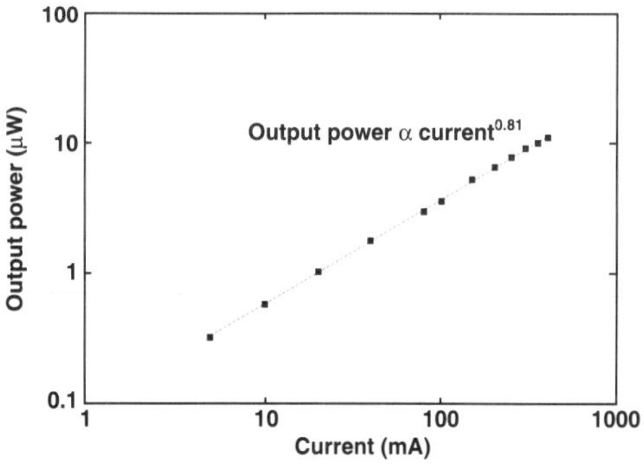


*Figure 4.* Output power as a function of forward current of an InSb electroluminescent diode at 294K.

## 5. Transistors

At 294K InSb has a bulk electron mobility of $7.8\times10^4 cm^2V^{-1}s^{-1}$, which is nearly ten times that of GaAs and fifty times that of silicon, and a saturation velocity of at least $5\times10^7 cms^{-1}$, which exceeds the wider-gap materials by a factor of five. Hence there is a potential for very high frequency operation in field-effect transistors with only a low voltage, and therefore low power requirement. The large intrinsic carrier concentration means that junctions can pass large currents with only a small voltage applied, so there is also scope for low power operation of high frequency InSb bipolar transistors. The large intrinsic carrier concentration also presents a problem to ambient temperature operation because it causes large leakage currents to flow which result in a very low output impedance or small dynamic range. One solution for field-effect transistors is to grow a very thin region of InSb on an insulating substrate so that the volume of conducting material is very small - a 'thin-film' transistor. This has been attempted [16] but with limited success as the device had only a 7dB dynamic range and a very low effective electron mobility of $3500 cm^2V^{-1}s^{-1}$.

The use of exclusion and extraction to reduce the free carrier concentrations enables us to design both field-effect and bipolar transistors which will operate successfully at ambient temperature. To date these have only been implemented in the case of the field-effect structures, and the first 5$\mu$m long gate, relatively low frequency MISFETs which verified the principles have been reported [17].

We have recently reported enhancement-mode MISFETs with a nominal 1$\mu$m gate length and a metal interconnect pattern better optimised for high frequency operation [18]. The growth sequence for the material was similar that used for the emitters and detectors, except that the active region was only 0.7$\mu$m thick. The channel of the MISFET was formed by dry etching through the $n^+$ layer into the $\pi$ region, followed by three anodize and strip stages to remove damage, see figure 5. The channel was then re-anodized and a thin $SiO_2$ layer sputtered to form the gate insulator with a total thickness of 70nm. The overall device shape was defined by etching to the $p^+$ region. An $SiO_2$ layer, up to 2.7$\mu$m thick, was then sputtered over part of this region to provide low capacitance isolation of the metal interconnects from the conducting, intrinsic substrate. Chromium/gold was used to form the source and drain contacts to the $n^+$ mesas and to provide a common 'substrate' contact to

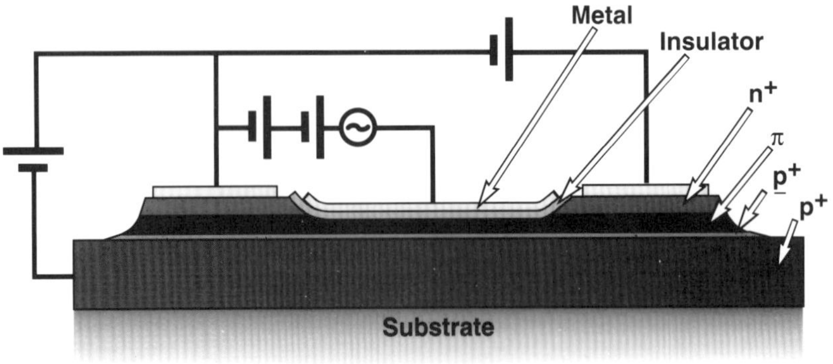


*Figure 5.* Schematic cross section of an InSb MISFET, showing bias arrangements.

the $p^+$ region. Aluminium/gold was used for the gate. In enhancement-mode devices the gate metal must completely span the channel, but overlap onto the source and drain regions should be avoided to minimise additional capacitance. In practice, the devices reported here have an actual channel length $0.3\mu m$ shorter than the nominal value and a gate metal length about $0.8\mu m$ longer, so there is an overlap of more than $1\mu m$.

The transistors are assessed in a grounded source configuration, with an additional dc negative bias applied to the $p^+$ contact so that the $\pi$ region, in which the channel is formed, can be extracted. The static output characteristics of a $140\mu m$ wide device with a substrate bias of -0.45V are shown in figure 6, and have a dynamic range of more than 20dB and output impedance of $45\Omega$ at a source-drain voltage ($V_{ds}$) of only 0.4V. If a negative bias is not applied to the $p^+$ contact, the minimum current which flows is increased by a factor of 28, so the dynamic range is only 1dB, and the output impedance is reduced to only $7\Omega$.

The high frequency performance was measured up to 26GHz on a vector network analyzer. The $140\mu m$ wide devices typically had a transconductance ($Y_{21}$) at 1GHz of 32mS ($230mSmm^{-1}$), and cut-off frequency where the current gain was unity ($f_T$) of 14GHz. Comparison with devices with a $40\mu m$ wide gate permits the capacitance due to the interconnects to be deduced and an 'internal' $f_T$ of 17GHz estimated. The total gate metal length, including that deposited on the sides of the source and drain mesas is up to $3.8\mu m$, compared with a channel length of only $0.7\mu m$. The 'intrinsic' frequency response can be inferred from the ratio of these lengths and is 92GHz. This extrapolation is supported by comparison of input capacitances on devices with different lengths.

The room temperature static electron mobility in the channel of $5\mu m$ long devices has been measured by magneto-resistance experiments. The transistors were placed in a perpendicular magnetic field of up to 3T and operated in the linear region with $V_{ds}=0.05V$. The transconductance ($g_m$) is then proportional to the effective mobility ($\mu_{eff}$), so we have

$$g_m(B)/g_{m,0} = 1/(1+\mu_0^2B^2)$$

and the zero field mobility ($\mu_0$) can be obtained by fitting to the variation in $g_m$ with

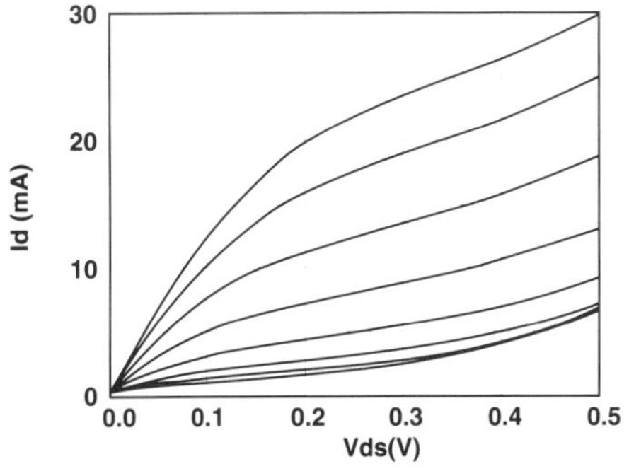


*Figure 6.* Output characteristics of a $140\mu m$ wide extracted InSb MISFET at 294K, gate voltages from -1.5V to +1.5V in increments of +0.5V

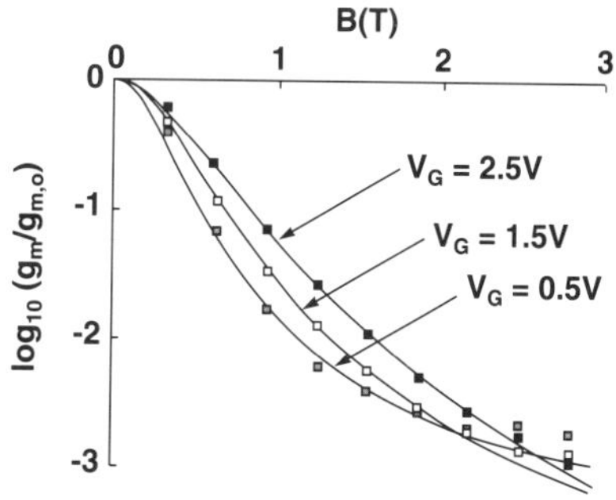


*Figure 7.* Transconductance vs magnetic field of an InSb MISFET at 294K for three gate voltages, showing experimental data (discrete) and theoretical fits (solid).

magnetic field, including the effects of magnetic quantisation. The results for several values of gate voltage are shown in figure 7, and values of mobility from $2.5x10^4cm^2V^{-1}s^{-1}$ just above threshold ($V_g$=+0.5V) to $1.6x10^4cm^2V^{-1}s^{-1}$ under heavy inversion ($V_g$=+2.5V) are deduced.

The high frequency data described above were taken under conditions of moderate inversion, therefore a mobility of $2.0x10^4cm^2V^{-1}s^{-1}$ is appropriate. The pinch-off voltage was 0.13V, so a carrier velocity of up to $3.7x10^7cms^{-1}$ might be attained, leading to a transit time limited frequency response of 84GHz, in reasonable agreement with the extrapolated microwave data.

## 6. Conclusion

We have demonstrated the feasibility of room temperature operation of a variety of devices made from InSb. These have included detectors with $D^*$ an order of magnitude higher than thermal detectors; infrared electroluminescence diodes operating over the 4$\mu$m to 7$\mu$m range with theoretical efficiency; and MISFETs which confirm the potential of InSb for ultra high frequency operation with low power requirement.



*Acknowledgements* I would like to thank colleagues at the DRA who contributed to this research - A B Dean, C T Elliott, N T Gordon, R S Hall, R Jefferies, A D Johnson, F. Khaleque, G J Pryce, A M White and H Willis. This work was supported by the Strategic Research Committee of the UK MoD.

**References**

[1] Ashley T, Elliott C T and Harker A T 1986 *Infrared Phys.* **26** 303
[2] Ashley T, Dean A B, Elliott C T, Johnson A D, Pryce G J, White A M and Whitehouse C R 1993 *Semicond. Sci. Technol.* **8** S386
[3] Williams G M, Whitehouse C R, Martin T, Chew N G, Cullis A G, Ashley T, Sykes D E, Mackay K, and Williams R H 1988 *J. Appl. Phys.* **63** 1526
[4] Ashley T, Beswick J A, Cockayne B. and Elliott C T 1995 *Proc. 7th Int. Conf. Narrow Gap Semiconductors, Santa Fe, 8-12 January 1995.*
[5] Lo W and Swets D E 1980, *Appl. Phys. Lett.* **36** 450
[6] Spanger B, Schiessl U, Lambrecht A, Bottner H, and Tacke M 1988, *Appl. Phys. Lett.* **53** 2582
[7] Tarry H A 1986, *Electron. Lett.* **22** 416
[8] Malinen J, Hannula T, Zotova N V, Karandashov S A, Markov I I, Matveev A, Stus N M and Talalakin G N 1993 *Proc. SPIE 'Optical methods for process control' Boston 7-10 Sept* **2069**
[9] Mao Y and Krier A 1994 *J. Elect. Materials* **23** 503
[10] Tang P J P, Pullin M J, Chung S J, Phillips C C, Stradling R A, Norman A G, Li Y B and Hart L 1995 *Appl. Phys. Lett.* in press
[11] Ashley T, Elliott C T, Gordon N T, Hall R S, Johnson A D, and Pryce G J 1994, *Appl. Phys. Lett.* **64** 2433
[12] Ashley T, Elliott C T, Gordon N T, Hall R S, Maxey C D and Matthews B E 1994, *Appl. Phys. Lett.* **65** 2314
[13] Morimoto T and Chiba M 1984 *Jpn. J. Appl. Phys.* **23** L821
[14] Bolgov S S, Malyutenko V K, Pipa V I and Savchenko A P 1991 *Sov. Tech. Phys. Lett.* **15** 641
[15] Bolgov S S, Kislyi V P, Malyutenko V K and Savchenko A P 1993 *Semiconductors* **27** 93
[16] Ohashi T, Bour D P, Itoh T, Berry J D, Jost S R, Wicks G W and Eastman L F 1986 *J. Vac Sci Technol.* **B4** 622
[17] Ashley T, Dean A B, Elliott C T, McConville C F, Pryce G J and Whitehouse C R 1991 *Appl. Phys. Lett.* **59** 1761
[18] Ashley T, Dean A B, Elliott C T, Pryce G J, Johnson A D and Willis H 1995 *Appl. Phys. Lett.* **66** 481

*Inst. Phys. Conf. Ser. No 144: Section 7*
*Paper presented at 7th Int. Conf. on Narrow Gap Semicond. Santa Fe 9–12 January*



# Modeling of multiple InGaAs/GaAs quantum well infrared photodetectors

**M Ershov and V Ryzhii**

Computer Solid State Physics Laboratory, University of Aizu, Aizu-Wakamatsu, 965-80 Japan

**Abstract.** InGaAs/GaAs Quantum Well Infrared Photodetectors (QWIP) are studied theoretically using a numerical modeling. It is shown that triangular emitter and collector barriers provide better transport conditions and higher QWIP performance as compared to rectangular barriers. The photoconductive gain differs considerably from the dark current gain, due to a redistribution of the potential in the QWIP under the influence of infrared radiation.

## 1. Introduction

Performance of intersubband Quantum Well Infrared Photodetectors (QWIPs), which are considered as an alternative to photodetectors based on narrow-gap semiconductors, is essentially influenced by the carrier injection and transport conditions [1, 2]. There has been a strong interest in GaAs/AlGaAs QWIP due to their mature growth and processing technology [1]. A potentially superior alternative to GaAs/AlGaAs materials system is a non-lattice matched $In_xGa_{1-x}As/GaAs$. Excellent quality strain layer $In_xGa_{1-x}As/GaAs$ heterostructures can be grown for lower In concentrations ($x < 0.2$), which makes this system very suitable for very long-wavelength ($\lambda > 15\ \mu m$) QWIPs. Binary GaAs barriers have better quality and lower electron effective mass than aluminium containing AlGaAs barriers. This provides better transport conditions and lower electron capture time, giving rise to higher photoconductive gain, responsivity and detectivity, as well as to faster intrinsic response time. Recently, excellent device characteristics of $In_xGa_{1-x}As/GaAs$ QWIPs have been reported [3, 4]. In these QWIPs, the emitter and collector barriers are of triangular shape, which, according to theoretical expectations [2, 5], are superior to rectangular barriers with respect to injection efficiency. $In_xGa_{1-x}As/GaAs$ QWIPs have also been studied by Kane *et al.* [6], who reported a very high responsivities, compared to those calculated theoretically, and normal incidence photodetection.

In this paper, we study theoretically the electrical and optical properties of $In_xGa_{1-x}As/GaAs$ QWIPs. A distributed model of the QWIP is proposed, which takes into account carrier injection and transport in QWIP structure, exchange between bound and continuum states electrons, and recharging of QWs. To make clear the role of the emitter and collector barriers, we compare systematically QWIPs with triangular and rectangular extreme barriers.

## 2. QWIP structure and numerical model

We consider $In_xGa_{1-x}As/GaAs$ QWIP, which comprises multiple QW structure of repeated n-doped thin narrow-gap $In_xGa_{1-x}As$ QWs, separated by undoped thick wide-gap GaAs barriers. It is supposed that there is only one bound level in QW. QW structure is sandwiched between the emitter and collector barriers followed by n-doped contacts. Two different types of contacts are considered. In the first type, contacts are made using GaAs, and the extreme, emitter and collector, barriers are of triangular shape, formed by band bending between heavily doped contacts and the first and last $In_xGa_{1-x}As$ QWs [3]. The contacts of the second type are supposed to be made of material with the band gap equal to that of $In_xGa_{1-x}As$, and extreme barriers have a rectangular shape.

Physical processes in QWIPs are studied using a distributed model [7], which incorporates the following equations:

(1) Poisson's equation for electric potential $\varphi$:

$$\frac{\partial^2 \varphi}{\partial x^2} = -\frac{e}{\varepsilon} \cdot \left[ N_d(x) - n_{QW}(x) - n(x) \right], \tag{1}$$

with the boundary conditions:

$$\varphi(x = 0) = 0, \qquad \varphi(x = L) = V.$$

Here $x$ is the coordinate in the direction normal to the plane of the layers, $e$ is the elementary charge, $\varepsilon$ is the dielectric constant, $N_d$ is the donor concentration in the QWs, $n_{QW}$ is the three-dimensional concentration of electrons confined in the QWs, $n$ is the concentration of the electrons in the continuum states (above the barriers), $L$ is the length of multiple QW structure, and V is the applied voltage.

(2) Current continuity equation for the continuum states electrons:

$$\frac{\partial n}{\partial t} + \frac{1}{e} \operatorname{div} j_n = G_O + G_T - R, \tag{2}$$

where the electron current density $j_n$ is given by a conventional expression in the drift-diffusion approximation:

$$j_n = en\mu_n E + eD_n \frac{\partial n}{\partial x}, \tag{3}$$

Here $G_O$ is the optical generation rate, $G_T$ is the generation rate due to thermal emission of bound electrons, $R$ is the capture rate of electrons into the QWs, $E(x) = -\partial\varphi/\partial x$ is the local electric field, $\mu_n = \mu_n(E)$ is the field-dependent electron mobility, and $D_n$ is the diffusion coefficient.

Electron current densities at the emitter and collector barriers, which form the boundary conditions for equation (2), are expressed as the differences of the tunneling assisted thermionic currents from the both sides of the barriers.

(3) The balance equation for electrons confined in the QWs:

$$\frac{\partial \Sigma^i}{\partial t} = L_W \cdot (r^i - g_O^i - g_T^i), \tag{4}$$

where $\Sigma$ is the sheet density of bound electrons, $L_W$ is the width of a QW, and index $i$ refers to the i-th well. The rate terms in the right side of equation (4), which have the same notation as for equation (2), are expressed as follows:

$$g_O^i = \sigma \, \Phi \, \Sigma^i / L_W, \tag{5a}$$

$$g_T^i = n_{eq}^i v_{QW} \exp(\Delta\varepsilon_F^i / kT) / L_W \tag{5b}$$

$$r^i = n^i v_{QW} / L_W, \tag{5c}$$

Here $\sigma$ is the photoionization cross section, $\Phi$ is the photon flux, $v_{QW}$ is a quantum well recombination velocity, $n_{eq}$ is the (field-dependent) equilibrium concentration of electrons above the wells, $\Delta\varepsilon_F^i$ is the difference of the bound electrons Fermi energies corresponding to sheet densities $\Sigma^i$ and $\Sigma_0 = N_d L_W$, respectively, $k$ is Boltzmann's constant and $T$ is the temperature.

The QW recombination velocity $v_{QW}$ is related to the QW capture probability [1] $p$ as $p = v_{QW}/v$, where $v$ is the average 3D electron velocity. Expressions for "generation-recombination" rates in equation (2) are obtained by renormalization of the respective formulas (5a)-(5c):

$$G_O = b \cdot g_O, \quad G_T = b \cdot g_T, \quad R = b \cdot r, \tag{6}$$

where $b = L_W/(L_W + L_B)$, and $L_B$ is the width of the barrier between QWs.

The self-consistent solution of equations (1)–(6) provides the distributions of all physical quantities (electric fields, potentials, electron concentrations etc.) in the QWIP structure, which allows us to calculate the electrical and optical properties of QWIPs. In particular, the dark current (noise) gain is evaluated as a ratio of the dark current to the current of thermoexcitation of bound electrons, and the photoconductive gain is given by a ratio of the photocurrent to the photoexcitation rate of bound electrons.

## 3. Results and discussion

We studied the $In_{0.2}Ga_{0.8}As$ QWIPs, operated at temperature $T = 50$ K, with the following parameters: $L_W = 60$Å, $L_B = 600$Å, number of QWs $N = 5$, QW doping $N_d = 5 \times 10^{17}$ cm$^{-3}$ and contact doping $N_c = 10^{17}$ cm$^{-3}$. We set $v_{QW} = 5 \times 10^5$ cm/s, low-field mobility $\mu_0 = 5 \times 10^3$ cm$^2$/V s and saturation velocity $v_s = 10^7$ cm/s. Physical parameters of $In_xGa_{1-x}As$ were taken from ref. [8].

Figure 1 shows the distributions of potential energy profile of continuum states electrons under different applied voltages. In QWIPs of the first type, the triangular barriers are formed by the band bending due to the electron accumulation in the first and last wells (see Fig. 2) and potential lowering of the QW region. With the increase of the bias voltage, the emitter barrier height is lowered, inducing the thermionic injection of electrons from the emitter contact. This is associated with the decrease by absolute value of the electric field at the emitter barrier and a further accumulation of electrons in the first QW (Fig. 2). The distribution of the electric field in the QW region is close to uniform, and corresponds to a negative charging of this region. In QWIPs of the second type, to the contrary, the first and last QWs are depleted by the electrons at low voltage. This results in an enhancement of the electric field at the extreme barriers and an increase of the QW region potential. With an increase of the bias voltage, the distribution of electric field in this QWIP becomes highly nonuniform, with the high-

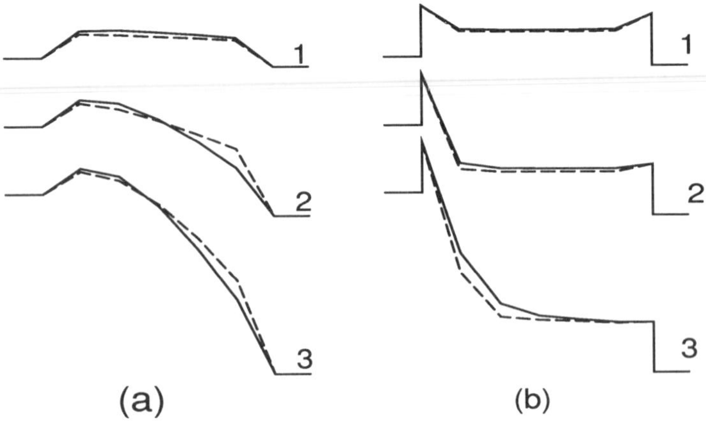


**Figure 1.** Potential energy profiles of the continuum states electrons in QWIPs with (a) triangular and (b) rectangular emitter and collector barriers under various applied voltages: 1 – $V$=0.02 V, 2 – 0.2 V, 3 – 0.4 V (solid lines – dark regime, dashed line – illuminated).

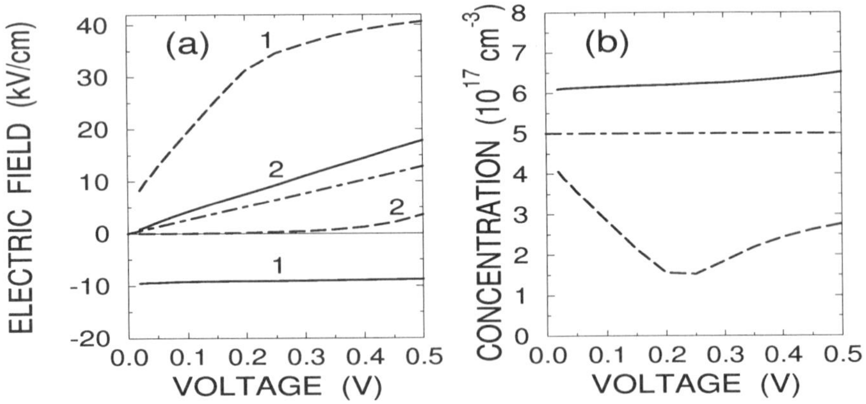


**Figure 2.** (a) Electric field ( 1 – at the emitter barrier, 2 – in the bulk of the QW region; dotted-dashed line – average electric field $E_0 = V/L$) and (b) electron concentration in the first QW (dotted dashed line – doping level) versus bias voltage. Solid and dashed lines correspond to QWIPs with triangular and rectangular extreme barriers, respectively.

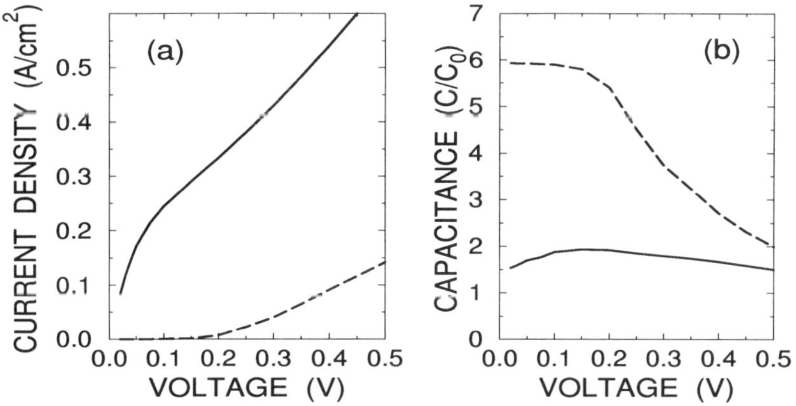


**Figure 3.** (a) Dark current density and (b) normalized capacitance of $In_{0.2}Ga_{0.8}As$ QWIPs with triangular (solid line) and rectangular (dashed line) emitter and collector barriers ($C_0 = \varepsilon/L$).

field domain near the emitter contact, and low uniform electric field domain in the rest of the structure (Fig. 1b). The high-field domain, due to strong depletion of the first well by the electrons (Fig. 2), is responsible for the tunnel electron injection through the emitter barrier. The injected electrons are transported through the multiple QW region owing to drift mechanism. Since the electric field in the low-field domain is much lower than the average electric field, $E_0 = V/L$, current densities in the QWIP of the second type are much lower than than those in QWIP with triangular extreme barriers (see Fig. 3a). The effects of the redistribution of the physical quantities in the QWIP structure with bias voltage are clearly reflected in the capacitance-voltage characteristics, illustrated in Fig. 3. In QWIP of the first type, the capacitance is nearly constant with voltage and ranges from $1.5\times C_0$ to $2\times C_0$ ($C_0$ is the geometric capacitance of QWIP). The capacitance of the second type QWIP is very high at low bias, and is close to the capacitance of the first barrier, because applied voltage drops predominantly on this barrier. At higher bias, potential is redistributed over the entire multiple QW region, and capacitance decreases and tends to the value of $C_0$ at high voltages.

Figure 4a shows the photoconductive gain and dark current gain versus bias voltage. Both quantities increase with voltage, due to the increase of the average electron velocity and decrease of the QW capture probability. It is interesting to note that the photoconductive gain can differ from the dark current gain, which is connected with the nonuniform electric field distribution and redistribution of the potential under the influence of infrared radiation (Fig. 1). For the QWIP with triangular emitter barrier, the photoconductive gain exceeds the dark current gain, due to the improvement of the carrier transport conditions under illumination. However, for the QWIP with the rectangular extreme barriers, the opposite inequality takes place.

It can be seen from Fig. 1 that the photodetection mechanism of the QWIPs is associated with the redistribution of the potential in the QWIP structure under illumination. The photocurrent of the QWIP with triangular emitter and collector barriers, induced by the decrease of the emitter barrier height, significantly exceeds that of the QWIP with rectangular barriers, produced by an increase of the electric field at the emit-

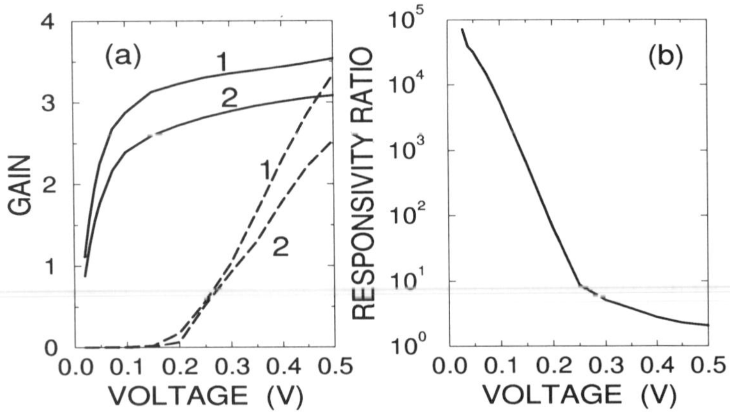


**Figure 4.** (a) Gain ( 1 – photoconductive gain, 2 – dark current gain; solid lines and dashed lines correspond to QWIPs with triangular and rectangular barriers) and (b) ratio of the responsivities of QWIPs with triangular and rectangular extreme barriers, versus voltage.

ter barrier. As a result, triangular barriers provide much higher detector responsivities than rectangular barriers (Fig. 4b).

## 4. Conclusion

A distributed model of intersubband QWIP has been proposed and used to study the $In_xGa_{1-x}As/GaAs$ QWIPs. It is found that the emitter and collector barriers of triangular and rectangular shapes give rise to strikingly different distributions of physical quantities inside the QWIP structure. Triangular barriers provide much better transport and injection conditions for continuum states electrons and higher QWIP performance as compared to rectangular barriers.

## References

[1] Levine B F 1993 *J. Appl. Phys.* **74** R1–R81

[2] Ryzhii V and Ershov M 1995 *Jpn. J. Appl. Phys.* **34** Part 1 (accepted for publication)

[3] Gunapala S D, Bandara K M S V, Levine B F, Sarusi G, Sivco D L and Cho A Y 1994 *Appl. Phys. Lett.* **64** 2288–90

[4] Gunapala S D, Bandara K M S V, Levine B F, Sarusi G, Park J S, Lin T L, Pike W T and Liu J K 1994 *Appl. Phys. Lett.* **64** 3431–33

[5] Ryzhii V 1994 *Jpn. J. Appl. Phys.* **33** 78–82

[6] Kane M J, Millidge S, Emeny M T, Guy D R P, Lee D and Whitehouse C R 1993 *J. Appl. Phys.* **73** 7966–68

[7] Ershov M and Ryzhii V 1994 *Proc. of the Int. Conf. on Optical Properties of Nanostructures*, Sendai

[8] *Properties of Lattice-Matched and Strained Indium Gallium Arsenide* 1993 No. 8 ed P Bhattacharya (INSPEC, EMIS Datareviews Series)

*Inst. Phys. Conf. Ser. No 144: Section 8*
*Paper presented at 7th Int. Conf. on Narrow Gap Semicond. Santa Fe 9–12 January*



# Electronic devices based on III-V/V heterostructures

**T. D. Golding, J. R. Meyer†, J. Huang, C. A. Hoffman†, E.G. Wang, J. H. Xu, and J.T. Zborowski**

Department of Physics, University of Houston, Houston, TX 77204, USA

† Naval Research Laboratory, Washington, DC 20375, USA

**Abstract.** Experimental results on the synthesis and transport properties of Sb/GaSb heterostructures and multilayers are presented, in which we have observed negative differential resistance in semimetal/semiconductor double barrier structures, and Aharanov-Bohm oscillations in a split-path Sb nanostructure. The ability to sandwich the semimetal between finite-barrier materials raises interesting possibilities for quantum confinement effects in the semimetal. Tight binding calculations of the band structure of Sb/Ga(Al)Sb superlattices indicate that an indirect narrow gap appears in such structures with thin (≈ 65Å) Sb layers. Such materials are of significant interest for non linear optical applications. Our studies indicate the possibility of incorporating Sb into the InAs/GaSb/AlSb family of heterostructures, raising numerous possibilities for new electronic devices. We discuss one possible structure with potential application in metal-base transistors and ultra-low-resistance interconnects based on a InAs/GaSb/Sb "Super-Type-II" heterostructure.

## 1. Introduction

Epilayer systems integrating (semi)metals and semiconductors have attracted growing attention because of the possibility of combining the transport properties of both materials to create attractive new classes of ultra-fast and novel electronic devices. An example is the metal-base transistor, in which the base region is made as conductive as possible in order to reduce the RC time constants, which limit switching speeds. However, previous studies of systems such as $Al_xGa_{1-x}As$/NiAl [1], and $GaAs/Sc_{1-x}Er_xAs$ [2], have found that while it is relatively easy to obtain an acceptable deposition of a metal film onto a semiconductor, redeposition of the semiconductor onto the metal results in inadequate overgrowth quality for devices.

In the present work, we report the successful synthesis of a semiconductor/semimetal system which displays high-quality epitaxial overgrowth: GaSb/Sb, which exploits a fortuitous match of the atomic surface nets in the (111) planes of each crystal structure - the rhombohedral crystal structure of the semimetal and the zincblende structure of the semiconductor.

In addition to devices employing the metallic properties of the semimetal we have also explored [3,4] further opportunities enabled by the large characteristic lengths in the semimetal, and by the indirect narrow-gap band alignment which can result if strong quantum confinement converts the semimetal into a semiconductor. Our findings also suggest [5] the feasibility of a more general integration of Sb into the InAs/GaSb/AlSb family of heterostructures, which opens a range of interesting device concepts.

In this paper we present experimental and theoretical results on Sb/GaSb heterostructures, multilayers and superlattices and discuss a promising structure based on a InAs/GaSb/Sb

"Super-Type-II" heterostructure for potential application in metal-base transistors and ultra-low-resistance interconnects.

## 2. V/III-V Heterostructures

Sb has a rhombohedral A7 crystal structure and an overlap between the conduction band L minima and the valance band H maxima of approximately 180 meV [6]. Despite the differing crystal symmetry of the semimetal and zincblende III-V semiconductors, the (111) plane of the bulk Sb forms a hexagonal surface net that is nearly identical to that of the III-V. For example, the in-plane lattice mismatches at the (111) interface are 0.06%, 0.71%, and 0.57% for Sb/GaSb, Sb/AlSb, and Sb/InAs respectively at room temperature. A typical Sb/III-V heterostructure is depicted in figure 1. Note that there can be two types of interfaces with bonding configurations of Sb-X=Sb-Sb or Sb-Sb=Sb-X (X = In, Al, Ga).

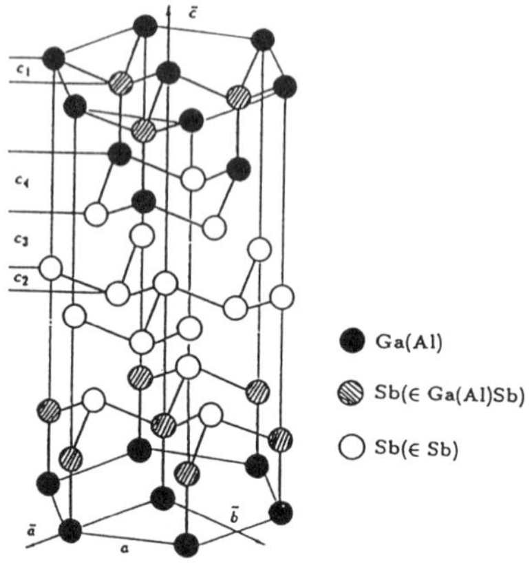


Figure 1. A III-V/V/III-V structure showing the two types of bonding configurations across the V/III-V interface. Note the atomic registry across the interface. Experimental results indicate the structure as shown.

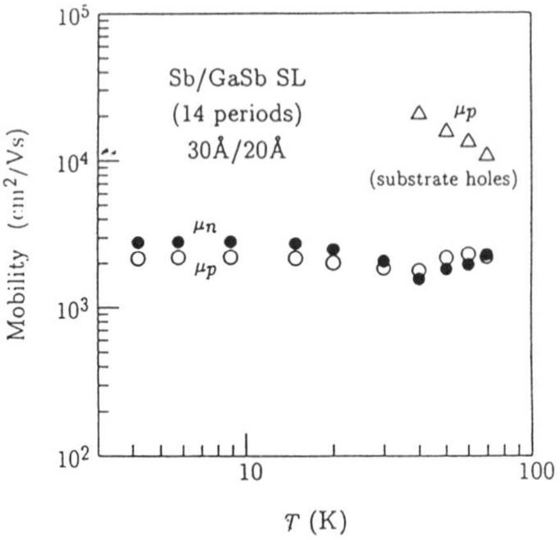


Figure 2. Electron (filled circles) and hole (open circles) mobilities in a 14-period Sb/GaSb superlattice. Substrate hole mobilities (triangles) are also shown.

## 3. Experimental results

Sb/GaSb heterostructures and multilayers, and Sb/AlSb and Sb/InAs heterostructures have been synthesized using molecular beam epitaxy (MBE) and migration enhanced epitaxy (MEE). Epitaxial growth of Sb is achieved by increasing the Sb residence lifetime on the III-V surface by lowering the temperature below a nucleation temperature, $T_n(J_{Sb})$. The critical III-V overgrowth is achieved at a temperature compatible with $T_n$ by using MEE. X-ray analysis [7] of Sb/GaSb multilayers show excellent epitaxial growth and minimal twin-related domains, and show that the bonding configuration across the heterostructure is as depicted in figure 1.

Magneto-transport measurements on a 1.9-μm-thick Sb/GaSb film yielded low-temperature electron and hole mobilities exceeding 30,000 $cm^2/Vs$, and pronounced Shubnikov-de Haas oscillations have been observed [8]. For further confirmation that the high overgrowth quality can be exploited in the fabrication of more complex structures, we have studied the mixed-conduction properties of Sb/GaSb superlattices. Figure 2 shows electron and hole mobilities for a 14-period superlattice (30Å/20Å) , in which the carrier concentrations are only slightly lower than those in bulk Sb. That both $\mu_n$ and $\mu_p$ exceed 2000 $cm^2/Vs$ in a sample having two interfaces every 50Å indicates that boundary scattering must be far weaker than in most thin heterostructures combining materials with different crystal structures [9,10]. Similar electron and hole mobilities were measured for a 37-period superlattice with somewhat thicker Sb layers (60Å/20Å).

Quantum confinement in Sb/GaSb multilayers has been indicated by the observation of negative differential resistance in a 35Å/50Å/35Å GaSb/Sb/GaSb double-barrier resonant tunneling diodes as shown in figure 3. To investigate the suitability of the semimetal-based structures for nanostructures and also to exploit the large mean free paths in the semimetal we have fabricated split-path Sb structures with lateral sizes below 1000Å using ion beam pattering. Aharanov-Bohm oscillations have been observed in a single ring loop (diameter 1μm, channel width 1000Å, channel height 1000Å) at T = 1.6 K.

## 4. Sb/Ga(Al)Sb superlattices: An indirect narrow bandgap material

The electronic structures of $(Sb_2)_m/(GaSb)_n$ and $(Sb_2)_m/(AlSb)_n$ (111) oriented ($m,n \leq 10$) superlattices have been calculated [11] using tight-binding theory including spin orbit interactions. A number of interesting quantum size effects have been predicted. Figure 4 shows the band structures and the densities of states of the (a) $(Sb_2)_2/(GaSb)_8$ and (b) $(Sb_2)_2/(AlSb)_8$ superlattices, where all energies are referenced to the valence band maximum of bulk GaSb (AlSb). As expected, the band overlap in Sb disappears due to the spacial quantization effect. An indirect narrow gap forms in the superlattice since the valence-band maximum ($H_v'$) and the conduction band minimum ($\Gamma_c$) are at different points in momentum space. This effect allows a strong optical nonlinearity in this artificially structured material [12].

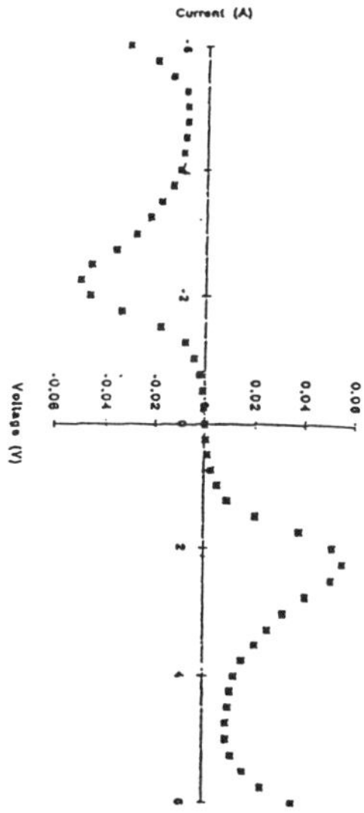


Figure 3. Experimental negative differential resistance for a Sb/GaSb/Sb/GaSb/Sb double-barrier resonant tunneling diode. The central Sb quantum well is 50Å thick, and the GaSb barriers are 35Å thick.

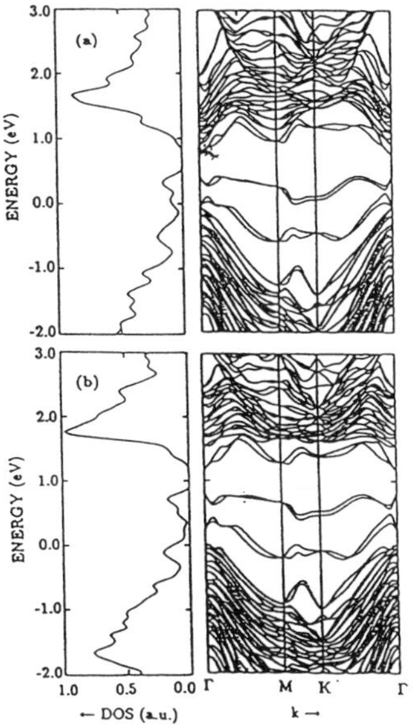


Figure 4. Band structures and densities of states of the (a) $(Sb_2)_2/(GaSb)_8$ and (b) $(Sb_2)_2/(AlSb)_8$ superlattices, where all energies are referenced to the valence band maximum of bulk GaSb and AlSb respectively.

## 5. Super-Type-II Heterostructures

While the finding that GaSb can be successfully overgrown onto Sb is significant in itself, we wish to emphasize the broader implication that a full integration of Sb into the InAs/GaSb/AlSb family of MBE-compatible heterostructures now appears feasible. As just one example, figure 5 illustrates the projected conduction (solid) and valence (dashed) band alignments for a GaSb/InAs/GaSb/Sb/GaSb heterostructure with a "Super-Type-II" energy overlap. Note from the unperturbed bands in fig. 5(a) that while the conduction-band minimum of InAs lies 100-150 meV below the valence band maximum of GaSb, we expect a far larger Type-II offset between the conduction-band minimum of InAs and the valence band maximum of Sb. Since our transport work on Sb/GaSb heterostructures confirms that the Sb valence and conduction bands both lie within the GaSb gap, it may be inferred that the negative offset for InAs/GaSb/Sb must be at least -300 meV and may be as much as -650 meV. Thus alignment of the Fermi energies will result in a massive transfer of electrons from the Sb layer into the InAs, as illustrated in fig. 5(b). Accounting for quantum confinement, nonparabolicity, the internal electric field induced by charge transfer, and the extremely high density of states in the Sb valence band, we estimate that in equilibrium the InAs conduction band will contain a sheet density $n_s$ of up to 1 x $10^{13}$ $cm^{-2}$ electrons in the absence of any doping. Furthermore, this population should maintain the high electron mobility usually associated with InAs/GaSb heterostructures ($\mu_n$ > 50,000 $cm^2/Vs$ has been reported for [111] growth [13]), since the InAs layer is grown onto the GaSb substrate and buffer layer before any Sb is deposited. Non-ideal Sb growth is then of less consequence, since the extremely low device resistance relies primarily on the conductivity by electrons in the InAs.

Conservatively taking $n_s \approx 5 \times 10^{12}$ $cm^{-2}$ and $m_n$ (300 K) $\approx$ 20,000 $cm^2/Vs$, we obtain a 2D resistivity of less than 100 $\Omega$ and an RC time constant of less than 100 fs, figures of merit which easily meet the requirements for high-speed metal-base transistors [14]. Further advantages of the Sb-based structure include the thinness of the base region ($\approx$ 200 Å) and the absence of intervalley hot-electron scattering (which is detrimental in devices with heavily-doped GaAs or $In_{1-x}Ga_xAs$ bases). One potential configuration would employ the combined InAs/GaSb/Sb substructure as the base and GaSb layers for the emitter and collector, with $Ga_{1-x}Al_xSb$ barriers separating the different regions Overgrowth quality is critical, since it will govern the unwanted reflection of injected electrons at the Sb/GaSb interfaces.

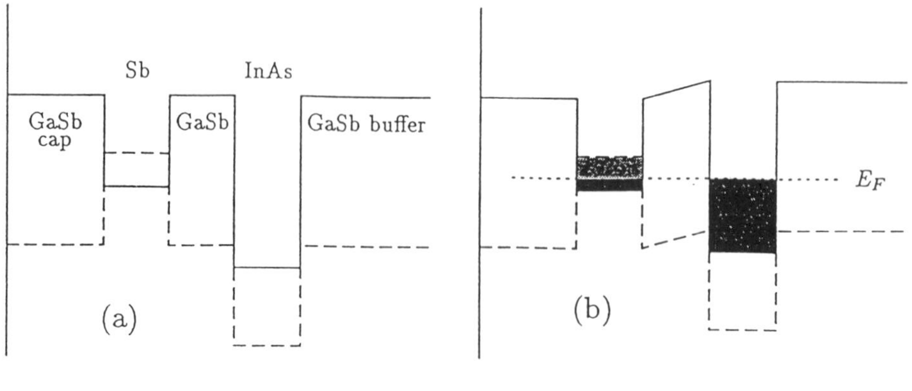


Figure 5. Schematic of the conduction (solid lines) and valence (dashed lines) band alignments in a GaSb/InAs/GaSb/Sb/GaSb "Super-Type-II" heterostructure. The unperturbed band edges are shown in (a), while (b) includes the internal electric field due to charge transfer and indicates the portions of the bands filled by electrons (dark shading) and holes (light shading).

## 6. Summary

In summary, we have demonstrated that high quality Sb/GaSb heterostructures and multilayers can be synthesized by molecular beam epitaxy and migration enhanced epitaxy. Transport results on vertical (double barrier resonant tunneling diodes, and superlattices) and lateral (sub-micron split-path loop) semimetal structures exhibit promising electronic properties, while also demonstrating the processing compatibility of the semimetal/semiconductor combination. Tight binding calculations of Sb/(Ga,Al)Sb superlattices indicate that spatial quantization effects will convert the semimetal into an indirect narrow bandgap semiconductor. This is unique among naturally-occurring and man-made semiconductors and is highly favorable for nonlinear optical applications, such as optical switching devices operating in the infrared. The results obtained to date on the Sb/GaSb combination point the way towards numerous opportunities based on the integration of Sb into the InAs/GaSb/AlSb family of materials. Studies of Sb/GaSb superlattices for infrared optical and nonlinear optical investigations requiring an indirect narrow bandgap material as well as the proposed "Super-Type-II" heterostructure are currently in progress.

## Acknowledgments

This work was supported by ARO Grant DAAH04-94-G-0075 and the Robert. A. Welch Foundation. We wish to thank J.A. Dura and C.S. Ting for many helpful discussions.

## References

[1] J.P. Harbison, T. Sands, N. Tabatabaie, W.K. Chan, L.T. Florez, and V.G. Keramidas, *Appl. Phys. Lett.* **53**, 1717 (1988).

[2] S.J. Allen, D. Brehmer, and C.J. Palmstrom, *Mat. Res. Soc. Symp. Proc.* **301** 307 (1993)

[3] T.D. Golding, J.A. Dura, W.C. Wang, J.T. Zborowski, A. Vigliante, J.H. Miller Jr. and J.R. Meyer. *Appl. Phys. Lett.*, **63** 1098 (1993).

[4] T.D. Golding, J.A. Dura, W.C. Wang, J.T. Zborowski, A. Vigliante H.C. Chen and J.R. Meyer. *Jour. Crys. Growth.* **127** 777 (1993).

[5] A. Vigliante, S.C. Moss, J.A. Dura, W.C. Wang, and T.D. Golding.*Mat. Res. Soc. Symp. Proc.* **281**, 721 (1993).

[6] J.-P. Issi, *Aust. J. Phys.* **32**, 585 (1979).

[7] J.A. Dura, A. Vigliante, T.D. Golding, and S.C. Moss, to be published *J. Appl. Phys.*

[8] J.H. Huang, T.D. Golding, J.C. Wolfe, C.A. Hoffman, and J.R. Meyer, to be published.

[9] N. Tabatabaie, T. Sands, J.P. Harbison, H.L. Gilchrist, L.T. Florez, and V.G. Keramidas, *Appl. Phys. Lett.* **54**, 2112 (1989).

[10] C.A. Hoffman, J.R. Meyer, F. J. Bartoli, A. DiVenere, XC.J. Yi, C.L. Hou, H.C. Wang, J.B. Ketterson, and G.K. Wong, *Phys. Rev. B* **48**, 11431 (1993).

[11] E.G. Wang, J.H. Xu, W.P. Su, and C.S. Ting, *J. Appl. Phys.* **76** (9), 5318 (1994).

[12] J. R. Meyer, F. J. Bartoli, E. R. Youngdale, and C. A. Hoffman, *J. Appl. Phys.* **70**, 4317 (1991)

[13] G.M. Sundaram, R.J. Warburton, R.J. Nicholas, G.M. Summers, N.J. Mason, and P.J. Walker, *Semicond. Sci. Technol.* **7**, 985 (1992).

[14] S. Luryi, in *High-Speed Semiconductor Devices,* ed. S.M. Sze, (Wiley, New York, 1990), Chapter 7.

*Paper presented at 7th Int. Conf. on Narrow Gap Semicond. Santa Fe 9–12 January*


# ECR-RIE of HgTe/CdTe and InAs/GaSb heterostructures patterned using nanochannel glass

C.R. Eddy, Jr., R.J. Tonucci†, D.H. Pearson†, J.R. Meyer‡ and C.A. Hoffman‡

Naval Research Laboratory, Code 6671, Washington, DC 20375-5345
‡Code 5613, †Code 5611

**Abstract.** This paper reports on both a novel lithographic process using nanochannel glass arrays and on the continued development of a dry etch process for narrow-gap semiconductors. Original and thin film replica nanochannel glass arrays have been used to pattern dot and/or anti-dot arrays on both HgTe/CdTe and InAs/GaSb heterostructures. Pattern transfer is achieved using reactive ion etching with methane/hydrogen/argon microwave plasmas generated under electron cyclotron resonance conditions. A desorption-limited, anisotropic etch process has been developed for both materials with exemplary etch rates of 200Å/min and 325Å/min for HgCdTe and InAs/GaSb, respectively. Features as small as 600 nm in diameter with nearly vertical sidewalls have been realized. Initial assessments on damage resulting from pattern transfer in both materials are also reported.

## 1. Introduction

The development of state-of-the-art pattern generation and pattern transfer technologies is critical to the ultimate success of narrow-gap semiconductor heterostructures in nanometer-scale devices. These include a wide range of promising mesoscopic electronic and optical devices that exploit the enhancement of quantum resonances in systems with small effective masses[1]. The photolithography used in most current-generation semiconductor processing can pattern thousands of devices in parallel with a single exposure, but is of low resolution. In contrast, nanometer-scale pattern generation has traditionally employed e-beam lithography, a serial lithographic process which provides very high resolution at the cost of very long write times for multiple devices. Thus, the development of a simpler, faster, and more parallel lithographic process for nanometer-scale pattern generation would have considerable technological and scientific impact.

In this work we employ NanoChannel Glass (NCG) arrays [2] and thin metal film replicas of these arrays [3] as lithographic masks for a massively parallel nanometer-scale lithography. The lithographic pattern is transferred using a desorption-limited etch process developed for HgCdTe nanometer-scale fabrication [4-6] and adapted for InAs/GaSb multilayer thin films using an electron cyclotron resonance microwave plasma reactive ion etching (ECR-RIE) system.

## 2. Experimental

### *2.1 Nanochannel glass fabrication*

Nanochannel glass lithographic masks are fabricated using a process that is outlined in Figure 1. In short, glasses with dissimilar etch characteristics (an etchable channel glass and an inert

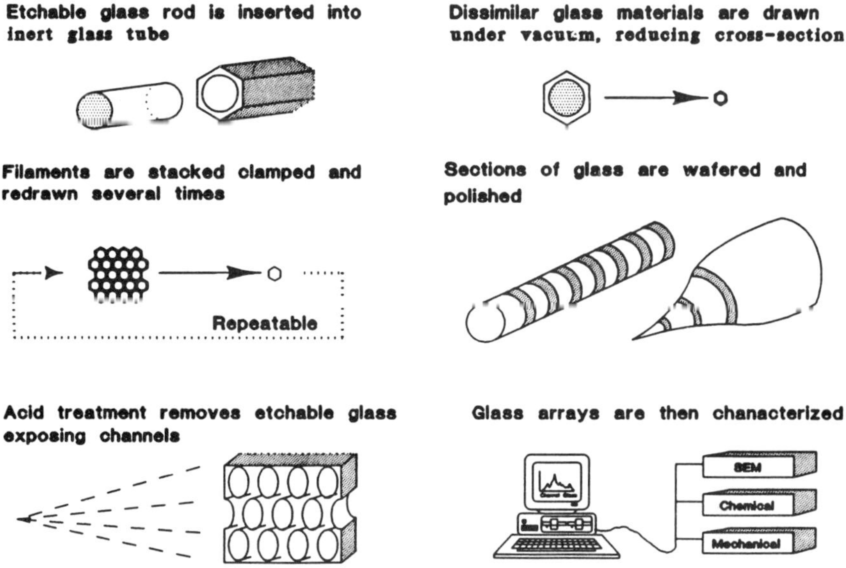


FIGURE 1. Outline of nanochannel glass lithographic mask fabrication process.

cladding glass) are packed in a hexagonal close-packed arrangement and drawn in a glass drawing tower in a manner similar to that used in the fabrication of optical fibers. This serves to reduce the diameter of the channel glass with each draw. Filaments are repacked and the process is repeated until the channel glass is of the desired diameter. To date, the smallest diameter realized is 17 nm. The final rod is then sliced and the NCG "wafers" are polished to a desired thickness (or aspect ratio). The channel glass is then etched away leaving a highly uniform, periodic array of holes in an otherwise continuous sheet of cladding glass.

At this point the NCG can be used directly as a lithographic mask for features > 1μm (determined by the minimum achievable glass thickness and the associated aspect ratio). For submicron lithography, the NCG is used as the base for generation of thin metal film replicas that can subsequently be used as a mask. In this process thin tungsten or platinum films (≈ 2000Å) are sputtered or evaporated onto the NCG, which has been coated with an easily dissolved buffer layer, such as aluminum. The buffer layer is then dissolved releasing the membrane which may be floated onto the desired substrate providing an anti-dot array mask. For dot arrays, this membrane serves as an evaporation mask through which a ≈ 1000Å thick metal film is deposited and then the replica membrane is removed.

### *2.2 Pattern transfer by ECR-RIE*

The NCG or replica pattern is transferred into the semiconductor using a $CH_4/H_2/Ar$ ECR plasma. Details of the plasma etching conditions for each semiconductor are given in Table I. HgCdTe samples used to demonstrate pattern transfer were alloy layers with nominal Hg concentrations of 20%. InAs/GaSb samples used to demonstrate pattern transfer were superlattice structures grown by molecular beam epitaxy. Samples used to demonstrate pattern transfer received no further treatment and were evaluated using scanning electron microscopy (SEM) to determine time-averaged etch rates and sidewall angle profiles.

| Semiconductor | Gas Mixture $CH_4/H_2/Ar$ (%) | Total Pressure (mTorr) | Substrate Bias (Volts) | Microwave Power (Watts) | Substrate Temperature (°C, ± 5°) |
|---|---|---|---|---|---|
| HgCdTe | 18/82/0 | 0.50 | -100 | 250 | 30 |
| InAs/GaSb | 22/53/25 | 1.00 | -300 | 400 | 60 |

TABLE 1. ECR-RIE process conditions for the etching of HgTe/CdTe and InAs/GaSb multilayer thin films.

## 3. Results and Discussion

### *3.1 Nanometer-scale fabrication in HgCdTe*

Figure 2 presents SEM micrographs of 600 nm dot and anti-dot arrays fabricated in HgCdTe using platinum thin film replicas. In each of the micrographs the metal film mask is present. As can be seen in the dot array, the pattern is very uniform and free of defects. Furthermore, the sidewall angles for both types of features is better than 88°. Measured etch rates were ~200Å/min.

### *3.2 Nanometer-scale fabrication in InAs/GaSb*

Figure 3 presents SEM micrographs of 600 nm anti-dot arrays fabricated into InAs/GaSb superlattice layers using tungsten thin film NCG replicas. In each of the micrographs the tungsten thin film is still present. Again, the features are very uniform throughout the arrays. The measured etch rate is ≈ 325Å/min and the sidewalls are ≈ 80°. The somewhat poorer sidewall angles likely result from the 1.0 mTorr total pressure during etching. Processing at lower pressures has been shown to improve the sidewall angles in HgCdTe and should apply to InAs/GaSb.

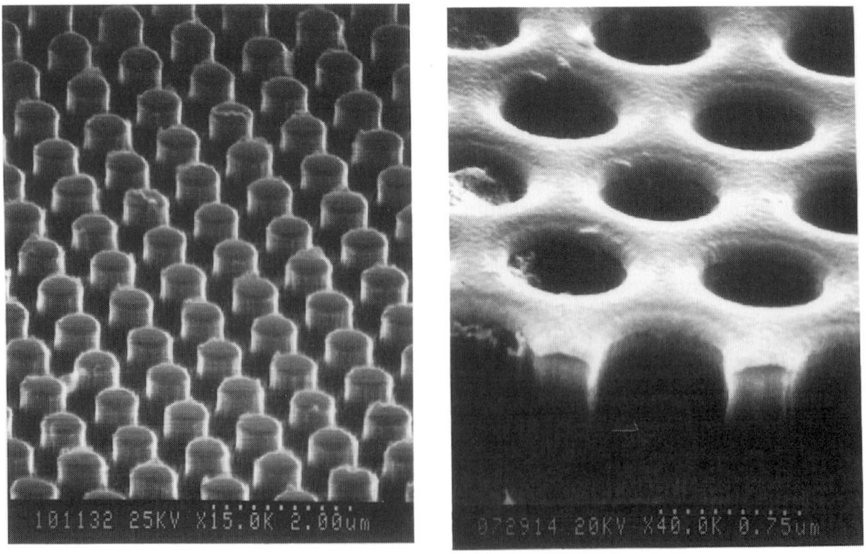


FIGURE 2. Arrays of 600 nm diameter dots and anti-dots fabricated in HgCdTe using platinum NCG replica masks and $CH_4/H_2$ ECR-RIE.

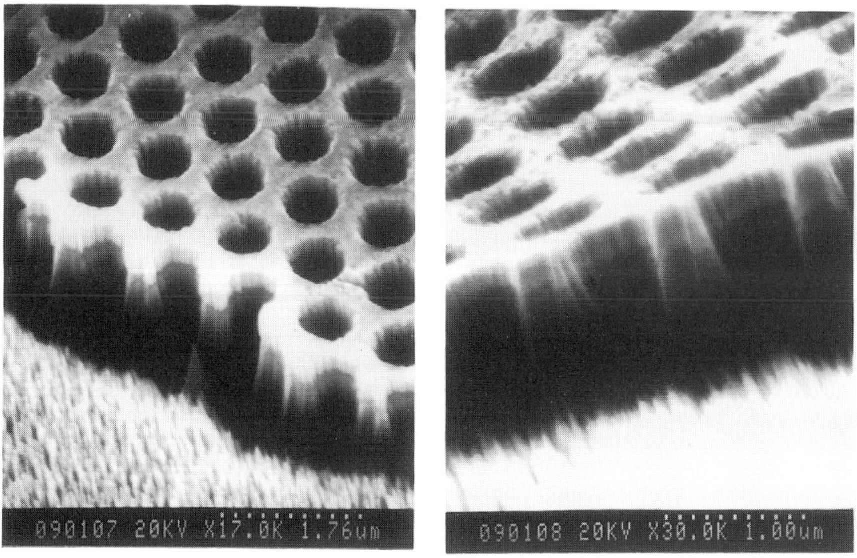


FIGURE 3. Arrays of 600 nm diameter anti-dots fabricated in a InAs/GaSb superlattice using tungsten NCG replica masks and $CH_4/H_2/Ar$ ECR-RIE.

### *3.3 Damage Assessment in HgTe/CdTe and InAs/GaSb processed by ECR-RIE*

In an initial experiment on Hg-based samples, photoresist was deposited on the top surface of a 6.4-μm-thick $Hg_{0.83}Cd_{0.17}Te$ alloy layer with net electron density 3.4 x $10^{15}$ $cm^{-3}$ and mobility 300,000 $cm^2/V•s$ at 4.2 °K. After being subjected to the lithographic and patterning processes discussed above (the etch exposure time was a relatively long 30 minutes), the carrier concentration fell to 1.1 x $10^{15}$ $cm^{-3}$ and the mobility decreased to 220,000 $cm^2/V•s$. These findings are consistent with the introduction of compensating acceptors, which are believed to result from hydrogen diffusion into the material from the plasma. Experiments to anneal out the hydrogen in a Hg overpressure are continuing. Further HgTe/CdTe samples were processed both with and without a thin (≈ 1000 Å) CdTe passivating layer, which was prepared by withdrawing the sample from the ECR-RIE system and then immediately removing the NCG mask and transferring the sample in air to a thermal evaporator for the film deposition. Characterization experiments on these samples are under way.

While the HgCdTe damage assessment remains somewhat ambiguous at this point, preliminary results for the InAs/GaSb system are quite encouraging. Before patterning, field-dependent Hall and resistivity measurements on a InAs/GaSb/GaSb single quantum well of width 150 Å yielded mobility spectra with 3 different electron peaks having values up to 100,000 $cm^2/V•s$. The sample was then RIE-patterned into an anti-dot matrix using an 8-μm-diameter NCG etch mask. The mobility spectra obtained following the patterning again showed 3 peaks with nearly identical mobilities but altered magnitude ratios. Furthermore, quantum transport measurements on the patterned sample at 1.6 °K displayed Shubnikov-de Haas oscillations with a period agreeing to within 4% with that obtained before patterning (implying a density of ≈ 8 x $10^{11}$ $cm^{-3}$), although fewer oscillations were observable in the anti-dot structure. Magneto-optical measurements were also performed both before and after the patterning. Most of the spectra, for FIR wavelengths between 70.6 and 118 μm and fields between 2 and 10 T, displayed multiple peaks which were quite similar in the two cases, and the resolution of the features was not appreciably different. Taken together, the magneto-transport and magneto-optical data confirmed that a substantial portion of the regions between the NCG/RIE-patterned anti-dots were essentially undamaged from the electronic standpoint. The electron concentrations and mobilities in those regions were virtually unaffected by the processing, and no additional carrier species appear to have been introduced due to damage near the anti-dot interfaces.

## 4. Conclusions

Micrometer- and nanometer-scale features have been fabricated in HgCdTe alloys and superlattices and in InAs/GaSb superlattices and single quantum wells. Features as small as 600 nm have been demonstrated in the form of both dot and anti-dot arrays using nanochannel glass-based lithography and methane-based ECR-RIE. NCG-based lithography has been used to demonstrate highly uniform, massively parallel patterning of these two narrow-gap semiconductors. The features resulting from the ECR-RIE pattern transfer are very uniform and demonstrate nearly vertical sidewalls. Etch rates of ≈ 200Å/min and ≈ 325Å/min have been realized for HgCdTe and InAs/GaSb, respectively. An unpassivated HgCdTe sample subjected to the etch conditions for 30 minutes showed slight compensation, presumably the result of hydrogen diffusion into the material from the plasma. InAs/GaSb samples subjected to the pattern transfer process demonstrate minimal damage in the remaining material between anti-dots from the electronic standpoint, with carrier densities and mobilities virtually unaffected. These experiments clearly demonstrate the feasibility of nanochannel glass-based lithography for nanometer-scale device fabrication and the feasibility of ECR-RIE as a low damage pattern transfer process for realizing mesoscopic devices in narrow gap semiconductors.

## 5. Acknowledgments

The authors would like to thank R. J. Wagner and J. P. Omaggio of NRL for performing the magneto-optical characterization of the InAs/GaSb sample.

## References

[1] Meyer J R, Bartoli F J , Hoffman C A and Ram-Mohan L R 1990 *Phys. Rev. Lett.* **64** 1963-1966

[2] Tonucci R J, Justus B L, Campillo A J and Ford C E 1992 *Science* **258** 783-785

[3] Pearson D H and Tonucci R J, to be published

[4] Eddy, Jr. C R, Dobisz E A, Hoffman C A, Meyer J R 1993 *Appl. Phys. Lett.* **62** 2362-2364

[5] Eddy, Jr. C R., Hoffman C A, Meyer J R, Dobisz E A 1993 *J. Electronic Materials* **22** 1055-1060

[6] Eddy, Jr. C R, Dobisz E A, Meyer J R, Hoffman C A 1993 *J. Vac. Sci. & Technol. A* **11** 1763-1767

*Inst. Phys. Conf. Ser. No 144: Section 8*
*Paper presented at 7th Int. Conf. on Narrow Gap Semicond. Santa Fe 9–12 January*



# Defects and Electric Property Characterization of Multi-electrode $Hg_{1-x}Cd_xTe$ Hall Devices

**Y. Cai[1,2] G.Z. Zheng[1] X.C. Zhu[2] J.H. Jiang[3] S.L. Guo[1] D.Y. Tang[1]**
[1]National Laboratory for Infrared Physics, Shanghai Institute of Technical Physics, Chinese Academy of Sciences, Shanghai 200083, P.R.China
[2]Kunming Institute of Physics, Kunming 650223, P.R.China
[3]Synchrotron Radiation National Laboratory, Institute of High Energy Physics, Chinese Academy of Sciences, Beijing 100039, P.R.China

**Abstract.** For fabricating $Hg_{1-x}Cd_xTe$ detectors effectively, an understanding of the defect behavior in $Hg_{1-x}Cd_xTe$ materials and devices is important. A multi-electrode $Hg_{1-x}Cd_xTe$ Hall device was specially prepared for investigating both material properties and detector performance, on which the synchrotron radiation Laue transmission topographs were taken for observation of lattice defects, the temperature dependent Hall effect, photoconductive decay and photocurrent spectra were measured for studying device. The results show that there exist many defects in the Hall device that include lattice strain, dislocation and substructure, etc.. These defects increase the concentration and decrease the carrier lifetime and cause abnormal photoconductive behavior, thus the detectivity of the Hall device as a detector is lowered.

## 1. Introduction

The infrared detecting capability of a photoconductive device is attribute to the excess carrier transport in the semiconductor, therefore, the transport properties of the semiconductors dominate the device performance. People are always interested in the correlations of the electrical properties and detector parameters. The electrical properties of the $Hg_{1-x}Cd_xTe$ wafers may be quite different from that of the device-size slices, which is due to poor homogeneity of $Hg_{1-x}Cd_xTe$ bulk crystals. Usually, there exist as-grown and process-induced extended lattice defects in $Hg_{1-x}Cd_xTe$ wafers, slices and detectors[1-4]. However, it is not clear how great the influence of the defects is, thus it is quite necessary to understand for growing material and fabricating device.

In this paper, With using a multi-electrode $Hg_{1-x}Cd_xTe$ Hall device to simulate the general fabrication process of photoconductive detectors, the behaviors of the as-grown and process-induced defects have been studied systematically by measuring the synchrotron radiation (SR) Laue transmission topography temperature dependent Hall effect, photoconductive decay and photocurrent spectra. The results show that the defects have an influence on the material and detector parameters.

## 2. Experiments and Discussions

The detector-grade $Hg_{1-x}Cd_xTe$ wafers were grown by solid state recrystallization technique and selected by density test, infrared absorption measurement, van der Pauw method and Schulz topography techniques. The composition of the material $x=0.214\pm0.005$, the electron concentration $n=3.4\times10^{14}cm^{-3}$ and Hall mobility $\mu=1.3\times10^5 cm^2/Vs$ were measured at 77K. By using techniques identical to those used for the fabrication of photoconductive $Hg_{1-x}Cd_xTe$ detectors, which include lap, polish, chemical etching, undulatory deposition, photoli-thography, ion-ecthing, sputtering gold in vacuum and wire bonding, a series of multi-electrode Hall devices were made on the wafer as shown in the Fig. 1.

Though the nondestructive detection of defects in $Hg_{1-x}Cd_xTe$ devices has not been solved at present, we think the high-resolution X-ray Laue topography may be an effective method. Compared with the conventional X-ray Laue topography, the synchrotron radiation (SR) Laue topography can give information of structure symmetry as well as the lattice perfection of the material[4-5], while the conventional X-ray Laue topography can only show the symmetry of the sample. By using the Beijing Electron Positron Collider, the high-resolution Laue patterns of 14 Hall devices were taken in transmission mode. The distances of the sample to the X-ray source spot and to the X-y film are set about 45 m and 45cm, respectively in the experiment. The electron bunch energy is about 2.6 GeV and beam current is about 44 mA in the storage ring. The exposure time is about a few seconds and the fine-particle AGFA X-ray films were used. The measured topographs of the Hall devices are shown in Fig.2 with a high resolution up to $1\mu m$.

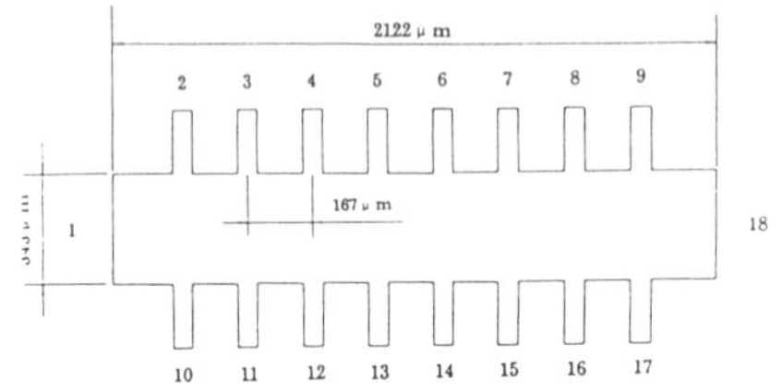


Fig.1. Schematic diagram of the 18-electrode $Hg_{1-x}Cd_xTe$ Hall device

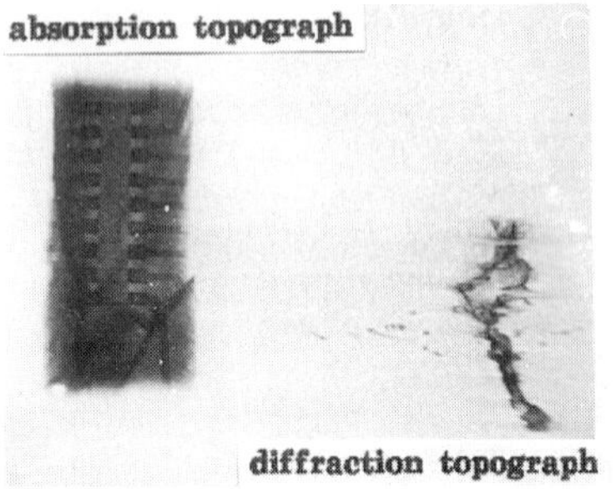


Fig.2a. A SR diffraction topograph of the $Hg_{1-x}Cd_xTe$ Hall device 607 10×

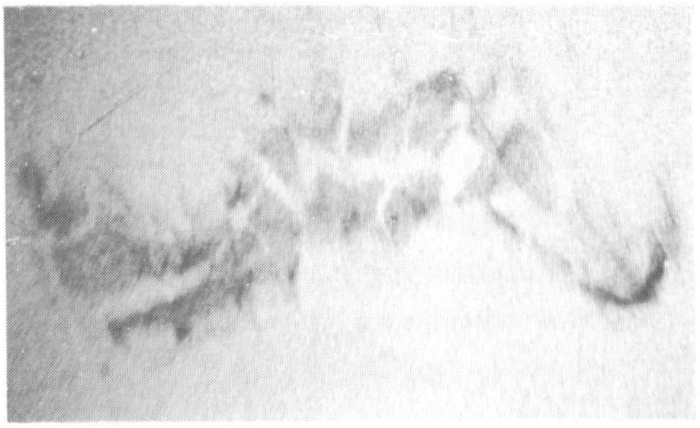

Fig.2b. A SR diffraction topograph of the $Hg_{1-x}Cd_xTe$ Hall device 601 43×

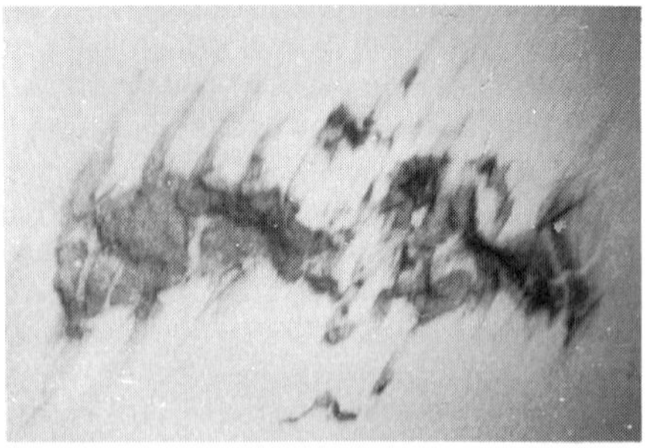

Fig.2c. A SR diffraction topograph of the $Hg_{1-x}Cd_xTe$ Hall device 602

Fig. 2a shows the X-ray absorption and diffraction topographs of the 18-electrode Hall devices, where the absorption topograph is formed by the difference of the $Hg_{1-x}Cd_xTe$ material density, while the diffraction topograph is attribute to the direct image and orientation contrast of defects. The diffraction topograph shows such a hazy lattice structure that the fine device shape almost can not be related with it. Both topographs demonstrate that there exist defects in the $Hg_{1-x}Cd_xTe$ crystal of the devices. The defects are result from the complicated states and distribution of dislocations and display mainly as the lattice distortion, subgrain boundary and polygonization structure where subgrain boundary angle has a typical value of $3\times10^{-4}$ rad. From the topographs, it can be seen that the SR topography is very useful to detect the defects of the devices.

It is well known that the material is always influenced by thermal and mechanical stresses during crystal growth and device fabrication process. Thus when the stresses are larger than the critical shear stress of the material, the stresses will be relaxed and dislocations be produced in it. By carefully studying, the high dislocation desity in $Hg_{1-x}Cd_xTe$ crystals is due to four factors:

(1) Low critical shear stress of $Hg_{1-x}Cd_xTe$ crystals. The typical value for the yield stresses are obtained about $2.6\times10^3$ N/cm$^2$ from the experiment[6], whereas the slip stress in the material is smaller than this value. So the dislocations are very easy to form.

(2). Many slip systems of $Hg_{1-x}Cd_xTe$ crystals. The ion bonds in the $< 111 >$ direction is the largest in them, therefore the lattice slip is switched to $\{110\}$ planes under any stress. But the highest line density of atoms in $< 110 >$ direction keeps the slip directions in $< 110 >$ orientations. As we know there are six $\{110\}$ equivalent planes in the $Hg_{1-x}Cd_xTe$ crystals and each plane has 0, 60 and 90 degree slip directions. Therefore, dislocations in $Hg_{1-x}Cd_xTe$ crystals have a total number of 18 slip systems.

(3). High residual stresses in $Hg_{1-x}Cd_xTe$ crystals. The residual stresses in $Hg_{1-x}Cd_xTe$ wafers grown by solid state recrystallization method have been determined from the X-ray diffraction experiment[7], which has a magnitude of 10~100 N/cm$^2$ and large enough to produce dislocations easily.

(4). Poor shape of the photoconductive $Hg_{1-x}Cd_xTe$ device. A material strength is related with its shape. For the $Hg_{1-x}Cd_xTe$ detectors designed with a thin thickness(about 10$\mu$m), dislocations are also easy to form in the fabrication process due to the thin thickness.

In the fabrication process of the Hall devices, it is found that defects are produced in the general lap, photoli-thography, ion-ecthing and sputtering gold in vacuum steps very easily using the X-ray topography. Based on these experimental resutls, it can be seen that the SR topography is a very useful tool to detect the defects of the devices, and the result shows that the defects in the Hall devices fabricated are not only result from the crystal growth but fabrication processes. In accordance with the failures in these steps, the fabricating techniques have been improved quite well. And the defects have a great influence in the electric properties of the devices as studied bellow.

The temperature dependent Hall effect were measured for the devices between the temperature of 25~300 K in the magnetic field of 800 Gauss with a cryogenic system, the results are shown in the Fig.3. Fig.4 shows the photocurrent spectra of the devices measured with a Fourier spectrometer at 80K. Fig. 5 shows a typical carrier lifetime measured with an abnormal peak observed.

As shown in the Fig.3, the behaviors of the electric properties are measured at different positions of the device, and marked from 601-1 to 601-8 downwords. It is obviously that they are similar to those of the normal $Hg_{1-x}Cd_xTe$ wafers, from which the different scattering mechanisms can be deduced. Compared with the starting wafers results measured at 77K, a obvious change of the electrical parameters is the great increase of the electron concentration with a magnitude up to $2\times10^{15}cm^{-3}$. From the measured electron concentration, the composition of Hall devices have been determinated by using formula given in Ref.[8] as shown in Fig.3a. The results show that composition of most samples become less than the starting wafer. The increase of the electron concentration may be result from the process-induced defects as we observed in the SR experiment.

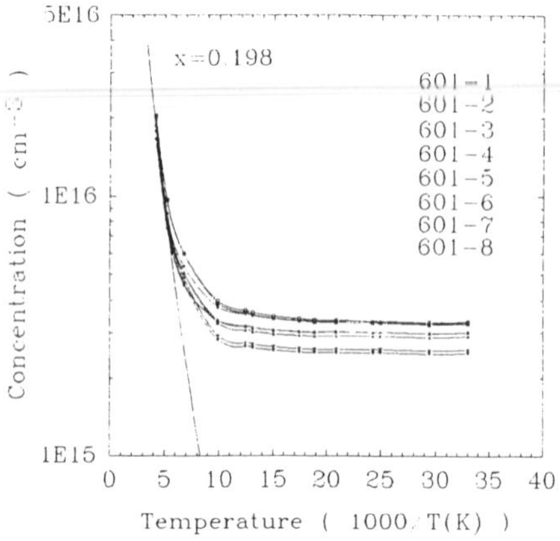


Fig. 3a. The temperature dependence of the carrier concentration of the $Hg_{1-x}Cd_xTe$ Hall device 601

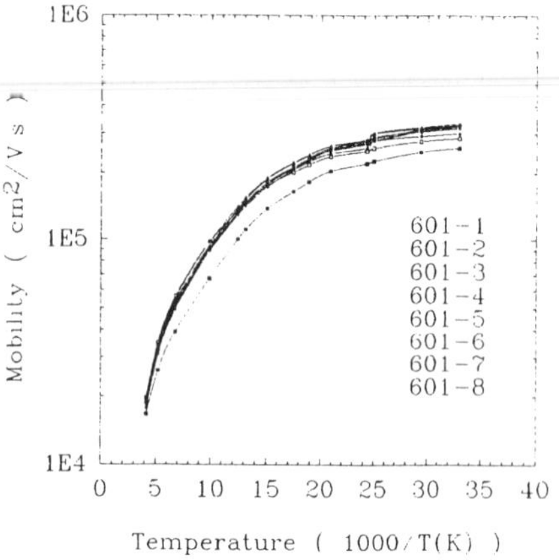


Fig.3b. The temperature dependence of the mobility of the $Hg_{1-x}Cd_xTe$ Hall device 601

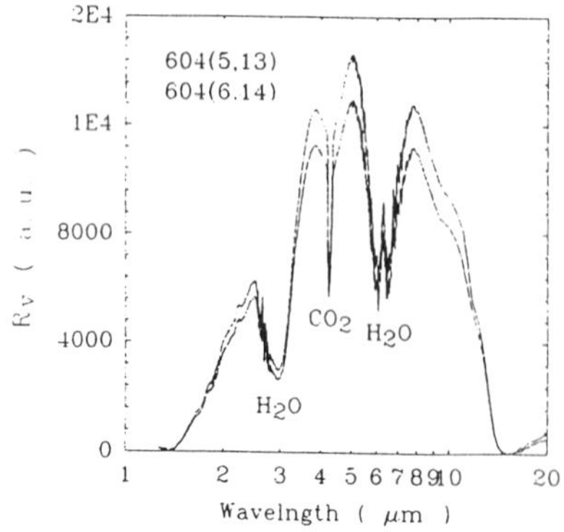


Fig.4. The photocurrent spectrum of the $Hg_{1-x}Cd_xTe$ Hall device 604

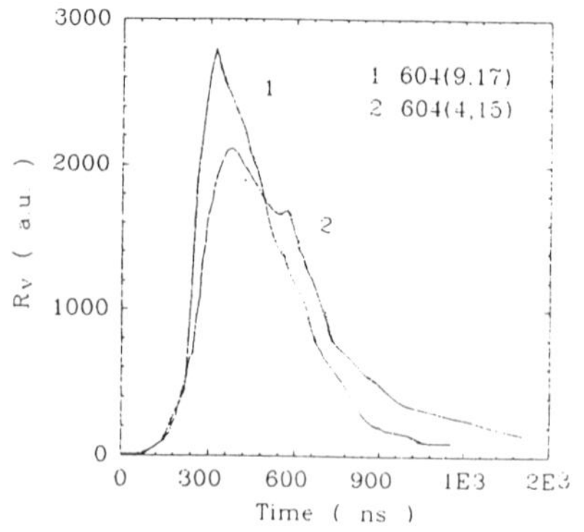


Fig.5. The photoconductive decay curve of the $Hg_{1-x}Cd_xTe$ Hall device 604

Due to the high electron concentration, the Burstein-Moss effect happens, and the optical energy gap becomes larger than the real one, which has been observed in the photocurrent experiment. The photocurrent spectra as shown in Fig.4. has the broad absorption edges measured. Based on composition of the starting wafer, the turning-point wavelength is calculated from 10.5$\mu$m at 80 K. While the experimental result shows the turning-point wavelength is shorter than 9.8 $\mu$m, this also indicates the degeneracy of the Fermi level and the existence of the BM effect.

Recombination process of the excess carriers has a great influence on the detector performance and the typical carrier lifetime is measured about 0.35$\mu$s in the devices with a high carrier concentrations and defect density. The measured result shows a two-energy-level decay process, which implies a high density of Shockley-Read recombination center in the devices. Just because of a high carrier codcentration and short lifetime, the detectivity of the Hall devices as a detector is measured lower. Discussions in detail will be published later.

## 3. Conclusions

The SR topography has been used to investigated the characteristics of $Hg_{1-x}Cd_xTe$ Hall device, and the results show that the device fabrication process can introduce defects in the forms of lattice distortion, dislocation, subgrain boundary and polygonization structure. This is the reason why the devices have more defects than the starting wafers usually. The electric transport experiments also show there exist many of defects in the devices that increase the electron concentration, decrease the lifetime of the carriers, and so the device performance is degenerated.

## References

1 Bye K L J, Mater. Sci., 1979; 14:619-625
2 Mirsky U, et al., J. Electron. Mater., 1980; (9):933-943
3 Cai Y, et al., Infra. Technol., 1993; 15(6):5-8 (in Chinese)
4 Dean B E, et al., J. Vac. Sci. Technol. B, 1991; 13(5); 1840-1846
5 Cai Y, et al., J. Infra. and Millim Waves, 1994; 13(5): 385-390
6 Ekpenuma S N, et al., J. Vac. Sci. Technol. B, 1992; 10(4):1451-1453
7 Wang Y, et al., Infra. Technol., 1990; 12(6): 1-5 (in Chinese)
8 Chu J H, Chin. IR Res., 1985;(1):39-45

*Inst. Phys. Conf. Ser. No 144: Section 8*
*Paper presented at 7th Int. Conf. on Narrow Gap Semicond. Santa Fe 9–12 January*



# Cyclotron Resonance and Hall Effect Studies of Ultra-high Mobility InSb Films

Y. Zhang, J. Su, J. Winesett, J.E. Furneaux, W.K. Liu, M.B. Santos, and R.E. Doezema

Department of Physics and Astronomy
and
Laboratory for Electronic Properties of Materials
University of Oklahoma, Norman, OK 73019

**Abstract.** A systematic study of homoepitaxial InSb films grown in a dedicated MBE system is reported. The films, grown on p-type substrates, have thicknesses up to 10 $\mu$m and are all found to be n-type. Electron cyclotron resonance at a wavelength of 119 $\mu$m is studied at temperatures from 4 K to 100 K. Depending on growth conditions, the carrier concentration and mobility are found to vary considerably. Impurity cyclotron resonance and conduction-band cyclotron resonance are resolved in the best films, where the carrier density is determined to be in the range of $10^{15}$ cm$^{-3}$ and the mobility in excess of 100,000 cm$^2$/Vsec. The observed cyclotron mass and high mobility indicate that the carriers are located in the bulk of the films and no evidence for surface-bound carriers is found. These conclusions from cyclotron resonance are corroborated by Hall effect measurements in the same films.

As a first step in a comprehensive program to create and investigate quantum-well systems based on InSb, we have studied homoepitaxial InSb films grown in a molecular beam epitaxy (MBE) machine dedicated to III-V antimonide growth. The structural properties of the films and the growth parameters in our INTEVAC Gen II system are described elsewhere [1-3]. Our purpose in growing and studying these InSb films was to provide a benchmark for future quantum-well work, in particular to obtain an upper limit for the carrier mobility and a lower limit for the carrier density achievable in our system (without deliberate compensation). Several previous studies of homoepitaxial, MBE-grown InSb films have been reported [4-9]. In these studies, unintentionally doped films were found to be n-type with carrier densities ranging from $10^{14}$ cm$^{-3}$ to $10^{17}$ cm$^{-3}$ and mobilities as high as 250,000 cm$^2$/Vsec as determined from Hall studies at 77 K. Because Hall measurements on films with conducting substrates are not without problems [7,10], we have used cyclotron resonance as our primary density and mobility probe along with Hall effect measurements.

The cyclotron resonance (CR) measurements were carried out using an optically-pumped far-infrared laser at a wavelength of 119 $\mu$m. The Faraday geometry with circularly polarized incident radiation is used and we monitor the transmission. The circular polarization enables us to discriminate between electrons and the holes, for which there is always a substrate contribution. In all films we observe electrons. Figure 1 shows a typical CR signal from a 1 μm-thick film on a p-type substrate together with the CR signal from an n-type bulk sample of thickness 0.051 cm and carrier density $n = 5.8 \times 10^{13}\ cm^{-3}$. In samples with sufficiently high mobilities the resonances from impurity-bound electrons and from free-carrier electrons are resolved. Several studies of the two resonances can be found in the literature [11]. Impurity cyclotron resonance (ICR) occurs at lower magnetic fields, therefore conduction-band cyclotron resonance (CCR) is the upper of the two lines. The position of the CCR line is determined by the effective mass at the bottom of the conduction band with appropriate corrections for non-parabolicity. Our observed CCR lines are consistent with the accepted effective mass to free electron mass ratio for InSb of 0.0139. It should be noted that a CR signal from surface-bound electrons would be considerably broader due to surface scattering and found at higher fields due to non-parabolicity [12], therefore the CR signals described here are due to electrons in the interior of the films. The integrated intensity of both lines is a measure of the electron sheet density, $N_{CR}$, and the width of the CCR line is a measure of the mobility.

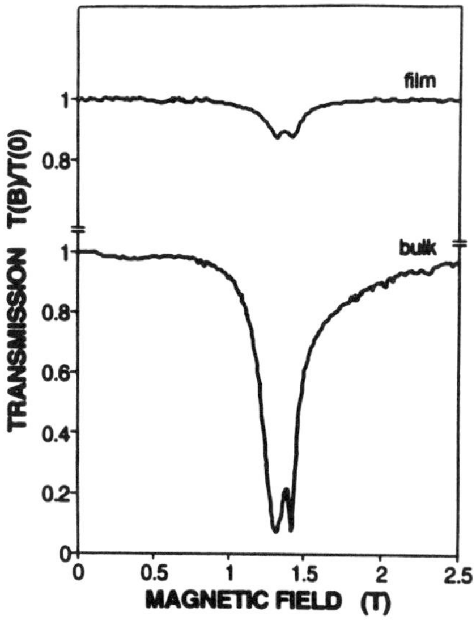


**Fig. 1** Comparison of cyclotron resonance at 119 μm in a homoepitaxial film (s11) and a bulk n-type sample of InSb at 4.2 K. The ratio of the transmission at magnetic field B to the transmission at zero field is plotted against B.

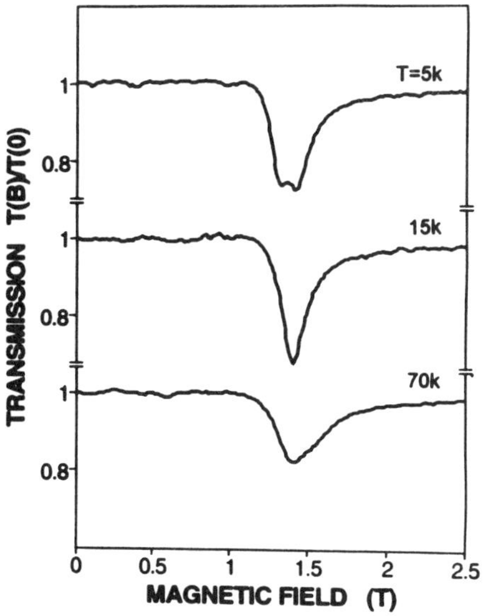


**Fig. 2** Cyclotron resonance at various temperatures for a homoepitixial InSb film (s35).

The 77 K Hall mobility of the bulk sample of Fig. 1 is $6.2 \times 10^5\ cm^2/Vsec$. Clearly, the mobility of the MBE-grown film in the figure is not quite as high as that of the bulk sample and its sheet density is considerably less. Using an oscillator model [13] to fit the observed transmission, we find for the film $N_{CR} = 1 \times 10^{11}$ cm$^{-2}$ and the mobility $\mu_{CR} = 2.5 \times 10^5$ cm$^2$/Vsec. Due to the proximity of the ICR line, the CR-determined mobility has a relatively high uncertainty at 4.2 K. To reduce this uncertainty and to make direct comparisons with 77 K data, we have studied the temperature dependence of the CR lines. The results on a second sample similar to that of Fig. 1 are shown in Fig. 2. One sees that in the 15 K trace, most of the impurity-bound carriers have been ionized and the CCR line dominates. As the temperature increases toward 77 K, the high-field side of the line broadens. This broadening is most likely due to higher-index Landau-level transitions which are shifted to higher magnetic fields by non-parabolicity. Therefore we fit the low-field side of the main CCR line which gives a nearly constant mobility between 4 K and 90 K of about $1.1 \times 10^5$ cm$^2$/Vsec. Thus we conclude that the 77 K mobilities and 4 K mobilities are comparable. The sheet density obtained by fitting the ICR and CCR lines below 25 K gives $N_{CR}$ a value of about $2.2 \times 10^{11}$ cm$^{-2}$. At 77 K, the sheet density is at most a factor of 3 higher depending on how one fits the high-field side of the main CCR transition.

Next we compare the results from our CR measurements to the results from the more traditional Hall effect performed on two samples, s69 (thickness 1 $\mu$m) and s84 (thickness 10 $\mu$m). These samples have a mobility difference characteristic of our MBE-produced samples. From the CR data as described above we find for s69 $N_{CR} = 3.2 \times 10^{12}$ cm$^{-2}$ and $\mu_{CR} = 0.5 \times 10^5$ cm$^2$/Vsec. For s84 we find $N_{CR} = 1.2 \times 10^{12}$ cm$^{-2}$ and $\mu_{CR} = 1.4 \times 10^5$ cm$^2$/Vsec. These two samples were also used in Hall studies with the standard van der Pauw geometry. At 77 K we find for s69 a Hall sheet density, $N_H = 3.2 \times 10^{12}$ cm$^{-2}$ and Hall mobility, $\mu_H = 0.36 \times 10^5$ cm$^2$/Vsec. For s84 we find $N_H = 2.0 \times 10^{12}$ cm$^{-2}$ and $\mu_H = 1.38 \times 10^5$ cm$^2$/Vsec. (These measurements are performed at fields below 0.1 T in order to avoid non-linearities.)

To translate the measured sheet densities from the CR and Hall studies into volume densities it is necessary to determine the thickness of the contributing electron layer. The most direct way to do this would be to use Shubnikov-de Haas measurements to directly determine the volume density. At densities of $10^{15}$ cm$^{-3}$ and lower, however, the observation of enough Shubnikov-de Haas oscillations to make such a measurement becomes problematic [14]. We therefore have used the nominal film thicknesses in order to calculate the volume densities. For the 1$\mu$m films this is not very accurate because the depletion thickness corresponding to $10^{15}$ cm$^{-3}$ is about 0.5 $\mu$m. Thus the volume densities we obtain are lower limits. For the thick films, however, using the nominal thickness should lead to accurate density values.

To compare the mobility value of our thick film (s84) with mobilities obtained in other work and in a theoretical calculation [15], we refer to Fig. 3. The data presented in this figure indicate that our films have mobilities comparable with the best obtained elsewhere. Comparison with theory indicates that our compensation ratio (the ratio of acceptor to donor concentration) is around 0.5. The considerable variation in our film mobilities and electron concentrations which we have observed suggests that further growth of (accidentally doped) homoepitaxial films could lead to higher mobilities and lower concentrations by a factor of two or so. Even at the present level of film quality, we are confident that we have achieved the substrate quality and near-optimum growth conditions for a successful program for InSb-based quantum-well and superlattice growth.

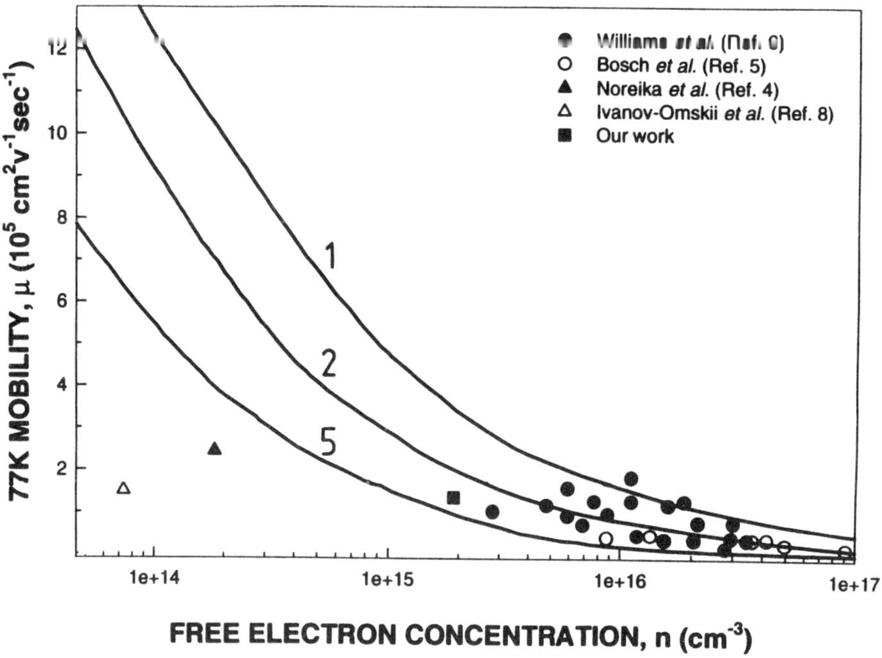


**Fig. 3** Mobility versus electron concentration for homoepitaxial InSb films. The solid curves are the theoretical results of Rode (Ref. 15) for different values of the ratio of the sum of the ionized donor and acceptor concentrations to the sum of the electron and hole concentrations: $(N_d + N_a)/(n + p)$.

It is a pleasure to thank Bruce Mason and Mansour Shayegan for helpful discussions. This work was partially supported by the National Science Foundation through grants OSR-9108771 and ECS-9410015.

## References

[1] W.K. Liu and M.B. Santos, Surf. Sci. **319**, 172 (1994).
[2] W.K. Liu and M.B. Santos, in this volume.
[3] M.B. Santos and W.K. Liu, in this volume.
[4] A.J. Noreika, J. Greggi, Jr., W.J. Takei, and M.H. Francombe, J. Vac. Sci. Technol. A **1**, 558 (1983).
[5] A.J. Bosch, R.G. van Welzenis, and O.F.Z. Schannen, J. Appl. Phys. **58**, 3434 (1985).
[6] G.M. Williams, C.R. Whitehouse, T. Martin, N.G. Chew, A.G. Cullis, and T. Ashley, D.E. Sykes, K. Mackey, and R.H. Williams, J. Appl. Phys. **63**, 1526 (1988).
[7] R.L. Williams, E. Skuras, R.A. Stradling, R. Droopad, S.N. Holmes, and S.D. Parker, Semicond. Sci. Technol. **5**, S338 (1990).

[8] V.I. Ivanov-Omskii, I.A. Petroff, V.A. Smirnov, Sh.U. Yuldashev, I.T. Ferguaon, P.J.P. Tang, C.C. Phillips, and R.A. Stradling, Semicond. Sci. Technol. **8**, 276 (1993).

[9] J. Heremans, D.L. Partin, D.T. Morelli, D.M. Thrush, G. Karczewski, and J.K. Furdyna, J. Appl. Phys **74**, 1793 (1993).

[10] R.A. Stradling, Phys. Scr. **T35**, 237 (1991).

[11] See for example, E. Gornik, T.Y. Chang, T.J. Bridges, V.T. Nguyen, J.D. McGee, and W. Müller, Phys. Rev. Lett. **40**, 1151 (1978) and references contained therein.

[12] U. Merkt, Festkörperprobleme **27**, 109 (1987).

[13] M.J. Chou, D.C. Tsui, and G. Weimann, Phys. Rev. B**37**, 848 (1988).

[14] See for example, M. Shayegan, V.J. Goldman, and H.D. Drew, Phys. Rev. B**38**, 5585 (1988).

[15] D.L. Rode, Phys. Rev. B**3**, 3287 (1971).

# Far-infrared studies of InAs quantum wells with Nb electrodes

E.L. Yuh[1], J.G.E. Harris[1], E.G. Gwinn[1], K.C. Wong[1,3] and H. Kroemer[2,3]

[1]Department of Physics, UC Santa Barbara, Santa Barbara, CA 93106
[2] Dept. of Electrical and Computer Engineering, UCSB, Santa Barbara CA
[3]QUEST, the NSF Center for Quantized Electronic Structures, Santa Barbara CA

Abstract

The development of InAs/AlSb quantum wells with high electron concentrations and high mobilities has recently made possible *super*current flow in SNS structures with a quasi-2D electron gas (2DEG) as the N channel. Previous studies of such superconducting 2DEG structures have focused on dc transport. Here we use the far-infrared transmission through a 2DEG that periodically contacts a superconducting grating to probe how the vanishing d.c. resistance of the 2DEG affects its excitation spectrum at frequencies near the superconducting grating's gap. The striking features we observe in transmission spectra indicate a more complex spectrum of excitations out of the ground state than in a uniform BCS film, and seem inconsistent with a conventional proximity effect model.

## 1. Introduction

The development of modulation-doped InAs/AlSb quantum wells with high electron concentrations and high mobilities [1] has recently enabled studies of quasi-two-dimensional electron gases (2DEGs) that carry a *super*current between superconducting banks [2-5]. Such 2DEG-based SNS structures promise to exhibit novel behavior associated with ballistic transport through N channels smaller than the mean free path [6], and have the potential to provide SNS structures with gate-modulated supercurrents.

Studies of S/N/S structures with sub-micron 2DEG widths have shown d.c. transport characteristics that are difficult to reconcile with a conventional proximity effect model. For example, substantial supercurrents have been observed across 2DEG gaps ~15 times wider than $\xi_N$, the proximity-induced coherence length, with critical voltages larger than predicted by theory [7]. This suggests that the conventional proximity effect [8], in which Cooper pairs penetrate into the normal metal over a distance on the order of $\xi_N$, is not the only mechanism that contributes to the supercurrent. In particular, theoretical work in the early 70's has shown that ballistic electron and hole-like quasiparticles can also carry supercurrents across N layers [9]. The quasiparticles transfer Cooper pairs across the N/S boundary in a scattering process called Andreev reflection, in which an incident electron reflects from the N/S interface as a hole, transferring a Cooper pair into the superconductor.

Recent experimental studies of induced superconductivity in quantum wells have suggested that multiple Andreev reflections between 2DEG/superconductor interfaces contribute to the supercurrent[3-5, 7]. However, the relevance of the two limiting pictures of supercurrent flow discussed above is not established, and the mechanism for induced superconductivity in quantum wells, and for its destruction by extremely small magnetic fields[3,5], is unresolved.

Here we use far-infrared (FIR) transmission through a 2DEG that periodically contacts a superconducting grating to probe how the 2DEG's vanishing d.c. resistance affects its excitation spectrum. FIR transmission is a well-established probe of the electronic excitations of thin films, and has been used to determine the energy gap in thin, uniform superconductors [10]. As shown below, the measured transmission though our SNSNS... grating structure differs qualitatively from that expected for a uniform BCS film, and seems inconsistent with a conventional proximity effect model of induced superconductivity.

## 2. Sample description

Figure 1a shows a schematic of the quantum well material used for our samples. A 150Å InAs layer, sandwiched between Te-doped AlSb barriers, forms a 1.35 eV potential well in the conduction band. A GaSb/AlSb superlattice lies above the well, and a layer of GaSb caps the structure. Electrons from Te donors fill the InAs well. Magneto-transport studies of samples without superconducting electrodes show that the Fermi energy is in the second subband, and give an electron sheet density $N_s = 7 \times 10^{12}$ cm$^{-2}$ and mobility $\mu = 90{,}000$ cm$^2$/V-sec. These parameters give two important lengths, the 2DEG's mean free path, $\ell \sim 3\mu$m, and its proximity-induced coherence length, $\xi_N$. For our 1.4 K experiments, the clean-limit coherence length, $\xi_{N,c} = (\hbar^2/2\pi m^* k_B T)(2\pi N_s{}^{1/2})$[8], is $\sim 1.5\mu$m. Since $\xi_{N,c} < \ell$, our samples fall between clean and dirty limits. The true $\xi_N$ should be $\sim \xi_{N,c}$, because the dirty-limit $\xi_N$ at 1.4 K is not much smaller than $\xi_{N,c}$.

Figure 1b shows a schematic of the SNSN... structure studied here, a superconductor grating that periodically contacts the underlying InAs well. To obtain the highly transmissive superconductor-2DEG interfaces required for induced superconductivity, we remove the semiconductor above the well, and deposit the superconductor onto the exposed InAs. Fermi-level pinning in the InAs conduction band allows a Schottky-barrier-free interface.

For the three samples studied here, the 1 μm-period superconductor grating was defined using holographic lithography. Samples A and B have a separation w ~ 3300 Å between edges of adjacent, 2000 Å-tall Nb strips, while sample C has w ~2300 A between 2000 Å-tall NbN strips (the NbN is deposited over 50 Å of Nb, to improve adhesion). Because w << $\ell$, we expect ballistic transport across the 2DEG strips.

Two patterns are etched onto each sample: a ~ 3mm x 3mm mesa sampled by the far-infrared beam, and a nearby, 80 μm-wide Hall bar used for monitoring the resistance. The small Hall bar avoids resistance due to imperfections in the superconducting grating, which affect d.c. transport results but should have little effect on the transmission measurements.

Figure 2 plots $R_\square$, the differential sheet resistance for transport perpendicular to the grating lines, versus temperature, T. Using the T at which $R_\square$ begins to drop as a rough estimate of $T_{cs}$, the critical temperature for the superconductor grating, Fig. 2 shows that sample C's NbN grating has $T_{cs} \sim 10$K; while the Nb gratings for A and B have $T_{cs} \sim 8.8$K and 7.5 K, respectively. The inset in Fig. 2 shows that $R_\square$ drops smoothly to zero at a temperature that we identify as the critical temperature, $T_{cN}$, of the 2DEG between adjacent

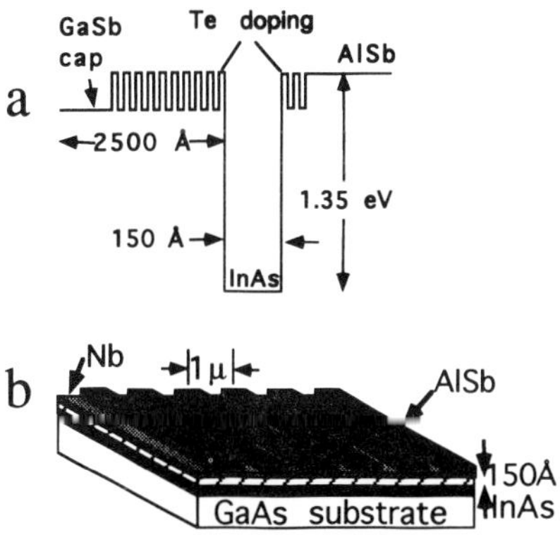


Figure 1. (a) InAs quantum well structure. (b) Sample schematic. A superconductor grating periodically contacts the underlying InAs well.

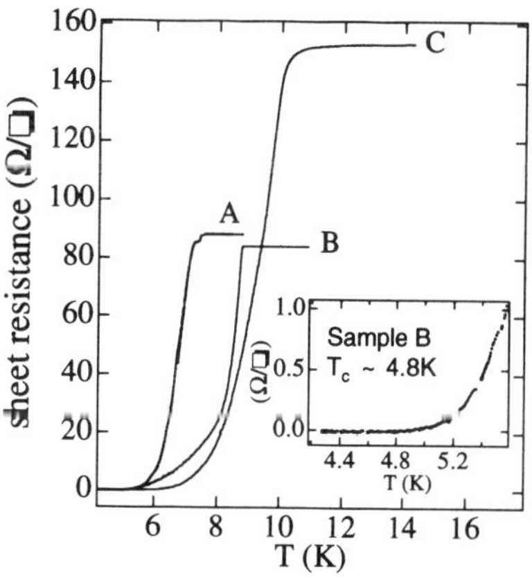


Figure 2. $R_\square(T)$ for samples A-C. Inset is exploded view for sample B.

superconductor strips. Sample C has the highest $T_{cN}$ ~ 5.5K, as well as the highest $T_{cs}$, while A and B have $T_{cN}$ ~ 4.2K and ~ 4.8K, respectively. Although sample A apparently has better quality Nb electrodes (higher $T_{cs}$) than sample B, $T_{cN}$ for sample B is *higher* than for sample A. This may be due to a better Nb-InAs interface in Sample B.

### 3. Far-infrared transmission data

To probe induced superconductivity in the quantum well, we measure the transmission spectrum, $T_S$, for the superconducting state of the 2DEG, and normalize $T_S$ to the transmission $T_N$ for the 2DEG's resistive state (for all spectra, the Nb or NbN grating remains superconducting). If induced superconductivity in the 2DEG did not change the high-frequency conductivity, which determines the transmission, $T_S/T_N$ would be one. Deviations away from one in $T_S/T_N$ indicate different electronic excitation spectra (and thus different high-frequency conductivity) in the superconducting and resistive states of the 2DEG.

We use a BOMEM DA3.002 Fourier transform spectrometer and $^3$He-cooled, Si bolometer to collect transmission data between 5 and 100cm$^{-1}$. The sample sits inside the detector dewar, directly in front of the bolometer. We control the sample temperature by setting the vapor pressure of the $^4$He bath.

For the data discussed here, a magnetic field, $H_{perp}$, applied normal to the sample plane switches the 2DEG between superconducting and resistive states. ($H_{perp}$ is the sum of the earth's field and the field from a coil near the sample, and is constant to ~3% over the area sampled by the far-infrared beam). The $H_{perp}$ that first produces electrical resistance in the quantum well, $H_{c,InAs}$, is tiny: at 1.4K, $H_{c,InAs}$ ~ 40 μTesla (0.4 Gauss). For all data shown below, the transmission $T_N$ is taken at $H_{perp}$ = 100μT, which is well above $H_{c,InAs}$, but orders of magnitude smaller than bulk critical fields for the superconductor electrodes.

Figure 3a shows $T_S/T_N$ from 5 - 25cm$^{-1}$ for Sample B, for several values of $H_{perp}$. Spectra were featureless from 25 - 100cm$^{-1}$. Figure 3b shows $R_\square$ vs. $H_{perp}$ for the Hall bar

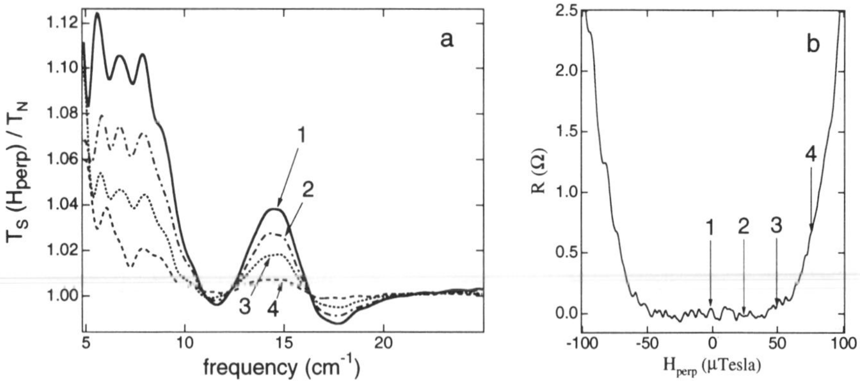


Figure 3. (a) Transmission spectra $T_S/T_N$ for four values of the perpendicular magnetic field, $H_{perp}$. Numbered arrows correspond to the following values of $H_{perp}$: 1) 0μT; 2) 24μT; 3) 49μT; and 4) 75μT. (b) Dependence of the sample resistance on $H_{perp}$. Arrows mark the field values used in Fig. 3a.

adjacent to the optically sampled area. Arrows in Fig. 3b show the $H_{perp}$ that correspond to the displayed $T_S(H_{perp})/T_N$.

The $T_S/T_N$ spectra in Fig. 3 show a distinct peak near 14.5 $cm^{-1}$, and a low-frequency rise below ~ 11 $cm^{-1}$. (Due to low signal levels below ~ 7 $cm^{-1}$, we could not determine whether this low-frequency structure is a broad peak. The oscillations below ~9 $cm^{-1}$ are residual Fabry-Perot interference between the faces of the wedged sample). Spectra for samples A and C have the same structure, with the distinct peak at slightly different frequencies, ~17.5 $cm^{-1}$ in sample A, and ~16.5 $cm^{-1}$ in sample C.

We have used a Nb grating with the same geometry as the sample gratings, but deposited onto a semi-insulating GaAs substrate, to show that all of the structure in $T_S/T_N$ arises from the 2DEG's interaction with the Nb. The Nb grating on semi-insulating GaAs has $T(H_{perp})/T(100\ \mu\text{Tesla}) = 1$, independent of frequency, for the same $H_{perp}$ as in Fig. 3 (this is as expected, since $H_{perp}$ is much smaller than Nb critical fields).

The dependence of $T_S(H_{perp})/T_N$ on $H_{perp}$ in Fig. 3 shows that the processes responsible for the distinct peak do not shift significantly in energy as $H_{perp}$ increases towards $H_{c,InAs}$. As shown, the peak frequency, $\omega_0$, does not shift, but the peak amplitude decreases smoothly as $H_{perp}$ approaches $H_{c,InAs}$. This behavior indicates that $H_{perp}$ may decrease the fraction of the area that a superconducting 2DEG occupies, by creating normal regions (vortices) in the 2DEG.

$T_S/T_N$ spectra measured between 1.4 K and 4.2 K show that the $\omega_0$ of the distinct peak in Fig. 3 has the same temperature-dependence as a BCS gap[11]. The best fit of $\hbar\omega_0(T)$ to the BCS form for $2\Delta(T)$ gives a $T_c$ of ~ 7.7K, close to sample B's Nb grating $T_{cS}$, ~ 7.5K. This suggests that the grating's gap, $\Delta_{Nb}(T)$, sets $\omega_0$.

This BCS-like temperature dependence of $\omega_0$ could be interpreted as evidence for a gap in the excitation spectrum of the 2DEG. However, this interpretation seems to contradict what one would expect from the penetration of Cooper pairs from the Nb banks into the 2DEG. The quantum well strips between the Nb grating lines, where we expect the absorption to occur, have width w << $\xi_N$, so that any induced gap, $\Delta_{2DEG}$, should be nearly constant across w. Since the Nb grating is opaque (and unchanged by $H_{perp}$), $T_S/T_N$ should be sensitive primarily to the these 2DEG regions between Nb strips. For these reasons, we expect that a conventional proximity effect model, applied to our grating samples, may predict $T_S/T_N$ similar to that of a uniform, thin BCS film, with gap $\Delta_{2DEG}$. (We note that the sample's response to applied fields is *non*local, in reality, so that this naive view is an oversimplification). However, the $T_S/T_N$ spectra in Fig. 3 differ qualitatively from that for a thin, uniform BCS film, which would have a *single* peak near $\hbar\omega = 2\Delta$, where $\Delta$ is the film's gap [12]. Further, in a conventional proximity effect picture [7] any $\Delta_{2DEG}$ induced in the quantum well should be *much* smaller than $\Delta_{Nb}$, so that peak structure associated with such a $\Delta_{2DEG}$ would lie at frequencies well below our range of observation. Thus, the $T_S/T_N$ structure we observe seems inconsistent with a conventional proximity effect model for the resistanceless state of the quantum well.

An alternate interpretation of the spectral features that we observe in $T_S/T_N$ is that they arise from excitations into quasiparticle states established by Andreev reflections at the Nb/2DEG interfaces. The temperature dependence and frequency of the features in $T_S/T_N$ seem roughly consistent with this interpretation: the temperature-dependent gap of the superconducting banks sets the maximum energy for Andreev bound states in the N region of SNS structures, so that excitations into the quasiparticle continuum might produce a feature in $T_S/T_N$ near $2\Delta_{Nb}$; while excitations into quasiparticle bound states could potentially produce the low frequency rise that we observe.

In conclusion, we have found that far-infrared transmission through a superconducting 2DEG shows striking features at frequencies up to ~ $2\Delta_{Nb}$, where $\Delta_{Nb}$ is the gap of the superconductor that contacts the 2DEG. Many interesting, unresolved questions on the origin for this structure remain, including the mechanisms for its extreme sensitivity to perpendicular fields, the influence of the samples' periodic SNSNS.. structure, and the role of interface transmittivity in determining characteristic frequencies.

This work was supported by NSF DMR-9314899. Material growth was partly supported by QUEST, the NSF Center for Quantized Electronic Structures.

## References

[1] See, for example, C. Nguyen *et al.*, J. Electr. Mat. **22**, 255 (1993).
[2] C. Nguyen, H. Kroemer and E.L. Hu, Phys. Rev. Lett. **69**, 2847 (1992).
[3] C. Nguyen, Ph.D. dissertation, UC Santa Barbara (1993).
[4] C. Nguyen, H. Kroemer, and E. Hu, Appl. Phys. Lett. **65**, 103 (1994).
[5] H. Kroemer, C. Nguyen, E.L. Hu, E.L. Yuh, M. Thomas, and K.C. Wong, to appear in Physica B.
[6] C.W.J. Beenakker and H. van Houten, Phys. Rev. Lett. **66**, 3056 (1991); T.M. Klapwijk, Physica B**197**, 481 (1994).
[7] D.A. Williams, A.M. Marsh and H. Ahmed, Phys. Rev. B **50**, 8118 (1994).
[8] G. Deutscher and P.G. de Gennes in *Superconductivity*, R. D. Parks, Ed., p. 1005 (Marcel Dekker, 1969).
[9] I.O. Kulik, Sov. Phys. JETP **30**, 944 (1970). For a recent discussion, see U. Schüssler and R. Kümmel, Phys. Rev. B**47**, 2754 (1993).
[10] See, for example, R. E. Glover III and M. Tinkham, Phys. Rev. **108**, 243 (1957).
[11] E.L. Yuh, J.G.E. Harris, E.G. Gwinn, K.C. Wong and H. Kroemer, to be published.
[12] S.W. McKnight, B.L. Bean, and S. Perkowitz, Phys. Rev. B**19**, 1437 (1979).

# Index